CISM COURSES AND LECTURES

The series presents lecture notes, monographs, edited works and proceedings in the field of Mechanics, Engineering, Computer Science and Applied Mathematics.
Purpose of the series is to make known in the international scientific and technical community results obtained in some of the activities organized by CISM, the International Centre for Mechanical Sciences.

INTERNATIONAL CENTRE FOR MECHANICAL SCIENCES

COURSES AND LECTURES - No. 450

MODELLING AND EXPERIMENTATION IN TWO-PHASE FLOW

EDITED BY

VOLFANGO BERTOLA
ECOLE NORMALE SUPÉRIEURE - PARIS

SpringerWienNewYork

This volume contains 234 illustrations

Printed in Italy
SPIN 10978504

In order to make this volume available as economically and as rapidly as possible the authors' typescripts have been reproduced in their original forms. This method unfortunately has its typographical limitations but it is hoped that they in no way distract the reader.

ISBN 3-211-20757-0 Springer-Verlag Wien New York

PREFACE

The study of two-phase flows, which include gas-solid, liquid-liquid, solid-liquid and gas-liquid flows, has great significance in several technological applications. As an example, gas-liquid flow, that also covers the whole subject of boiling and condensation, is a topic of the utmost importance and can be encountered in a wide range of industrial applications including evaporators, boilers, distillation towers, chemical reactors, condensers, oil pipelines, nuclear reactors, etc. Although not exhaustive, these few examples suggest how two–phase flow represents an interdisciplinary subject, which is of interest for mechanical, chemical, nuclear and environmental engineers, as well as for physicists and applied mathematicians.

During the last few decades, research in two-phase flows made great advancements, which are essentially related to the development of sophisticated experimental techniques (such as process tomography) and of computational fluid dynamics resources (such as direct numerical simulation). In spite of these advancements, the knowledge of two-phase flows is still somewhat limited, so that they continue to represent an extremely rich and appealing research domain for both the fundamental and the application oriented scientist. Even modelling of an apparently simple two-phase flow regime, such the stratified gas-liquid flow in a horizontal pipe, requires nontrivial closure relationships in order to give trustable results.

This volume collects contributions on different topics in the field of two-phase flow, with emphasis on gas-liquid and liquid-liquid systems. Specifically, this book contains selected printouts based on the lectures given at the advanced course "Modeling and Control of Two-Phase Flow Phenomena", held at the International Centre of Mechanical Sciences in Udine, Italy. The lectures aimed at presenting in a concise but comprehensive manner some of the main topics arising in the study of two-phase flows, and were delivered to an audience of graduate students, young researchers and professionals. The main objective of the course was to give participants the tools to carry out independent research, as well as to solve specific problems, fostering at the same time the exchange between the scientific community and the industrial world.

The contributions cover the description of flow patterns and flow pattern maps (Azzopardi and Hills), modelling of stratified and slug flow (Fabre), models for pressure drops in straight pipes and fittings (Azzopardi and Hills), liquid-liquid systems (Brauner), experimental techniques (Bertola). Two examples of state of the art two-phase flow modelling are also presented: the transport of solid particles in turbulent boundary layers (Soldati), which represents an important environmental issue, and the phenomenon of critical heat flux in boiling flows (Celata), which is relevant for several industrial applications, which include safety in nuclear power plants.

I would like to thank all the contributors, as well as the members of the International Centre of Mechanical Sciences, in particular Prof. M. Velarde, who was able to create a friendly and convivial atmosphere during the course week, and Prof. C. Tasso, for his precious assistance during the editing of this volume. A special thank is due to Prof. A. Soldati, who since the beginning fostered and encouraged the organization of this course, giving an essential contribution to make it a successful one.

Volfango Bertola

CONTENTS

Flow Patterns, Transitions and Models for Specific Flow Patterns

Barry Azzopardi and John Hills

Multiphase Flow Research Group, School of Chemical, Environmental and Mining Engineering, University of Nottingham, U.K.

Abstract. This chapter presents flow patterns, the configurations into which gas/liquid flows can arrange themselves in pipes and other geometries. Simple, mainly empirical, graphical methods for predicting the flow patterns occurring are then presented. The effect of flow patterns on heat transfer and of heat transfer on flow patterns are then considered. Models for the individual flow pattern transitions for vertical and horizontal (and near horizontal) pipes are then introduced. Important background areas such as flooding are covered. The extension of the models to annuli is also presented. Finally models for individual flow patterns are considered. In vertical flow slug and annular flow are examined. In addition, the latest material for churn flow, for which there is not a well-developed model, is presented. For horizontal flows, stratified, annular and slug flow are studied.

1 Flow Pattern and Flow Pattern Maps

An inherent complication in two-phase gas-liquid flows is caused by the interface between the phases. This produces a wide range of configuration of the phases in the channel with consequences both for the hydrodynamics and for heat and mass transfer. The groupings of similar configurations are important as studies have shown that a single correlation for the whole range of gas/liquid flows from 100% liquid to 100% gas is inadequate (Hewitt, 1983). In addition, advanced codes for transient analysis such as TRAC use descriptions specific to individual flow patterns (or configurations).

Here, flow patterns inside tubes and annuli and on the shell side of bundles are considered. After a brief presentation of the methods of flow pattern identification, the most common flow patterns are discussed and their importance identified. Graphical methods for identifying when flow patterns occur (flow pattern maps) are examined and the interaction between flow patterns and phase change is considered.

2 Flow Patterns

2.1 Definition and importance of flow patterns

In describing the configurations taken up by gas and liquid flowing together, researchers have used a very large number of descriptions. Some of these are alternative names for the same flow pattern, whilst others are subdivisions of more major groupings. Much of this confusion has arisen from the subjective way in which flow patterns are characterised.

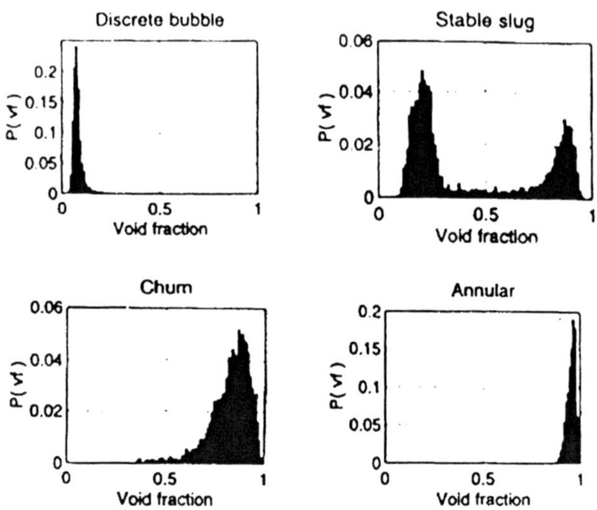

Figure 1. PDF of void fraction – Reprinted from International Journal of Multiphase Flow, vol. 23, Costigan, G. and Whalley, P.B., Slug flow regime identification from dynamic void fraction measurements, pp 263-282, Copyright (1997), with permission from Elsevier.

Initially, identification of flow pattern was by visual observation of the flow in transparent tubes, and this remains the primary definition. Given the almost infinite number of possible shapes and states of subdivision of the phases, a large number of reported flow patterns was inevitable. However, the number of flow patterns used in any description should be limited if the descriptions are to be of any use, and a small number of major patterns have been agreed, as described below. Suggested sub-divisions of these major patterns, where they may be important, are indicated in the descriptions.

Yet, the problem remains of assigning a given flow to one of the agreed patterns. In narrow tubes, at moderate fluid velocities, visual observation is reasonably reliable, but at high velocities it is difficult to see anything, and in wider tubes and the shell side of heat exchangers only the flow near the wall can be seen. Photographs with a high-speed flash, or high speed video studies, can help to overcome the problem of high fluid velocities, although it may be noted that some steam-water studies at Harwell in the 1960s using flash photographs had to be analysed by majority vote among a team of experts!

Jones and Zuber (1975) were among the first to suggest a more objective way of flow pattern determination. Using X-rays, they measured the time-varying mean void fraction (fraction of cross-section occupied by gas) at one cross-section in the tube, and plotted the probability density function (PDF) of these measurements. They found significant differences in the PDF "signatures" of different flow patterns, and suggested that this could be used for flow pattern identification. Thus, bubbly flow gave a single peak at low void fractions, annular flow a single peak at high void fractions, and intermittent flow gave two peaks. Subsequent workers have analysed pressure fluctuations (Matsui, 1984), or used other techniques, such as electrical resistance, for determining time-varying void fraction, and more sophisticated statistical analyses of time-varying data have also been suggested. A recent comprehensive study by Costigan and Whalley (1997) of flow patterns in vertical upflow used segmental impedance electrodes to determine void fraction, combined with the PDF technique of Jones and Zuber. Typical PDF curves from their work are shown in Figure 1 (see Figure 2 for explanation of the flow patterns). These more objective techniques are becoming increasingly popular, although there can still be disagreements between different investigators.

The difficulty of assigning a given flow to one of the accepted patterns must be borne in mind when we come to look at models for flow pattern transition in the third lecture. Such models are tested against data banks, which consist mainly of data determined by visual observation and which cannot, therefore, be regarded as totally reliable.

Bearing the above caveats in mind, this section defines the major flow patterns for vertical up- and down-flow, horizontal flow and flow at other inclinations and considers some implications arising from the characteristics of these patterns.

2.2 Vertical Upflow in Pipes

In vertical upflow four main patterns can be seen, as shown in Figure 2. **Bubbly flow:** In this configuration, there is a continuous liquid phase and the gas phase is dispersed as bubbles within the liquid continuum. The bubbles travel with a complex motion within the flow, may be coalescing and are generally of non-uniform size. In some situations, they congregate mainly at the pipe centre, in others, near the pipe walls, and the **wall-peaking** and **core-peaking** flows have sometimes been treated as sub-patterns of bubbly flow (Serizawa and Kataoka, 1988). At lower liquid velocities, the small bubbles must be generated either at the gas distributor or in the process of nucleate boiling, whereas at higher liquid velocities they can be formed by turbulent breakup of larger bubbles. Some workers treat these as two sub-patterns called **discrete bubbly** (or just **bubbly**) and **dispersed bubbly** flow respectively. The concentration of bubbles is not uniform but there are waves in concentration which travel along the pipe. **Plug flow:** This flow pattern, which in vertical systems is often referred to as **slug flow**, occurs when coalescence begins, and the bubble size tends towards that of the channel. Characteristic bullet-shaped bubbles, often called Taylor bubbles, flow up the pipe surrounded by a thin film of liquid. The liquid slug between the Taylor bubbles often contains a dispersion of smaller bubbles. Recent work has shown that this flow pattern does not occur in larger diameter pipes (0.15 and 0.2 m), where there is a direct transition from bubble to churn, Cheng *et al.* (1998), Ohnuki and Akimoto (2000). **Churn flow:** At higher velocities, the Taylor bubbles in slug flow break down into an unstable pattern in which there is a churning or oscillatory motion of liquid in the tube. Churn flow with its characteristic oscillations is an important pattern, often covering a fairly wide range of gas flow rates. At the lower end of the range, it may be regarded as a breaking up of plug flow with occasional bridging across the tube by the liquid phase; whilst at the higher range of gas flow rates it may be considered as a degenerate form of annular flow with the direction of the film flow changing and very large waves being formed on the interface. In the latter range the term **semi-annular flow** has sometimes been used. The Plug and Churn flow patterns, which both show large fluctuations in void fraction and pressure drop, are often grouped together as **intermittent flow**, particularly in shell-side flows. **Annular flow:** This configuration is characterised by liquid travelling as a film on the channel walls. Part of the liquid can also be carried as drops in the central gas core. In fact, for certain flow rates, the majority of the liquid travels as drops, leading to the term **mist**

flow being applied to this flow pattern in some industries. However, only in heat transfer systems where walls can become too hot to be wetted is there flow with no liquid film, since in adiabatic systems a minimum film flow is needed before drops can be generated. Interchange of liquid occurs between the film and the drops. Under some circumstances bubbles of gas can be entrained in the liquid film. At very high liquid flow rates liquid concentrations in the gas core are so high that "wisps" are observed instead of droplets. This is the **wispy annular flow** defined in certain flow pattern maps.

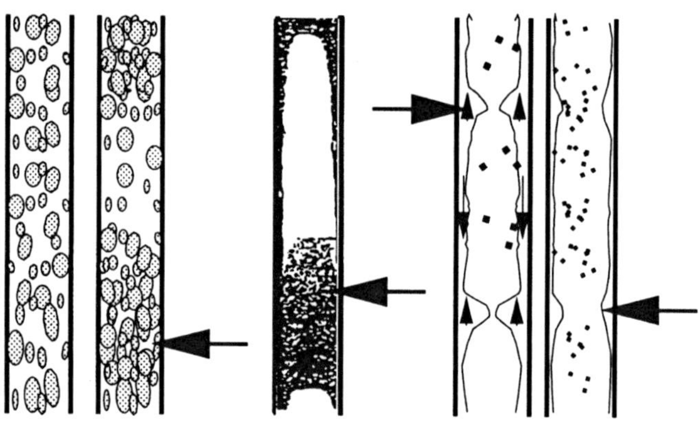

Bubbly Plug Churn Annular

Figure 2. Flow patterns in vertical upflow.

2.3 Horizontal Flow in Pipes

When gravity acts perpendicularly to the tube axis separation of the phases can occur. This increases the possible number of flow patterns, as seen in figure 3. **Bubbly flow:** This, like the equivalent pattern in vertical flow, consists of gas bubbles dispersed in a liquid continuum. However, except at very high liquid velocities when the intensity of the turbulence is enough to disperse the bubbles about the cross section, gravity tends to make bubbles accumulate in the upper part of the pipe as illustrated. **Stratified flow:** In this flow pattern liquid flows in the lower part of the pipe with the gas above it. The interface is smooth. **Wavy flows:** An increase of gas velocity causes waves to form on the interface of stratified flow. **Plug flow:** This flow pattern is characterised by bullet shaped gas bubbles as seen in vertical flow. However here they travel along the top of the pipe. **Slug flow:** This flow pattern, like plug flow, is intermittent. The gas bubbles are bigger whilst the liquid slugs contain many smaller bubbles. When the slugs are very aerated, the term **frothy** surges has been used (Coney, 1974). Some workers have used the term **semi-slug** to describe cases where the surges do not fill the pipe completely (Sakaguchi *et al.*, 1979). However, this might be more correctly considered as part of wavy flow. **Annular flow:** A continuous gas core with a wall film characterises this flow pattern. As with the equivalent pattern in vertical flow, liquid can be entrained as drops in the gas core. Gravity causes the film to be thicker on the bottom of the pipe but as the gas velocity is increased the film becomes circumferentially more uniform.

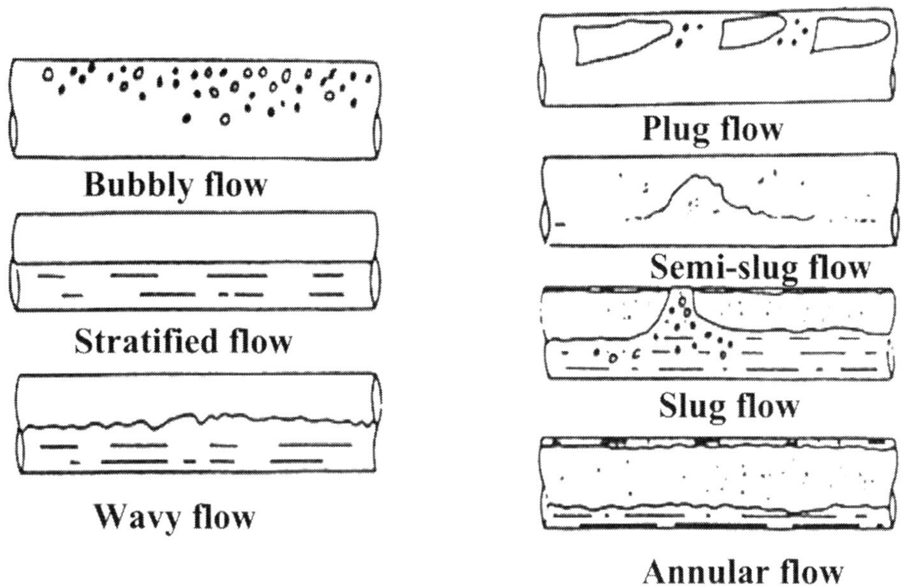

Figure 3. Flow patterns in horizontal flow.

2.4 Pipes at Other Inclinations

In **vertical downflow**, flow patterns are very similar to upflow with bubble, plug, churn and annular flows being reported. However, these patterns occur over different ranges of flow rates. Thus, low gas and liquid rates, which would yield bubble or slug flow in vertical upflow, produce a falling film flow, which resembles annular flow.

In general, for **inclined upflow**, the range of conditions occupied by slug-type flows increases considerably, starting at even small inclinations from the horizontal. For **inclined downflow** the range of conditions for slug-type flows diminish considerably. Studies into the effect of inclination can be divided into those which consider small deviations from the horizontal, and those which look at the complete range of angles from vertical upflow to vertical downflow. In both cases the flow patterns reported are similar to those described above, though some of the names used might differ.

2.5 Vertical Upflow in an Annulus

Two phase flow in annuli can occur in a variety of industrial equipment. An example in heat transfer equipment is the flow in double pipe heat exchangers. In addition, this geometry is sometimes used in experiments to provide data and understanding for the more complex rod bundle geometries used in nuclear reactors.

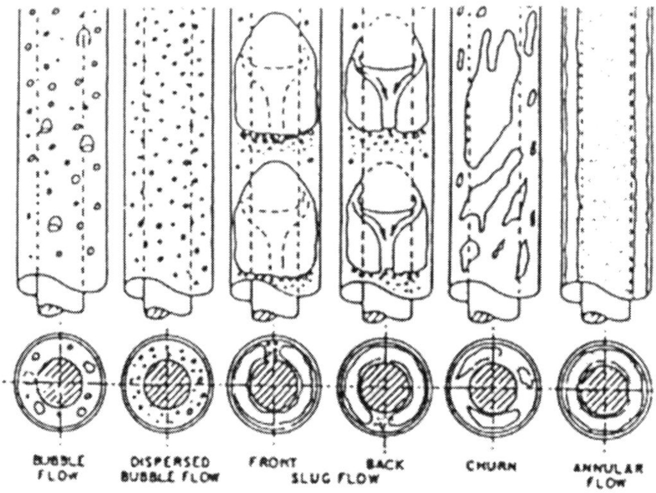

Figure 4. Flow patterns observed in a concentric annulus – Reprinted from Journal of Energy Resources Technology, vol. 114, Caetano, E.F., Shoham, O. and Brill, J.P., Upwards vertical two-phase flow through an annulus. Part I: Single-phase friction factor, Taylor bubble rise velocity and flow pattern prediction, pp 1-13, Copyright (1992), with permission from ASME.

In the oil/gas production industries, produced fluids are sometimes brought up the well in the annulus between the casing and the tubing. Though these "casing flow" wells are few in number, they are usually high producers and account for a significant part of world oil production.

The annulus geometry immediately increases the geometric parameters from one to three, for in addition to the inside and outside diameters, the eccentricity of the centre rod must be taken into account; the value of this parameter can vary from 0 (concentric) to 1 (rod touching outside tube).

Data on flow patterns in annuli have been published by Sadatomi *et al.* (1982) [concentric]; Caetano *et al.* (1992) [concentric and fully eccentric]; Kelessidis and Dukler (1989) [concentric and partially eccentric; e =0.5] and Das *et al.* (1999) [concentric]. They report flow patterns very similar to those seen in round tubes. Kelessidis and Dukler (1989) describe similar flow patterns but at higher gas and liquid flows they report a pattern which they label "annular flow with lumps". Their description indicates that this corresponds to the wispy annular flow pattern considered above. The major difference between these flow patterns and those found in round tubes occurs in slug flow where the Taylor bubble does not occupy the complete annulus and liquid can flow down the remaining sector. There is still a liquid film falling between the Taylor bubble and the inside and outside tubes. The effect persists even when the inside tube diameter is quite small: Hills & Chéty (1998) found it in isolated slugs in still liquid with an annulus of 22 mm inside diameter and 140 mm outside diameter. Sketches produced by Caetano *et al.* (1992) are shown in

Figure 4 for the concentric case; in the fully eccentric case the same patterns are found, with the liquid bridge in slug flow occupying the narrowest part of the annulus.

Horizontal flow in an annulus has been studied by Osamasali & Chang (1988). Patterns were similar to those in horizontal flow in a pipe, with transitional regimes in which waves on the bottom of annular flow touched the central rod, and hence produced liquid slugs. In narrow annuli, a similar transitional regime occurs where waves on the surface of stratified flow can also create liquid slugs by touching the central rod (Ekberg *et al.*, 1999)

2.6 Axial Flow in Bundles

Flows in nuclear reactor fuel bundles (normal operation in Boiling Water Reactors and accident conditions in pressurised Water Reactors) are the most important occurrences of axial flows. In spite of the safety analysis necessary for such reactors very little data is available for such flows. For **vertical bundles**, experiments at high pressure have been carried out by Bergles (1969) (4 rod bundle with square array) and Williams and Peterson (1978) (4 rod bundle with an in-line array). Both identified bubble and annular flow patterns together with an intermittent regime which Bergles termed slug flow whilst Williams and Peterson used the term froth flow. A subdivision of the intermediate flow has been recorded by Venkaseswararao *et al.* (1982) who carried out experiments on a 24 rod bundle using air/water at near atmospheric pressure. In some cases the Taylor bubble was confined to one or two subchannels between rods. Under other circumstances the gas plug occupied the entire channel with the rods penetrating the Taylor bubble. **Horizontal bundles**, found in the CANDU reactor, have been studied by Aly (1981) and Krishnan & Kowalski (1984). Observed patterns are similar to those found in horizontal tubes.

2.7 Cross Flow Through Bundles

Data for two-phase flow in cross flow over tube bundles is important in the development of design methods for shell and tube heat exchangers where two-phase flow occurs on the shell side. In spite of the wide occurrence of such duties, the available information is very sparse. Grant (1975) observed flow patterns in a segmentally baffled shell with four shell passes. He studied the flow through the transparent tube-plates, and identified a series of flow patterns, which are illustrated in Figure 5. In horizontal flow, they are **Stratified, Spray, Stratified Spray** and **Bubbly flow.** These flow patterns are very similar to those found in horizontal tubes, though the spray-stratified pattern, unobtainable in tubes, occurs because film on upper tubes can be stripped off as spray. The Spray and Bubbly flows are also found in vertical flows, together with an intermittent pattern which Grant called "**Chugging**". Although his 4-pass shell would have had downflow in the fourth pass, he does not appear to make any distinction between the patterns in upflow and downflow. However, Xu *et al.* (1998), working with a six-pass shell in the vertical orientation, did find such differences. In **upflow**, they added a fourth regime, which they call **churn flow**, that occurs at low gas and liquid flows and consists of large, irregular bubbles in a continuous liquid. In **downflow**, the low flow-rate pattern became **falling film** flow, in which liquid films covered the tubes and ran down the shell and baffle walls between them. **Bubbly flow**, in downflow, was more like the falling film flow, but with thicker, aerated films.

2.8 Consequences of Flow Patterns

At the start of this chapter, it was suggested that identification of flow patterns was important as the wide range of physical processes that they incorporate make it unlikely that a single model could accurately describe the flow.

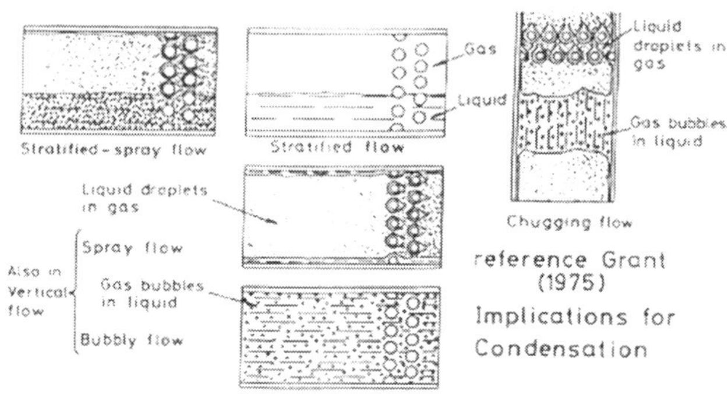

Figure 5. Flow Patterns in horizontal and vertical crossflow as defined by Grant (1975).

Having defined the flow patterns we can now see other aspects in which flow patterns are important. For example, we see that bubbly flows produce a large interfacial area and thus are most appropriate for processes where mass transfer is required. Hence bubble columns are used for absorption and reaction.

The intermittent flow patterns (plug/slug flow) can cause vibration damage particularly at bends and fittings as the momentum of the gas- and liquid-dominated packages can be significantly different. Alves (1974) discusses this point. The occurrence of intermittent flows can also be important if the two phases are to be separated. Separators designed to cope with time averaged gas and liquid flow rates can fail if slug flow occurs and the feed becomes all liquid for periods. The effects of flow patterns on phase change and of phase change on flow patterns are considered in section 4 below.

3 Flow Pattern Maps

3.1 Vertical Upflow in Pipes

Flow pattern data are often represented on a two dimensional diagram in terms of system variables. The most common variables used are the liquid and gas superficial velocities (volumetric flow rate/cross sectional area of the pipe). Since variables other than the superficial velocities are known to affect the flow pattern, maps of this kind are specific to a particular combination of fluids and geometry. However, they are simple to use, and unlike the case of single-phase Newtonian flow where the single parameter of Reynolds number brings all flows together, it is by no means clear exactly which other variables should be included. No reliable universal flow map has yet been produced. The commonest way of constructing a flow map is to identify the flow pattern at a set of conditions covering the field, and then to sketch in boundary lines separating the different patterns. An example is shown in Figure 13 for flow in an annulus. Because of problems in correctly identifying flow patterns, it often happens that a few experimental points lie on the wrong side of these lines, and the lines would be better regarded as transition zones, of indeterminate width. This should always be remembered when using maps on which only the boundary lines appear.

For **vertical upflow**, flow pattern maps based on superficial velocities have been published since the 1960's (Sterling, 1968 and Wallis, 1969) and are still being produced. Figure 6 shows a recent example, for air/water in a 82.6 mm pipe, due to Zhang *et al.* (1997). Here, the boundary lines are based on various objective criteria, and not the "sketching in" process described in the previous paragraph. While the position of the various lines may be open to discussion, there is general agreement about the shape of the map, as may be seen by comparing Figure 6 with the widely-accepted semi-theoretical map of Taitel *et al.* (1980), shown in Figure 9. Other workers have presented maps where the superficial velocities are modified by factors in the form of ratios of actual physical properties to standard values raised to different powers. The horizontal flow pattern map of Baker (1954) shown in Figure 11 is an example of the type. A popular approach, which tries to incorporate some physical reality, is that of Hewitt and Roberts (1969), shown in Figure 7. The data are plotted as $\rho_g u^2_{gs}$ against $\rho_l u^2_{ls}$ (*ie* gas momentum flux against liquid momentum flux), and data for air/water at 3 bar and steam/water at 35 and 70 bar were brought together by this approach. The Hewitt and Roberts map remains popular with researchers. However, a new version by Owen (1986), using the same axes, but with significantly different boundary curves, appears to give better results particularly in evaporating flows (Frankum *et al.*, 1997). It should be noted that neither version is in particularly close agreement with the Taitel *et al.* (1980) map.

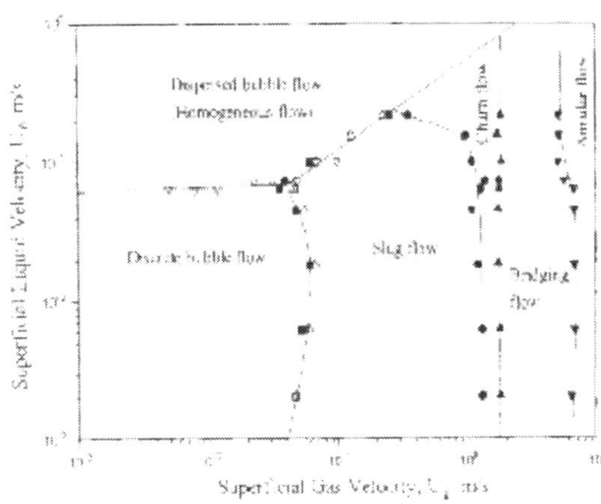

Figure 6. Flow pattern map of Zhang et al. (1997) - vertical upflow Reprinted from Chemical Engineering Science, vol. 52, Zhang, J.-P., Grace, J.R., Epstein, N. and Lim, K.S., Flow regime identification in gas-liquid flow and three-phase fluidised beds, pp 3979-3992, Copyright (1997), with permission from Elsevier.

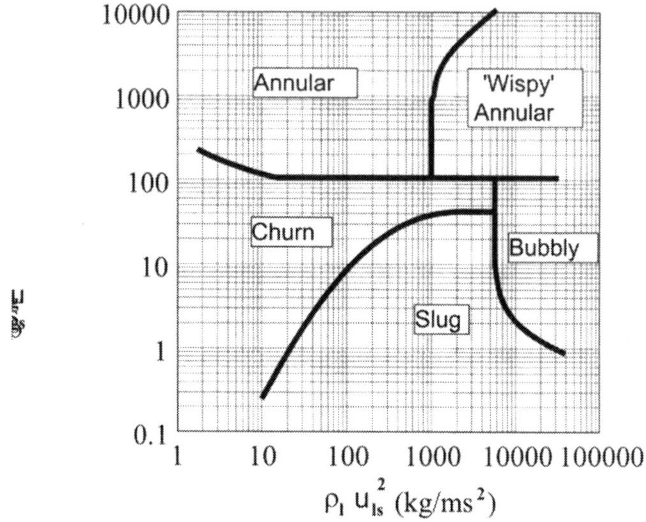

Figure 7. Flow pattern map of Hewitt and Roberts (1969) - vertical upflow.

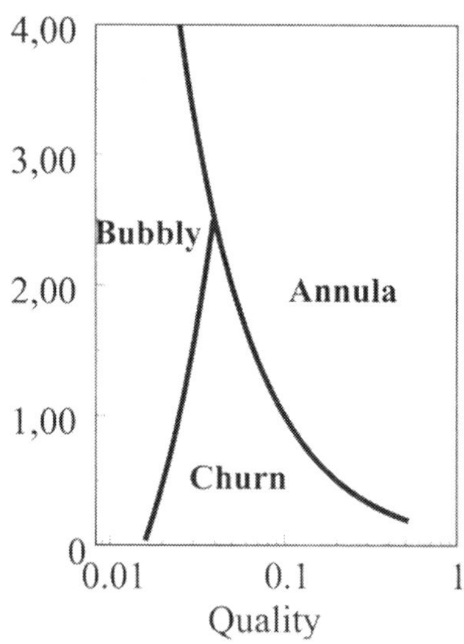

Figure 8. Flow pattern map after Bennett et al. (1965) and Mayinger and Zetzmann (1976) - vertical upflow - water.

Another approach, put forward by Bennett *et al.* (1965), is to plot data as mass flux *versus* quality. This is useful for evaporating systems, where liquid and vapour contain the same component, as the map quickly reveals the distribution of patterns in an evaporator tube (quality increasing at constant mass flux). Mayinger and Zetzmann (1976) have suggested a generalisation in which the quality is corrected by a multiplier (a power of the reduced pressure) which, to a large extent, brings together data from different fluids, though differences still exist. Figure 8 shows typical boundaries for a steam /water system at one pressure.

Perhaps the most popular map with researchers is a semi-theoretically based one, due to Taitel et al. (1980). Some of the transitions on this map have a tube diameter dependence, and the authors regard churn flow as an entrance effect, so the slug/churn boundary also depends on the tube length. The map is shown in Figure 9 for the air/water system, with the common tube diameter of 30 mm, and a length of 1 metre. It is further discussed in lecture 3, where models for the transitions are considered.

Vertical Downflow in Pipes being less prevalent than upflow, has received less attention. Available studies are almost all for air/water in small diameter pipes. Figure 10 shows a map published by Barnea *et al.* (1982). There has not yet been a systematic comparison of the divers flow patterns maps published for vertical downflow.

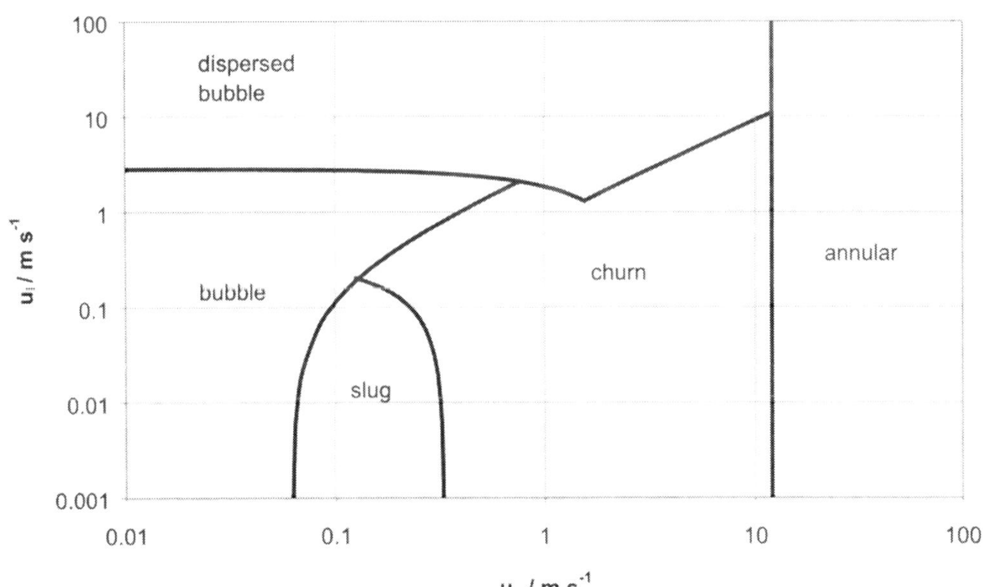

Figure 9. Vertical flow pattern map based on models of Taitel *et al.* (1980).

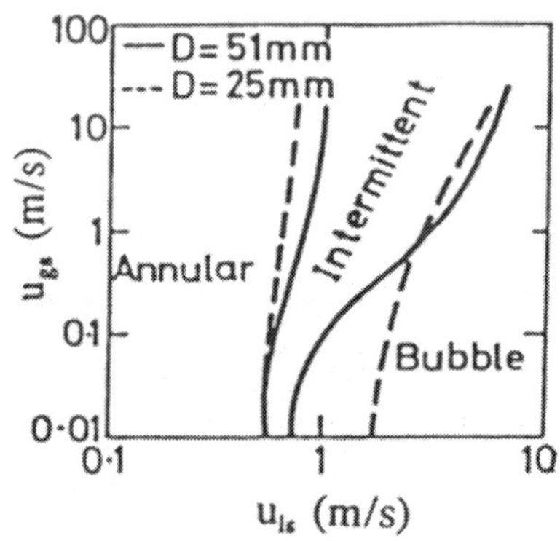

Figure 10. Flow pattern map for vertical downflow - after Barnea *et al.* (1982).

3.2 Horizontal Flow in Pipes

In **horizontal flows** the flow pattern map of Baker (1954) (Figure 11) still has great popularity. To its credit, it is simple and based on industrially relevant data (gas/condensate flows at high pressure in 5"-10" lines). Subsequent work has shown some of its transition boundaries to be poor. Taitel and Dukler (1976) made a significant step forward. They concluded that a single pair of axes was not sufficient, and produced the map of figure 12. The approach, described in detail in the third lecture, was to produce models for the individual transitions and from these determine both appropriate axes and the transition curve.

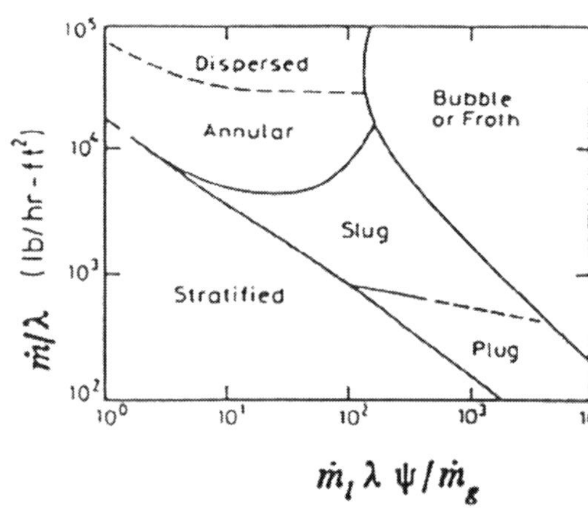

Figure 11. Flow pattern map of Baker (1954).

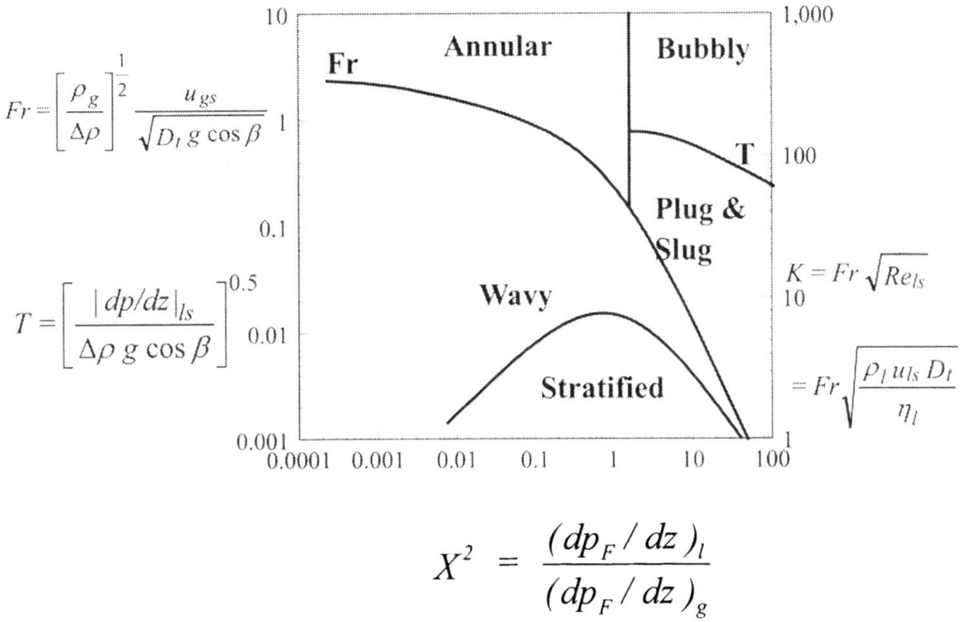

Figure 12. Flow pattern map constructed from models of Taitel and Dukler (1976) - horizontal flow.

For **inclined flows** Spedding and Nguyen (1980), Gould (1972) and Mukerjee and Brill (1985) give maps for steep inclination. In addition the Taitel and Dukler (1976) approach can handle small deviations from the horizontal.

Information available on flows in horizontal pipes with **non-Newtonian liquids** and liquids with suspended solids has been collected by Chhabra and Richardson (1985). They present specific flow pattern maps for both vertical and horizontal flows.

3.3 Annuli

The conditions at which different flow patterns occur are shown in Figure 13 (air/water, concentric annulus). This data is from Caetano *et al.* (1992), who also present maps for air/kerosene in a concentric annulus and for air/water in a fully eccentric annulus. The differences between the three cases are not great. The maps of Kelessidis and Dukler (1989) and of Das *et al.*(1999) are also very similar.

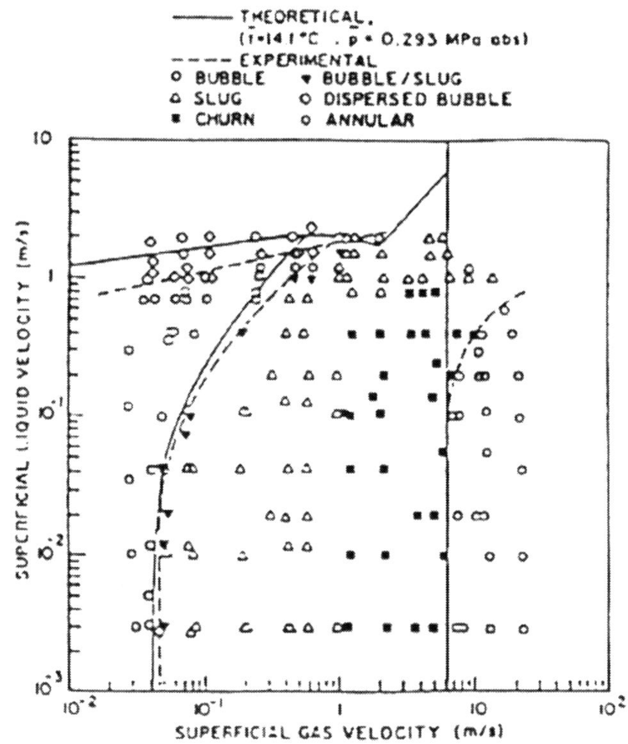

Figure 13. Flow pattern map for a concentric annulus - air/kerosene - – Reprinted from Journal of Energy Resources Technology, vol. 114, Caetano, E.F., Shoham, O. and Brill, J.P., Upwards vertical two-phase flow through an annulus. Part I: Single-phase friction factor, Taylor bubble rise velocity and flow pattern prediction, pp 1-13, Copyright (1992), with permission from ASME.

3.4 Cross Flow Through Bundles

Grant (1975) provided flow pattern maps for flow across tube banks for both horizontal and vertical flow. These maps are in terms of superficial velocities of the phases (defined as the volumetric flow rate divided by the minimum flow area) multiplied by factors to take account of physical properties. However, as Grant only worked with air-water mixtures, these maps should be used with caution for other fluids.

Ulbrich and Mewes (1994) gathered a great deal more data for the upflow case with an in-line arrangement of ten rows of five tubes in a rectangular shell. They covered much wider ranges of superficial velocities (both gas and liquid) than those employed in the original work of Grant (1975). In addition they assembled data published since Grant's original paper. The map of Grant was tested against the bank of data and found to be accurate for only 45% of the points.

Similarly, when the model of Taitel *et al.* (1980), originally devised for in-tube flows but often used in shell side flows for want of anything better, were applied to shell side cross flow, only 48% of the points were correctly predicted

Based on their data bank, Ulbrich and Mewes suggested their own purely empirical Flow Pattern Map which predicts 89% of the data correctly. As the great majority of the data in the bank used by Ulbrich and Mewes were from air/water tests, the map should be used with caution for other fluids. A more recent study, by Noghrehkar *et al.* (1999), conflicts with the work of Ulbricht and Mewes (1994) although it was carried out in similar equipment over a similar range of flow rates. They studied both in-line and staggered tube arrangements in vertical upflow, with 25 rows of 5 tubes. Their in-line data are shown in Figure 14, where it may be seen that the intermittent flow pattern persists up to the highest liquid rates considered (1.0 m/s), unlikethe Ulbrich and Mewes curves, where it vanishes above 0.4 m/s. In the staggered arrangement, the range of intermittent flow was smaller, but it still persisted at 1.0 m/s liquid superficial velocity. Noghrehkar *et al.* attribute this difference to the fact that they measured the void fraction with an electrical impedance probe deep within the tube bundle, and used its PDF to give an objective measure of flow pattern, whereas Ulbricht and Mewes, like Grant before them, used visual observation at the transparent tube plate. However, Ulbricht and Mewes also measured the fluctuations of pressure drop across their bundle, and claim that the region of large fluctuations corresponds to their visually observed intermittent flow pattern. This conflict cannot at present be resolved.

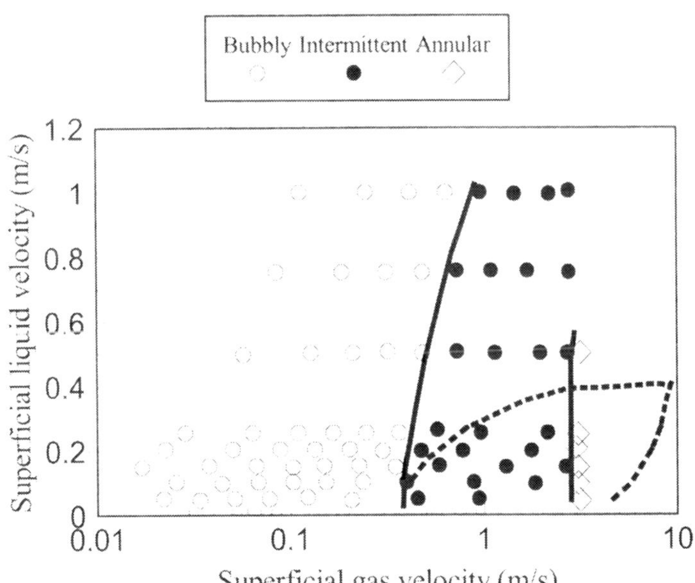

Figure 14. Map for two-phase crossflow - vertical upflow - Noghrehkar *et al.* (1999) ...... boundaries of Ulbrich and Mewes (1994); ——— boundary proposed by Norgrehkar *et al.* (1999).

4 Flow Patterns with Phase Change

In flows with phase change, flow patterns are important because they affect the heat and mass transfer as well as the hydrodynamics. In addition, the change of phase has an effect on the conditions at which flow patterns occur. This section considers the implication of flow patterns on condensation and evaporation and discusses a study which allows for the effect of phase change on transitions.

4.1 Condensation

The importance of flow patterns for interpreting and predicting condensation heat transfer was noted by Bell *et al* (1970). There have been many other publications on the subject since. Soliman (1985) has carried out a thorough review, collecting most of the available data and comparing these with flow pattern prediction methods. Here we merely seek to illustrate the types of flow patterns which occur, and the consequent heat transfer implications.

Figure 15, taken from Palen *et al* (1979) shows patterns and the condensation paths occurring in horizontal flow at high and low vapour feed rates on a modified form of the Baker (1954) map. At high rates there is a wavy flow which differs from the usual wavy flow by the existence of a thin draining condensing layer on the upper part of the tube. This soon changes to annular flow and as more vapour is turned into liquid, slug and plug flow occur. At lower flows, annular flow, if it occurs, will always be asymmetric with a thicker film on the bottom. Thereafter, the flow will be wavy and, when the vapour velocity has fallen sufficiently, stratified smooth flow occurs.

The implication for heat transfer of these flows can be seen in Figure 16 prepared by Sardesai *et al* (1981). They measured the heat transfer coefficients at points around the tube circumference. They then considered the ratio of the coefficient at the bottom to that at the top and plotted this against a parameter β which they obtained from

$$Fr = \frac{\beta}{0.7\,X^2 + 2\,X + 0.85} \tag{1}$$

where Fr and X are defined by

$$Fr = \left[\frac{\rho_g}{\Delta\rho}\right]^{0.5} \frac{u_{gs}}{\sqrt{d\,g\cos\beta}} \tag{2}$$

$$X^2 = \frac{(dp_F/dz)_l}{(dp_F/dz)_g} = \frac{u_{ls}^2}{u_{gs}^2}\frac{\rho_l}{\rho_g}\frac{f_{ls}}{f_{gs}} \tag{3}$$

Equation (1) with β equal to 1.0 gives a reasonable approximation to the stratified/annular transition suggested by Taitel and Dukler (1976) and plotted in Figure 12. Values of $\beta>1$ would correspond to annular flow, and of $\beta<1$ to stratified flow. Figure 16 shows that for values of β above 1.75, the ratio of coefficients is approximately constant and close to 1.0.

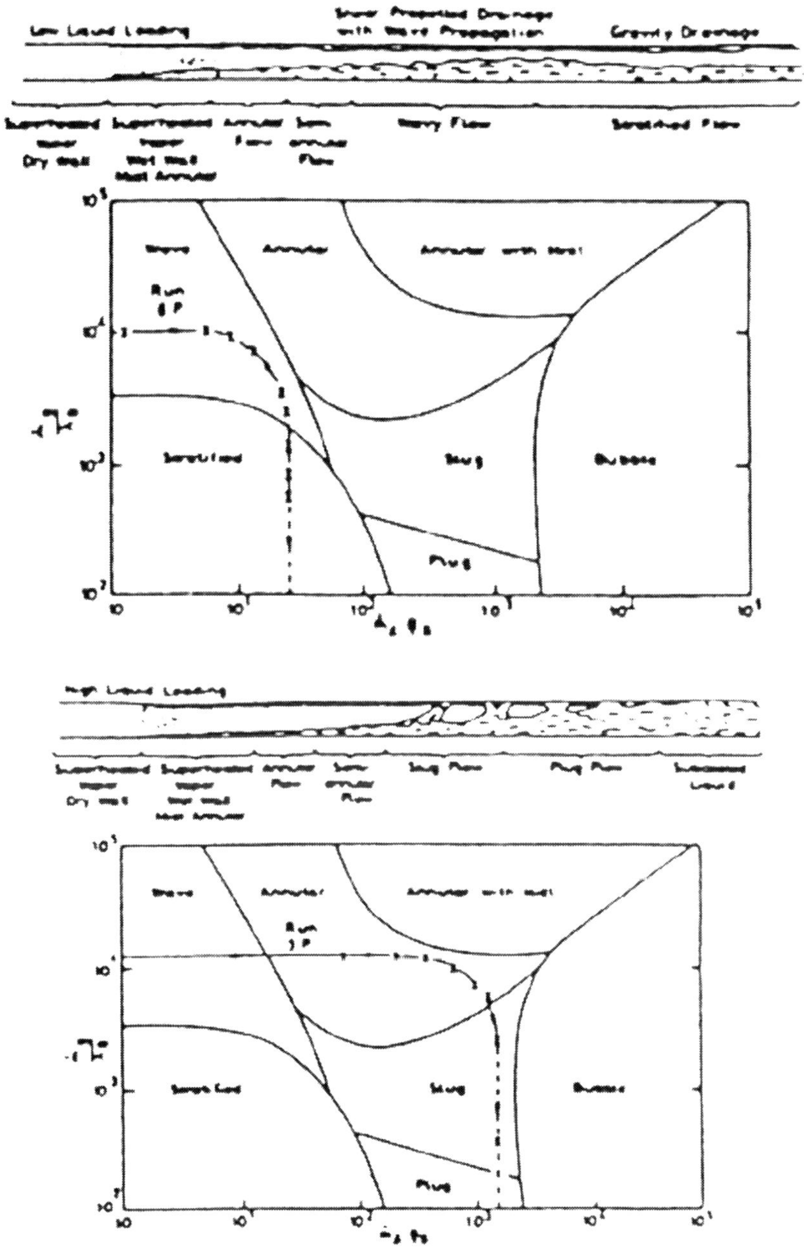

Figure 15. Flow patterns in condensation in a horizontal tube - Palen *et al* (1979).

In this case, a single heat transfer coefficient can be used for the whole tube. For values of β below 1.75, account must be taken of the circumferential variation of heat transfer coefficient with high heat transfer occurring in the thin draining film on the upper side of the tube and poorer heat transfer in the stratified layer at the bottom.

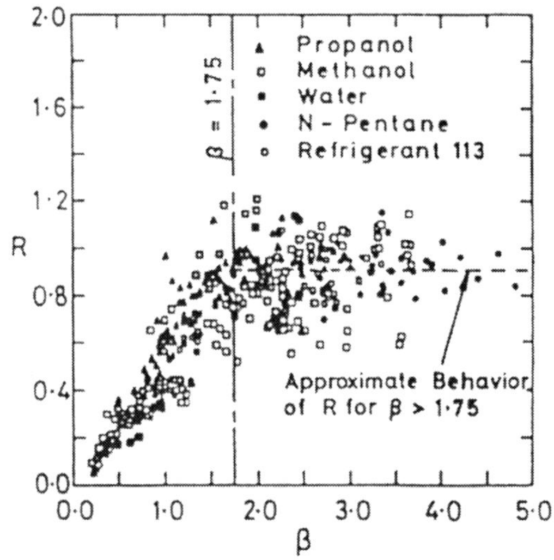

Figure 16. Variation of ratio of heat transfer coefficient at tube bottom and tube top with parameter β - Reprinted from Chemical Engineering Science, vol. 36, Sardesai, R.G., Owen, R.G. and Pulling, D.J., K.S., Flow regimes for condensation of a vapour in a horizontal tube, pp 1173-1180, Copyright (1981), with permission from Elsevier..

4.2 Evaporation

Most evaporating flows are vertical upwards and the flow patterns observed are similar to those described above, but all occurring sequentially in the same tube, as shown in Figure 17. The extent of each flow pattern depends on the combination of heat input, flow rate and subcooling of the liquid entering the pipe. At high flows and low heat flux there might only be bubbly flow, whilst at low flows and higher heat fluxes the whole range of flow patterns shown in Figure 17 would appear.

One important factor to be considered in evaporating flows is the possibility of dryout or critical heat flux. If the wall becomes covered with a layer of vapour then the heat transfer coefficient deteriorates. For temperature-controlled systems, such as a steam-heated reboiler, this leads to a loss of efficiency. If the system is heat flux controlled, such as a nuclear reactor or the radiant section of a fossil-fuelled boiler, the wall temperature will rise to remove the heat and damage may occur. Examination of Figure 17 leads to insight into the possible mechanisms of dryout. At low qualities, in bubbly flow, the vapour layer could build up if bubbles are produced at a rate faster than they can be removed from the boundary layer and into the liquid core. At higher qualities, in annular flow, the liquid film could be so depleted by evaporation and entrainment of liquid as drops that the film could dry out. This would occur before all the liquid was evaporated as the vapour effectively insulates the liquid travelling as drops. It is also possible that in the slug flow region, the liquid film around the Taylor bubble could dry out. However, this would require very rapid evaporation or very long vapour bubbles as the hot surface sees liquid plugs at regular intervals.

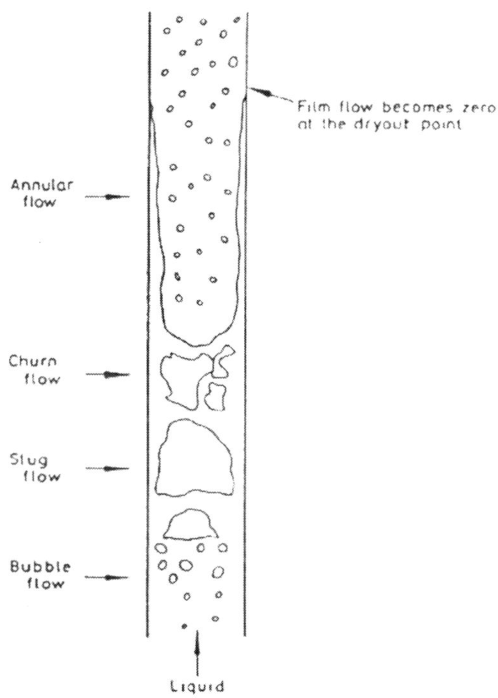

Figure 17. Flow patterns during evaporation in vertical upflow.

In those cases where the evaporation is taking place in horizontal tubes, stratification of the flow occurs, Figure 18. If the conditions are opportune, dryout will occur at the top of the tube when the flow passes through wavy flow. As more vapour is produced, the flow could revert to annular and the top surface is rewetted. Further evaporation could thin the film and once again cause dryout at the top of the tube where the film is thinnest. The fact that dryout usually occurs at the top of the tube has lead to the practice of passing the heating medium upwards over the horizontal tube bank. In this way, the hottest part, the tube bottom, has most chance of being wetted. Equations have been proposed to identify when strong stratification effects are present. Chawla (1968) recommends

$$\left[\frac{\dot{m}^2 (1-x_g)^2}{\rho_l^2\, g\, D_t}\right]\left[\frac{\dot{m}(1-x_g)\, D_t}{\eta_l}\right] > 109 \tag{4}$$

whilst Shah (1976) gives

$$\frac{\dot{m}^2}{\rho^2\, g\, D_t} > 0.04 \tag{5}$$

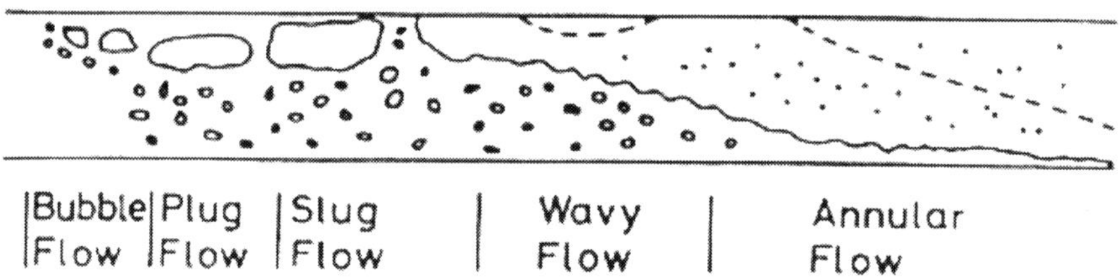

Figure 18. Flow patterns during boiling in horizontal tubes.

4.3 Effect of Phase change on Flow Pattern

Phase change, whether evaporation or condensation can affect the interface of the two-phase flow and hence its flow patterns. The production of new vapour with low momentum or the removal of high momentum vapour causes this. Thus boiling will result in reduced shear stress at the interface whilst in condensation it will be increased.

No Flow Pattern Maps have been derived specifically for flows with phase change, so the only option is to test the existing maps against data where flow change is taking place.

For vertical upflow with evaporation, Frankum *et al.* (1997) have recently carried out such an exercise. They considered both empirical flow pattern maps as well as models for individual transitions. As well as the data of Bennett et al. (1965) for steam/water at 35 and 70 bar, they considered the steam/water data of Hoesler (1967) and the refrigerant 12 data of Celata *et al.* (1991). This exercise showed that all the models gave similar levels of accuracy. The models derived by McQuillan and Whalley (1985) predicted 75% of the data points correctly. Interestingly, the maps of Owen (1986) and Mayinger and Zetzmann (1976) gave similar results whilst that of Fair (1960) was accurate 88% of the time.

Dukler and Taitel (1984) have produced an extension of the Taitel and Dukler (1976) model for flow pattern transitions in horizontal flow which allows for the direct effect of momentum exchange on interfacial shear stress. They use a dimensionless group

$$\Phi = \frac{\phi\, \pi\, u_g}{\Delta h_v\, D_t\, (dp/dz)_g} \tag{6}$$

where q is the heat flux and Δh_v is the latent heat of evaporation. The effect of Φ on the flow pattern transitions can be seen in Figure 19. It appears that this modification has not been tested against experimental data.

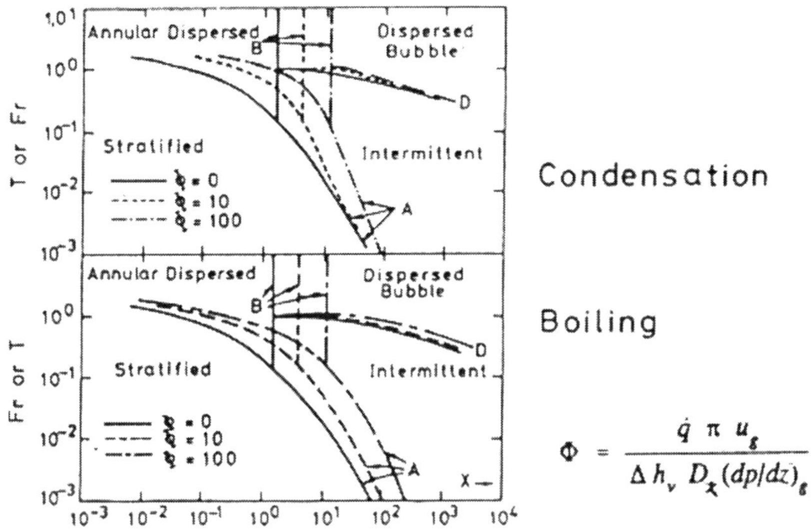

Figure 19. Effect of heat flux on flow patterns in a horizontal tube (Dukler and Taitel, 1984).

5 Introduction to Flow Pattern Transition Models

The sections above on Flow Patterns introduced flow pattern maps, which are graphical, usually empirical methods for the identification of the conditions at which different flow patterns occur. However, this empiricism limits their applicability to the ranges of parameters on which the maps were developed, and the lecture also presented a number of more theoretically derived maps. This paper presents some of the physically based models which describe the transitions between flow patterns, using the theoretical maps of the flow pattern lecture as a starting point, but presenting other models for individual transitions. So far, the theoretical approach has only been developed for flows in tubes and annuli.

One important caution to be borne in mind is that most of the models are based on a steady-state equilibrium description of the flow. In fact, the development lengths for establishment of fully-developed two-phase flow patterns can be significantly in excess of 100 pipe diameters – figures up to 200D are quoted. 100D in a 50 mm pipe (a typical size) is already 5 meters, which exceeds the length of many experimental rigs, and is a non-negligible fraction of many industrial installations, apart from oil/gas wells and pipelines. Thus, much real gas/liquid flow may exist in a developing state, rather than as a fully developed flow, and this will influence the position of the transition boundaries.

6 Vertical Upflow in Pipes

Taitel *et al.* (1980), and Mishima and Ishii (1984) presented the earliest flow maps based on theoretical models of these transitions, and both models remain popular with researchers. They provide the framework for this section, which discusses each transition in turn, including other, more recent models as appropriate, and providing systematic tests, such as that of McQuillan and Whalley (1985b) whenever possible.

6.1 Transititions Involving Bubbly Flows

A number of sub-regimes of bubbly flow have been described; some are best regarded as transition regions between the bubble and slug regimes, such as the "bubble cluster" and "cap bubble" regimes of Cheng *et al.* (1998), the "coalesced bubble" regime of Zhang *et al.* (1997), and probably the "churn-turbulent" regime of Zuber and Findlay (1965) also. All involve the presence of bubbles significantly larger than in the bubble regime proper, and in some cases it is uncertain as to when, or if, the bubble-to-slug transition has occurred. For example, Hills (1976) observed large bubbles, or regions of high voidage, at high gas rates in his 150 mm column, which led him to call the regime "slug flow", although the Taylor bubbles of classical slug flow in narrower columns were not present. Subsequent papers by Hills and his co-workers have re-classified the regime as "churn flow", while Ohnuki and Akimoto (2000) described "churn-froth", "churn-slug" and "churn bubbly" regimes in their 200 mm column, in conditions where traditional slug flow would be expected. Slug flow was not observed in these wider columns.

Bubble size is a parameter whose importance has only been recognised comparatively recently. In any bubbly flow, two opposing processes are at work: bubble coalescence as a result of collisions between bubbles, and bubble break up as a result of turbulence in the liquid phase. At low liquid velocities, turbulence is small, so coalescence dominates and the equilibrium bubble size is large; these larger bubbles have distorted, constantly varying shapes, and rise in spiral or zigzag motion, which itself promotes collision. At higher velocities turbulence is increased, and the equilibrium bubble size is smaller; these smaller bubbles are essentially spherical, and rise rectilinearly, reducing collision. This leads to the two sub-regimes described above as "Discrete Bubbly" (low turbulence) and "Dispersed Bubbly" (high turbulence). The following sections treat these two sub-regimes and their transitions separately.

Discrete bubble to slug transition. In discrete bubbly flow, turbulence forces are insufficient to break up larger bubbles, so if such bubbles are present at the start of the flow, slug flow is more likely to occur. In narrow pipes (less than about 50 mm for air/water) the Taylor bubbles of slug flow have a lower velocity than smaller bubbles, so any smaller bubbles will overtake a Taylor bubble and tend to coalesce with it. This was pointed out by Taitel *et al.* (1980), who concluded that air/water discrete bubbly flow could not exist in tubes narrower than 50 mm – a conclusion manifestly untrue in view of the fact that many studies of bubbly flow have been carried out in tubes below this diameter. A more likely conclusion for such narrow tubes is that bubbly flows may develop into slug flow as the gas velocity increases, but that slug flow will never break down into bubbly flow as it falls; *i.e.*, a configuration which generates Taylor bubbles at the pipe inlet will lead to slug flow even at minimal gas flows. Such conditions could arise if the phases enter separately through different legs of a junction, or if a vertical pipe is preceded by a bend and a section of horizontal pipe, where gas can accumulate at the bend and only pass into the vertical pipe in large packets (as in the experiments of Nishikawa *et al.*, 1968, for example).

It is clear that entrance conditions, in particular initial bubble size, are critical in the study of this transition. Bubbles generated from porous sinters or small drilled holes, as well as those arising from nucleate boiling on the surface of the tube, are likely to be small and hence lead initially to discrete bubbly flow. Since turbulent break up is insignificant, the transition to slug flow will be determined mainly by the processes of bubble coalescence.

As non-spherical bubbles have random paths they should collide and coalesce; thus it might be argued that if tubes are long enough, coalescence will always lead to slug flow, so that discrete bubbly flow is an unstable entrance effect. It is hard to test this hypothesis in ambient pressure studies, since the decrease in hydrostatic pressure along the tube causes an expansion of the bubbles and an increased likelihood of coalescence for that reason.

Radovcich and Moissis (1962) assumed the hypothesis, and modelled collision by assuming that bubbles arrange themselves in a cubic lattice and move randomly with respect to other bubbles in the lattice with a mean fluctuating velocity. They determined a bubble collision frequency:

$$f_{BC} = \frac{\bar{c}}{D_t\left[\left(\frac{0.74}{\varepsilon_g}\right)^{1/3} - 1\right]^5} \tag{7}$$

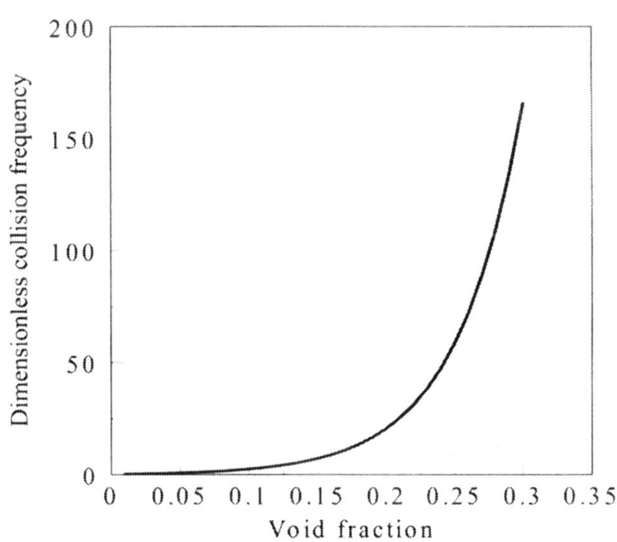

Figure 20. Variation of bubble collision frequency with void fraction as proposed by Radovcich and Moissis, 1962.

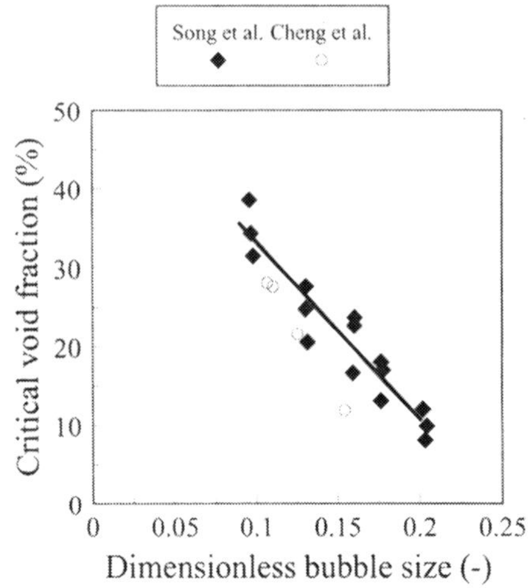

Figure 21. Effect of bubble size on critical void fraction.

The dimensionless frequency, $f_{BC}D_t/c$ is shown in figure 20 as a function of the void fraction, ε_g. It can be seen that at low void fractions the frequency of collision is low and hence bubbly flow is probable; however at void fractions of 0.2 - 0.3 collision frequencies have risen sharply and collision and plug flow will result. Radovcich and Moissis assumed the coalescence of bubbles to form slugs occurred randomly and uniformly throughout the flow. More recent studies suggest that the coalescence is preceded by void fraction (density) waves in which there are regions of higher and lower bubble concentration (Song *et al.*, 1995, Matuszkiewicz *et al.*, 1997). At low void fractions, these waves are attenuated, but above a critical void fraction, they grow, leading to eventual coalescence and slug flow. This approach is not yet fully developed, but it does reinforce the idea that the transition to slug flow will occur at a critical void fraction, ε_{gcr}. Following Radovcich and Moissis (1962), Taitel *et al.* (1980) suggested taking $\varepsilon_{gcr} = 0.25$, and Mishima and Ishii (1984) $\varepsilon_{gcr} = 0.3$.

However, a number of studies of void fraction waves have found that there is a bubble size dependence of the critical void fraction which is not included in Equation 7. Figure 21 presents data from Song *et al.* (1995) in a 39 mm column, and Cheng *et al.* (2002) in a 29 mm column, plotted as ε_{crit} as a function of the dimensionless bubble diameter, D_b/D_t. The trendline, fitted to Song's data, has the equation:

$$\varepsilon_{crit} = 0.55 - 2.37(D_b/D_t) \qquad (8)$$

Song *et al.* do not justify their use of D_b/D_t on the ordinate of Figure 21, but the fact that it brings together the two sets of data in different tubes suggests it is the correct parameter.

Some experiments by Prasser *et al.* (2000) in a 51.2 mm column dramatically indicate this bubble diameter effect. At the same air rate of 0.125 m/s and water rate of 1 m/s, the use of 4mm holes in the gas distributor (bubble diameter > 10 mm) led to rapid coalescence and slug flow after 3.03m (60D), whereas the use of 0.8mm capillaries led to uniform bubbly flow with almost no coalescence over the same distance.

As has already been said, true slug flow, in which Taylor bubbles effectively fill the cross-section, does not occur in larger columns. Taylor bubbles can be formed in still liquids in tubes up to 300 mm in diameter (Hsu and Simon, 1969), but liquid-phase turbulence and the presence of smaller bubbles causes them to break up, and the upper limit for slug flow in air/water systems is thought to be around 100 mm. Above this size, bubbly flow gradually degenerates into a highly turbulent regime with many large bubbles, but none large enough to bridge the tube. There is no clear critical void fraction or transition in this case (Cheng *et al.*, 1998).

Taitel *et al.* (1980) use the concept of a critical void fraction to define the transition in terms of superficial velocities. Actual velocities for the liquid u_l, and bubbles, u_B, are defined as:

$$u_l = \frac{u_{ls}}{(1 - \varepsilon_g)} \qquad u_B = u_g = \frac{u_{gs}}{\varepsilon_g} \tag{9}$$

For the velocity of bubbles relative to the liquid, known as the slip velocity, they use the equation suggested by Harmathy (1960):

$$u_{Br} = u_B - u_l = 1.53\left(\frac{g\sigma\Delta\rho}{\rho_l^2}\right)^{1/4} \tag{10}$$

This equation is for an isolated bubble in an infinite medium. For a bubble rising in a swarm of bubbles, Zuber and Hench (1962) suggested that an additional term $(1 - \varepsilon_g)$ would be required, while other authors suggest other functions of the void fraction. Combining (9) and (10) yields

$$u_{ls} = 3\,u_{gs} - 1.15\left(\frac{g\sigma\Delta\rho}{\rho_l^2}\right)^{1/4} \tag{11}$$

Substituting for u_{br} and setting ε equal to the critical value of 0.25 (Taitel *et al.*, 1980) gives:

$$u_{ls} = 3\,u_{gs} - 1.15\left(\frac{g\sigma\Delta\rho}{\rho_l^2}\right)^{1/4} \tag{12}$$

Use of a different critical void fraction will clearly change both the coefficients of Equation (11), while use of a different relative velocity will only change the second term. This is only significant at very low flow rates, as the term only has a value of around 0.2 m/s.

It may be noted that the derivation of equation (11) is *one dimensional*, in that it does not allow for variations of gas or liquid flows across the tube. The drift flux model of Zuber and Findlay (1965) presented in lecture makes allowances for this effect. The actual mean bubble velocity given by equation (9) is expressed relative to the mixture velocity, $(u_{gs} + u_{ls})$, rather than the actual liquid velocity, and correction terms are applied to both mixture velocity and relative velocity to allow for radial non-uniformity:

$$\frac{u_{gs}}{\varepsilon} = C_0(u_{gs} + u_{ls}) + C_1 \tag{13}$$

This equation can be rearranged in a form similar to Equation (11) to give

$$u_{ls} = 3\,u_{gs} - 1.15\left(\frac{g\sigma\Delta\rho}{\rho_l^2}\right)^{1/4} \tag{14}$$

This is the approach favoured by Mishima and Ishii (1984), using the equations of Ishii (1977) for C_0 and C_1:

$$C_0 = 1.2 - \sqrt{(\rho_g/\rho_l)} \qquad u_{Br} = u_B - u_l = 1.53\left(\frac{g\sigma\Delta\rho}{\rho_l^2}\right)^{1/4} \tag{15}$$

together with a value of $\varepsilon_{gcr} = 0.3$ to give, at low pressures,

$$u_{ls} = 3\,u_{gs} - 1.15\left(\frac{g\sigma\Delta\rho}{\rho_l^2}\right)^{1/4} \tag{16}$$

Neither Equation (12) nor (16) allows for the sort of bubble-size dependence of ε_{gcr} shown in Figure 21, but the vast majority of published data on the transition does not include information on bubble size anyway, and it would be rare for a designer to be able to predict this quantity. It is probable that a lot of the scatter found in the database can be attributed to the lack of control over bubble size.

None of the above models make any allowance for the effect of liquid viscosity on the transition, but the work of Furukawa and Fukano (2001) showed that the transition line moves to higher u_{ls} at a given u_{gs} as the viscosity increases from 0.001 to 0.0125 Pa s. There was no obvious change in initial bubble size, but coalescence occurred more readily in the more viscous liquids (glycerol solutions). Void fractions were not measured, so it is not certain that it is the critical void fraction which is affected by liquid viscosity, but their data can be fitted by Equation (11) with ε_{gcr} falling from 0.39 in water to 0.15 in the most viscous solution. The very early data of Oshinowo and Charles (1974) also show an increase in u_{ls} with viscosity, although to a lesser extent.

With water-like liquids, provided the bubbles in a given flow are produced by fine sinters or during flow boiling, they are likely to be of a size where $0.2 < \varepsilon_{gcr} < 0.3$, and Equations (12) or (16) will make adequate predictions. Since slug flow is to be avoided, designers will probably choose the more conservative (12).

It should also be noted that the above discussion is limited to pure liquids. The effect of surface contaminants can be important, as they hinder bubble coalescence, and thus increase ε_{gcr}. For example, Whalley *et al.* (1972) found that an air/water mixture containing small amounts of hexanol or butanol could produce bubbly flow at a void fraction up to 0.6.

Dispersed bubble to slug transition. Here, the liquid phase turbulence is sufficient to cause breakup of larger bubbles, and it is possible to calculate theoretically the maximum stable bubble size in any flow. Hinze (1955) and Sevik and Park (1973) considered the process of bubble breakup by turbulence in the continuous phase. The theory results in a critical Weber number for bubble breakup

$$We_c = \frac{\rho_l \overline{v^2} r_B}{\sigma} \tag{17}$$

where r_B is the bubble radius and v^2 is the spatial average value of the square of the velocity difference over a distance r_B. Sevik and Park (1973) assumed that breakup would occur when resonant frequency of the bubble was equal to a characteristic turbulence frequency. This led to good agreement with experiment and an expression for the maximum possible stable bubble size.This has been extended by Taitel *et al.* (1980) and Barnea and Taitel (1985)

$$d_{MAX} = We_c \left(\frac{\sigma}{\rho_l} \right)^{3/5} E^{-2/5} \tag{18}$$

where $We_{crit} = 1.17$ (both theoretically and by experiment) and E is the rate of energy dissipation per unit mass in the continuous phase. E is determined in pipe flow from single-phase pressure drop, assuming turbulent energy dissipation is the major cause of pressure loss.

$$E = \left| \frac{dp}{dz} \right| \frac{u_m}{\rho_m} \tag{19}$$

and

$$\frac{dp}{dz} = \frac{2f}{d} \rho_m u_m^2 \tag{20}$$

Barnea and Taitel (1985) argued that plug flow cannot be formed if the bubbles are small enough to remain spherical, since such small bubbles do not have the random, zigzag motion which leads to coalescence in larger bubbles. They used the correlation of Migayi (1925) for the maximum size at which bubbles remain spherical:

$$d_{crit} = 2\sqrt{\frac{2\sigma}{5\,\Delta\rho\, g}} \tag{21}$$

They also note that $We_{crit} = 1.17$ only holds for a single isolated bubble. For a bubble swarm, they suggest a void fraction based equation developed from the work of Calderbank (1958).

$$\left[4.15 \left(\frac{u_{gs}}{u_{gs} + u_{ls}} \right)^{0.5} + 0.725 \right] = 2\sqrt{\frac{2\sigma}{5\,\Delta\rho\, g}} \left(\frac{\rho_l}{\sigma} \right)^{0.6} \left(\frac{f}{2D_t} \right)^{0.4} \left(u_{gs} + u_{ls} \right)^{1.2} \tag{22}$$

Assuming that the void fraction can be described by the homogeneous model, these can be combined to yield an implicit equation in u_{ls} as a function of u_{gs}. Taitel *et al.* (1980) also argued that there must still be a critical void fraction above which bubbles are so close-packed that coalescence is inevitable no matter how small they are; they took this to be a cubic lattice, for which $\varepsilon =$

0.52. Relative velocities can be ignored in these high velocity flows, so the flow is homogeneous, and Equation (11) becomes:

$$u_{ls} = 3\,u_{gs} - 1.15\left(\frac{g\sigma\Delta\rho}{\rho_l^2}\right)^{1/4} \tag{23}$$

Note that Equations (21) and (22) represent, according to this analysis, the boundary of the dispersed bubbly flow regime: what lies on the other side of the boundary could be either discrete bubbles, slugs, or, in larger tubes the highly turbulent churn-like flows described earlier.

Chen *et al.* (1997) criticise Equation (23) because it suggests that as u_{gs} increases, the critical u_{ls} needed to maintain dispersed bubbles *decreases*. This is counter-intuitive, and does not fit the experimental results of Shoham (1982) and others. They propose a different model for this transition, in which the turbulent kinetic energy of the liquid phase is balanced against the surface energy of the dispersed gas phase to determine an equilibrium bubble size: this bubble size is then equated to the maximum size for bubbles to remain spherical (half of Equation 21) as in the analysis of Taitel *et al.* (1980).

They write the rate of supply of turbulent kinetic energy by the liquid as:

$$E_T = \frac{3}{4} f\rho_l u_{ls}^{\;2} A u_{ls} \tag{24}$$

while the total surface energy of the generated gas bubbles is given by:

$$E_S = \frac{6\sigma}{d_B} A u_{gs} \tag{25}$$

Equating these two terms, and using the Blasius formula for *f*, they derive

$$d_B = \frac{8\sigma u_{gs}}{0.046(D_T u_{ls}/\nu_l)^{-0.2}\rho_l u_{ls}^{\;3}} \tag{26}$$

Equations (21) and (26) are then combined to give the equation of the transition from dispersed bubbly flow:

$$u_{ls} = 4.927\left[\frac{\Delta\rho g\sigma}{\rho_l^{\;2}}\right]^{0.357}\left(\frac{D_t}{\nu_l}\right)^{0.0714} u_{gs}^{\;0.357} \tag{27}$$

This comparatively simple equation has the correct trend, and Chen *et al.* (1997) suggest it should apply for all pipe orientations, testing it against the data of Shoham (1982) to prove the point. Its major difference from the Taitel *et al.* model is in the use of u_{ls}^2 rather than u_m^2 in the kinetic energy term, and it is this which allows it to predict a monotonously increasing transition curve.

However, in presenting her own model, valid for liquid/liquid as well as gas/liquid systems, Brauner (2001) implies that Chen *et al.* (1997) are wrong. Because of the homogeneous model assumption, the actual velocity of the continuous phase, $u_{cs}/(1-\varepsilon_d)$ is equal to that of the dispersed phase, and both are equal to the mixture velocity, u_m. This explains the use of u_m in

Equations (19) and (20), but also suggests that Equation (24) should read:

$$E_T = \frac{3}{4} f \rho_l u_m^2 A u_{ls} \tag{28}$$

Brauner's model combines those of Taitel *et al.* (1980) and Chen *et al.* (1997) by using the former for low concentrations of dispersed phase, and the latter (with Equation (28) replacing (18)) at higher concentrations. This leads to two different equations for d_{MAX}, which are as follows: For low concentrations:

$$\left(\frac{d_{MAX}}{D_t}\right)_0 = 0.833\left(\frac{\rho_c u_m^2 D_t}{\sigma}\right)^{-0.6}\left(\frac{\rho_m}{\rho_c(1-\varepsilon_d)} f\right)^{-0.4} \tag{29}$$

(Note that her paper mistakenly gives the numerical constant as 0.55 and not 0.833, but her figures have been drawn using the correct value.)
For high concentrations:

$$\left(\frac{d_{MAX}}{D_t}\right)_\varepsilon = 2.22 C_H^{0.6}\left(\frac{\rho_c u_m^2 D_t}{\sigma}\right)^{-0.6}\left(\frac{\rho_m}{\rho_c(1-\varepsilon_d)} f\right)^{-0.4}\left(\frac{\varepsilon_d}{1-\varepsilon_d}\right)^{0.6} \tag{30}$$

where f is the friction factor, given by the Blasius formula, and C_H is a proportionality factor; Brauner wrote $E_T = C_H \cdot E_S$, although she suggests using $C_H = 1$, which corresponds to the assumption of Chen *et al.* (1997). She notes that, for bubbly gas/liquid flows, the term $\rho_m/\rho_c(1-\varepsilon_d)$ is very close to unity, while the term $\varepsilon_d/(1-\varepsilon_d)$ in (30) corresponds quite closely to the function of ε_g in (22). Brauner takes whichever of the two values of d_{MAX} is the larger, and equates it to the d_{crit} of Equation (21) to obtain the transition line. Like the Barnea and Taitel (1985) model, the final equation is implicit in u_{cs}. The Hinze model which lies behind all these studies only applies for $d/D_t < 0.1$, which, using (21) for the critical bubble diameter, means

$$D_t > 20\sqrt{\frac{2\sigma}{5\Delta\rho g}}$$

For air/water, this yields $D_t > 34$ mm, but for liquid/liquid systems, the critical tube diameter may be significantly larger. For smaller tubes, Brauner (2001) suggests using a model due to Kubie and Gardner (1977) which replaces (29) and (30) with:

$$\left(\frac{d_{MAX}}{D_t}\right)_0 = 1.38\left(\frac{\rho_c u_m^2 D_t}{\sigma}\right)^{-1.0}(f)^{-1.0} \tag{31}$$

and

$$\left(\frac{d_{MAX}}{D_t}\right)_\varepsilon = 8 C_K\left(\frac{\rho_c u_m^2 D_t}{\sigma}\right)^{-1.0}\left(\frac{\rho_m}{\rho_c} f\right)^{-1.0}\left(\frac{\varepsilon_d}{1-\varepsilon_d}\right) \tag{32}$$

For capillary tubes, if the above equations predict $d_{MAX}/D_t > 0.5$, she suggests using:

$$\left(\frac{d_{MAX}}{D_t}\right)_0 = \left(\frac{d_{MAX}}{D_t}\right)_\varepsilon = 0.5 \tag{33}$$

Figure 22 compares the predictions of the models of Taitel *et al.* (1980), Barnea and Taitel (1985), Chen *et al.* (1997) and Brauner (2001) for the dispersed bubble transition for air/water in a 25 mm vertical pipe. There is insufficient published data which clearly identifies the dispersed bubble regime for a thorough test of these models.

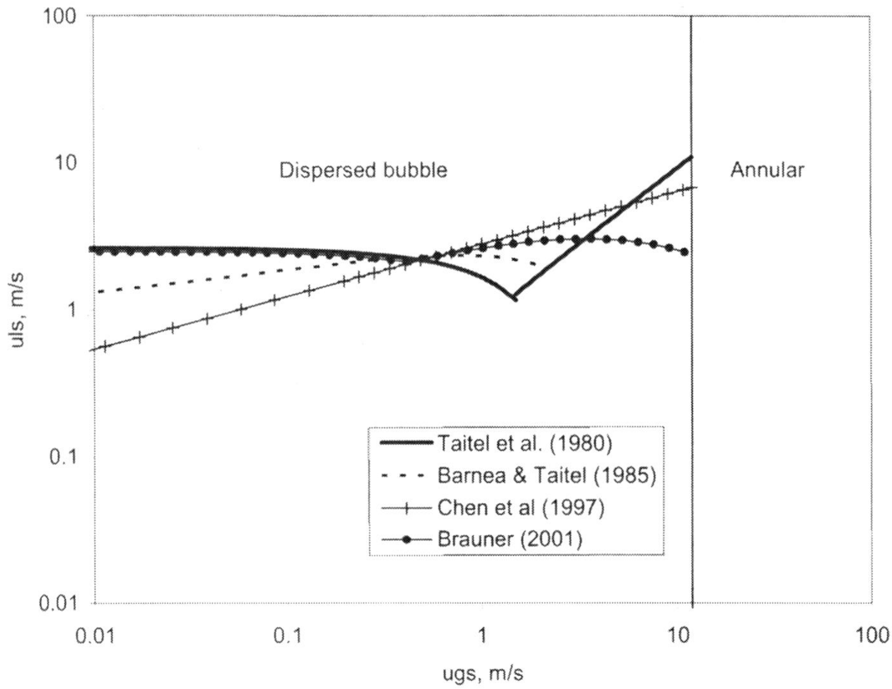

Figure 22. Models for the transition to dispersed bubbly flow.

6.2 Transitions at Higher Gas Velocity

Before considering the plug/churn and churn/annular transitions it is useful to give some consideration to the phenomena of **flooding** and **flow reversal** which are used in many of the models of these transitions. Flooding occurs in counter-current flow when the flow rates are so large that one of the phases (usually the liquid phase in gas/liquid systems) can no longer maintain its flow, and is forced back out of the equipment by the other phase. It is familiar to Chemical Engineers because of its occurrence in packed distillation and absorption columns, but it also occurs in other industrial processes, such as reflux condensation.

In two-phase flows in tubes, the phenomena are sometimes referred to collectively as the **counter-current flow limitation**. They are best illustrated by reference to Figure 23. The following description is a composite of the papers by Govan *et al.* (1991), Watson and Hewitt (1998) and Vijayan *et al.* (2001), which are in broad agreement with the earlier work of McQuillan (1985) and others; the letters in brackets refer to Figure 23. Liquid is introduced part-way down a vertical tube through a porous section or is formed in the tube by condensation. If the upflow of gas or vapour is low the film travels downward with small ripples on the interface, (a). Increasing the gas velocity causes the size of the waves to increase and their velocity to decrease, but they remain incoherent. A further increase in the gas velocity can, suddenly, cause the formation of coherent ring waves at the liquid outlet, which are carried up repeatedly and periodically to above the liquid inlet, where they establish a churn flow, (b). This is the **flooding** point, and it is marked by a significant increase in the mean pressure drop, as shown in Figure 24, and also in the amplitude of the fluctuations of pressure drop. Further increase in the gas rate carries more liquid up above the liquid inlet, leaving a smaller downflow below it, until all the liquid is flowing upwards, (c). If the gas flow is now reduced again, liquid first starts to fall below the inlet at the point of **flow reversal**, (d). Continued reduction of the gas flow eventually restores the pure counter-current flow, at the point called "**deflooding**" by Govan *et al.* (1991), (e). At all but the lowest liquid feed rates, there is a hysteresis effect, with deflooding occurring at lower gas rates than flooding.

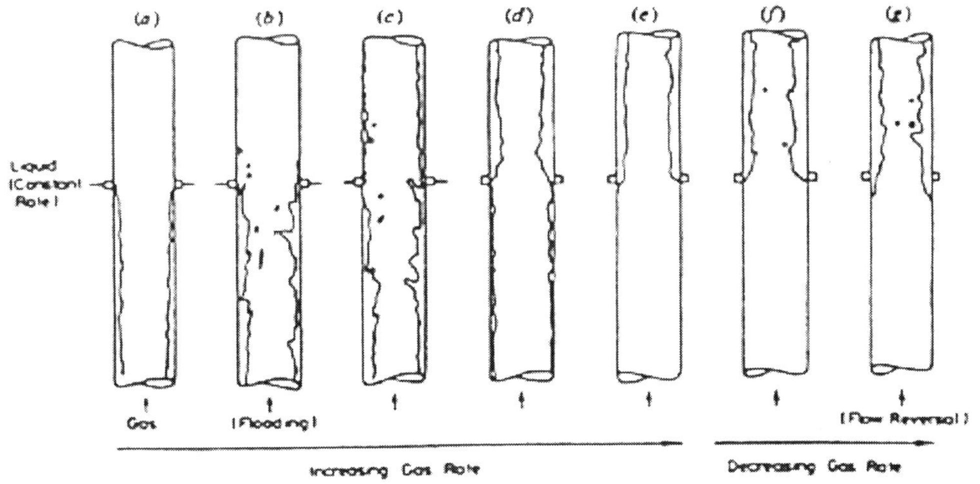

Figure 23. Counter-current flow limitation.

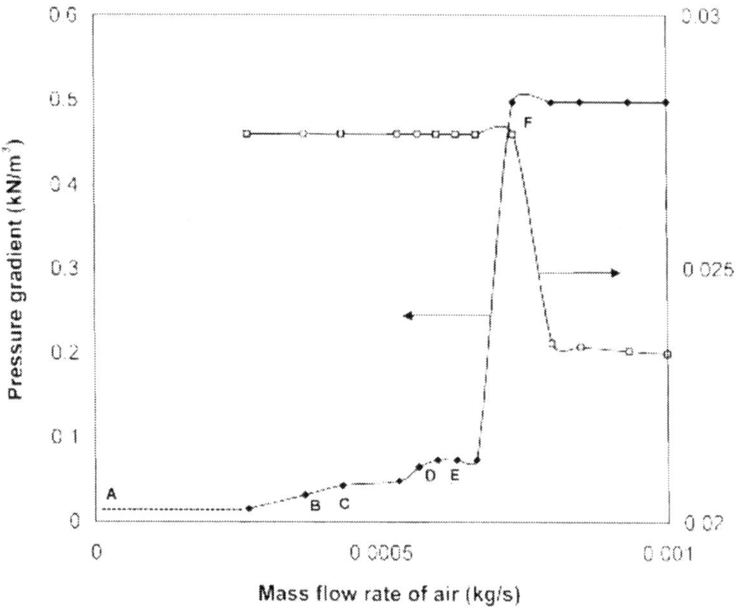

Figure 24. Pressure drop and liquid downflow in a 25 mm tube. Reprinted from International Journal of Multiphase Flow, vol. 27, Vijayan, M., Jayanti, S. and Balakrishnan, A.R., Effect of tube diameter on flooding, pp 797-816, Copyright (2001), with permission from Elsevier.

The description above refers to narrow tubes (the three cited papers used tubes of 32, 26 and 25 mm), in which the liquid outlet is also a porous sinter, with sufficient empty tube below it to establish a fully developed gas flow, and the instability is due to a large wave being carried up the tube. Instability starts at the bottom of the tube, where the waves are largest, and this mechanism is borne out by the fact that, for a given liquid rate, the flooding gas velocity decreases with increasing separation of liquid inlet and outlet (waves have more chance to grow), and also falls sharply if the outlet sinter is replaced by the square-edged end of the tube (the vena contracta in the gas flow promotes instability). A bell-shaped outlet, which reduces the vena contracta effect, gives intermediate results. The square-edged outlet shows no tube length effect. These results are shown in Figure 25, redrawn from Govan *et al.* (1991), which is plotted using the dimensionless velocities defined in equation (35) below.

Wider tubes show a different mechanism, as found by Watson and Hewitt (1998) in a 82 mm tube, Vijayan *et al.* (2001) in 67 and 99 mm tubes and Zabaras and Dukler (1988) in a 50 mm tube. Here, instead of the large waves formed at the liquid outlet being carried up past the tube inlet, they rise a short distance before breaking down into a high concentration of droplets, which are swept up in the gas stream and re-deposited higher up the tube, increasing the liquid flow rate to the point where it becomes unstable in turn. The observed result is a churn-type flow which gradually spreads up the tube beyond the liquid inlet. Flooding, with the characteristic sharp increase in pressure drop, occurs when the churn-type flow reaches the liquid inlet.

Attempts to predict flooding conditions have ranged from empirical correlations to theoretical models. A popular description is due to Wallis (1961).

$$u_g^{*\frac{1}{2}} + u_l^{*\frac{1}{2}} = c \tag{34}$$

where u_g^* and u_l^* are dimensionless superficial velocities given by:

$$u_g^* = \frac{u_{gs}\sqrt{\rho_g}}{(g\, D_t\, \Delta\rho)^{1/2}}; \qquad u_l^* = \frac{u_{ls}\sqrt{\rho_l}}{(g\, D_t\, \Delta\rho)^{1/2}} \tag{35}$$

and c is a constant of order unity. Dukler and Smith (1979) have also used this description. Other authors have included a second constant to allow for more tuning of the curve to experimental results, although $c_1 = 1$ and $c_2 = 0.88$ are the usual recommended values.

$$u_g^{*\frac{1}{2}} + u_l^{*\frac{1}{2}} = c \tag{36}$$

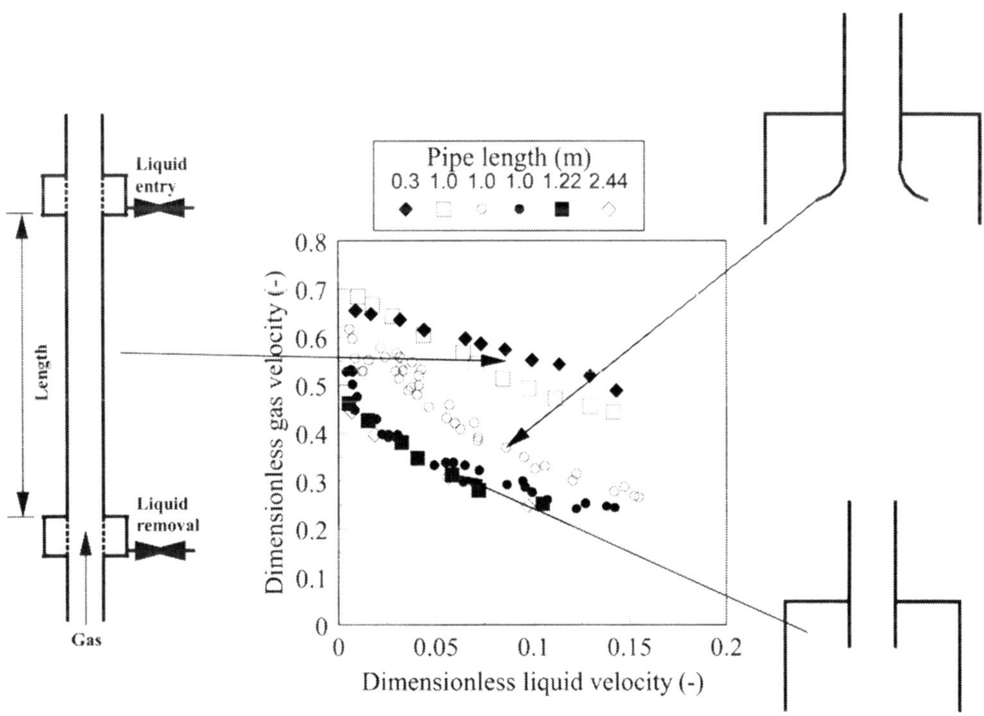

Figure 25. Flooding in a 32 mm tube. (after Govan *et al.*, 1991).

Since there is a large effect of film length on the flooding point when liquid is removed through a porous wall, Jayanti and Hewitt (1992) suggested making c_l a function of L/D as follows:

$$c_1 = 0.1928 + 0.01089\left(\frac{L}{D}\right) - 3.754\times10^{-5}\left(\frac{L}{D}\right)^2 \qquad \text{for (L/D)< 120} \tag{37}$$

For larger tube diameters, the equation of Pushkina and Sorokin (1969) has proved popular.

$$Ku_g = \frac{u_g\sqrt{\rho_g}}{(g\sigma\Delta\rho)^{0.25}} = 3.2 \tag{38}$$

This does not allow for any effect of liquid rate on flooding, although experimentally the flooding gas rate falls as the liquid rate rises. Sun (1979) suggested an equation analogous to (30) but using Kutateladze numbers as the dimensionless velocities, which does allow for an effect of liquid rate:

$$\sqrt{Ku_g} + c_3\sqrt{Ku_l} = c_4 \tag{39}$$

where Ku_g is defined in Equation (38) and Ku_l uses u_{ls} and ρ_l in place of u_{gs} and ρ_g. The recommended values of the constants are c_3 = 1.0 and c_4 = 1.79, making Equation (39) agree with Equation (38) as $u_{ls} \rightarrow 0$. An equation of this type was also proposed by Chung *et al.* (1980), based on a Kelvin-Helmholz instability analysis, in which the constant c_4 was replaced by c_5 $\tanh(c_6 Bo^{1/8})$, where the Bond number, *Bo*, is defined as

$$Bo = \frac{D^2 g(\rho_l - \rho_g)}{\sigma} \tag{40}$$

and c_5 and c_6 are constants dependent on tube geometry.

It may be noted that Equation (36) predicts an increase in flooding velocity with increasing tube diameter, whereas Equation (39) predicts no effect. This could be associated with the change in mechanism of flooding noted above. Jayanti *et al.* (1996) devised a model to represent the effect of tube diameter on flooding, in which the upwards drag force on a large wave due to the air flow is equated to the gravitational force. It was found that the air flow needed to generate flooding increased with tube diameter, and they argued that, ultimately, the gas rate needed to carry up a roll wave would exceed that needed to create massive entrainment from the wave, so that the large tube mechanism detailed above would occur first. Interestingly, Equations (39) and (36) almost coincide for air and water at ambient conditions in a 50 mm tube, as pointed out by Watson and Hewitt (1998), and the experimentally noted switch in flooding mechanism occurs at about this tube diameter.

The effect of physical properties of the fluids was investigated by Zapke and Kröger (1996) in a 30 mm tube. They clearly showed that it is gas phase momentum flux, $\rho_g u_{gs}^2$ (or its square root, in u_g^*) and not gas phase Reynolds number which is the appropriate correlating parameter, and they suggested that liquid properties could be incorporated into equation (28) by writing

$$u_g^{*\frac{1}{2}} + u_l^{*\frac{1}{2}} = c \tag{41}$$

where Oh, the Ohnesorge number, is given by

$$Oh = \sqrt{\frac{\eta_l^2}{\rho_l D \sigma}} \tag{42}$$

In a later paper (Zapke and Kröger, 2000) they propose a plot of Fr_g against $Fr_l^{0.2} Oh^{0.3}$ as a way of correlating data with a range of different liquids and gases in a given apparatus. For circular tubes, the Froude numbers Fr_g and Fr_l are simply the squares of u_g^* and u_l^*. The plots vary with the geometry of the apparatus, and there is no indication as to whether the additional dependence on D introduced via the Ohnesorge number gives a better description of the effects of tube diameter than equation (4.30). However, as an empirical way of predicting non-aqueous flooding data from air/water experiments in the same apparatus, the method seems promising.

For vertical tubes with square ends, they suggest the following equation

$$Fr_g = \frac{0.0055}{Fr_l^{0.2} Oh_l^{0.3}} \tag{43}$$

McQuillan and Whalley (1985a) carried out a systematic comparison of flooding equations against a bank of 2762 experimental data points available at the time. They concluded that a modified version of the equation of Alekseev *et al.* (1972) was the most accurate.

$$Ku = 0.286\ Bo^{0.26}\ Fr^{-0.22}\ (1 + \frac{\eta_l}{\eta_w})^{-0.18} \tag{44}$$

Note that Fr in this equation is <u>not</u> a standard Froude number! A certain amount of manipulation transforms it into the following:

$$Fr_g = \frac{0.1505}{Fr_l^{0.22} Bo^{0.31} (\rho_l / \Delta\rho)^{0.22} (1 + \eta_l / \eta_w)^{0.36}} \tag{45}$$

which has close similarities with Equation (37) above.

Plug/Churn Transition. A number of approaches have been suggested to delineate the plug/churn transition. Taitel *et al* (1980) consider that churn flow is an entry effect associated with plug flow and have produced a model including a length effect. This approach implies that for infinitely long tubes only bubble, plug and annular flow should occur and thus a direct transition mechanism between plug and annular flows is required. However, in plug flow the film can be flowing downwards whilst in annular flow it flows upward, so an intermediate state might be expected between the film downflow and upflow, particularly from the evidence provided by flooding experiments. For this reason the Taitel *et al* approach is not considered to be appropriate.

Mishima and Ishii (1984) postulated that the transition between plug and churn flow results from a void fraction limitation. They calculated the mean void fraction over the length of a gas plug by a potential flow analysis and the mean void fraction over the total plug unit from a drift flux analysis. The transition was defined as the condition at which the void fractions were equal. Though their transitions showed good agreement with experiment it is not considered reliable because: (i) there is an apparently incorrect application of Bernoulli's equation; and (ii) they assumed that the plug length is determined when the film thickness reaches that calculated by Nusselt (1916) for a falling film and that the length of the liquid slug is approximately zero. Both these are unreasonable assumptions and this approach is not considered suitable. Brauner and Barnea (1986) also used a limiting void fraction approach, and proposed that slug flow would change to churn flow when the void fraction in the liquid plugs between the Taylor bubbles became equal to the maximum packing void fraction (defined above as ε_g = 0.52). Thus the transition line can be obtained from Equation (22) replacing $\varepsilon_g = u_{gs}/(u_{gs} + u_{ls}) = 0.52$.

Nicklin and Davidson (1962) suggested that churn flow occurred when the gas flow in the plug is sufficient to cause flooding in the film surrounding it. This approach is realistic as one can see instabilities (waves) appear on the film towards the end of the gas plug at conditions below the transition. Obviously this is the first indication of the instability which at higher flow conditions results in flooding. Subsequently, McQuillan and Whalley (1985b) and McQuillan (1985) followed this suggestion and produced a transition criterion. The approach was modified by Jayanti and Hewitt (1992) who argued that, since there is a significant effect of film length on the flooding point in counter-current flow, the length of the Taylor bubbles should be taken into account. They used their own correlation, Equation (37) to include this effect.

Recently, Watson and Hewitt (1999) have compared the predictions of the correlations of Mishima and Ishii (1984), McQuillan and Whalley (1985b), Brauner and Barnea (1986) and Jayanti and Hewitt (1992) with some extensive air/water data at 1.2, 3 and 5 bara in a 32 mm tube. One of the comparisons is shown in Figure 26, and it can be seen that only the equation of Jayanti and Hewitt (1992) predicts the correct trend.

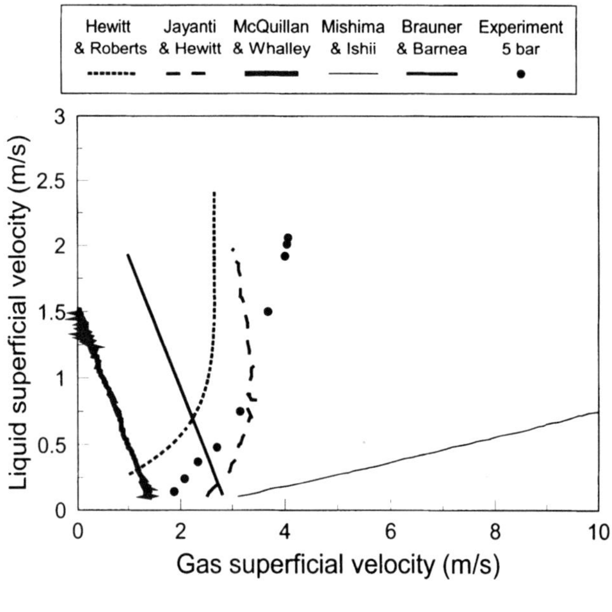

Figure 26. Comparison of slug/churn predictions with air/water data at 5 bara (Watson & Hewitt, 1999).

Churn/annular transition. For this transition a number of approaches have been published. Taitel *et al.* (1980) suggest that the transition can be determined when the gas velocity is just sufficient to suspend a drop. They used a drag equation

$$\frac{1}{2} c_D \pi \frac{d^2}{4} \rho_g u_g^2 = \frac{\pi d^3}{6} g \Delta\rho \tag{46}$$

The pertinent drop size was defined as the largest stable drop which was determined from the equation of Hinze (1955)

$$d = \frac{We_c \sigma}{\rho_g u_g^2} \tag{47}$$

Turner *et al.* (1969) suggested values of 30 and 0.44 for We_c and c_D. Equations (46) and (47) can then be combined to give

$$Ku = \frac{u_{gs} \sqrt{\rho_g}}{[\sigma g \Delta\rho]^{0.25}} = 3.1 \tag{48}$$

Now, whilst the churn/annular transition is accompanied by a large amount of liquid entrainment, it seems more likely that the transition is associated with wave growth mechanisms and unlikely that the relationship of Equation (48) can be applied generally. A more realistic approach links the transition with the flow reversal transition discussed above. Different workers have defined the transition in a number of ways. some have used flow reversal experiments, others have used the associated minimum in pressure drop or visual observations. Their result can be defined in terms of the Wallis parameter

$$u_g^* = \frac{u_{gs} \sqrt{\rho_g}}{(g D_t \Delta\rho)^{1/2}} = C \tag{49}$$

Values of C have been suggested by a number of workers. Figure 27 shows some examples of the pressure drop minimum reported by Willetts *et al* (1987). This shows that there are some differences in u_g^* at the minimum for different fluids but that the value of about 1.0 is a reasonable representation. Chaudry (1967) reported low values, which could be due to his method of detection.

Table 1: Comparison of predicted and experimental flow pattern

		Experimental Flow Pattern			
		Bubble	Plug	Churn	Annular
Pred Flow Pattern	Bubble	128	36	3	4
	Plug	48	198	30	0
	Churn	13	108	296	97
	Annular	4	13	12	355

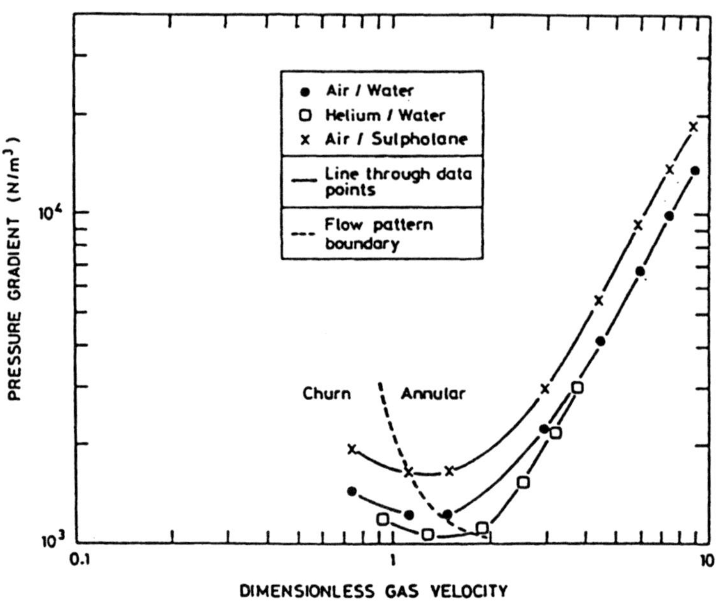

Figure 27. Minimum in pressure drop. Reprinted from Proceedings of the 3rd International Conference on Multi-phase Flow, The Hague, Netherlands, Willetts, I.P., Azzopardi, B.J. and Whalley, P.B., The effect of gas and liquid properties on annular two-phase flow, pp 85-95, Copyright (1987), with permission of BHR Group.

Latest observations using photochromic dye markers have shown that though the film appears to be moving upward the liquid can be oscillating with large waves carrying up the liquid whilst the liquid in the film between the waves flows downwards. A value of C = 1 is suggested to describe the transition.

1.2 Comparison with Experiment.

The accuracy of the flow pattern transition equations selected above, Equations (12), (22) and (44) and $\varepsilon_{gmax} = 0.74$, has been tested against a bank of data by McQuillan (1985) and McQuillan and Whalley (1985b). They used a data bank of 1399 points taken from n sources and covering a wide range of parameters. The results of the tests are given in Table 1 show that 73.7% of the data were correctly predicted using the selected equations. The greatest problem appears to be at the plug/churn transition. If we consider the flow as continuous (annular or bubble) or intermittent (plug and churn) the accuracy rises to 84.1%.

7 Horizontal Flow in Pipes

The transitions for horizontal and near horizontal flows have been considered by a number of workers, but the most popular model-based map remains that of Taitel and Dukler (1976). They considered that the stratified/wavy, wavy/slug and wavy/annular transitions depend on the growth of waves on a smooth stratified liquid layer caused by transfer of energy from the faster moving gas phase. In considering this they started from a model of smooth stratified flow, and calculated the equilibrium height of the liquid layer.

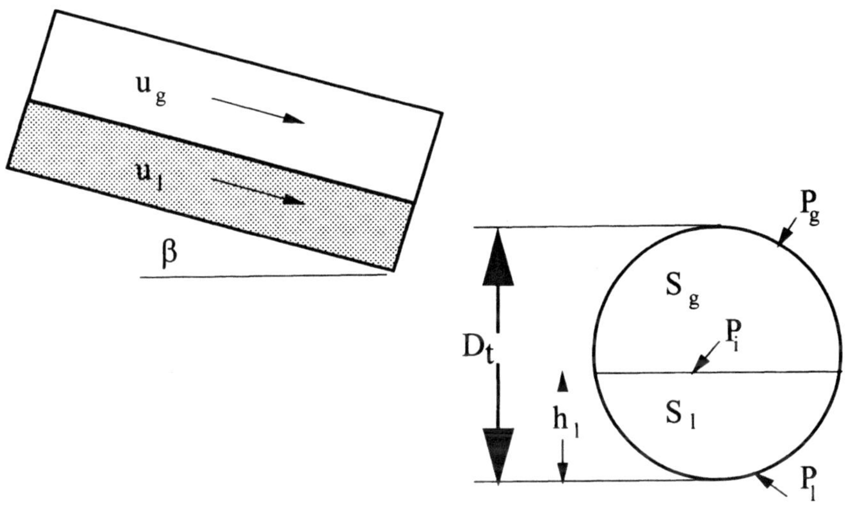

Figure 28. Parameters of stratified gas-liquid flow.

7.1 Stratified Flow

They assumed a uniform steady state flow in which there was no hydraulic gradient (h_l constant) and in which there was no acceleration. For this case, the momentum equations for the liquid and the gas are (Figure 28).

$$-S_l \frac{dp}{dz} + \rho_l\, g \sin\beta - \tau_l P_l + \tau_i P_i = 0 \tag{50}$$

$$-S_g \frac{dp}{dz} + \rho_g\, g \sin\beta - \tau_g P_g - \tau_i P_i = 0 \tag{51}$$

where β is the angle of inclination of the channel, τ_l and τ_g are the wall shear stresses adjacent to the liquid and gas phases, which act over the peripheries P_l and P_g, τ_i is the interfacial shear stress, which acts over the interface periphery P_i and S_l and S_g the flow areas of the respective phases. Eliminating the pressure gradient term (dp/dz) between Equations (50) and (51) we have:

$$\tau_g \frac{P_g}{S_g} - \tau_l \frac{P_l}{S_l} + \tau_i P_i \left(\frac{1}{S_l} + \frac{1}{S_g} \right) + (\rho_l - \rho_g) g \sin\beta = 0 \tag{52}$$

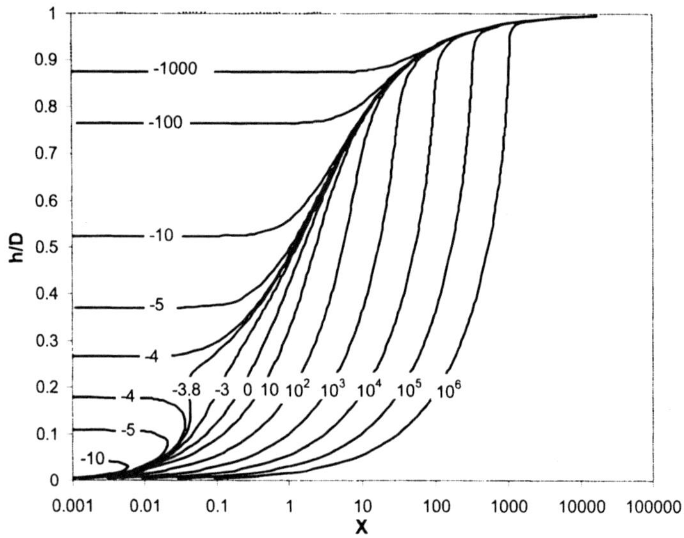

Figure 29. Dimensionless liquid height as a function of X with Y as parameter.

In order to calculate the shear stress terms Taitel and Dukler invoked the following relationships involving friction factors

$$\tau_l = \frac{f_l \rho_l u_l^2}{2}; \tau_g = \frac{f_g \rho_g u_g^2}{2}; \tau_i = \frac{f_i \rho_g (u_g - u_l)^2}{2} \tag{53}$$

where u_g and u_l are the gas and liquid velocities. Assuming a smooth interface, we have $f_i = f_g$. and if in addition u_g is much greater than u_l, then $\tau_i = \tau_g$

Taitel and Dukler related the friction factors to Reynolds numbers as follows:

$$f_l = C_l \left(\frac{\rho_l u_l D_l}{\eta_l} \right)^{-n}; f_g = C_g \left(\frac{\rho_g u_g D_g}{\eta_g} \right)^{-m} \tag{54}$$

where D_l and D_g are equivalent hydraulic diameters and are given by

$$D_l = \frac{4 S_l}{P_l}; D_g = \frac{4 S_g}{(P_g + P_i)} \tag{55}$$

For turbulent flow, Taitel and Dukler assumed that $C_g = C_l = 0.046$ and $n = m = 0.2$. Similarly for laminar flow they assume that $C_g = C_l = 16$ and $n = m = 1$.

Taitel and Dukler non-dimensionalised areas by the pipe diameter squared, lengths by the pipe diameter and velocities by the respective superficial velocities.

$$X^2(\tilde{u}_l\tilde{d}_l)^{-n}\tilde{u}_l^2\frac{\tilde{P}_l}{\tilde{S}_l} - (\tilde{u}_g\tilde{d}_g)^{-m}\left(\frac{\tilde{P}_g}{\tilde{S}_g}+\frac{\tilde{P}_i}{\tilde{S}_l}+\frac{\tilde{P}_i}{\tilde{S}_g}\right) - 4Y = 0 \tag{56}$$

where the Martinelli parameter X is defined as:

$$X^2 = \frac{(dp_F/dz)_l}{(dp_F/dz)_g} = \frac{u_{ls}^2}{u_{gs}^2}\frac{\rho_l}{\rho_g}\frac{f_{ls}}{f_{gs}} \tag{57}$$

and the dimensionless parameter Y is defined as

$$Y = \frac{(\rho_l - \rho_g)\, g \sin\beta}{(dp_F/dz)_g} \tag{58}$$

With the exception of X and Y, the terms in Equation (56) are all functions of the dimensionless liquid height h_l/D_t, and can be found by simple trigonometry. Equation (56) can thus be solved to give the dimensionless height in terms of X, Y. Figure 29 illustrates the result when all flows are turbulent ($n = m = 0.2$). Note that Y has the same sign as $\sin\beta$ and is positive for downflow, negative for upflow, and zero for horizontal flow.

Before looking at some criticisms of the details of this model, it is worth pointing out some features of Figure 29. For horizontal flow, $Y = 0$, the liquid height increases as X increases, which makes sense since the ratio u_{ls}/u_{gs} increases. A constant non-zero value of Y may be taken as representing the effect of slope at a given gas rate. Positive Y (downwards slope) leads to decreased liquid height at given liquid and gas rates, which also makes sense. Negative Y (upwards slope) leads to more interesting effects. For $-3<Y<0$, the curve has a similar shape to the horizontal case, but shifted to greater liquid heights, but for $Y < -3.7$, there are three solutions for liquid height at a given value of X, and for $Y \leq -4$, X^2 becomes negative over part of the range. This sort of behaviour, which was pointed out by Landman (1991), suggests a hydraulic jump, with the intermediate root unstable.

This ties in well with some recent experiments on air/water flow in a 76.3 mm pipe by Simmons and Hanratty (2001). These authors found, for upward inclinations of 0.2°, 0.4° and 1.2°, that there was a critical gas velocity below which slugs always formed at the tube entrance no matter how small the liquid velocity. Gas forced these slugs up the pipe, but between the passage of the slugs, the liquid flowed back down towards the entrance. If we substitute their measured critical gas velocities and inclinations into Equation (58), we can calculate a critical value of Y. The results are Y = -17.1, -14.5 and -23.1, which are reasonably constant, but about 5 times larger than the value of -3.7 found in the Taitel and Dukler analysis. The discrepancy could well be due to the wavy interface structure and consequent increased friction factors discussed below; Simmons and Hanratty found interfacial friction factors greater by a factor of between 5 and 9 than those predicted with $f_i = f_g$. If the Taitel and Dukler analysis is run with $f_i/f_g = 5$, the critical Y value becomes -14, which is much closer to what Simmons and Hanratty found. For an upward inclination of 0.05°, Simmons and Hanratty (2001) found no critical gas velocity, which they attribute to the fact that, at this angle, the bottom of the test section at the exit lies below the top of the test section at inlet ($\sin^{-1}(D/L) = 0.19°$). Simon (1998) looked at flow boiling in slightly inclined tubes, and found that slugging always occurred in upward inclined flows above a critical angle even though the horizontal flow was stratified at the same conditions. He quotes the critical angle as lying between 0.1° and 0.4°: the criterion of Simmons and Hanratty gives an angle of 0.18° for his test section.

The major criticisms of the Taitel and Dukler model centre round their assumptions of a smooth interface, and the modelling of the various frictional stresses. Experimentally, a smooth interface is only found at very low gas flow rates; at higher rates, various types of waves are formed. They will be looked at in more detail later, as they eventually lead to the various flow regime transitions, but their major importance here is that they lead to increased interfacial turbulence, so that $f_i = f_g$ is no longer valid.

Andritsos and Hanratty (1987) used the work of Kowalski (1987) and Andritsos (1986) to suggest that the gas/wall shear stress τ_g can still be calculated from the above. Using this calculated value together with measured pressure drops and h/D values in Equation (51) leads to a value of τ_i which can be compared with the calculated τ_g. Andritsos and Hanratty make the comparison in terms of f_i/f_g and present the following equations:

$$\frac{f_i}{f_g} = 1 \qquad \text{for } u_{gs} \le u_{gs,t} \tag{59}$$

$$\frac{f_i}{f_g} = 1 + 15\left(\frac{h}{D}\right)^{0.5}\left(\frac{u_{gs}}{u_{gs,t}} - 1\right) \qquad \text{otherwise} \tag{60}$$

where $u_{gs,t}$, the transition velocity for the formation of large irregular waves, is given approximately by:

$$u_{gs,t} \approx 5m/s\left(\frac{\rho_{g0}}{\rho_g}\right)^{0.5}$$

where ρ_{g0} is the gas density at atmospheric pressure.

Having calculated τ_i, Andritsos and Hanratty (1987) were then able to calculate τ_l from Equation (50).. They found that at high gas rates, equations (52) and (54) gave acceptable results: for low gas rates, they proposed a complex procedure involving a dimensionless liquid film height h^+ rather than f_l.. Since equation (60) involves h/D, an iterative procedure is needed when using it. Spedding and Hand (1997) proposed an equation for f_i which only involves superficial velocities and hence should be easier to implement:

$$\frac{f_i}{f_{gs}} = 1.76\frac{u_{gs}}{6} + 2.7847\,log_{10}\left(\frac{u_{ls}}{6 + u_{ls}}\right) + 7.8035 \tag{61}$$

where f_{gs} is the friction factor for gas alone flowing in the pipe.

At high gas and liquid rates, the interface becomes curved, as liquid starts to climb up the walls, eventually leading to annular flow. This curvature leads to different values for the various perimeters, as well as to the friction factors. The most sophisticated of the models based on this interface structure is the MARS model (Modified Apparent Rough Surface) by Grolman and Fortuin (1997), which covers a range of gas and liquid rates, physical properties and pipe inclinations. They propose a correlation for the fraction of wall wetted.

Stratified to Slug or Annular Transition. Observations of horizontal pipe flow have shown that waves grow on initially stratified flows if the liquid flow rate is increased. At low gas flow rates the wave fills the tube and slug flow occurs. At higher gas flow rates, there can be too little liquid to fill the pipe. In this case, Taitel and Dukler (1976) argued that the wave is swept round the pipe to form annular flow. They note that there is some atomisation of liquid. Lin and Hanratty (1987) has shown that atomisation can precede the formation of annular flow at lower liquid flow rates. Butterworth (1967) did observe the waves being swept round the periphery.

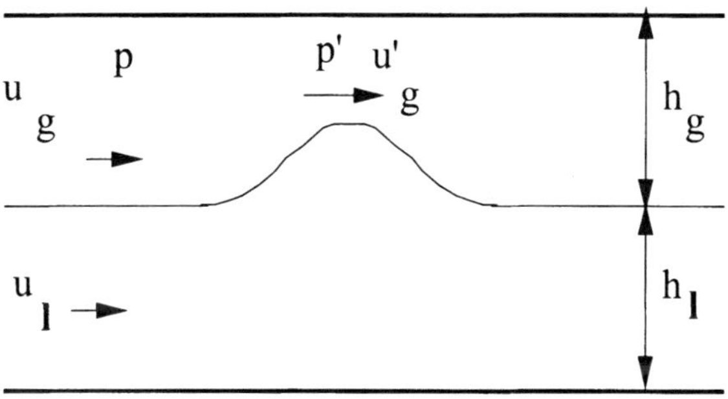

Figure 30. Growth of a wave.

Taitel and Dukler (1976) analysed the roll-wave instability by a Kelvin-Helmholtz approach with a finite wave amplitude. For flow in a rectangular channel, Figure 30, neglecting the motion of the wave, growth of a wave will occur if

$$p - p' = (h_g - h'_g)(\Delta\rho\, g) \tag{62}$$

But, from Bernoulli,

$$p - p' = 1/2\, \rho_g (u'^2_g - u^2_g) \tag{63}$$

Thus, the instability criterion is

$$u_g > \left[\frac{2}{(h_g / h'_g)(h_g / h'_g + 1)}\right]^{1/2} \left[\frac{g\, \Delta\rho\, h_g}{\rho_g}\right]^{1/2} \tag{64}$$

They note that classical (linear) Kelvin-Helmholz analysis with a wave of infinitesimal amplitude ($h'_g \rightarrow h_g$) replaces Equation (58) by its limiting form:

$$u_g > \left[\frac{g\,\Delta\rho\,h_g}{\rho_g} \right]^{1/2} \tag{65}$$

The finite amplitude (non-linear) approach ($h'_g < h_g$) gives Equation (64) with the first bracket having a value of less than 1.0. This suggests a solution to the problem experienced by Wallis and Dobson (1973), who used the classical theory, Equation (65), but were forced to introduce a factor of 0.5 to obtain agreement with experiment. Equation (64) can be extended to a pipe geometry to give

$$u_g > \left[\frac{2\,g\,\Delta\rho\cos\beta(h'_l - h_l)}{\rho_g} \right] \left[\frac{S_g^2}{(S_g^2 - S'^2_g} \right] \tag{66}$$

If we consider small but finite disturbances we can expand S'_g in a Taylor series about S_g:

$$u_g > \left[\frac{2(s'_g/S_g)^2}{(1+S'_g/S_g)} \right]^{1/2} \left[\frac{g\,\Delta\rho\cos\beta\,S_g}{\rho_g(dS_l/dh_l)} \right]^{1/2} \tag{67}$$

Taitel and Dukler (1976) suggest that the first term in Equation (67) can be approximated by $1 - h_l/D_t$. Equation (67) can then be rearranged to give

$$Fr^2 \left[\frac{1}{(1 - h_l/D_t)^2} \frac{\tilde{u}_g\, d\tilde{S}_l/d\tilde{h}_l}{\tilde{S}_g} \right] \geq 1 \tag{68}$$

where Fr, the Froude number is defined by

$$Fr = \left[\frac{\rho_g}{\Delta\rho} \right] \frac{u_{gs}}{\sqrt{D_t\, g\cos\beta}} \tag{69}$$

Now, $\tilde{h}_l$ was defined in terms of the Lockhart-Martinelli parameter, X, above. From this, Equation (68) can be used to give the transition line shown in Figure 12. This transition has been found to give a reasonable description of the experimentally observed flow patterns. However, the theory behind the equation can be criticised on several counts. First, it only applies to long-wavelength, inviscid waves, whereas it has been observed experimentally that it is the shorter wavelength roll waves which appear to form slugs. Inviscid Kelvin-Helmholtz (IKH) instability in a channel is governed by:

$$(u_g - u_l)^2 = \left[\frac{g}{k}\frac{\Delta\rho}{\rho_g}\cos\beta + \frac{\sigma k}{\rho_g} \right] \left[\tanh(kh_g) + \frac{\rho_g}{\rho_l}\tanh(kh_l) \right] \tag{70}$$

where k is the wave number and σ the interfacial tension. For $\rho_l \gg \rho_g$ and h_g large enough so that $kh_g > 3$, the second bracket on the RHS of Equation (70) approximates to unity. Instability will occur first for that wave number which minimises the value of the first bracket, leading to:

$$k_{cr} = \left[\frac{\rho_l g \cos\beta}{\sigma}\right]^{0.5} \quad \text{and} \quad (u_g - u_l)^2_{cr} = 2\left[\frac{\sigma g \cos\beta}{\rho_l}\right]^{0.5} \frac{\rho_l}{\rho_g} \tag{71}$$

For air/water in a horizontal pipe, Equation (71) gives the critical wave number as 369 m^{-1}, corresponding to a critical wavelength of 17 mm and a critical relative velocity of 6.6 m/s. The long wave Equation (65) can be obtained from Equation (70) by setting $k \rightarrow 0$.

Lin and Hanratty (1986) and Wu *et al.* (1987) introduce wall and interfacial drag terms into the analysis in what they call a Viscous Long Wave (VLW) model. The stability criterion for circular pipes is shown to be:

$$(u_g - C)^2_{cr} = \frac{\Delta\rho g \cos\beta}{\rho_g} \frac{S_g}{(dS_l / dh_l)} - \frac{\rho_l S_g}{\rho_g S_l}(C - u_l)^2_{cr} \tag{72}$$

where C, the kinematic wave velocity, is a complex function of pipe geometry and frictional drag terms. For inviscid flow, $C = u_l$, and Equation (72) reverts to the Taitel and Dukler Equation (61) with $Sg' \rightarrow Sg$. The second term represents the destabilising effect of liquid inertia, and causes a transition at lower gas velocities when $C \neq u_l$.

Funada and Joseph (2001) present an analysis based on the theory of Viscous Potential Flow (VPF). Their equation can be written as:

$$(u_g - u_l)^2_{cr} = \frac{[\hat{\eta}\tanh(kh_l) + \tanh(kh_g)]^2}{(\hat{\eta}/\hat{\rho})\hat{\eta}\tanh(kh_l) + \tanh(kh_g)}\left[\frac{\Delta\rho g \cos\beta}{\rho_g k} + \frac{\sigma k}{\rho_g}\right] \tag{73}$$

where k is the wave number as in Equation (70) and

$$\hat{\eta} = \frac{\eta_g}{\eta_l} \quad \text{and} \quad \hat{\rho} = \frac{\rho_g}{\rho_l}$$

When $\hat{\eta} = \hat{\rho}$, Equation (73) reduces to the IKH equation (64), and for $\hat{\eta} \leq \hat{\rho} << 1$, with $kh_g > 3$, Equation (71) results. For air/water at atmospheric pressure, $\hat{\eta} = 0.018$ and $\hat{\rho} = 0.0012$, so $\hat{\eta} = 15\,\hat{\rho}$, and for large kh_g and kh_l the first term of Equation (73) equals 0.816 rather than 1.0. Only for more viscous liquids will the inviscid analysis apply. This is supported by the observation of Hurlburt and Hanratty (2002) that the critical *h/D* data of Andritsos et al. (1989) for glycerol/water-air flows (η_l = 0.07 and 0.1 Pa.s) in two different pipe diameters is well predicted by the IKH model, whereas that for water-air is not. The latter is better predicted by the VLW model.

The IKH, VLW and VPF analyses, Equations (71-73) all lead to values of the actual gas velocity, u_g, at the transition. Since $u_g = u_{gs}/\varepsilon$, they can be used to derive a plot of u_{gs} (or *Fr*) as a function of ε (or *h/D*) to compare with the model of Taitel and Dukler (1976). To get the critical u_{ls} from critical *h/D*, it is necessary to solve the momentum Equations (50) and (51). Unfortunately, the results are extremely sensitive to friction factors; in view of the uncertainties discussed in section (3.1), it is unsurprising that predictions are poor.

Though as it will be shown below, the predictions of this analysis agrees well with data for air/water flows in 1 to 2 inch diameter pipes, there are questions as to its applicability to other systems and larger pipe diameters. Wu *et al* (1987) and Fernschnieder *et al* (1985) have carried out a more elaborate stability analysis and concluded that the transition was related to the conditions of the kinematic waves being faster than dynamic waves. Watson (1989) presents an alternative approach, which examines the growth of disturbance waves.

7.2 Other Transitions

Slug/Annular Transition. This is one of the most difficult transitions to describe, as the mechanism controlling it has not been identified. Taitel and Dukler (1976) suggest that if the liquid level predicted by their analysis of stratified flow is $0.5D_t$ or greater then, if the criterion suggested above indicates wave growth, slugs will occur. This corresponds to a value of X of 1.6 as shown in Figure 12 above. Barnea *et al* (1982) note that there can be significant gas hold up in the predominantly liquid part of the slugs and that the transition criteria should account for this. They suggest $h_l/D_t = 0.5\,(1-\varepsilon_g)$. From measurements in slug flow a value of $\varepsilon_{gs} = 0.3$ was chosen leading to $h_l/D_t = 0.35$ or $X = 0.6$. However, comparison with experimental data shows these criteria to give a poor description.

An alternative approach has been produced by Reimann *et al* (1981). That produced an empirical equation based on air/water, steam/water and refrigerant data. This predicts that annular flow will occur when

$$u_{gs} > 0.0285\,(g\,D_t\,)^{1/6}\left(\frac{\sigma^2}{\eta_g\,\eta_l}\right)^{1/3}\left[\frac{\Delta\rho}{\rho_g}\right]^{2/5} \tag{74}$$

Stratified/Wavy Transition. Here Taitel and Dukler (1976) follow a suggestion by Jeffreys (1925, 1926) who proposed that waves will be generated when

$$(u_g - C\,)^2\,C > \frac{4\,\eta_l\,g\,\Delta\rho}{s\,\rho_g\,\rho_l} \tag{75}$$

with C being the wave velocity and s a sheltering coefficient. Jeffreys suggested a value of 0.3 for s but subsequent work has favoured lower values. Taitel and Dukler select a value of 0.01, they also suggest that $u_g \gg C$ and $C = u_l$. If these approximations are used equation (75) can be rewritten as

$$u_g \geq \left[4\frac{\eta_l\,g\,\Delta\rho\cos\beta}{s\,\rho_g\,\rho_l\,u_l}\right]^{1/2} \tag{76}$$

which in dimensionless form is

$$K \geq \frac{2}{\tilde{u}_g\sqrt{\tilde{u}_l\,s}} \tag{77}$$

where

$$K = Fr \sqrt{Re_{ls}} = \frac{u_{gs} \sqrt{\rho_g}}{[\Delta\rho\, d\, g \cos\beta\,]^{1/2}} \sqrt{\frac{\rho_l u_{ls} D_t}{\eta_l}} \tag{78}$$

Bubble/Slug Transition. At high liquid and low gas rates bubble flow might be expected. However, the effect of gravity will tend to cause bubbles to accumulate at the top of the pipe where they would coalesce to form slug (or plug) flow. High liquid velocities could provide turbulent forces, which would disperse bubbles. Following Taitel and Dukler (1976) we could write the buoyancy force (per unit length) as

$$F_B = \Delta\rho\, g \cos\beta\, S_g \tag{79}$$

and the force due to turbulence as

$$F_T = \frac{1}{2} \rho_l \overline{v'^2} P_i \tag{80}$$

approximating v', the root mean square radial velocity fluctuation to the friction velocity

$$\overline{v'^2} = u_+^2 = u_l^2 \frac{f_l}{2} \tag{81}$$

Dispersion of gas will occur when $F_T >= F_B$ or

$$u_l > \left[\frac{4\, S_g\, g \cos\beta}{P_i f_l} \left(1 - \frac{\rho_g}{\rho_l} \right) \right]^{1/2} \tag{82}$$

or in dimensionless form

$$T^2 \geq \left[\frac{8\, \tilde{S}_g}{\tilde{P}_i \tilde{u}_l^2 (\tilde{u}_l \tilde{D}_l)^{-n}} \right] \tag{83}$$

where

$$T = \left[\frac{|\,dp/dz\,|_{ls}}{\Delta\rho\, g \cos\beta} \right]^{0.5} \tag{84}$$

7.3 Comparison with Experiment.

For horizontal flow no systematic comparison has been carried out. However, there has been more restricted comparison for individual transitions. Figure 31 shows the behaviour of data for the stratified /slug annular transition. The data indicates that the theory shows the correct trends though there is some variation in absolute values. Data from high-pressure natural gas/condensate taken from a 0.2 m diameter pipe are shown in figure 32. The Taitel and Dukler (1976) predictions can be seen to be low whilst the more elaborate analysis of Wu *et al* (1987) gives much better agreement. Figure 33 shows a test of the Taitel and Dukler (1976) and Barnea *et al* (1982) criteria for the slug/annular transition. The agreement is not good but it is noted that this method of plotting exaggerates small differences.

Figure 34 shows a comparison carried out by Pearce (1982) of the Reimann *et al* (1981) slug/annular transition against a wide range of data. Equation (74) gives good predictions. However, examination of experimental data shows some dependence on the liquid flow rate. Equation (74) indicates the transition to depend only on the gas flow rate.

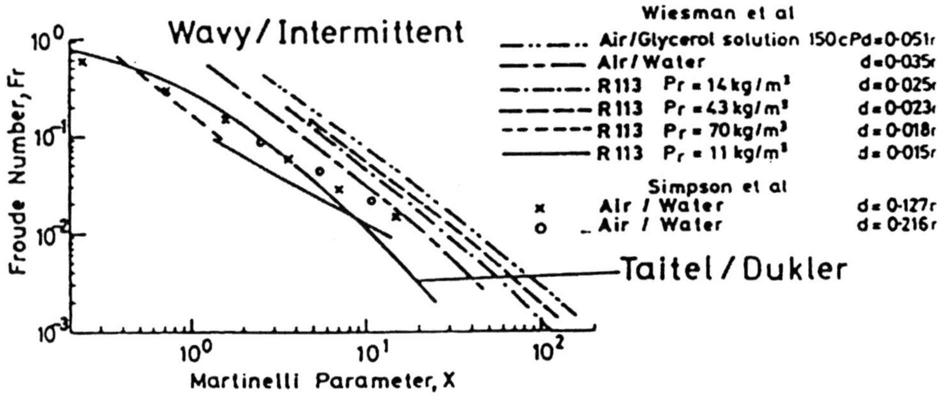

Figure 31. Test of stratified/annular-slug transition proposed by Taitel and Dukler (1976).

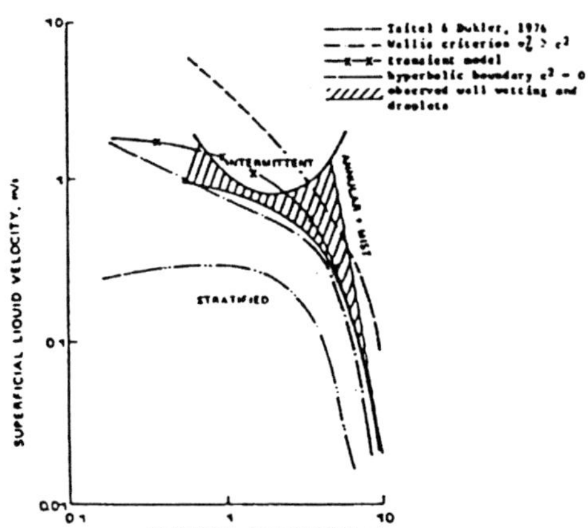

Figure 32. Comparison of stratified/slug-annular transition of Taitel and Dukler (1976) with experiments of Wu *et al* (1987). Reprinted from Proceedings of the 3rd International Conference on Multi-phase Flow, The Hague, Netherlands, Wu, H.L., Pots, B.F.M, Hollenberg, J.F. and Meerhoff, R., Flow pattern transitions in two–phase gas/condensate flow at high pressure in an 8-inch horizontal pipe, pp 13-21, Copyright (1987), with permission of BHR Group.

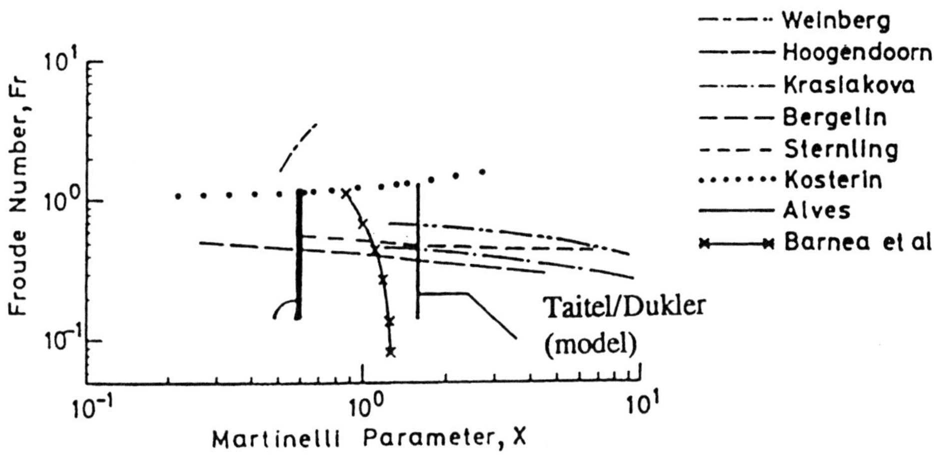

Figure 33. Test of prediction for the slug flow / annular flow transition of Taitel and Dukler (1976) and Barnea *et al* (1982).

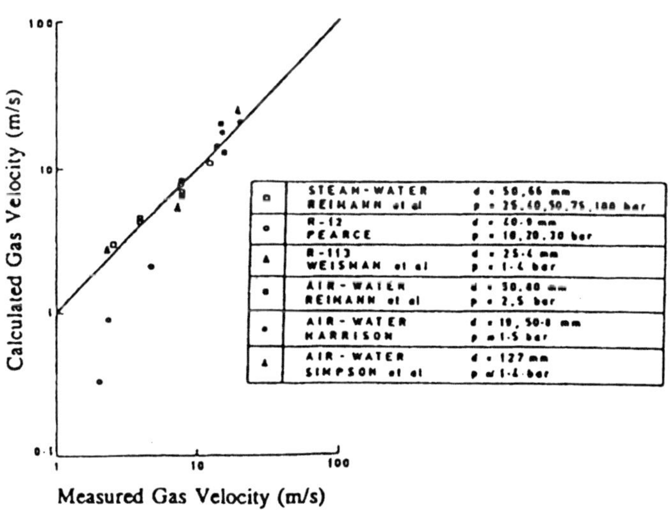

Figure 34. Test of slug flow /annular flow transition proposed by Reimann *et al* (1981) carried out by Pearce (1982).

8 Models for Specific Flow Patterns

In section 2 it was noted that a gas/liquid flow in pipes can have very different characteristics according to the flow pattern present. Yet many methods are available for pressure drop and void fraction, which have been developed to be applicable across all flow patterns without any explicit links to different flow patterns.

There have been significant developments in the modelling of individual flow patterns. However, only recently have these individual models been drawn together produce an overall prediction scheme. The following sections report recent work using this approach. For example, Holt *et al.* (1999) who considered vertical upflow, Kaya *et al.* (1999) and Gomez *et al.* (1999) who describe methods which are aimed at the full range of inclinations from vertical to horizontal and Tribbe and Muller-Steinhagen (2000) who focused on horizontal flows. An exercise looking at models for horizontal slug flow and their comparison with experimental data is described. They all consider methods for flow pattern identification, the individual flow pattern models and a comparison of the predictions of the combination of models with experimental data. From this the successes and limitations of the approach can be identified.

9 Vertical Upflow

9.1 Flow pattern identification

In the material that follows the flow patterns are identified as bubbly, slug, churn and annular. The transition models employed are those discussed in section 6. For the bubble/slug transition the approach introduced by Taitel *et al.* (1980) is used. Slug flow occurs if

$$u_{ls} < 3\,u_{gs} - 1.15\left(\frac{g\sigma\Delta\rho}{\rho_l^2}\right)^{1/4} \tag{85}$$

For slug/churn we first consider the transition from slug to dispersed bubble. Brauner and Barnea (1986) considering the balance between dispersive forces to surface tension forces which results in

$$2\left[\frac{0.4\,\sigma}{\Delta\rho\,g}\right]^{1/2}\left(\frac{\rho_l}{\sigma}\right)^{3/5}\left[\frac{2}{D_t}0.046\left(\frac{D_t}{\nu_l}\right)^{-0.2}\right](u_{gs}+u_{ls})^{1.12} = 0.725 + 4.15\left(\frac{u_{gs}}{u_{gs}+u_{ls}}\right)^{1/2} \tag{86}$$

This can be solved for the slug-dispersed bubble boundary when $\varepsilon_g = u_{gs}/(u_{gs}+u_{ls}) = 0.52$.

There is clearly an upper limit on the possible void fraction in bubbly flow due to the close packing of bubbles. Taitel *et al.* (1980) take this value as 0.52. Assuming homogeneity in the liquid slug results in churn flow rather than dispersed bubble flow if

$$u_{ls} < 0.92\,u_{gs} \tag{87}$$

For the churn/annular transition the approach employed by McQuillan and Whalley (1985) is used. This predicts annular flow if

$$u_{gs} > \frac{\sqrt{(g\, D_t\, \Delta\rho)}}{\sqrt{\rho_g}} \tag{88}$$

9.2 Models for specific flow patterns

Bubbly flow. For this flow pattern the drift flux approach is used for void fraction

$$\varepsilon_g = \frac{u_{gs}}{C_0\,(u_{gs} + u_{ls}) + v_{gd}} \tag{89}$$

Zuber and Findlay (1965) suggested a value of 1.13 for C_0 and $(1.4\,[\sigma g \Delta\rho/\rho_l^2]^3)$ for v_{gd}. The gravitational pressure drop is then

$$\left(\frac{d\,p}{d\,z}\right) = \left[\rho_g\,\varepsilon_g + \rho_l\,(1 - \varepsilon_g)\right] g \tag{90}$$

For the frictional component of pressure drop, the equation suggested by Wallis (1969) is used

$$\left(\frac{d\,p}{d\,z}\right)_f = -\frac{4\,f\,\rho_l\,u_{ls}\,(u_{gs} + u_{ls})}{2\,D_t} \tag{91}$$

where f is given a value of 0.005.

Slug flow. Here the model developed by de Cachard and Delhaye (1994) is used. There have been complex models derived for vertical slug flow, e.g., Fernandes *et al.* (1983). In these, steady and fully developed slug flow is described by a succession of identical unit cells each of which consists of a cylindrical bubble (Taylor bubble), surrounded by a falling liquid film, and of a liquid slug. The complexity in the models arises from the way in which the gas contained as bubbles in the liquid slug is described. Several workers have noted that there are conditions under which the liquid slugs contain no bubbles. Ros (1961) and Fitremann (1977) specify the condition for "non-aerated" slugs by

$$Bo = \frac{(\rho_l - \rho_g)\,g\,D_t^2}{\sigma} < 140 \tag{92}$$

Under such conditions the flow is as illustrated in Figure 35.

The bubble fraction, β, is defined as the fraction of the unit cell transit time corresponding to the Taylor bubble, L_B/L. If ε_B is the void fraction of the Taylor bubble part of the cell, then

$$\beta\,\varepsilon_B\,u_g = u_{gs} \tag{93}$$

$$(1 - \beta)\,u_{lSl} + \beta\,(1 - \varepsilon_B)\,u_{lB} = u_{ls} \tag{94}$$

where u_g is the actual averaged velocity in the Taylor bubble, u_{lSl} is the liquid velocity in the slug, u_{lB} the liquid velocity in the Taylor bubble region and u_{ls}, u_{gs} are the liquid and gas superficial velocities.

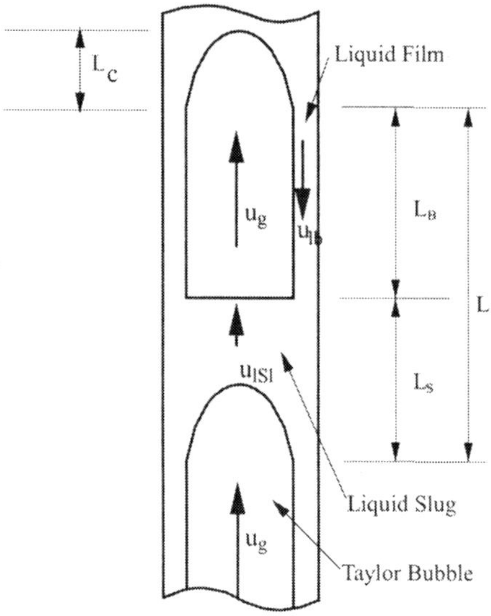

Figure 35. Schematic of slug flow showing major features.

The gas or Taylor bubble velocity is described by the equation proposed by Nicklin *et al.* (1962)

$$u_g = C_0(u_{gs} + u_{ls}) + u_0 \tag{95}$$

where for turbulent flow $C_0 = 1.2$ and u_0 is the rise velocity of a Taylor bubble in stagnant liquid obtained from

$$u_o = \Gamma \sqrt{g\ D_t}$$

$$\Gamma = 0.345\left(1 - exp\left[-\frac{0.01\ \eta^*}{0.345}\right]\right)\left(1 - exp\left[\frac{3.37 - Bo}{\delta}\right]\right)$$

$$\delta = 10 \qquad \eta^* > 250 \tag{96}$$

$$\delta = 69(\eta^*)^{-0.35} \quad 18 < \eta^* < 250$$

$$\delta = 25 \qquad \eta^* < 18$$

With $\eta^{*2} = \dfrac{\rho_l(\rho_l - \rho_g)\ g\ D_t^3}{\eta_l^2} < 3.0\ 10^{-5}$.

The film around the Taylor bubble is modelled as a falling film without interfacial shear. For laminar flow

$$\delta \left(\frac{g}{\nu_l^2} \right)^{\frac{1}{3}} = 3 \ Re_f^{\frac{1}{3}} \tag{97}$$

where, Re_f, the Reynolds number for the liquid film is defined by

$$Re = -\frac{u_{lB} \ \delta}{\nu_l} \tag{98}$$

For turbulent flow the expression suggested by Belkin *et al.* (1959) and Wallis (1969) is used

$$\delta \left(\frac{g}{\nu_l^2} \right)^{\frac{1}{3}} = 0.159 \ Re_f^{\frac{2}{3}} \tag{99}$$

These can be expressed in terms of u_{lB} as

$$u_{lB} = -0.333 \ g \ \frac{\delta^2}{\nu_l} \quad Re_f < 750 \tag{100}$$

$$u_{lB} = -15.8 \ (g \ \delta)^{\frac{1}{2}} \quad Re_f > 750 \tag{101}$$

Now as the geometrical relationship between the film thickness and void fraction is

$$\delta = \frac{D_t}{2} \left(1 - \varepsilon_B^{\frac{1}{2}} \right) \tag{102}$$

equations can be written as

$$u_{lB} = -0.333 \ \frac{g \ D_t^2}{\nu_l} \left(1 - \varepsilon_B^{\frac{1}{2}} \right)^2 \quad Re_f < 750 \tag{103}$$

$$u_{lB} = -11.2 \left[g \ D_t \left(1 - \varepsilon_B^{\frac{1}{2}} \right) \right]^{\frac{1}{2}} \quad Re_f > 750 \tag{104}$$

The slug flow parameters u_g, u_{lB}, u_{lS}, ε_B, β can be determined from u_{ls} and u_{gs} using

$$u_g = C_0 \ (u_{gs} + u_{ls}) + u_0 \tag{105}$$

$$\varepsilon_B \ u_g + (1 - \varepsilon_B) \ u_{lB} = u_{gs} + u_{ls} \tag{106}$$

$$u_{lB} = f \ (\varepsilon_B) \tag{107}$$

$$\beta \; \varepsilon_B \; u_g = u_{gs} \tag{108}$$

The average void fraction can be expressed as

$$\varepsilon_g = \frac{u_{gs}}{u_g} = \frac{u_{gs}}{C_0 \; (u_{gs} + u_{ls}) + u_0} \tag{109}$$

The gravitational pressure drop can be expressed as

$$\left(\frac{d\,p}{d\,z}\right) = -\,\rho_l \; g \; [(1 - \beta) + \beta(1 - \varepsilon_B)] \tag{110}$$

The friction term for the liquid slug is

$$\left(\frac{d\,p}{d\,z}\right)_{fS} = -\,\frac{4 \; f \; \rho_l \; (u_{gs} + u_{ls})^2}{2 \; D_t} \tag{111}$$

where the friction factor is obtained from the Blasius equation using a Reynolds number based on $u_{gs} + u_{ls}$.

In the Taylor bubble region, for a fully developed film, the wall friction force balances the weight of the film. The friction term is thus the opposite of the gravitational term

$$\left(\frac{d\,p}{d\,z}\right)_{fB} = \rho_l \; g \; (1 - \varepsilon_B) \tag{112}$$

and the average friction component is

$$\left(\frac{d\,p}{d\,z}\right)_f = (1 - \beta)\left(\frac{d\;p}{d\;z}\right)_{f\,S} + \beta\left(\frac{d\;p}{d\;z}\right)_{f\,B} \tag{113}$$

$$\left(\frac{d\,p}{d\,z}\right)_a = \alpha\left[\left(\frac{d\;p}{d\;z}\right)_g + \left(\frac{d\;p}{d\;z}\right)_f\right] \tag{114}$$

where

$$\alpha = \left[1 - \rho_l \; u_{ls}^2 \frac{u_{gs} \; (C_0 \; u_{ls} + u_0)}{p \; [(C_0 - 1) \; u_{gs} + C_0 \; u_{ls} + u_0]^2}\right]^{-1} \tag{115}$$

is an acceleration correction factor.

For larger diameter pipes the assumption that there are no bubbles the liquid slugs is no longer correct. For this models have been published by, inter alia, Fernandes et al. (1983) and Sylvetster (1978).

Churn flow. Churn flow is probably the least understood of the flow patterns. de Cachard and Delhaye (1994) used the same approach as for slug flow to determine the void fraction in churn flow. For the frictional component a modified form of Equation (111) is used with the combined superficial velocity replaced by the liquid film velocity. The friction factor is calculated using this velocity. The accelerational term is calculated using Equations (114) and (115). However, Holt *et al.* (1999) found that the churn flow pressure drops were better predicted by an annular flow model.

Annular flow. A schematic view of the processes occurring in annular flow is shown in Figure 36 and illustrates the waves on the interface, source of drops, which are entrained into the gas core and subsequently, redeposited. The phenomenological annular flow model takes entrainment, deposition and evaporation of the liquid film into account. This approach, first produced by Whalley *et al.* (1974), calculates the variation of film flow rate with axial position in a (hydrodynamic) non-equilibrium flow. It treats the flow as three interacting fields, vapour, liquid film and entrained drops. This flow can be described by the following mass balance equations.

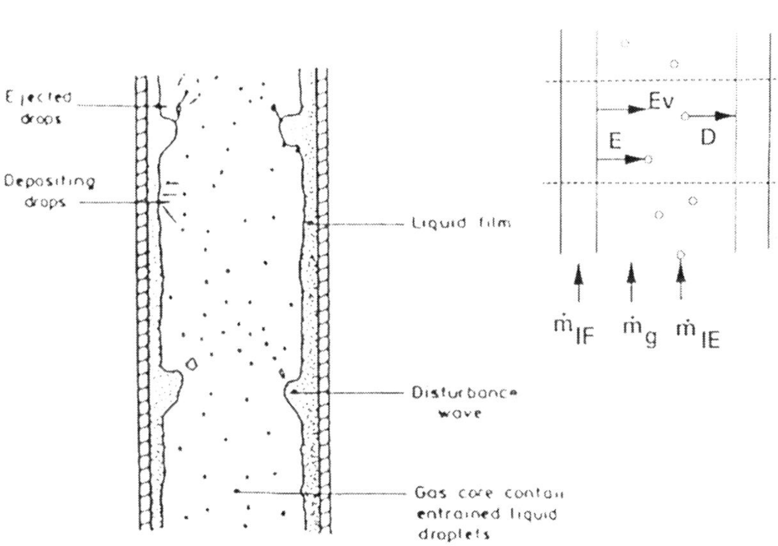

Figure 36. Major features of annular flow.

$$\frac{d\dot{m}_{lF}}{dz} = \frac{4}{D_t}(D - E) \tag{116}$$

$$\frac{d\dot{m}_{lE}}{dz} = \frac{4}{D_t}(E - D) \tag{117}$$

where D is the deposition flux onto the film, and E is the entrainment flux from the film. It is noted that an extra term must be added in the brackets of Equation (116) when there is heat addition. Drop evaporation is only expected to occur when the film dries out.

The deposition flux is described by

$$D = k\,c \tag{118}$$

where k is a mass transfer coefficient and c is the concentration of drops calculated assuming that they are travelling at the same velocity as the gas. The deposition mass transfer coefficient was then determined from the description of Hewitt and Govan (1990)

$$k\sqrt{\frac{\rho_g D_t}{\sigma}} = 0.18 \qquad for\ \frac{c}{\rho_g} < 0.3$$

$$k\sqrt{\frac{\rho_g D_t}{\sigma}} = 0.083\left(\frac{c}{\rho_g}\right)^{-0.65} \qquad for\ \frac{c}{\rho_g} > 0.3 \tag{119}$$

In predicting the rate of entrainment, it was considered that there are conditions (very thin films) at which no entrainment occurs. This limiting film flow rate is specified by

$$\dot{m}_{lFC} = \frac{\eta_l}{D_t}\exp\left(5.8504 + 0.4249\frac{\eta_g}{\eta_l}\sqrt{\frac{\rho_l}{\rho_g}}\right) \tag{120}$$

For values greater than this

$$E = 5.75\ 10^{-5}\ \dot{m}_g\left([\dot{m}_{lF} - \dot{m}_{lFC}]^2\frac{D_t\,\rho_l}{\sigma\,\rho_g^2}\right)^{0.316} \tag{121}$$

These equations are still being refined. Some recent work is given by Chong *et al.* (2001).

There is some evidence that there is additional entrainment caused by any added heat flux, Miloshenko *et al.* (1989). There is also extra entrained if there is a constriction.

For mass fluxes less than 100 kg/m^{2}s, the interfacial friction factor is calculated from the expression suggested by Ambosini *et al.* (1990)

$$\frac{f_i}{f_s} = 1 + 13.8\ We_D^{0.2}\ Re_g^{-0.6}\left(\delta_g^+ - 200\sqrt{\frac{\rho_g}{\rho_l}}\right) \tag{122}$$

where $\delta_g^+ = \dfrac{\delta\ u_g^*}{\nu_g}$ $\quad Re_g = \dfrac{u_{gs}\ D_t}{\nu_g}$ and $We_D = \dfrac{\rho_g\ u_{gs}^2 D_t}{\sigma}$

At higher mass fluxes, a new equation suggested by Holt *et al.* (1999) is employed

$$\frac{f_i}{f_s} = 1 + \frac{13.8\ We^{0.175}\ m_g^+}{Re_g^{0.7}} \tag{123}$$

The film thickness is obtained from

$$\delta_l^+ = 0.34\ Re_{lF}^{0.6} \quad \mathrm{Re_{lF}} < 1000 \tag{124}$$

$$\delta_l^+ = 0.0512\ Re_{lF}^{0.875} \quad \mathrm{Re}_{lF} > 1000 \tag{125}$$

where $\delta_l^+ = \frac{\delta\ u_l^*}{\nu_l}$ and $u_l^* = \sqrt{\frac{\tau_i}{\rho_l}}$, $\delta_g^+ = \delta_l^+ \frac{\eta_l}{\eta_g} \sqrt{\frac{\rho_g}{\rho_l}}$ and $Re_{lF} = \frac{u_{ls}\ D_t}{\nu_l}$.

Void fraction is calculated from the film thickness and entrained fraction and the gravitational pressure drop calculated from Equation (90). The frictional component of pressure drop is given by

$$\left(\frac{d\ p}{d\ z}\right)_f = \frac{4\ f_i\ \rho_g\ u_{gs}^2}{2\ D_t} \tag{126}$$

Comparison of predictions with experimental data. The above combination of models has been tested against available data. The comparisons are in terms of a correction factor and a range factor.

F The "correction factor": this is the average number by which the predicted value must be multiplied to give the experimental value. Values of F greater than unity mean that the given correlation under-predicts the mean density.

R The "range factor": this gives the factor by which the corrected result must be multiplied and divided to give the true mean density with 99% probability and 95% confidence.

The results are shown in Table 2 where the information is considered flow pattern by flow pattern. As so shown are the equivalent information for the correlations of Friedel (1979) for frictional pressure drop and C.I.S.E. for void fraction and hence gravitational pressure drop. The table shows that the flow pattern specific models give good prediction. Indeed, they do as well as Friedel/CISE. It is noted that, in these cases, the gravitational component is dominant. For churn flow it is the annular flow model that does best. Wallis (1969) had already remarked on this. The agreement with annular flow data is shown in figure 37.

Table 2: Comparison of data with flow pattern specific model for vertical upflow - summary of results

Flow pattern	Model/ correlation	No. of data points	Correction factor	Range factor
Bubbly	Bubbly flow model	181	1.003	1.162
	Friedel/CISE	181	1.008	1.165
Slug	Slug flow model	1495	0.854	1.720
	Friedel/CISE	1495	1.031	1.828
Churn	Churn flow model	524	1.155	3.370
	Annular flow model	524	0.975	1.693
	Friedel/CISE	514	0.809	1.861
Annular	Annular flow model	3556	0.540	5.010
	Friedel/CISE	3545	1.0171	1.788

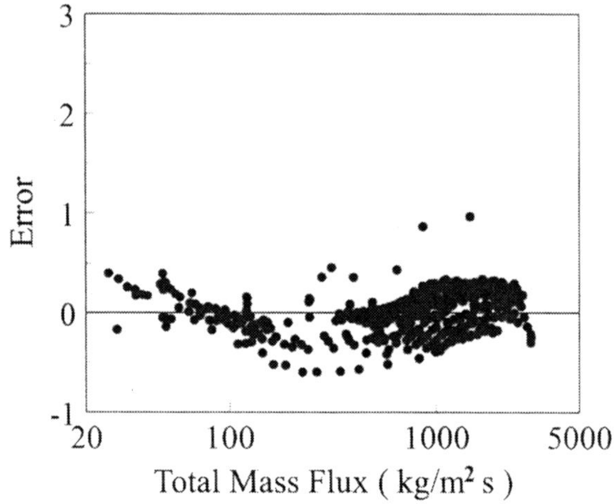

Figure 37. Error in prediction: annular flow model using combination of interfacial friction factor equations. Reprinted from Chemical Engineering Research and Design, vol. 77, Holt, A.H., Azzopardi, B.J. and Biddulph, M.W., Calculation of two-phase pressure drop through a flow pattern specific model, pp 7-15, Copyright (1999), with permission of the Institution of Chemical Engineers.

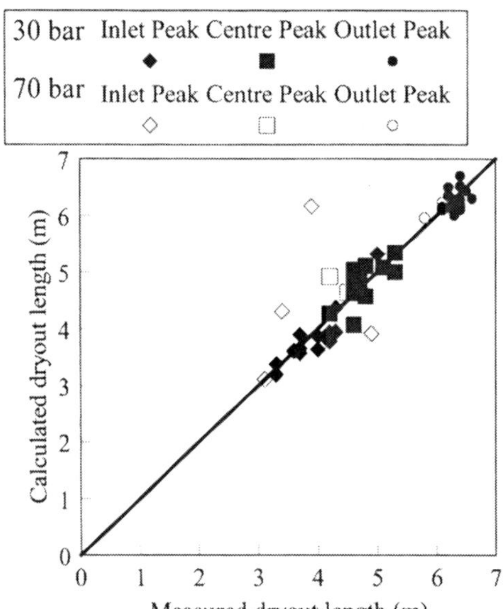

Figure 38. Accuracy of prediction of dryout length.

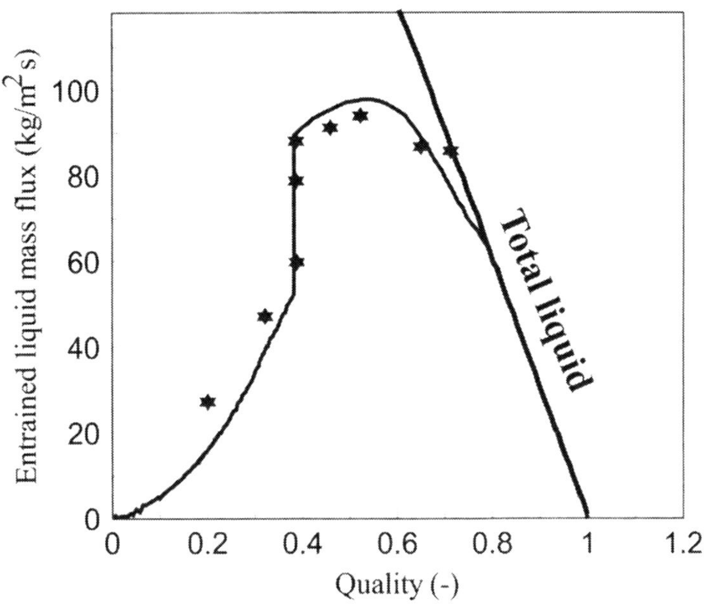

Figure 39. Comparison of prediction of entrained liquid flow rate with experiments – Mass flux = 297 kg/m^2s; heat flux = 617 $kW.m^2$; pressure = 3.77 bar.

The annular flow model can also be used to determine when the film dries out in heated systems. The heat flux at which the film dries out is known as the critical heat flux. Systems with axially varying heat fluxes are very difficult to predict. The annular flow model is capable of providing such predictions. Azzopardi (1996) has shown that the model accurately predicts the length to dryout in experiments with a number of different axial heat flux profiles. The success of the model at predictiong the dryout length is shown in Figure 38.

Hewitt and Govan (1990) have carried out predictions on a case where the heat flux is uniform at the entry and exit of the vertical pipe but with a "cold patch" 0.61 m long near the middle. The annular flow model successfully predicts the variation in film flow rate along the pipe, Figure 39. No other prediction method can handle this case.

The model can be extended to more complex cases. Azzopardi (2000) gives details of the application of the methodology to Venturis. Chong *et al.* (2002) have applied the annular flow model to fired reboilers. These have serpentine arrangements with vertical pipes and 180° bends. The effect of the bends was assumed to be to cause drops to deposit. This was incorporated into the model. Typical predictions are show in Figure 40.

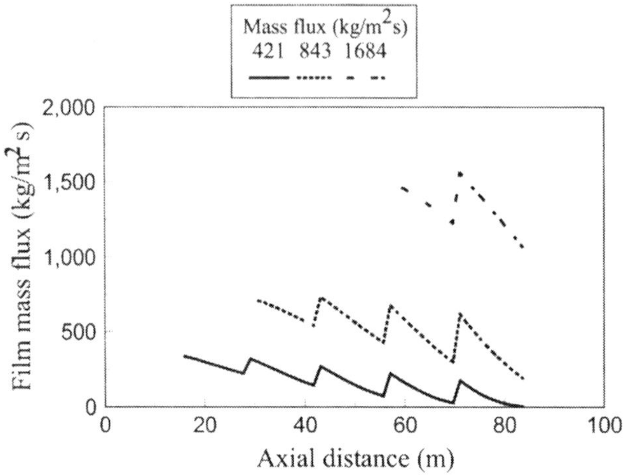

Figure 40. Effect of mass flux on axial film mass flux variation. Minimum heat flux = 20 kW/m^2, maximum heat flux =40 kW/m^2.

10 Horizontal Flow

For horizontal flows Tribbe and Muller-Steinhagen (2000) carried out a very systematic exercise in which he tested a large number of pressure drop methods against a bank of ~7,000 data. He considered both empirical equations as well as more mechanistic methods. However, he carried out tests using specific methods for the individual flow patterns. He lists and gives full details of the methods used. He we consider his "best buys" but confine ourselves to the more mechanistic models. and Muller-Steinhagen (2001 found that he obtained different results according to the way flow pattern was identified. He used the flow pattern definitions suggested by Taitel and Dukler (1976), stratified smooth, stratified wavy, annular, intermittent and dispersed bubble. If the flow pattern present was identified by using the Taitel and Dukler model, as described in section 4, then the method of Agrawal *et al.* (1973) is used for stratified smooth, that of Hashizume *et al.* (1985) for stratified wavy and annular and the models of Nicholson *et al.* (1978) for intermittent and dispersed bubbly flow. An alternative approach was determined by looking at the variation of error across the flow pattern map. This resulted in the Hart *et al.* (1989) method being used for stratified wavy. S will be seen below, the latter considers the fraction of wall wetted as a parameter. When this reaches a value of 1.0 is considered the lower limit of annular flow and the approach then chooses the method of Hashizume *et al.*. The intermittent/dispersed bubbly transition is as specified by Nicholson *et al.*. This section considers these methods in detail and where appropriate gives details of the methods and discusses their limitation pointing out the direction that more complex modelling might take. It is considered in terms of flow patterns.

10.1 Stratified smooth flow

The analysis of this flow pattern is outlined in section 7.1. The method of Agrawal *et al.* (1973) differs in two way from that material. Firstly, different friction factor equations are used. Secondly, velocity profile effect have to be taken into account in the liquid.

10.2 Stratified wavy flow

Though models similar to those described for stratified smooth flow are applicable to stratified wavy flow, the model of Hart *et al.* (1989), which was found to be best by Tribbe and Muller-Steinhagen (2000) in his systematic testing of methods, has a slightly different approach. In this, the interface is not considered flat but is curved: liquid is climbing up the walls. One of the first things to be determined is the fraction of wall wetted, θ.

$$\theta = 0.52\left(1-\varepsilon_g\right)^{0.374} + 0.26 Fr^{0.52} \tag{127}$$

where

$$Fr = \frac{\rho_l u_{sl}^2}{\left[g D_t \left(\rho_l - \rho_g\right)\left(1-\varepsilon_g\right)^2\right]} \tag{128}$$

and

$$1-\varepsilon_g = \frac{\dfrac{\varepsilon_g}{1-\varepsilon_g}\dfrac{u_{sl}}{u_{sg}}\left[1+\left(10.4\, Re_{sl}^{-0.363}\left\{\dfrac{\rho_l}{\rho_g}\right\}^{0.5}\right)\right]}{1+\dfrac{\varepsilon_g}{1-\varepsilon_g}\dfrac{u_{sl}}{u_{sg}}\left[1+\left(10.4\, Re_{sl}^{-0.363}\left\{\dfrac{\rho_l}{\rho_g}\right\}^{0.5}\right)\right]} \tag{129}$$

Obviously if θ>1, the flow is annular. As this equation is implicit, it has to be solved iteratively. This angle is then used to predict the two-phase friction factor

$$f_{TP} = \left(1-\theta\right) f_g + \theta f_i \tag{130}$$

where

$$\frac{1}{2\sqrt{f_g}} = -0.86\left(\frac{\varepsilon / D_t}{3.7} + \frac{1.2505}{Re_g \sqrt{f_g}}\right) \quad \textit{for} \quad Re_g > 2100 \tag{131}$$

and $f_g = 16/Re_g$ otherwise. The interfacial friction factor is given by

$$f_i = \frac{0.0625}{\left[\log_{10}\left(\dfrac{15}{\mathrm{Re}_g} + \dfrac{k}{3.715 D_t}\right)\right]^2} \tag{132}$$

where

$$\frac{k}{D_t} = \frac{2.3(1-\varepsilon_g)}{4\theta} \tag{133}$$

The pressure drop is then calculated from

$$\frac{dp}{dz} = \frac{1}{\varepsilon_g}\left[\frac{4f_{TP}}{D_t}\frac{\rho_g u_{sg}^2}{2\varepsilon_g^2} - \frac{16\theta f_i}{D_t}\frac{\rho_g\left(\frac{2u_{sg}u_{sl}}{\varepsilon_g(1-\varepsilon_g)} - \left(\frac{u_{sl}}{\{1-\varepsilon_g\}}\right)^2\right)}{2}\right] \tag{134}$$

10.3 Annular flow

The method identified as the best by the systematic testing of Tribbe and Muller-Stienhagen (2000), that of Hashzume et al. (1985), which assumes a film uniform around the circumference and ignores entrainment. Reality indicates that there can be very significant entrainment and that, particularly for large pipe diameters, low gas velocities and higher liquid flow rates, the film flow rate and film thickness can be highly asymmetric. This is illustrated in Figure 41 which shows values of the ratio of mean film thickness to that at the bottom of the pipe calculated from a correlating equation by Hurlburt and Newell (2000).

The following is taken from Azzopardi and Rea (1999). One of the problems in modelling this asymmetric flow is that there is not agreement in what causes the film to remain at the top of the pipe given the draining effect of gravity. Butterworth (1973) suggested that transfer of liquid by entrainment and deposition, circumferential secondary gas flow and wave spreading were possible mechanisms. Butterworth (1973), Fisher and Pearce (1978, 1993), James *et al.* (1987) and Peng and Shoukri (1997) developed models employing entrainment and deposition. Laurinat *et al.* (1985) invoked the circumferential secondary gas flow whilst Fukano and Ousaka (1989) utilised wave spreading in their model.

Though entrainment and deposition are believed to be important, and indeed shown by Jepson (1988) to be the dominant, there as yet very rudimentary methods to describe these features. There are important and fundamental developments in describing deposition e.g., Mols and Oliemans (1998), much more work is required for entrainment. Though the circumferential secondary gas flow invoked by Laurinat *et al.* (1985) receives some support from the experiments of Flores *et al.* (1995), detailed computations by Jayanti *et al.* (1990) indicate that this secondary flow is an order of magnitude smaller than required to maintain the liquid film at the top of the pipe. The wave spreading mechanism utilised by Fukano and Ousaka (1989) is supported by early visualisation studies by Butterworth and Pulling (1972) and by more recent studies using photochromic dye tracing by Sutharshan *et al.* (1995)

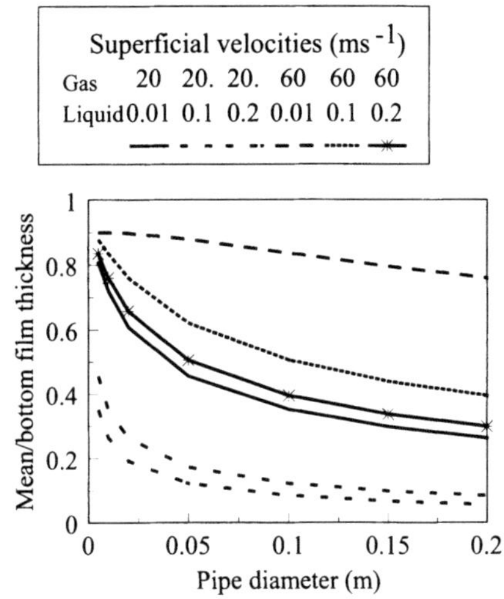

Figure 41. Illustration of the effect of pipe diameter and superficial velocities on the degree of asymmetry - air/water at 1.5 bar(a). Reprinted from Chemical Engineering Research and Design, vol. 77, Azzopardi, B.J. and Rea, S., Modelling the split of horizontal annular flow through a T-junction, pp 713-720, Copyright (1999), with permission of the Institution of Chemical Engineers.

Fukano and Ousaka (1989) published a model that introduced a description of wave spreading through momentum balances. This required two constants to be fitted. One could be optimised via a mass balance to allow for the effect of entrainment. Roberts *et al.* (1997) provided a simple correlation for the second, the film thickness at the bottom of the pipe, which included a simple diameter correction factor. Further examination of the database employed by Roberts *et al.* (1997) which has been augmented in the present work by additional data to cover a wider range of pipe diameters, shows that a better description is achieved if the correction is to the pipe diameter squared.

$$h_o^+ = 846\left(\frac{D}{D_{ref}}\right)^2 \frac{Re_{LF}^{0.44}}{Re_{GS}^{0.59}} \tag{135}$$

The most recent model for horizontal annular flow is by Hurlburt and Newell (2000) who started from the axial, circumferential and radial momentum balances developed by Laurinat *et al.* (1985).

$$\frac{\partial \tau_{yz}}{\partial y} + \frac{2}{D_1}\frac{\partial \tau_{xz}}{\partial \bar{x}} = 0 \tag{136}$$

$$-\frac{2}{D_1}\frac{\partial p}{\partial \bar{x}} + \frac{\partial \tau_{yz}}{\partial y} + \frac{2}{D_1}\frac{\partial \tau_{xx}}{\partial \bar{x}} - \rho_l g \sin \bar{x} = 0 \tag{137}$$

$$-\frac{\partial p}{\partial y}-\rho_l g \cos\bar{x}=0 \tag{138}$$

These were derived based on the assumptions of: negligible surface tension effects, thin turbulent films, steady state, effects of waves handled indirectly through the axial shear model, a constant pressure is imposed by the gas phase on the surface of the liquid film, τ_{xx} and τ_{xz} can be modelled as independent of radial position and the normal stress in the liquid scales with the liquid axial velocity. Integration of the radial momentum equation and differentiation with respect to the circumferential coordinate enables the pressure gradient term to be eliminated from the circumferential equation.

The equations can be non-dimensionalised with respect to the gas friction velocity based on a smooth pipe. Further simplifications result from assuming that entrainment and deposition and circumferential shear due to secondary flows in the gas contribute little to the system. Also, the static pressure due to circumferential film thickness variation is assumed small compared to the normal stress term and the dispersion term (second term in Equation (137)) is taken to be small relative to the axial shear. This results in

$$\frac{\partial\tau_{xx}^{+}}{\partial\bar{x}}-\frac{1}{Fr_{\tau_s}}\sin\bar{x}=0 \tag{139}$$

$$\frac{\partial\tau_{yz}^{+}}{\partial y^{+}}=0 \tag{140}$$

Equation (140) implies that the axial shear stress is constant in the radial direction and can thus represented by the interfacial shear stress. Hurlburt specified this by a new correlation that represents the relationships identified for both thin and thick films.

$$\frac{\tau_i}{\tau_s}=10\left[1-exp\left(-\frac{\phi h_i^{+}}{250}\right)\right] \tag{141}$$

The normal stress term is specified assuming that there are two regions, within each of which there is a different relationship with h^+, linear at lower thickness and exponential for thicker films. Using this in Equation (139) results in

$$-\frac{1}{24C}exp\left(-\frac{h^{+}}{12}\right)\frac{dh^{+}}{d\bar{x}}-\frac{1}{Fr_{\tau_s}}\sin\bar{x}=0 \tag{142}$$

This can be integrated from the value h^+_0 at the pipe bottom to yield

$$h^{+}=-12\,ln\left[exp\left(-\frac{h_0^{+}}{12}\right)+CFr_{\tau_s}\left(\cos\bar{x}-1\right)\right] \tag{143}$$

The axial film velocity is determined from flow rate/film thickness relationships developed by Asali and Hanratty (1985) and Henstock and Hanratty (1976) for thin and thicker films respectively. These can be combined to give a composite equation

$$u_{av} = \frac{u_i^*}{\sqrt{\left[\frac{1}{\left(1.5h_i^{+0.666}\right)^2} + \frac{1}{\left(9.5h_i^{+0.111}\right)^2}\right]}} \tag{144}$$

This leaves an unspecified constant C. Hurlburt determined the value of this by comparing the integrated film flow rate determined by from the above, Equation (145), to that expected from an entrained fraction prediction method,. Although Hurlburt used the experimental values for the film thickness at the pipe bottom, in our calculations, the values specified by Equation (135) is employed.

$$\frac{\rho_l D_1}{w_l} \int_0^{\pi} h u_{av} d\bar{x} = E - 1 \tag{145}$$

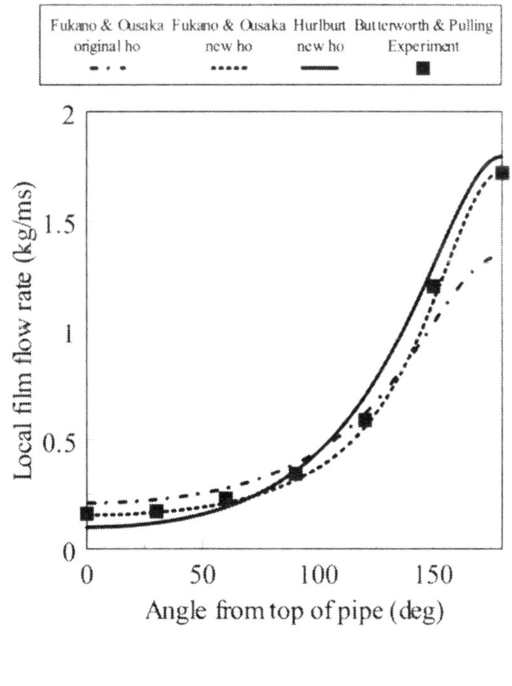

Figure 42. Validation of the horizontal annular flow model against the experimental data of Butterworth & Pulling (1973). Reprinted from Chemical Engineering Research and Design, vol. 77, Azzopardi, B.J. and Rea, S., Modelling the split of horizontal annular flow through a T-junction, pp 713-720, Copyright (1999), with permission of the Institution of Chemical Engineers.

This version of the model has been validated against the circumferential film flow data of Butterworth and Pulling (1973). Figure.42 shows that there is excellent agreement between model and experimental data and that this model gives better predictions than the model of Fukano and Ousaka (1989) as employed by Roberts *et al.* (1997). However, the prediction of Fukano and Ousaka's model are significantly improved if the film thickness at the pipe bottom is specified by Equation (51) rather than the corresponding equation of Roberts *et al.*.

10.4 Slug flow

The main features of horizontal slug flow are shown diagrammatically in Figure 43. This shows that periodically the pipe is filled with liquid, which may or may not contain some of the gas in the form of bubbles. This liquid slug travels rapidly along the pipe at velocities greater than the liquid film surrounding the gas bubble in front of it. At the back of the slug, liquid is ejected which initially has a fairly high velocity (similar to that of a slug) but soon decelerates to the typical velocity of the film. These physical features are used to varying degrees in the models presented.

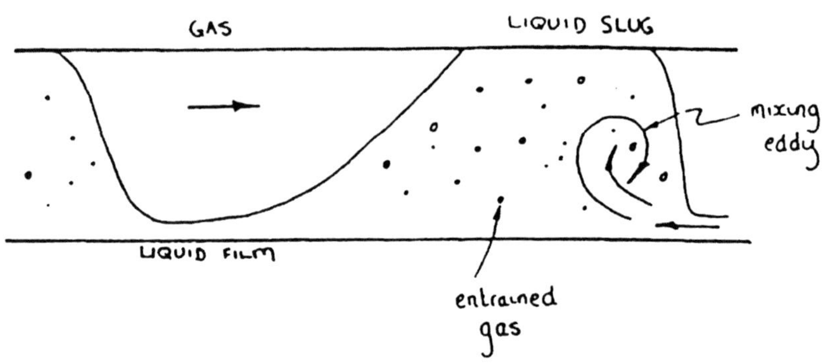

Figure 43. Main features of slug flow.

The simplest model is that of Bonnecaze *et al.* (1971) who assume that there was no slip between the liquid and the gas (and, therefore, ignored the effect of the film) and that the main cause of pressure drop was the wall friction acting on the liquid in the slugs. The pressure gradient was given by

$$\frac{dp}{dz} = \frac{4\, f\, \lambda\, \rho_l\, u_m^2}{2\, D_t} \tag{146}$$

where f is the friction factor correlated in terms of the slug Reynolds number, $Re_s = \rho_l u^2_m D/\eta_l$,, an equation for which is given by Bonnecaze *et al.*. Here u_m is the sum of the gas and liquid superficial velocities and λ is the liquid volume fraction given by

$$\lambda = \frac{u_{ls}}{u_m} \tag{147}$$

The model of Vermeulen and Ryan (1971) also considered the effect of the liquid and assumed that the pressure drop arose from (i) the wall friction acting on the slug and (ii) the acceleration of the liquid entering the front of the slug. In this case the deceleration of the liquid emerging from the back of the slug is neglected. Thus,

$$\frac{dp}{dz} = \frac{4\, f\, \lambda\, \rho_l\, u_m^2}{2\, D_t} + v_s\, \rho_l\, R_f\, u_m \tag{148}$$

where the first term accounts for the frictional losses and the second for the acceleration losses.

A Blasius type relationship was used for the friction factor, f. The slug is assumed to be all liquid and the ratio of slug length/distance between slugs is taken to be equal to λ in this equation. vs is the slug frequency and R_f the film hold up. Vermeulen and Ryan suggest that

$$R_f = 1 - \varepsilon_g - \lambda \tag{149}$$

with the void fraction calculated from the correlation of Lockhart and Martinelli (1949). Azzopardi *et al.* (1985) also tested this model using the HTFS correlation for void fraction. The correlation of Gregory and Scott (1969) was used to calculate v_s.

A much more complete model is that of Dukler and Hubbard (1975). In contrast to Vermeulen and Ryan, Hubbard and Dukler did not assume that the slug was entirely liquid, that is, that the void fraction in the slug was not equal to 0. Thus, ρ_s is not equal to ρ_l. In addition, they allowed for the fact that not all of the liquid is contained in the slug. Their model results in

$$\frac{dp}{dz} = \frac{1}{z_t}\left[\dot{m}_s\,(u_m - V_f) + \frac{4\, f\, \rho_s\, u_m^2}{2\, D_t}\right] \tag{150}$$

where z_t is the total length of a slug unit, that is a slug and the accompanying gas bubble. In applying the model, the correlation of Gregory *et al.* (1978) can be used to determine R_s as a function of u_m. The rate of mass pick up by a slug, as a mass flux, m_s, is calculated from the ratio, C', of the apparent slug velocity to the mixture velocity, this ratio being correlated in terms of the slug Reynolds Number. Dukler and Hubbard gave values of C' ranging from 1.25 to 1.3. Values of u_f and z_s were obtained by solving mass and momentum balances on the liquid film. In applying the momentum balance any two-dimensional effects associated with the flow around the nose of the gas bubble are ignored.

Normally, the Dukler and Hubbard momentum balance and mass balance equation for the liquid film between two slugs (their equations (40b) and (44)) are solved iteratively to give R_{fe}, the liquid hold up at the onset of the film region behind the slug. However, their iteration does not converge at very small and very large mixture velocities (this is due to the simplification introduced in the model and has also been noted by Nicholson *et al.* (1978)). For these limiting cases, R_{fe} was taken as $R_sC/(1 + C)$ and $(R_sC + V_L/u_m)/(1 + C)$ respectively which are the minimum and maximum conceivable values for R_{fe}. It was found that the difference in overall average prediction error obtained, omitting the non-convergent cases and alternatively including them with the limiting values of R_{fe}, was not large. For slug frequency, the Gregory and

Scott correlation was used as for the Vermeulen and Ryan model. The mixing length z_m is a correction to allow for the penetration of the liquid film into the slug. Dukler and Hubbard calculated it from

$$z_m = \frac{0.15}{g} \left(u_m - V_f\right)^2 \tag{151}$$

However, in many cases this gave values of z_m greater than z_s, in which case z_m was set equal to z_s, thus making the second term of Equation (150) equal to zero. The case with $z_m = 0$ and with $R_s = 1$ were also considered. The method by Nicholson et al. (1978) which was identified by Tribbe and Muller-Steinhagen (2000) bas the best is an extension of the model of Dukler and Hubbard (1975).

Comparison of predictions with experimental data. The data used in this comparison were taken from a large bank of pressure drop measurements obtained from the literature. The data have been thoroughly checked for correctness and consistency. Information corresponding to slug flow was selected according to two different criteria:

(i) The observation of the experimenter - flows described as either plug flow or slug flow have been selected.

(ii) the criterion of Weisman *et al.* (1979) for the stratified/intermittent transition

$$u_{ls} = 0.284 \left(g \, D_t \right)^{0.455} u_{gs}^{0.091} \tag{152}$$

and the criterion of Pearce (1982) for the intermittent annular transition

$$\frac{\rho_g \, u_{gs}^2 \, g \, D_t}{(\rho_l - \rho_g)} = 2 \tag{153}$$

Method (i) gave a total of 1798 data points (all for air-water and air-oil systems) while (ii) gave 3489 points. Of these 3489 points it is of interest that 470 points had been identified as being a flow pattern other than slug or plug flow by the experimenters concerned.

The overall results of the comparison are presented in Table 4. Here we can see that amongst the three slug models tested the predictive capability appears to improve with the complexity of the model. However, the scatter of the data is still fairly large. If the more accurate holdup equation of HTFS equation is used in the Vermeulen and Ryan (1971) model there is little change in mean error but the scatter is reduced significantly. If we examine the variants on the model of Dukler and Hubbard (1975) we see that the inclusion of the correction z_m improves the predictions but the use of the Gregory correlation for R_s appears to make them worse.

The effect of mass flux has been examined using the "best" Dukler and Hubbard model (case 2) and is shown in Table 5. This illustrates that the model overpredicts at low mass flux but underpredicts at higher fluxes. The scatter appears to decrease with mass flux. In the case of diameter, shown in Table 6 for the same model, it can be seen that the model underpredicts for small diameter tubes but is also showing an overprediction as the diameter increases. The scatter also decreases as the diameter increases.

Table 4: Overall comparison between horizontal slug flow models and data

Model	Data selected using (i) - 1798 points	Data selected using (ii) - 3489 points	Notes
Bonnecaze et *al.*	0.51/7.0	0.7/7.0	1
Vermeulen and Ryan	0.66/3.5	0.8/4.1	2
Vermeulen and Ryan (HTFS void fraction)	0.74/3.0	0.8/3.4	
Dukler & Hubbard (simplest case: $R_s = 1$, $z_m = 0$)	0.81/2.8	0.9/2.9	
Dukler & Hubbard (more complex case)	1.15/2.9	1.22/3.3	
HTFS	0.88/2.0	0.93/2.2	

1 Worse at low quality/ low mass flux 2 Worse at low mass flux, distinct diameter effect

Table 5: Effect of mass flux on accuracy of slug flow model

Mass flux range (kg/m^{2}s)	100-250	250-500	500-750	750-1000	1000-2000	2000-3000	3000-4000	>4000
Number of data points	202	599	442	446	995	362	212	231
Correction factor	0.80	0.84	0.91	0.97	1.25	1.19	1.15	1.1
Range factor	8.35	4.82	3.92	3.3	2.64	2.19	1.98	1.64

Table 6: Effect of pipe diameter on accuracy of slug flow model

Diameter (m)	0.0032-0.01	0.01-0.032	0.032-0.1	>0.1
Number of data points	17	1892	1356	222
Correction factor	1.9	1.16	0.91	0.804
Range factor	4.6	3.7	2.9	2.8

References

Agrawal, S.S., Gregory, G.A., and Govier, G.W. (1973) An analysis of horizontal stratified two-phase flow in pipes. *Canadian Journal of Chemical Engineering* 51:280-286.

Akelseev, V.P., Poberezkin, A.E., and Gerasimov, P.V. (1972) Determination of flooding rates in regular packings. *Heat Transfer Soviet Research* 4:159-163.

Alves, G.E. (1974) Experience with industrial co-current liquid-gas pipelines. *Institution of Chemical Engineers Symposium Series*. No.38, Paper F1.

Aly, A.M.M. (1981) Flow regime boundaries for an interior subchannel of a horizontal 37-element bundle. *Canadian Journal of Chemical Engineering* 59:158-163.

Ambrosini, W., Andreussi, P., and Azzopardi, B.J. (1990) A physically based correlation for drop size in annular flow. *International Journal of Multiphase Flow* 17:497-507.

Andritsos, N. (1986) Effect of pipe diameter and liquid viscosity on horizontal stratified flow. PhD thesis, Univ. Illinois, Urbana.

Andritsos, N., and Hanratty, T.J. (1987) Influence of interfacial waves in stratified gas-liquid flows. *American Institute of Chemical Engineers Journal* 33:444-454.

Andritsos, N., Williams, L., and Hanratty, T.J. (1989) Effect of liquid viscosity on the stratified-slug transitions in horizontal pipe flow. *International Journal of Multiphase Flow* 15:877-892

Asali, J.C., Hanratty, T.J., and Andreussi, P. (1985) Interfacial drag and film height for vertical annular flow. *AIChE J* 31:895-902.

Assad, A., Jan, C., Lopez de Beltodrano, M., and Beus, S., (1998) Scaled experiments in ripple annular flow in a small tube. *Nuclear Engineering Design* 184:437-447.

Azzopardi, B.J., Govan, A.H., and Hewitt, G.F. (1985) Slug flow in horizontal pipes. *Symposium on Pipelines*, Utrecht, I.Chem.E., European Branch Symposium Series No 4, 2:213-225.

Azzopardi, B.J. (1996) Prediction of dryout and post-burnout heat transfer with axially non-uniform heat input by means of an annular flow model. *Nuclear Engineering and Design* 163:51-57.

Azzopardi, B.J., and Rea, S. (1999) Modelling the split of horizontal annular flow at a T-junction. *Tranaction of the Institution of Chemical Engineers* 77A:713-720.

Azzopardi, B.J., and Zaidi, S.H. (2000) Determination of entrained fraction in vertical annular gas/liquid flow. *Journal of Fluids Engineering*. 122:146-150.

Azzopardi, B.J. (2000) Multiphase flow in Venturis. *7th International Conference on Multiphase Flow in Industrial Plants,* Bologna, 13-15 September.

Azzopardi, B.J., and Wren, E. (2002) What is entrainment in vertical two-phase churn flow? *40th European Two-Phase Flow Group Meeting*, Stockholm, June.

Azzopardi, B.J., Conte, G., and Wren, E. (2002) The split of two-phase slug and churn flow at a vertical regular T-junction. Submitted.

Baker, O. (1954) Simultaneous flow of oil and gas. *Oil and Gas Journal* 53:185-195.

Barbosa, J., Richardson, S., and Hewitt, G.F. (2001) Churn flow: myth, magic and mystery. *39th European Two-Phase Flow Group Meeting*, Aveiro, Portugal, 18-20, June.

Barnea, D., Shoham, O., and Taitel, Y. (1982) Flow pattern transition for downward inclined two-phase flow: Horizontal to vertical. *Chemical Engineering Science* 37:735-740.

Barnea, D., and Taitel, Y. (1985) Flow pattern transition in two-phase gas-liquid flows. In *Encyclopedia of Fluid Mechanics*, Volume 3 (ed. N. Cheremisinoff), Gulf Publishing Co.

Barnea, D. (1986) Transition from annular flow and from dispersed bubble flow - unified models for the whole range of pipe inclinations. *International Journal of Multiphase Flow* 12:733-744.

Barnea, D., and Taitel, Y. (1992)

Belkin, H.H., Macleod, A.A., Monrad, C.C., and Rothfus, R.R. (1959) Turbulent liquid flow down vertical walls. *American Institute of Chemical Engineers Journal* 5:245-248.

Bell, K.J., Taborek, J., and Fenoglio, F. (1970) Interpretation of horizontal in-tube condensation heat transfer correlations with a two-phase flow regime map. *American Institute of Chemical Engineers Symposium Series* No. 102, 66:150-163.

Bennett, A.W., Hewitt, G.F., Kearsey, H.A., Keeys, R.K.F., and Lacey, P.M.C. (1965) Flow visualisation studies of flow boiling at high pressures. *Proceedings of the Institution of Mechanical Engineers* 180: Paper no 5.

Bergles, A.E. (1969) Two-phase flow structure observations for high pressure water in a rod bundle. *ASME Winter Annual Meeting, Los Angeles, Two Phase Flow in Rod Bundles*, :47-55.

Bonnecaze, R.H., Erskine, W., and Greskovich, E.J. (1971) Holdup and pressure drop for two-phase slug flow in inclined pipelines. *American Institute of Chemical Engineers Journal* 17:1109-1113.

Brauner, N., and Barnea, D. (1986) Slug/churn transition in upward gas-liquid flow. *Chem.Eng.Sci.* 40:159-163.

Brauner, N. (2001)The prediction of dispersed flow boundaries in liquid-liquid and gas-liquid systems. *International Journal of Multiphase Flow* 27:885-910.

Butterworth, D. (1967) A visual study of mechanisms in horizontal air water flow. UKAEA Report, AERE M 2556.

Butterworth, D., and Pulling, D.J., (1972) A visual study of mechanisms in horizontal annular, air-water flow. UKAEA Report AERE M2556.

Butterworth, D. (1973) An analysis of film flow for horizontal flow and condensation in a horizontal tube. UKAEA Report AERE R7575.

Butterworth, D., and Pulling, D.J. (1973) Film flow and film thickness measurements for horizontal annular air-water flow. UKAEA Report AERE R7576.

Caetano, E.F., Shoham, O., and Brill, J.P. (1992) Upward vertical two-phase flow through an annulus. Part I: Single-phase friction factor, Taylor bubble rise velocity and flow pattern prediction. *Journal of Energy Resources Engineering* 114:1-13.

Calderbank, P.H. (1958) Physical rate processes in industrial fermentation. Part I: The interfacial area in gas-liquid contacting with mechanical agitation. *Transactions of the Institution of Chemical Engineers* 36:443-463.

Celata, G.P., Cumo, M., Farello, G.E., Mariani, A., and Solimo, A. (1991) Flow pattern recognition in heated vertical channels: steady state and transient conditions. *Experimental Thermal and Fluid Science* 4: 737-746.

Chaudry, A.B. (1967) A study of the flow of air and water in vertical tubes. PhD Thesis, University of Edinburgh.

Chawla, J.M. (1967) Waermeubergang und druckabfall in waagerechten rohren fur der stromung von verdampfenden. *VDI Forschungs Heft* 523.

Chen, X.T., Cai, X.A., and Brill, J.P. (1997) A general model for transition to dispersed bubble flow. *Chemical Engineering Science* 52:4373-4380.

Cheng, H., Hills, J.H., and Azzopardi, B.J. (1998) A study of the bubble-to-slug transition in vertical gas-liquid flow in columns of different diameter. *International Journal of Multiphase Flow* 24:431-452.

Cheng, H., Hills, J.H., and Azzopardi, B.J. (2002) Effects of initial bubble size on flow pattern transition in a 28.9 mm diameter column. *International Journal of Multiphase Flow* 28:1047-1062.

Chhabra, R.P., and Richardson, J.F. (1985) Co-current horizontal and vertically upward flow of gas and non-Newtonian liquid. In *Encyclopedia of Fluid Mechanics*, Volume 3 (ed. N. Cheremisinoff), Gulf Publishing Co, Houston.

Chong, L.Y., Azzopardi, B.J., and Bate, D.J. (2002) Modelling the process side of fired reboilers. 40th European Two-Phase Flow Group Meeting, Stockholm, June.

Chong, L.Y., Azzopardi, B.J., and Hankins, Hankins, N.P. (2001) Entrainment rate in annular two-phase flow. *7th U.K. National Heat Transfer Conference*, Nottingham, September

Chung, K. S., Liu, C. P., and Tien, C. L. (1980) Flooding in two-phase counter-current flows – II Experimental investigation. *PhysicoChemical Hydrodynamics* 1:209-220.

Coney, M.W.E. (1974) The analysis of a mechanism of liquid replenishment and draining in horizontal two-phase flow. *International Journal of Multiphase Flow* 1:647-670.

Costigan, G., and Whalley, P.B. (1997) Slug flow regime identification from dynamic void fraction measurements in vertical air-water flows. *International Journal of Multiphase Flow* 23:263–282.

Das, G., Das, P.K., Purohit, N.K., and Mitra, A.K. (1999) Flow pattern transition during gas liquid upflow through vertical concentric annuli. *Journal of Fluids Engineering* 121:pp 895-907.

de Cachard, F., and Delhaye, J.M. (1996) A slug-churn model for small-diameter airlift pumps. *International Journal of Multiphase Flow* 22:627-649.

Dukler, A.E., and Hubbard, M.G. (1975) A model for gas-liquid slug flow in horizontal and near horizontal tubes. *Industrial and Engineering Chemistry Fundamentals* 14:337-347.

Dukler, A.E., and Smith, L. (1979) Two-phase interactions in counter-current flow: studies of the flooding mechanism. USNRC Report NUREG/CR- 0617.
Dukler, A.E., and Taitel, Y. (1984), "Flow pattern transitions in gas-liquid systems: Measurement and modelling", In *Multiphase Science and Technology*, Volume 2, Hemisphere Pub. Corp.
Ekberg, N.P., Ghiaasiaan, S.M., Abdel-Khalik, S.I., Yoda, M., and Jeter, S.M. (1999) Gas-liquid two-phase flow in narrow horizontal annuli. *Nuclear Engineering and Design* 192:59-80.
Fair, J.R. (1960) What you need to know to design thermo-siphon re-boilers. Petroleum Refiner 39:
Fernandes, R.C., Semiat, R., and Dukler, A.E. (1983) Hydrodynamic model for gas-liquid slug flow in vertical tubes. . *American Institute of Chemical Engineers Journal* 29:981-989.
Fernschneider, G., Lagiere, M., Bourgeois, T., and Fitremann, J.M. (1985) How to calculate two-phase flow of gas and oil in pipelines. *Pipe Line Industry* 63:33.
Fisher, S.A., and Pearce, D.L. (1978) A theoretical model for describing horizontal annular flows. *International Seminar on Momentum, Heat and Mass Transfer in Two-Phase Energy and Chemical Systems*, Dubrovnik, Yugoslavia.
Fisher, S.A., and Pearce, D.L. (1993) An annular flow model for predicting liquid carryover into austenitic superheaters. *International Journal of Multiphase Flow* 19:295-307.
Fitremann, J.M. (1977) Ecoulements diphasiques: théorie et applications à l'étude de quelques régimes d'écoulements verticaux ascendants d'un mélange gaz-liquide. Thèse d'état, Univ. P.& M. Curie, Paris VI.
Flores, A.G., Crowe, K.E., and Griffith, P. (1995) Gas-phase secondary flow in horizontal stratified and annular two-phase flow. *International Journal of Multiphase Flow* 21:207-221.
Frankum, D.P., Wadekar, V.V., and Azzopardi, B.J. (1997) Two-phase flow patterns for evaporating flow. *Experimental Thermal and Fluid Science* 15:183-192.
Friedel, L. (1979) Improved friction pressure drop calculations for horizontal and vertical two-phase pipe flow. *European Two-phase Flow Group Meeting.*
Fukano, T., and Ousaka, A. (1989) Prediction of the circumferential distribution of film thickness in horizontal and near-horizontal gas-liquid annular flow. *International Journal of Multiphase Flow* 15:403-419.
Funada, T., and Joseph, D.D. (2001) Viscous potential flow analysis of interfacial instability in a channel. *Journal of Fluid Mechanics* 445:263-283.
Furukawa, T., and Fukano, T. (2001) Effects of liquid viscosity on flow patterns in vertical upward gas-liquid two-phase flow. *International Journal of Multiphase Flow* 27:1109-1126.
Gomez, L.E., Shoham, O., Schmidt, Z., Chohski, R.N., Brown, A., and Northug, T. (1999) A unified mechanistic model for steady-state two-phase flow in wellbores and pipelines. *SPE 56520, Proceedings of the SPE Annual Technical Conference and Exhibition*, Houston, Π1:307-320.
Govan, A. H., Hewitt, G. F., Richter, H. J., and Scott, A. (1991) Flooding and churn flow in vertical pipes. *International Journal of Multiphase Flow* 17:27-44.
Gould, T.L. (1972) Vertical two-phase flow in oil and gas wells. PhD Thesis, University of Michigan.
Grant, I.D.R. (1975) Flow and pressure drop with single-phase and two-phase flow on the shell side of segmentally baffled shell-and-tube heat exchangers. NEL Report 590.
Gregory, G.A., and Scott, D.S. (1969) Correlation of liquid slug velocity and frequency in horizontal co-current gas-liquid slug flow. . *American Institute of Chemical Engineers Journal* 15:933-935.
Gregory, G.A., Nicholson, M.K., and Aziz, K. (1978) Correlation of the slug liquid volume fraction in the slug for horizontal gas-liquid slug flow. *International Journal of Multiphase Flow* 4:33-39.
Grolman, E., and Fortuin, J.M.H. (1997) Gas-liquid flows in slightly inclined pipes. *Chemical Engineering Science 52:4461-4471.*
Harmathy, T.Z. (1960) Velocity of large drops and bubbles in media of infinite or restricted extent. *American Institute of Chemical Engineers Journal* 6:281-288.

Hart, J., Hamersma, P.J., and Fortuin, J.M.H. (1989) Correlations predicting frictional pressure drop and liquid holdup during horizontal gas-liquid pipe flow with a small liquid holdup. *International Journal of Multiphase Flow* 15:947-964.

Hashizume, K., Ogiwara, H., and Taniguchi, H. (1985) Flow pattern void fraction and pressure drop of refrigerant two-phase flow in a horizontal pipe – II analysis of frictional pressure drop. *International Journal of Multiphase Flow* 11:643-658.

Henstock, W.H., and Hanratty, T.J. (1976) The interfacial drag and height of the wall layer in annular flows. . *American Institute of Chemical Engineers Journal* 22:990-1000.

Hewitt, G.F. (1983) Two-phase flow and its applications: past, present and future. *Heat Transfer Engineering* 4:67-79.

Hewitt, G.F., and Roberts, D.N. (1969) Studies of two-phase patterns by simultaneous x-ray and flash photography. UKAEA Report AERE M2159.

Hewitt, G.F., and Govan, A.H. (1990) Phenomenological modelling of non-equilibrium flow with phase change. *International Journal of Heat Mass Transfer* 32:229-242.

Hewitt, G.F., Gill, L.E., Roberts, D.N., and Azzopardi, B.J. (1990) The split of low inlet quality gas/liquid flow at a vertical T - Experimental data. UKAEA Report AERE M3801.

Hills, J.H. (1976) The operation of a bubble column at High Throughput - I Gas holdup measurements. *Chemical Engineering Journal* 12:89-99.

Hills, J.H., and Chéty, P. (1998) The rise velocity of a Taylor bubble in an annulus. Transactions of the Institution of Chemical Engineers 76A: 723-727.

Hinze, J.O. (1955) Fundamentals of the hydrodynamic mechanism of splitting of dispersion processes. *American Institute of Chemical Engineers Journal* 1:289-295.

Holt, A.J., Azzopardi, B.J., and Biddulph, M.W. (1999) Calculation of two-phase pressure drop for vertical upflow in narrow passages by means of a flow pattern specific model. *Transactions of the Institution of Chemical Engineers* 77A:7-15.

Hosler, E.R. (1967) Flow pattern in high pressure (steam-water) flow. Westinghouse AEC R&D Report No. WAPD-TM-658.

Hsu, Y.Y. and Simon, F.F. (1969) Stability of cylindrical bubbles in a vertical pipe. ASME pap. 69-HT-28.

Hurlburt, E.T., and Newell, T.A. (2000) Prediction of the circumferential film thickness distribution in horizontal annular gas-liquid flow. *Journal of Fluids Engineering* 122:1-7.

Hurlburt, E.T., and Hanratty, T.J. (2002) Prediction of the transitions from stratified to slug and plug flow for long pipes. *International Journal of Multiphase Flow* 28:707-729.

Ishii, M. (1977) One-dimensional drift-flux model and constitutive equations for relative motion between phases in various two-phase flow regimes. ANL Report ANL-77-47.

James, P.W., Wilkes, N.S. Conkie, W., and Burns A. 1(987) Developments in the modelling of horizontal annular two-phase flow. *International Journal of Multiphase Flow* 13:173-198.

Jayanti, S. Wilkes, N.S. Clarke, D.S., and Hewitt, G.F., (1990) The prediction of turbulent flows over roughened surfaces and its application to interpretation of mechanisms of horizontal annular flow. *Proceedings of the Royal Society A* 431:71-88.

Jayanti, S, and Hewitt, G.F. (1992) Prediction of the slug-to-churn transition in vertical two-phase flow. *International Journal of Multiphase Flow* 18:847-860.

Jayanti, S, Tokarz, A., and Hewitt, G.F. (1996) Theoretical investigation of the diameter effect on flooding in countercurrent flow. *International Journal of Multiphase Flow* 22:307-324.

Jeffreys, H. (1925) On the formation of water waves by wind. *Proceeding of the Royal Society (London)* A107:189-206.

Jeffreys, H. (1926) On the formation of waves by wind. *Proceeding of the Royal Society (London)* A110:241-247.

Jepson, W.P. (1988) Liquid film thickness variation in horizontal annular flow in large diameter pipes. AERE Report R12991.

Jones Jr, O.C., and Zuber, N. (1975) The interaction between void fraction fluctuations and flow patterns in two-phase flow. *International Journal of Multiphase Flow* 2:273 - 306

Kaya, A.S., Sarica, C., and Brill, J.P. (1999) Comprehensive mechanistic modeling of two-phase flow in deviated wells. SPE 565220, *Proc. SPE Annual Technical Conf. and Exhib.*, Houston, 1:331-342.

Kelessidis, V.C., and Dukler, A.E. (1989) Modelling flow pattern transitions for upward gas-liquid flow in vertical concentric and eccentric annuli. *International Journal of Multiphase Flow* 15:173-191.

Kowalski, J.E. (1987) Wall and interfacial shear stress in stratified flow in a horizontal pipe. *American Institute of Chemical Engineers Journal* 33:274-281.

Krishnan, V.S., and Kowalski, J.E. (1984) Stratified-slug flow transition in a horizontal pipe containing a rod bundle. *American Institute of Chemical Engineers Symposium Series* 80 (236):282-289

Kubie, J., and Gardner, G..C. (1977) Drop sizes and drop dispersion in straight horizontal tubes and helical coils. *Chemical Engineering Science* 32:195-202.

Landman, M.J. (1991) Non-unique holdup and pressure drop in two-phase stratified inclined pipe flow. *International Journal of Multiphase Flow* 17:377-394.

Laurinat, J.E. Hanratty, T.J., and Jepson, W.P. (1985) Film thickness distribution for gas-liquid annular flow in a horizontal pipe. *PhysicoChemical Hydrodynamics* 6:179-195.

Lin, P.Y., and Hanratty, T.J. (1986) Prediction of the initiation of slugs with linear stability theory. *International Journal of Multiphase Flow* 12:79-98.

Lin, P.Y., and Hanratty, T.J. (1987) Effect of pipe diameter on flow patterns for air-water flow in horizontal pipes. *International Journal of Multiphase Flow* 13:549-563.

Lockhart, R.W., and Martinelli, R.C. (1949) Proposed correlation of data for isothermal, two-phase, two-component flow in pipes. *Chemical Engineering Progress* 45:39-48.

Matsui, G. (1984) Identification of flow regimes in vertical gas-liquid two-phase flow using differential pressure fluctuations. *International Journal of Multiphase Flow* 10:711-720.

Matuszkiewicz, A., Flamand, J.C., and Bouré, J.A. (1987) The bubble-slug flow pattern transition and instabilities of void fraction waves. *International Journal of Multiphase Flow* 13:199-217.

Mayinger, F., and Zetzmann, K. (1976) Flow pattern of two-phase flow in inside-cooled tubes; a generalised of flow pattern map based on investigation in water and freon. *Advanced Study Institute in Two-phase Flow and Heat Transfer*, Istanbul, Turkey, 16-27 August.

McQuillan, K.W. (1985) Flooding in annular two-phase flow. DPhil Thesis, University of Oxford.

McQuillan, K.W., and Whalley, P.B. (1985a) A comparison between flooding correlations and experimental flooding data for gas-liquid flow in vertical circular tubes. *Chemical Engineering Science* 40:1425-1440.

McQuillan, K.W., and Whalley, P.B. (1985b) Flow patterns in vertical two-phase flow. *International Journal of Multiphase Flow* 11:161-176.

Miloshenko, V.I., Nigmatulin, B.I., Petukhov, V.V., and Tribunkin, N.I. (1989) Burnout and distribution of liquid in evaporative channels of various lengths. *International Journal of Multiphase Flow* 15:393-402.

Mishima, K., and Ishii, M. (1984) Flow regime transition criteria for upward two-phase flow in vertical tubes. *International Journal of Heat and Mass Transfer* 27:723-736.

Miyagi, O. (1925) On air bubbles rising in water. *Philosophical Magazine* 50:112-140.

Mols, B., and Oliemans, R.V.A. (1998) A turbulent diffusion model for particle dispersion and deposition in horizontal tube flow *International Journal of Multiphase Flow* 24:77-92.

Mukerjee, H., And Brill, J.P. (1985) Empirical equations to predict flow patterns in two-phase inclined flow. *International Journal of Multiphase Flow* 11:299-315.

Nicholson, M.K., Aziz, K., and Gregory, G.A. (1978) Intermittent two-phase flow in horizontal pipes. *Canadian Journal of Chemical Engineering* 56:653-663.

Nicklin, D.J., and Davidson, J.F. (1962) The onset of instability in two-phase slug flow. *Institution of Mechanical Engineers Symposium on Two-Phase Flow*, London.

Nicklin, D.J., Wilkes, J.O., and Davidson, J.F. (1962) Two-phase flow in vertical tubes. *Transaction of the Institution of Chemical Engineers* 40:61-68.

Nishikawa, K., Sekoguchi, K., and Fukano, T. (1968) Characteristics of pressure pulsation in upward two-phase flow. *International Symposium on Research in Co-current Gas-Liquid Flow*, Univ. of Waterloo, paper A2

Noghrehkar, G.R., Kawaji, M., and Chan, A.M.C. (1999) Investigation of two-phase flow regimes in tube bundles under cross-flow conditions. *International Journal of Multiphase Flow* 25:857-874.

Nusselt, W. (1915) Die oberflachenkondensation des wasserdampfes. *VDI Zeitschrift* 60:541-546 and 569-575.

Ohnuki, A., and Akimoto, H. (2000) Experimental study on transition of flow pattern and phase distribution in upward air-water two-phase flow along a large vertical pipe. *International Journal of Multiphase Flow* 26:367-386

Oshinowo, T., and Charles, M.E. (1974) Vertical two-phase flow: Part I. Flow pattern correlations . *Canadian Journal of Chemical Engineering* 52:25-35.

Osmasali, S.I., and Chang, J.S. (1988) Two-phase flow regime transition in a horizontal pipe and annular flow under gas-liquid two-phase flow *ASME FED* 72:63-69

Owen, D.G. (1986) An experimental and theoretical analysis of equilibrium annular flows. PhD Thesis, University of Birmingham.

Palen, J.W., Breber, G., and Taborek, J. (1979) Prediction of flow regimes in horizontal tube-side condensation. *Heat Transfer Engineering* 1:47-57.

Pearce, D.L. (1982) An experimental investigation of flow regimes in R12. *European Two-phase Flow Group Meeting*, Paris, 2-4 June, Paper A24.

Peng, F., and Shoukri, M. (1997) Modelling the phase redistribution of horizontal annular flow divided in T-junctions. *Canadian Journal of Chemical Engineering* 75:264-270.

Prasser, H.-M., Krepper, E., and Lucas, D. (2000) Fast wire mesh sensors for gas-liquid flows and decomposition of gas fraction profiles according to bubble size classes. *2nd Japanese-European Two-Phase Flow Group Meeting*, Tsukuba, Japan, September 25-29.

Pushkina, O.L., and Sorokin, Y.L. (1969) Breakdown of liquid film motion in vertical tubes. *Heat Transfer Soviet Research* 1:56-64.

Radovcich, N.A., and Moissis, R. (1962) The transition from two-phase bubble flow to slug flow. MIT Report No. 7-7673-22.

Reimann, J., John. G., and Seeger, W. (1981) Transition to slug and annular flow in horizontal air-water and steam-water flow. Report No. KfK3189.

Roberts, P.A., Azzopardi, B.J., and Hibberd, S. (1997) The split of horizontal annular at a T-junction. *Chemical Engineering Science* 52:3441-3453.

Ros, N.C.J. (1961) Simultaneous flow of gas and liquid as encountered in well tubing. *Journal of Petroleum Technology* 13:1037-1049.

Sadatomi, M., Sato, Y., and Saruwatari, S. (1982) Two-phase flow in vertical non-circular channels. *International Journal of Multiphase Flow* 8:641-655.

Sakaguchi, T., Akagawa, K., Hamaguchi, H., Imoto, M., and Ishida, S. (1979) Flow regime maps for developing steady air-water two-phase flow in horizontal tubes. *Memoirs of the Faculty of Engineering of Kobe University* 25:191-202.

Sardesai, R.G., Owen, R.G., and Pulling, D.J. (1981) Flow regimes for condensation of a vapour in a horizontal tube. *Chemical Engineering Science* 36:1173-1180.

Sawai, T., and Kaji, M. (2001) Flow structure and pressure gradient in churn flow. *Experimental Heat Transfer, Fluid Mechanics and Thermodynamics 2001* (Ed. G.P. Celata, P. Di Marco, A. Goulas and A. Mariani) Editzioi ETS, Pisa, 2:1791-1796.

Sekoguchi, K., and Mori, K., (1998) New development of experimental study on interfacial structures in gas-liquid two-phase flow. *Proeedings of the Conference on Experimental Heat Transfer, Fluid Mechanics and Thermodynamics* :1177-1188.

Serizawa, A., and Kataoka, I. (1988) in *Transient Phenomena in Multi-phase flow*, Afghan, N.H. (ed), Hemisphere, New York, pp. 179-224.

Sevik, M., and Park, S.H. (1973) The splitting of drops and bubbles by turbulent fluid flow. *Journal of Fluids Engineering* 95:53-60.

Shah, M.M. (1976) A new correlation for heat transfer during flow boiling in pipes" *ASHRAE Transactions* 82:60-86.

Shoham, O. (1982) Flow pattenr transitions and characterization in gas-liquid flow in inclined pipes. PhD Tesis, Tel-Aviv University, Israel.

Simmons, M.J.H., and Hanratty, T.J. (2001) Transition from stratified to intermittent flows in small angle upflows. *International Journal of Multiphase Flow* 27:599-616.

Simon, M (1998) On the effects of inclination on non-adiabatic gas/liquid two-phase flow. 3rd Internationasl Conference on Multiphase Flow, Lyon, 8-12 June.

Soliman, H.M. (1985) Flow pattern transitions during horizontal in tube condensation. In *Encyclopedia of Fluid Mechanics* (ed. N. Cheremisinoff), Gulf Publishing Co, Houston.

Song, C.H., No, H.C., and Chung, M.K (1995) Investigation of bubble flow developments and its transition based on the instability of void fraction waves. *Int. J. Multiphase Flow* 21:381-404.

Spedding, P.L., and Nguyen, V.T. (1980) Regime maps for air-water two-phase flow. *Chemical Engineering Science* 35:779-793.

Spedding, P.L., and Hand, N.P. (1997) Prediction in stratified gas-liquid co-current flow in horizontal pipelines. International Journal of Heat and Mass Transfer 40:1923-1935.

Sterling, V.C. (1985) Two-phase flow theory and engineering decision. Lecture Presented at AIChE Annual Meeting.

Sun, K.H. (1979) Flooding correlations for BWR bundle upper tieplates and bottom side-entry orifices. in Veziroglu T.N. (*ed*) *Proceedings of Multiphase Flow and Heat Transfer Symposium Workshop*, Miami Beach, Florida :1615 – 1635.

Sutharshan, B., Kawaji, M., and Ousaka, A. (1995) Measurement of circumferential and axial film velocities in horizontal annular flow. *International Journal of Multiphase Flow* 21:193-206.

Sylvester, N.D. (1987) A mechanistic model for two-phase vertical slug flow in pipes. *Journal of Energy Resources Technology* 109:206-213.

Taitel, Y., and Dukler, A.E. (1976) A model for predicting flow regime transitions in horizontal and near-horizontal gas-liquid flow. *American Institute of Chemical Engineers Journal* 22:47-55.

Taitel, Y., Barnea, D., and Dukler, A.E. (1980) Modelling flow pattern transitions for steady upward gas-liquid flow in vertical tubes. *American Institute of Chemical Engineers Journal* 26:345-354.

Tribbe, C., and Muller-Steinhagen, H.M. (2000) An evaluation of the performance of phenomenological models for predicting pressure gradient during gas-liquid flow in horizontal pielines. *International Journal of Multiphase Flow* 26:1019-1036.

Turner, R.G., Hubbard, M.G., and Dukler, A.E. (1969) Analysis and prediction of minimum flow rates for continuous removal of liquid from gas wells. *Journal of Petroleum Tech.* 21:1475-

Ulbrich, R., and Mewes, D. (1994) Vertical, upward gas-liquid two-phase flow across a tube bundle. *International Journal of Multiphase Flow* 20:249-272.

Venkaseswararao, P., Semiat, R., and Dukler, A.E. (1982) Flow pattern transition for gas-liquid flow in a vertical rod bundle. *International Journal of Multiphase Flow* 8:509-524.

Vermeulen, L.R., and Ryan, J.T. (1971) Two-phase slug flow in horizontal and inclined tubes. *Canadian Journal of Chemical Engineering* 49:195-201.

Vijayan, M., Jayanti, S., and Balakrishnan, A.R. (2001) Effect of tube diameter on flooding. *International Journal of Multiphase Flow* 27:797-816.

Wallis, G.B. (1961) Flooding velocities for air and water in vertical tubes. UKAEA Report AEEW R123.

Wallis, G.B. (1969) *One-dimensional Two-phase Flow*, McGraw-Hill.

Wallis, G.B., and Dobson, J.E. (1973) The onset of slugging in horizontal stratified air-water flow. *International Journal of Multiphase Flow* 1:173-193.

Watson, M. (1989) Wavy stratified and the transition to slug flow. *4th International Conference on Multiphase Flow*, Nice, France, 19-21 June (BHRA pub.).

Watson, M.J., and Hewitt, G.F. (1998) Effect of diameter on the flooding initiation mechanism. *3rd International Conference on Multiphase Flow*, Lyon, France, 8-12 June.

Watson, M.J., and Hewitt, G.F. (1999) Pressure effects on the slug to churn transition. *International Journal of Multiphase Flow 25:1225-1241.*

Weisman, J., Duncan, D., Gibson, J., and Crawford, T. (1979) Effects of fluid properties and pipe diameter on two-phase flow patterns in horizontal lines. *Int. J. Multiphase Flow* 5:437-462.

Whalley, P.B., Hedley, B.D., and Davidson, J.F. (1972) Gas hold-up in bubble columns with liquid flow, *VDI Berichte* 182:57-61.

Whalley, P.B., Hutchinson, P., and Hewitt, G.F. (1974) The calculation of critical heat flux for forced convection boiling, *5th International Heat Transfer Conference* Tokyo, paper B6.11.

Whalley, P.B., Azzopardi, B.J., Hewitt, G.F., and Owen, R.G. (1982) A physical model for two-phase flows with thermodynamic and hydrodynamic non-equilibrium. *7th International Heat Transfer Conference*, Munich, Paper CS29.

Willetts, I.P., Azzopardi, B.J. and Whalley, P.B. (1987), The effect of gas and liquid properties on annular two-phase flow, *3rd Int. Conf. on Multiphase Flow*, The Hague, The Netherlands, 18-20 May (BHRA pub.).

Williams, C.L., and Peterson, A.C. (1978) Two-phase flow patterns with high pressure water in a heated four-rod bundle. *Nuclear Science and Engineering* 68:155-169.

Wu, H.L., Pots, B.F.M., Heelenberg, J.F. and Meerhoff, R. (1987), "Flow pattern transitions in two-phase gas/condensate flow at high pressure in an 8-inch horizontal pipe", *3rd International Conference on Multiphase Flow*, The Hague, The Netherlands, 18-20 May (BHRA pub.).

Xu, G.P., Tso, C.P., and Tou, K.W. (1998) Hydrodynamics of two-phase flow in vertical up- and down-flow across a horizontal tube bundle. *International Journal of Multiphase Flow* 24:1639-1648.

Zetzmann, K. (1984) Phase separation of air-water flow in a vertical T-junction. *German Chemical Engineering* 7:305-312.

Zabaras, G.J., and Dukler, A.E. (1988) Countercurrent gas-liquid annular flow including the flooding state. *American Institute of Chemical Engineers Journal 34:389-396.*

Zapte, A., and Kroger, D. G. (1996) The influence of fluid properties and inlet geometry on flooding in vertical and inclined tubes. *Int. J. Multiphase Flow* 22:461-472.

Zapte, A., and Kroger, D. G. (2000) Countercurrent gas-liquid flow in inclined and vertical ducts – I: Flow patterns, pressure drop characteristics and flooding. *Int. J. Multiphase Flow* 26:1439-1455. Countercurrent gas-liquid flow in inclined and vertical ducts – II: The validity of the Froude-Ohnesorge number correlation for flooding. *Int. J. Multiphase Flow* 26:1457-1468.

Zhang, J.-P., Grace, J.R., Epstein, N., and Lim, K.S. (1997) Flow regime identification in gas-liquid flow and three-phase fluidised beds. *Chemical Engineering Science* 52:3979-3992.

Zuber, N., and Findlay, J.A. (1965) Average volumetric concentration in two-phase flow systems. *Journal of Heat Transfer* 87:453-468.

Zuber, N., and Hench, J. (1962) Steady state and transient void fraction of bubbling systems and their operating limits, Part I: Steady state operation. General Electric Report 62GL100.

Modelling of Stratified Gas-Liquid Flow

Jean Fabre

Institut de Mécanique des Fluides, Toulouse, France

Abstract. This chapter focuses on the behaviour of stratified gas-liquid flow in horizontal or nearly horizontal pipes. Conservation equations and closure relationships are described for stratified smooth and stratified wavy flows. Details on the local structure, surface waves, wall and interfacial shear stresses are also provided.

1 Introduction

When gas and liquid flow in a pipe or a channel, there exist some particular conditions for which the two phases are separated from each other by a continuous interface. This pattern is dominated by the gravity force that causes the liquid to stratify at the bottom (Figure 1). This flow pattern can be observed in horizontal or slightly inclined pipelines. It is characterised by the structure of the interface that may be smooth or wavy according to the gas flow rate (Figure 2). At low gas velocity, the interface is smooth or may be rippled by small capillary waves of a few millimetres length. With increasing gas velocity, small amplitude regular waves appear. At high enough velocity of gas, droplets can be entrained from the large amplitude irregular waves and deposited at the wall or at the interface; however this atomisation phenomenon is out of the scope of this chapter.

In stratified flow, we are first interested in pressure drop and liquid hold-up for given flow conditions and geometry. Although this appears as a very simple problem, it requires an accurate prediction of the friction at the wall and at the interface. In particular, the difference of velocity between phases can be high, suggesting that the momentum transfer between phases through interfacial friction is little effective for the gas to drive the liquid.

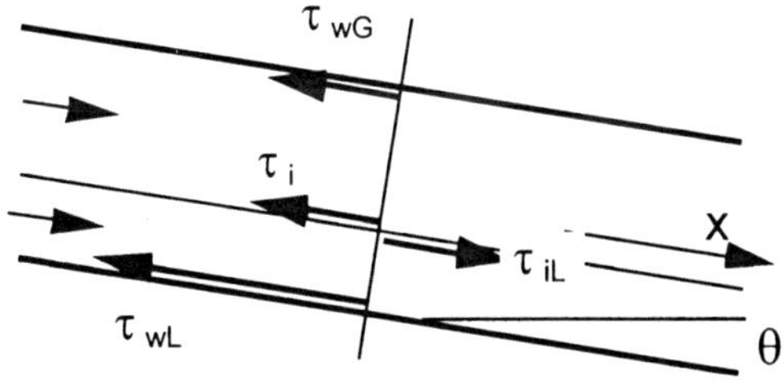

Figure 1. Gas-liquid stratified flow

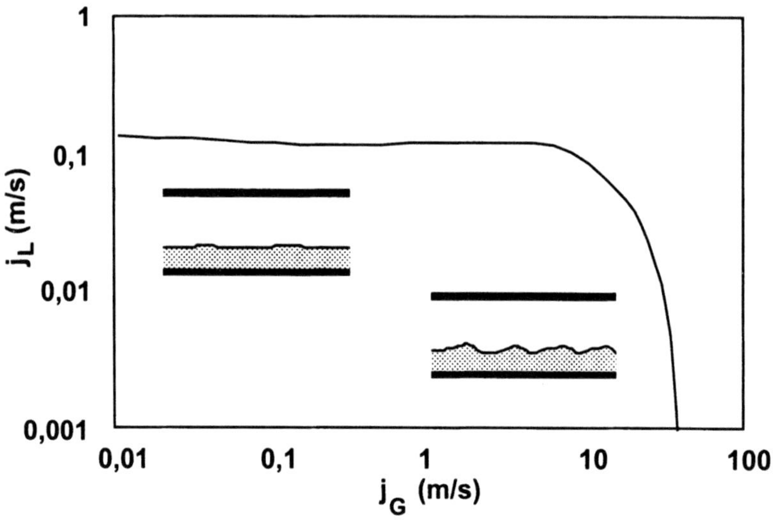

Figure 2. Occurrence of the stratified flow in horizontal pipe of 5 cm inside diameter: the coordinates of the map are the gas and liquid fluxes, the phases are air and water.

However, the transfer between phase is subtler: the higher the interfacial friction, the higher the pressure drop in the gas. In fully developed flow, as the pressure drop is identical in both phases, it can be understood that the pressure gradient is a driving force for the liquid. This remark proves that friction plays a central role in the flow modelling and especially interfacial friction that can be dominant if the interface is wavy.

Indeed if at low enough gas velocity the interface is maintained flat by gravity and surface tension, at increasing gas velocity, waves develop at the surface due to viscous instability and/or Kelvin-Helmholtz instability of the perturbations that appear on this surface. We will see that the gas "sees" the interface as if it was motionless: thus for the gas flow, waves are nothing else than roughness. As waves are produced by the work done by the fluctuations of the rate of strain against the deformation of the interface, it can be understood that the interfacial friction is strongly coupled to the motion of both phases. On the one hand, it results from the energy transfer near the interface, on the other hand it controls the momentum transfer between these phases. This is a twofold problem that is considered as the central issue of stratified flow modelling.

Different approaches have been explored to solve this problem. An empirical but effective approach correlates the interfacial stress to the mean phase velocities and fluid properties: by similarity to single-phase turbulent flow over rough surfaces, it is possible to correlate the interfacial friction factor to the wave roughness experienced by the gas. A more modern approach involves numerical simulation to predict the wave drag over simplified interfacial shapes, the monochromatic wave being the simplest case to be studied. As pointed out by Hanratty and McCready (1992), "a critical physical problem is to reconcile these approaches so as to produce a unified theory" of interfacial transfer of momentum in stratified two-phase flows.

Before considering this question, it is useful to start with a very simple case that leads to an analytical solution, the case of stratified flow with a smooth interface: this case allows to understand some of the important question related to friction modelling. Then the case of real flows will be considered to understand in detail in what the presence of the other phase modifies the flow structure and what are the consequence on the frictions at the wall and at the interface. Finally the central question of interfacial behaviour will be discussed.

2 A simple case: stratified smooth flow in 2D channel

2.1 Presentation of the problem

We start with the very simple case of a stratified 2D flow that is both steady and full developed (Figure 3). This case has the pleasant feature of being the addition of two single-phase flow problems rather than a two-phase flow problem. This makes its solution analytical at least under some restrictive assumption indicated further.

Each single-phase flow problem may be viewed as indicated in Figure 4. The layer of thickness h contains gas or liquid that flows between two moving plate. If it contains liquid, the velocity U_0 of the lower plate is zero, if it contains liquid then the upper plate velocity U_h is zero. The angle of the channel with respect to the horizontal is $\theta = (\mathbf{OX}, \mathbf{Ox})$: note that in Figure 4, this angle is negative. Let U be the mean velocity of the fluid:

$$U = \frac{1}{h}\int_0^h u\, dz\text{'}$$

where u is the x-component of the local velocity. Let τ_0 and τ_h be the shear stresses exerted by each plate over the fluid.

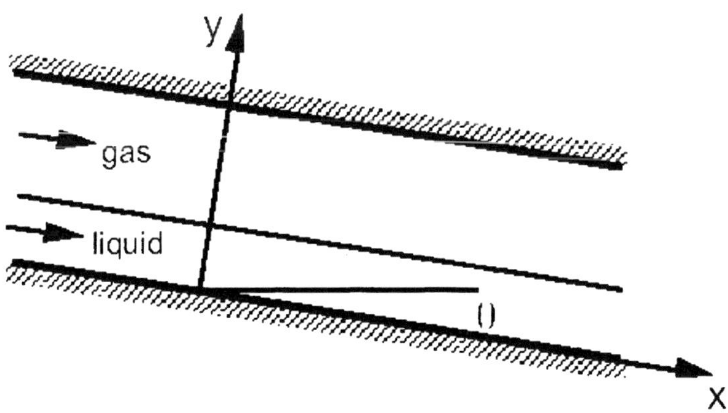

Figure 3. Stratified flow in 2D channel.

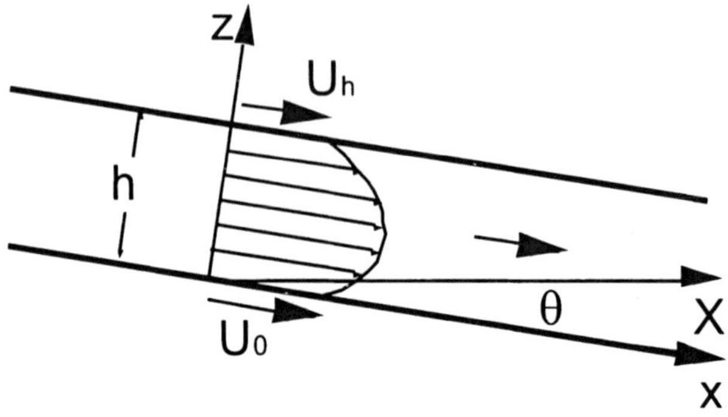

Figure 4. Scheme of each layer.

2.2 The single layer problem: laminar Couette-Poiseuille flow

Taking into account that the local velocity depends neither on x (fully-developed flow) nor on the time t (steady flow), the momentum equation projected on x (resp. z) becomes:

$$0 = -\frac{\partial p}{\partial x} - \rho g \sin\theta + \frac{d\tau}{dz}, \quad (1)$$

$$0 = -\frac{\partial p}{\partial z} - \rho g \cos\theta, \quad (2)$$

where g is the acceleration of gravity, p the pressure and τ the shear stress exerted over a surface of fluid oriented by a unit normal $\mathbf{n_z}$. The derivative of Eq. (2) with respect to is zero. Thus $\partial p/\partial x$ is constant. Then, the force opposed to the friction is the sum of the pressure gradient and the driving contribution of gravity. It is the constant a, such as:

$$a = -\frac{\partial p}{\partial x} - \rho g \sin\theta. \quad (3)$$

With this definition, it can be seen that Eq. (1) leads to a linear relation for the shear stress:

$$\tau = -az - \tau_0. \quad (4)$$

The minus sign for τ_0 accounts for the fact that the outer normal at the lower wall is equal to $-\mathbf{n}_z$. If the condition at the upper wall is used, then:

$$a = -\frac{\tau_0 + \tau_h}{h}. \quad (5)$$

It must be pointed out that Eq. (4) holds whatever the relation between the stress and the rate of strain. This relation is valid in laminar flow. It is also valid for turbulent flow provided the total shear stress (viscous stress + turbulent stress) is considered. This point will be useful further.

In what follows, we restrict the analysis to the case of laminar flow. This assumption is needed for simple analytical treatment. The flow under consideration is a Couette-Poiseuille flow.

The constitutive relation reads:

$$\tau = \mu \frac{du}{dy}, \tag{6}$$

where μ is the fluid viscosity. The solutions for the velocity field and its space average are readily obtained by using the no-slip conditions at both walls:

$$u = -\frac{az^2}{2\mu} + \left(\frac{ah}{2\mu} + \frac{U_h - U_0}{h} \right) z + U_0, \tag{7}$$

$$U = -\frac{ah^2}{12\mu} + \frac{1}{2}(U_0 + U_h). \tag{8}$$

As shown by Eq. (7), the parabolic velocity distribution is recovered. Moreover, Eq. (8) demonstrates that the fluid is moved by two different mechanisms: the motion of each plate to which the mean velocity is proportional and the contribution of pressure that opposes to the friction.

From (7), the shear stresses exerted by the plate may be determined. Then by eliminating a with Eq. (8) leads to:

$$\tau_0 = -\frac{\mu}{h}(6U - 4U_0 - 2U_h), \tag{9}$$

$$\tau_h = -\frac{\mu}{h}(6U - 2U_0 - 4U_h). \tag{10}$$

These two expressions are symmetrical with respect to the indices 0 and h. They also show that the force exerted on a specific plate is proportional to the fluid viscosity and to the difference between the mean velocity U and a wall fictitious velocity. This fictitious velocity is equal to 2/3 of the velocity of that plate, plus 1/3 of the velocity of the other plate.

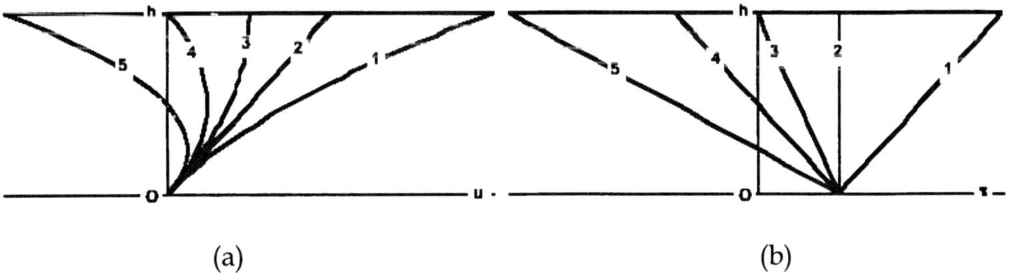

Figure 5. Vertical distribution of velocity (a) and shear stress (b) for a given shear stress and zero velocity at the lower plate. For case 1 $a < 0$, for case 2 (Couette flow) $a = 0$, for cases 3, 4 (Poiseuille flow) and 5 $a > 0$.

They contain two particular solutions: the case of Poiseuille flow (U_0=U_h=0) and the case of free surface flow (U_0=0, τ_h=0):

$$\tau_0 = -6\frac{\mu}{h}U \quad \text{(Poiseuille flow)}$$
$$\tau_0 = -3\frac{\mu}{h}U \quad \text{(free surface flow)} \qquad (11)$$

The solutions (9)-(10) show that it is hopeless to get a simple closure law between the mean velocity and the wall shear stress as it the case is in single-phase flow. This remarks also applies to the shear stress at the interface.

Figure 5 shows the velocity and shear stress distribution for a given shear stress and zero velocity at the lower plate and for different values of the effective pressure gradient a. As pointed out before, the velocity profile is parabolic, the shear stress is linear. This figure contains the particular cases of Couette and Poiseuille flows.

The flow is driven by the pressure (i) from the left to the right in cases 3, 4 and 5, (ii) from the right to the left in case 1. It is driven by the upper plate, from the left to the right in cases 1, 2 and 3, from the right to the left in case 5. It can be noted if a maximum velocity exists, it corresponds to the point where the shear stress is zero. Moreover, there exist some conditions for which the sign of velocity changes (e.g. case 5).

2.3 The two-layer laminar problem

Two-phase stratified flow is obtained by superimposing two similar layers as pictured in Figure 6. Let us focus our attention on the case of two-phase flow in a channel: in this case both the upper and the lower walls have zero velocity. In contrast, the velocity U_i of the interface as well as its position with respect to each plate are unknown.

The solution of this problem will provide a simple and pedagogic illustration of a typical problem of stratified two-phase flow in ducts. Suppose that the flow rates by unit width of the channel, q_L and q_G are fixed. The channel height $H = h_L + h_G$ is also given as well as its slope sin θ.

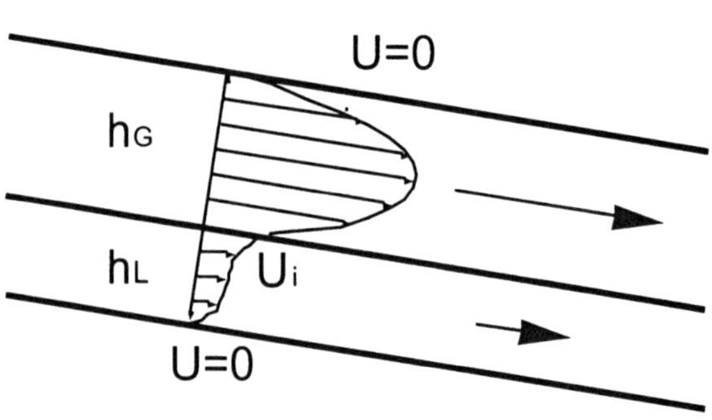

Figure 6. Schematic representation of 2D stratified flow.

Conservation equations. The conservation of mass and momentum, integrated over the transverse direction, can be written for each phase k:

$$q_k = U_k h_k \,, \tag{12}$$

$$0 = h_k\left(\frac{dP}{dz} + \rho_k g \sin\theta\right) + \tau_{wk} + \tau_{ik} \,, \tag{13}$$

with

$$\sum_{k=L,G} \tau_{ik} = 0 \,. \tag{14}$$

Eq. (13) is obtained from the integral formulation of the momentum balance projected on the x-axis. Eq. (14) is the jump condition at the interface. The pressure gradient may be eliminated between the two equations (13) written for $k=L$ and G. By using Eq. (14), it yields:

$$(\rho_L - \rho_G) g \sin\theta = \frac{\tau_{wL} - \tau_{iG}}{h_L} - \frac{\tau_{wG} - \tau_{iG}}{h_G} \,. \tag{15}$$

Closure problem: expressions for the shear stresses. The two equations of mass conservation (12) plus the momentum balance (15) forms a system that contains 6 unknown quantities: U_L, U_G., τ_{wL}, τ_{wG}, τ_{iG}, h_L (or h_G). The constitutive relations (9) and (10) suggests that it is possible to link the shear stresses to the mean velocities. These relations can be written for the present case:

$$\tau_{wk} = -\frac{\mu_k}{h_k}\left(6U_k - 2U_i\right), \tag{16}$$

$$\tau_{ik} = -\frac{\mu_k}{h_k}\left(6U_k - 4U_i\right). \tag{17}$$

To be useful, the unknown velocity U_i at the interface must be eliminated. Combining Eqs. (17)-(14) leads to:

$$U_i = \frac{3}{2}\frac{\frac{\mu_L}{h_L}U_L + \frac{\mu_G}{h_G}U_G}{\frac{\mu_L}{h_L} + \frac{\mu_G}{h_G}} \,. \tag{18}$$

The velocity of the interface is 1.5 times the sum of the phase velocities weighted by the ratio μ_k/h_k. As the gas viscosity is usually much smaller than the liquid one the interface velocity becomes:

$$U_i \approx \frac{3}{2}U_L \qquad \text{if} \qquad \mu_G << \mu_L \,. \tag{19}$$

To be complete, this condition requires also that $h_G/h_L >> \mu_G/\mu_L$ and $U_G h_G/U_L h_L >> \mu_G/\mu_L$: this is true in stratified flows. In this case, the interface velocity depends only on the liquid velocity, as in single-phase flow. The small viscosity ratio breaks the symmetry of the solution. This fact will be confirmed below.

With the above expression of U_i, one may simplify the relations (16)-(17) that express the shear stresses:

$$\tau_{iG} = -\tau_{iL} = -6\frac{U_G - U_L}{\frac{h_G}{\mu_G} + \frac{h_L}{\mu_L}}, \tag{20}$$

$$\tau_{wk} = -3\frac{\mu_k}{h_k}U_k + \frac{\tau_{ik}}{2}. \tag{21}$$

As previously noted, it appears again that the low viscosity of the gas phase breaks the symmetry in the expression of the shear stresses. Indeed, it yields:

$$\left.\begin{aligned} \tau_{iG} &\approx -6\frac{\mu_G}{h_G}(U_G - U_L) \\ \tau_{wG} &\approx -6\frac{\mu_G}{h_G}U_G\left(1 - \frac{U_L}{2U_G}\right) \\ \tau_{iG} &\approx -6\frac{\mu_L}{h_L}U_L \underbrace{-\tau_{iG}}_{negligible} \end{aligned}\right\} \text{if } \mu_G << \mu_L \tag{22}$$

In its approximate expression (22), the interfacial shear stress depends neither on the viscosity of the liquid phase nor on the thickness of the liquid layer: the symmetry of the solution has been lost. It is only preserved through the difference of the phase velocities. Roughly speaking, it may be said that the gas imposes the friction at the interface.

The shear stress exerted by the wall on each phase, given by Eq. (21), is the sum of two terms: (i) the contribution of the motion of the phase itself, expressed like in single-phase flow, (ii) the contribution of the other phase through the friction at the interface. To understand the effect of the other phase, let us take the practical case of Figure 6 in which $U_G > U_L > 0$: then $\tau_{iG} < 0$ ($\tau_{iL} > 0$). Since the first term of the right hand side of Eq. (21) is negative, the effect of the interfacial shear stress is to increase $|\tau_{iG}|$ and to decrease $|\tau_{iL}|$. In this particular case, the shear stress at the interface is a driving force for the liquid and a resisting force for the gas. It is understandable that it contributes to reduce the friction at the wall wetted by the liquid phase and to increase it at the other wall. However, the modification of the wall shear stress in the liquid is weak and it has been neglected in the approximate expression (22). This is not the case for the shear stress at the wall in the gas phase for which the two terms are nearly the same. Indeed, in its approximate expression (22), if the U_G high enough compared to U_L the liquid velocity may be ignored. As a consequence, the gas flows like if it was in a channel of height h_G, ignoring the presence of the liquid phase. The shear stresses at the wall and at the interface are nearly equal (see Eq. 11): this is a Poiseuille flow. In the lower layer, the wall shear stress contains two contributions. However, in the limit of vanishing gas viscosity, the second term is negligible. Again in this case, the wall friction is not influenced by the presence of the other phase. The flow behaves as a free surface flow (see Eq. 11).

Similarity analysis. The solution for the phase distribution may be found by solving Eq. (15). At this stage, it is convenient to introduce the fraction R_k and the flux j_k of each phase:

$$R_k = \frac{h_k}{h}, \tag{23}$$

$$j_k = R_k u_k . \tag{24}$$

Note that the flux, often referred to as "superficial velocity"), is the flow rate by unit height of the channel (Eq. 12). R_L is the liquid hold-up and the gas fraction satisfies $R_G=1-R_L$. Using the expressions (22) of the shear stresses at the wall and at the interface, Eq. (15) becomes:

$$-3j_L\left(\mu_L R_G^3 + 3\mu_G R_L R_G\right) + 3j_G \mu_G R_L^2\left(3+R_L\right) - \Delta\rho g h^2 \sin\theta\, R_L^3 R_G^3 = 0 . \tag{25}$$

This 6th degree equation in R_L can be solved for given flow conditions to determine the hold-up: this is why we will refer to it as the hold-up equation. Its general solution can be put under the form $F(R_L, j_L, j_G, \mu_L, \mu_G, \Delta\rho, g, \theta)=0$. Rather than solving this equation for the whole set of the flow parameters, one may reduce its complexity by introducing the dimensionless parameters:

$$\mathrm{Y} = \frac{(\rho_L - \rho_L)gh^2 \sin\theta}{\mu_L j_L}, \qquad \mathrm{N}_\mu = \frac{\mu_G}{\mu_L}, \qquad \mathrm{X} = \frac{\mu_G j_G}{\mu_L j_L} . \tag{26}$$

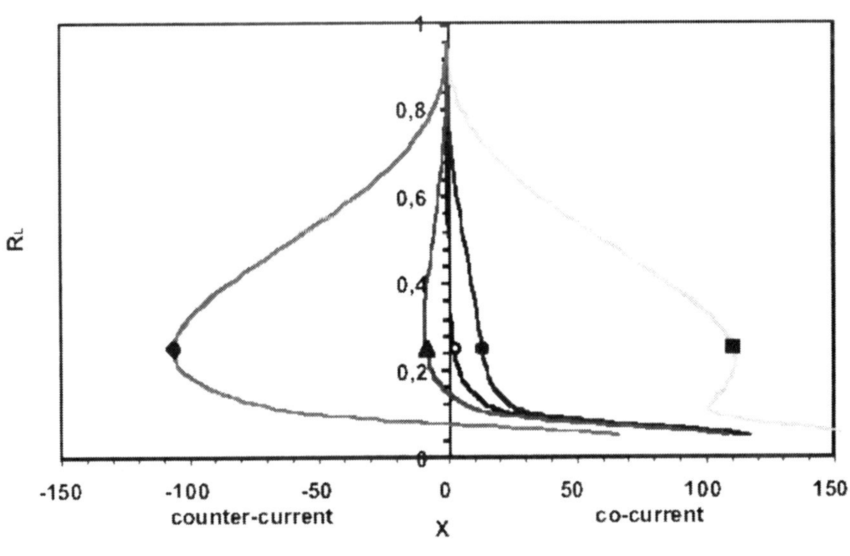

Figure 7. Solution of the laminar problem in the limit $\mu_G/\mu_L \to 0$ at fixed Y: –10000, –1000, 0, 1000, 10000 from left to right.

Y relates the gravity force to the viscous force in the liquid, N_μ is a ratio of viscosities that is supposed small in gas-liquid flow, X is the ratio of frictions force. Note that X is also the ratio of the single-phase pressure drops of the gas and the liquid: this dimensionless number is the root of the Martinelli parameter.

Introducing these definitions in the equation (25) leads to the dimensionless hold-up equation:

$$-3\left(R_G^3 + 3\mathrm{N}_\mu R_L R_G\right) + 3\mathrm{X}R_L^2(3 + R_L) - \mathrm{Y}R_L^3 R_G^3 = 0\,. \tag{27}$$

The general solution for the hold-up can be put under the form: $R_L = F(\mathrm{Y}, \mathrm{X}, \mathrm{N}_\mu)$. However it simplifies to $R_L \approx F(\mathrm{X}, \mathrm{Y}$ if the gas viscosity is neglected with respect to that of the liquid. The solution is plotted in Figure 7 for different values of X.

The graph may be viewed as the representation of the hold-up versus the gas flux at constant liquid flux and for different channel slopes. Indeed, if j_L as well as the fluid properties are kept constant, X is proportional to the gas flux and Y is proportional to sinθ. Thus negative (resp. positive) values of X correspond to counter-current (resp. co-current) flow. Negative (resp. positive) values of X correspond to ascending (resp. descending) channel.

The graph shows that in descending channel, counter-current flow cannot occur. The two different situations are pictured in Figure 8a-b according to the sign of the velocity difference. For Y=10000, the curve presents some particular behaviour if 100<X<110: two solutions may exist.

Counter-current flow may occur in ascending channel as pictured in Figure 8. But for a given channel slope (i) there exist two solutions at given fluxes, (ii) above a certain value of X there is no solution. If –X is greater than a critical value that depends on the channel slope, the gas cannot move the liquid up.

Figure 7 shows another important feature. In the bottom of the graph all the curves have the same asymptotic behaviour. They tend towards the same curve whatever the gravity parameter: the flow is dominated by gravity that balances the friction in the liquid. The hold-up equation reduces to an algebraic equation the 3rd degree:

$$R_G^3 - \mathrm{X}R_L^2(3 + R_L) \approx 0\,.$$

For the other part of the domain, the flow is dominated by gravity.

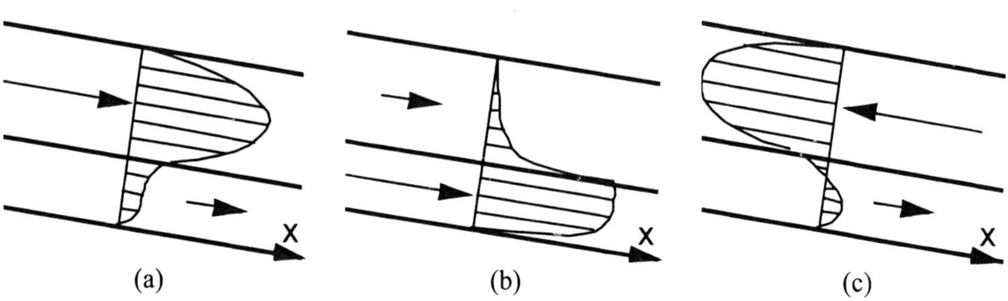

Figure 8. Different cases of stratified gas-liquid flow.

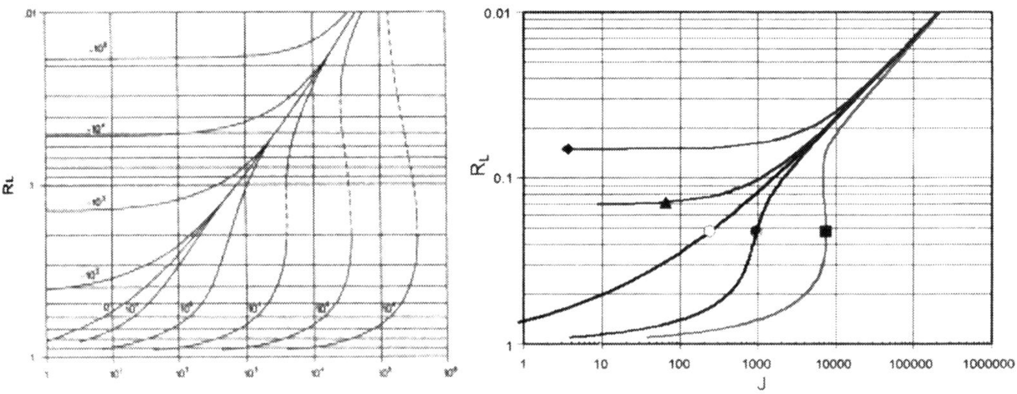

(a) (b)

Figure 9. R_L vs. J for different values of Y Comparison between (a) pipe flow (Rosant, 1984) and (b) 2D flow with Y= -10^{-4},, -10^{-3}, 0,, 10^{-3}, 10^{-4}.

2.4 Remark on 3D effect.

For pipe flow, the exact solution may be found by solving the Navier-Stokes equations in the circle within two domains separated by a chord. The basic equation to solve in each domain is a Poisson equation. The details of the method can be found in Rosant (1984) and since then other researchers have solved this problem by using various numerical method. We have plotted in **Figure 9** a comparison between the 3D and 2D cases by using for this comparison the same coordinates than Rosant: the flux ratio J and the liquid hold-up R_L. It must be pointed out that choosing J rather X limits the validity of the graph to a given value of the viscosity ratio. The choice of a logarithmic scale for J precludes also from considering counter-current flow.

For co-current flow, the 3D effects are more quantitative than qualitative. Indeed the same trends are observed, even the existence of two possible solutions when Y=10^4.

2.5 Friction factors.

It is useful to write the closure laws (20)-(21) in a generic dimensionless form. The dimensionless shear stress is called friction factor: it is defined as $2\tau/(\rho U^2)$ where ρ (resp. U) is a density (resp. velocity) scale. The choice of these scales is suggested by the form of the closure laws. The laminar case is useful to extend the results to other geometries (e.g. pipe) or other flow regimes (e.g. turbulent). Let us limit the analysis to the case $\mu_G << \mu_L$. Eqs. (22) suggest to scale:

- The shear stress at the interface with ρ_G and U_G–U_L.
- The shear stress at the wall in the gas phase with ρ_G and U_G.
- The shear stress at the wall in the liquid phase with ρ_L and U_L.

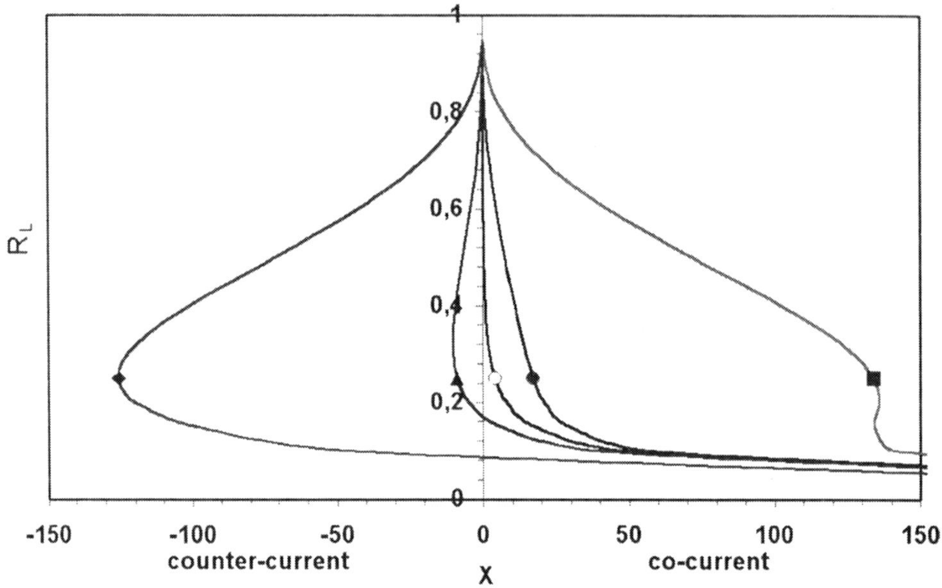

Figure 10. Solution of the turbulent problem at fixed Y (–1000, –100, 0, 100, 1000 from left to right).

We have chosen to express the friction exerted on each phase by the wall or the other phase. With this choice, the force is opposed to the velocity as can be seen on Eqs. (22). The friction factor may be viewed as a transfer coefficient whose sign is always positive. To fulfill this condition, the friction factor will be $f = 2\tau/\rho U|U|$. With this definition and the scaling density and velocity, the friction factors are:

$$\left.\begin{aligned} f_i &= \frac{\tau_{iG}}{\frac{1}{2}\rho_G(U_G-U_L)|U_G-U_L|} \\ f_G &= \frac{\tau_{wG}}{\frac{1}{2}\rho_G U_G|U_G|} \\ f_L &= \frac{\tau_{wL}}{\frac{1}{2}\rho_L U_L|U_L|} \end{aligned}\right\} \tag{28}$$

There is an interesting consequence of this definition. Let us apply these definitions to the case where $|U_G| >> |U_L|$. This is indeed the most interesting case where the gas has some influence on the liquid motion. If we use the expressions (22), the friction factors take the same form:

$$f_i = \frac{24}{\mathrm{Re}_i}, \qquad f_G = \frac{24}{\mathrm{Re}_G}, \qquad f_L = \frac{24}{\mathrm{Re}_L} \tag{29}$$

provided that the Reynolds numbers are defined as:

$$\left.\begin{aligned} \mathrm{Re}_i &= \frac{2\rho_G h_G |U_G - U_L|}{\mu_G} \\ \mathrm{Re}_G &= \frac{2\rho_G h_G |U_G|}{\mu_G} \\ \mathrm{Re}_L &= \frac{4\rho_L h_L |U_L|}{\mu_L} \end{aligned}\right\} \tag{30}$$

The above definitions are nearly the same. The only difference lies in the value of the numerical coefficient. Why is it so? The Reynolds number is $\mathrm{Re} = \rho \cdot D_h \cdot U/\mu$ with $D_h = 4A/S$, the hydraulic diameter, S the wetted perimeter, A the wetted cross-section. To recover the expressions (30) of the Reynolds numbers one must consider that both the upper wall and the interface must be taken into account for the wetted perimeter whereas for the liquid phase, only the lower wall must be taken into account. This choice is a logical consequence of the analytical results although it has been systematically done in literature without justification.

2.6 The two-layer turbulent problem

A similar procedure may be followed in turbulent flow to solve the hold-up equation (15). The shear stresses are replaced by their expressions (28). To simplify the solution we assume that $U_G - U_L \approx U_G$. In channel flow, we not need to substitute the expressions of the friction factors by Eqs. (29)-(30). Indeed, with the foregoing assumption, the Reynolds numbers are not function of the solution: they expresses only with the liquid and gas fluxes ($\mathrm{Re}_L = 4\rho_L j_L/\mu_L$, $\mathrm{Re}_G = \mathrm{Re}_i = 2\rho_G j_G/\mu_G$). To put Eq. (28) in dimensionless form we introduce similar dimensionless numbers:

$$\mathrm{Y} = \frac{(\rho_L - \rho_G) g h^2 \sin\theta}{f_L \rho_L j_L}, \qquad \mathrm{X} = \frac{f_G \rho_G j_G^2}{f_L \rho_L j_L^2}. \tag{31}$$

As in laminar case, Y is the ratio of the gravity to the friction force in the liquid and X is the ratio between the single-phase pressure drops of the gas and the liquid, also defined as the Martinelli multiplier. Then the hold-up equation reads:

$$-2R_G^3 - \mathrm{X} R_L^2 (3 + R_L) - 4\mathrm{Y} R_L^3 R_G^3 = 0. \tag{32}$$

Eq. (32) is similar to Eq. (27) in the limit of vanishing gas viscosity and the results plotted in Figure 10 look the same as Figure 7. They would have been exactly identical if we had taken care of expressing X and Y as force ratios.

3 Modelling of stratified two-phase flow: closure problem

We have got a taste of the modelling of two-phase stratified flow through the extensive analytical treatment of 2D laminar flow. Thus we want to follow the same road to treat the case of 3D flow in pipe and channels. Suppose that we know the specific flow conditions, namely:

- The pipe size, i.e. its diameter D or its cross-section area $A = \pi D^2/4$, the inclination θ.
- The fluid properties, i.e. the densities ρ_k ($k = L,G$), the kinematic viscosities ν_k and the surface tension σ.
- The flux —or superficial velocity— of each phase, $j_k = Q_k/A$, where Q is the volumetric flow rate.

A useful model of stratified flow would produce at least the following information:

- The mean velocities of both gas and liquid U_k.
- The phase fractions of both gas and liquid R_k.
- The wall and interfacial stresses.
- The pressure gradient dP/dx.

An advanced model should also give some detailed information on the flow structure:

- The local velocity of each phase
- The characteristic of waves (height, length, velocity).

For the sake of simplicity, let us restrict the study to the case of fully-developed steady flow and assume that the gas is incompressible. From the conservation equations integrated over the pipe cross-section we get the mass and momentum balances. The mass balance for each phase reduces to:

$$Q_k = Aj_k = AR_kU_k \tag{33}$$

where the phase fractions are related by:

$$R_G + R_L = 1\,. \tag{34}$$

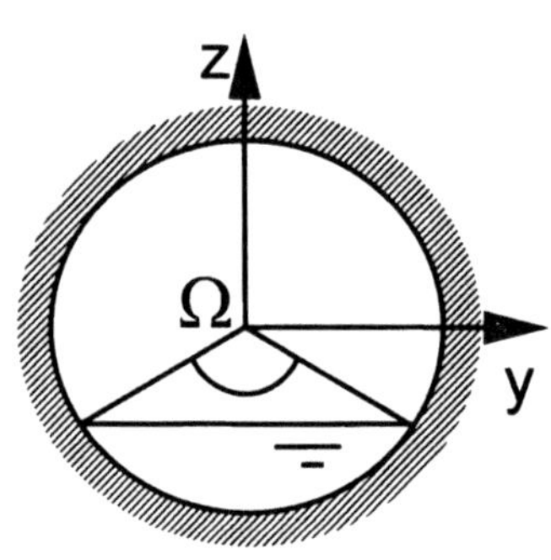

Figure 11. Representation of the smooth pattern in a cross-section of the pipe.

The momentum balance in each phase and across the interface may be written:

$$-R_k \frac{dP}{dx} = \frac{S_{wk}\tau_{wk} + S_i\tau_{ik}}{A} + \rho_k R_k g \sin\theta \tag{35}$$

$$\tau_{iG} + \tau_{iL} = 0 . \tag{36}$$

where S_{wk} and S_i are the perimeters of the wall and the interface, τ_{wk} and τ_i the shear stresses at the wall and the interface, and θ the pipe inclination. The pressure gradient hold-up equations are obtained from the two momentum equations:

$$-\frac{dP}{dx} = \frac{S\tau_w}{A} + \rho_M g \sin\theta , \tag{37}$$

$$\frac{S_{wG}\tau_{wG}}{AR_L} - \frac{S_{wL}\tau_{wL}}{AR_L} \frac{S_i\tau_{iG}}{AR_L R_G} - \Delta\rho \; g \sin\theta = 0 , \tag{38}$$

where $S = S_{wL} + S_{wG}$ is the total perimeter wetted by the phases, $\tau_w = (S_{wL}\tau_{wL} + S_{wG}\tau_{wG})/S$ is the total shear stress and $\rho_M = \rho_G R_G + \rho_L R_L$ the density of the mixture. The wall and interfacial perimeter can easily be related to the phase fractions. For pipe of circular cross-section, they are easily expressed versus the angle Ω that intercepts the interface (Figure11):

$$R_L = \frac{\Omega - \sin\Omega}{2\pi} , \tag{39}$$

$$S_{wL} = \frac{1}{2} D\Omega , \qquad S_{wG} = \frac{1}{2} D(2\pi - \Omega), \qquad S_i = D \sin\frac{1}{2}\Omega . \tag{40}$$

The shear stresses at the wall and at the interface may be put under dimensionless form by introducing the friction factors at the wall and at the interface. Although this introduction is no more than a definition that brings nothing new for modelling the shear stresses, it is important to take the form that is the most suitable. The 2D analytical analysis has suggested such a form (Eqs. 28). Thus we extend the previous definition by putting:

$$f_k = \frac{\tau_{wk}}{\frac{1}{2}\rho_k |U_k| U_k} \qquad f_i = \frac{\tau_{iG}}{\frac{1}{2}\rho_G |U_G - U_L|(U_G - U_L)} . \tag{41}$$

The shear stresses as they were defined represent in fact the mean momentum transfer across certain surfaces (wall wetted by each phase, interface). With respect to the simple situation discussed in Section 1, the real flows present several additional features that make more complex the modelling:

- The 3D character of the flow. 2D flow is an ideal geometry that is seldom encountered in practical situations.
- The turbulence in one or in both phases. The momentum exchanges are drastically modified by the presence of turbulent eddies.

- The waves at the interface. The interaction between the gas and the liquid across the interface may produce waves. To develop, these waves are fed by the energy they take from the liquid.

These remarks show that it is hopeless to get the answer from an analytical treatment of the problem: it only put some light on it and serves as a guide. One may think to the numerical simulation: it could be efficient for solving the questions related to the 3D aspects and the turbulence, but its capabilities are limited to explain how the waves develop and by which mechanism they transfer momentum to the liquid. To understand the real questions we need to discuss in detail the local structure of each phase and of the interface in stratified flow.

4 Local structure of stratified gas-liquid flow

4.1 Stratified flow in rectangular channel

The results presented in this section were obtained by a group of researchers in a rectangular channel of 10 cm height and 2/1 aspect ratio in turbulent regime (Suzanne, 1985; Fernandez-Flores 1984; Fabre *et al* 1982, 1983, 1987).

Kinematic structure. In a first step, let us focus on the kinematic structure of the flow (velocity and Reynolds stress) on the vertical axis of symmetry of the channel. Two experiments will be discussed. The first (referred to as Run 2) corresponds to Re_L=13600, Re_G=15100. The second (referred to as Run 3) corresponds to Re_L=13600, Re_G=25100. In the first case, the interface was smooth, in the second, wind waves of large amplitude covered it.

The components of the mean velocity are plotted in Figure 12: in (a), the *x*-component, in (b) the *z*-component. Note that different scales have been used for the gas and the liquid. Also two components of the Reynolds stress have been plotted: in (c), the turbulent shear stress that plays a special role as the signature of the momentum diffusion, in (d) the turbulent kinetic energy (tke).

There are significant differences between the two cases. When the interface is smooth, the flow appears to be a superimposition of two single-phase flow layers:

- The gas velocity is nearly symmetrical, suggesting that the upper wall and the interface play a similar role. This was suggested by the analytical study of the laminar 2D case (see Section 1). As expected, the flow is parallel as proved by the zero value of the *z*-component.
- The velocity in the liquid presents some similarity with the turbulent Couette single-phase flow (Wang and Gelhar, 1974). Like in the gas, the flow is parallel.
- The turbulent shear stress is linear in both phases as it must be in parallel flow: this point has been already mentioned in Section 1 (see Eq. 4). There is only a small deviation to the linearity close to the interface that is due to LDA limitation close to the surface.
- The tke possesses the same features that it has in single-phase flow. In particular it presents an important gradient near the wall and the interface where it is produced.

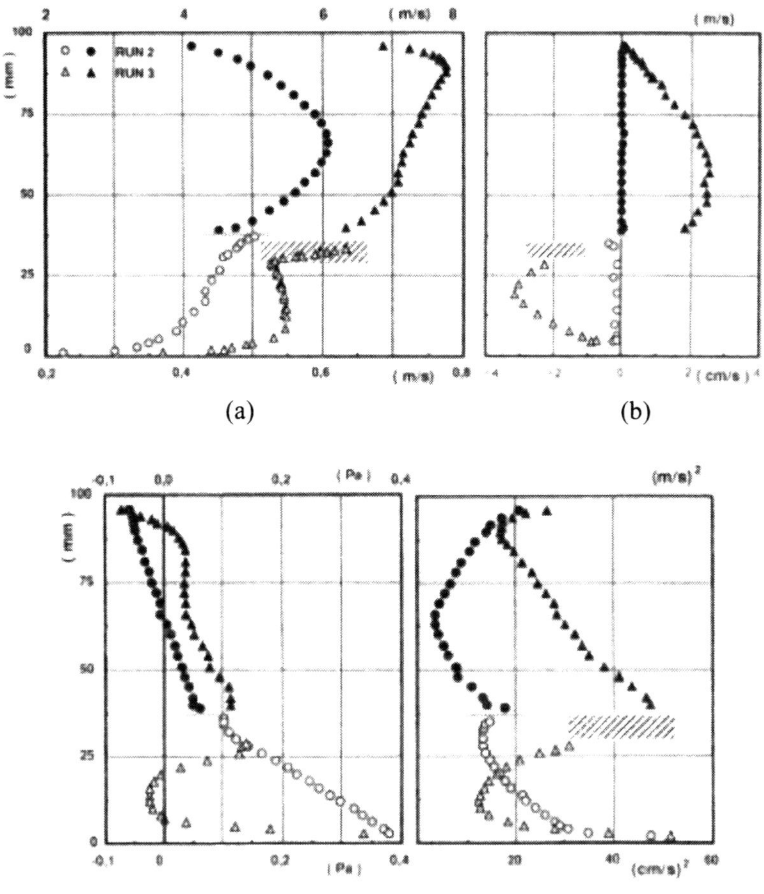

Figure 12. Vertical distribution of the longitudinal (a) and vertical (b) components of the mean velocity. Vertical distribution of the turbulent shear stress (c) kinetic energy (d) (Suzanne, 1985; see also Fabre *et al* 1982, 1983, 1987).

Let us consider now the second case (Run 3). As already mentioned, the interface is wavy. We will prove further that these waves are responsible for the deviation with respect to the smooth interface even far from it. The main features that can be put in evidence are as follows (Figure 12):

- In the gas, the symmetry of the x-velocity profile is lost. The maximum is drastically shifted towards the upper wall. The flow is not parallel as proved by the z-velocity profile: although weak the vertical component demonstrate clearly an upward motion from the interface to the upper wall. This point will be discussed further.

- In the liquid, a maximum appears in the x-velocity profile. Under the waves, the velocity gradient is slightly negative whereas it becomes positive within this region. A similarity exists with the gas flow since the flow is not parallel. In this phase however, the z-velocity on the channel axis is negative suggesting a downward motion from the interface to the lower wall.
- In both phases, the shear stress lost its linearity. This feature is stronger in the liquid than in the gas. On each side of the wavy region, it is hard to say if the shear stress keeps the same value. However the two-phase shear stress $\tau = \alpha_L\tau_L + (1 - \alpha_L)\tau_G$ remains continuous across it even if at a given location $\tau_L \neq \tau_G$.
- The tke reflects the presence of the vertical motion that contributes to transport the turbulence up in the gas and down in the liquid. As a consequence the tke profile in the gas lost its symmetry and its maximum is shifted to the bottom in the liquid.

From these results we can certainly learn about the turbulence in each phase and by extension about the law at the wall.

To discuss these points, the x-velocity has been plotted in Figure 13 and Figure 14 using viscous coordinates. These coordinates are defined as:

$$z^+ = \frac{zu^*}{\nu}, \qquad U^+ = \frac{U}{u^*}, \tag{42}$$

where $u^*=(\tau_w/\rho)^{1/2}$ is the friction velocity and ν is the kinematic viscosity. The local wall shear stress τ_w is adjusted so that the near wall distribution fit at best the logarithmic law:

$$U^+ = \frac{1}{\kappa}\text{Ln}y^+ + B \tag{43}$$

This fitting method does not guarantee that the coefficients κ and B take the same values as in single-phase flow where κ=0.4 (von Karman constant) and $B \approx 5.5$ (for smooth regime).

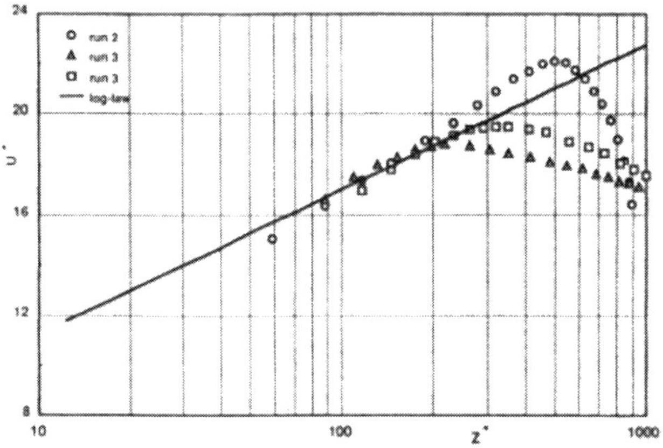

Figure 13. Vertical distribution of liquid velocity using viscous coordinates.

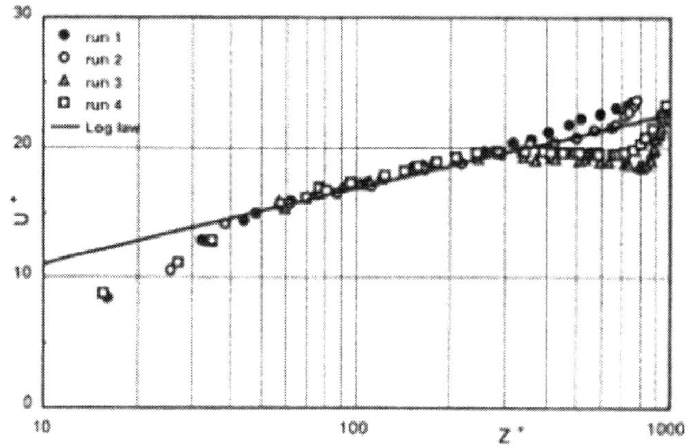

Figure 14. Vertical distribution of liquid velocity using viscous coordinates.

The conclusion that can be drawn is twofold. On the one hand, from Figure 13 and Figure 14 it can be seen that the data are closely grouped around the log-law distribution used in single-phase flow in a wide range of the domain. The results in the liquid phase (Figure 14) even display the typical behaviour of the viscous sub-layer in the range $y^+ < 30$. On the other hand, from Figure 12 it is shown that, for wavy flow there is a region above the interface in the liquid where positive shear stress is associated with negative velocity gradients. Therefore the effective viscosity concept does not apply for momentum transport, and "turbulent-viscosity" models fail to predict velocity profiles.

This conclusion is of major importance for the turbulence modelling of two-phase flow. It may be summarized as follows:

- In the region near wall regions where the production of turbulence by the work of the shear stress is dominant, the classical single-phase flow is observed and the eddy viscosity concept may be used. Moreover, the wall law seems to apply without restriction. In these regions the classical turbulence models are expected to provide acceptable results.
- When the interface is wavy, the eddy viscosity concept does not apply in a region below the waves. The classical turbulence model fails to predict the behaviour of the liquid phase.

Influence of waves upon the flow structure. When the waves are present at the interface, we have seen that the flow does not remain parallel. Suzanne (1985) carried out extensive measurements over the whole cross-section. The results are reported in Figure 15 where the projection of the mean velocity over a cross-sectional plane is plotted at various locations. The picture becomes clear: the flow in each phase is made of two counter rotating cells symmetrical with respect to the vertical axis. Magnaudet (1989) has shown that the secondary motions have a specific origin in each phase.

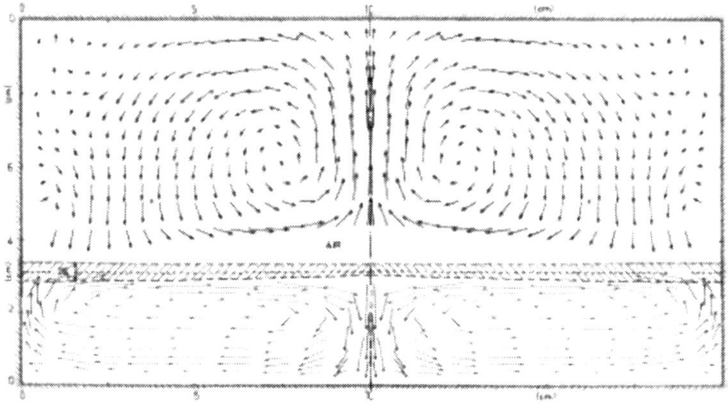

Figure 15. Secondary flow from the transverse components of the mean velocity, the velocity in the gas (upper layer) and in the liquid (lower layer) are in the ratio 100/1.

Concerning the upper layer, one must remember this rough picture that the gas flows as if the interface was a solid wall. However if waves develop on this surface, the gas will experience the interface as hydraulically rough. Because the wave amplitude is not homogeneous in the transverse direction, the gradient of the roughness, $\partial k/\partial y$, may prevent the gas flow to remain parallel. In the results of Figure 15 the interface roughness was higher close to the wall than at the channel axis. Concerning the lower layer, the mechanism of formation of the secondary motion is of the same kind as the Langmuir circulations observed at the ocean surface in windy situations.

To end with this particular structure that is obviously of two-phase nature, a simple experiment has been carried out to relate clearly the secondary motion to the wave structure. This experiment consists in flowing a thin oil film at the interface to prevent the waves to form. In doing so, the flow conditions remain identical, only the waves at the interface are suppressed. The results are given in Figure 16. In the absence of waves, the flow in the liquid recovers the classical behaviour that was pictured in smooth flow. We may even certify that secondary flow was cancelled by the presence of oil film at the surface. For fully-developed steady flow, the Reynolds equations on the channel axis read:

$$\overline{w}\frac{\partial \overline{u}}{\partial z} = -\frac{1}{\rho}\frac{\partial \overline{p}}{\partial x} + \frac{\partial \tau}{\partial z} - g\sin\theta$$

$$\overline{w}\frac{\partial \overline{w}}{\partial z} = -\frac{1}{\rho}\frac{\partial \overline{p}}{\partial x} - \frac{\partial}{\partial z}\overline{w'^2} - g\cos\theta$$

Taking the derivative with respect to x of the second equation proves that the derivative of the pressure with respect to x does not depend on z. As was done in Eq. (3), we can introduce the constant a defined as:

$$a = -\frac{\partial \overline{p}}{\partial x} - \rho\ g\sin\theta .$$

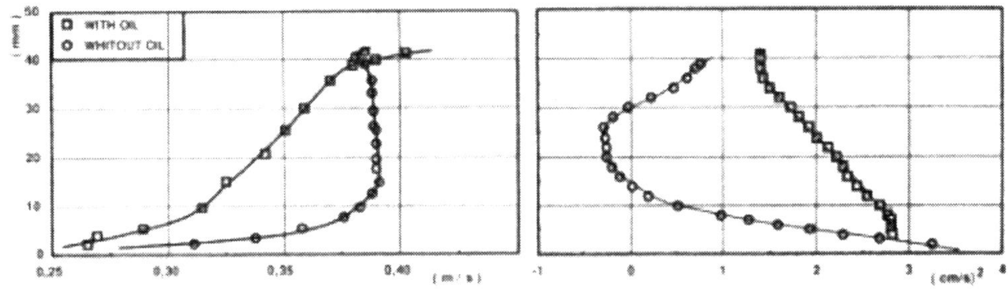

(a) (b)

Figure 16. Experimental evidence of the influence of wave motion on the structure of the liquid phase: (a) mean velocity (b) shear stress (O) with waves (□) without waves suppressed by an oil film.

Then the first equation writes:

$$\frac{\partial \tau}{\partial z} = -\frac{a}{\rho} + \overline{w}\frac{\partial \overline{u}}{\partial z}. \tag{44}$$

As a consequence, the shear stress τ does not remain linear as soon as the flow is not parallel. With a downward secondary motion, the gradient of τ is positive when the gradient of u is negative and *vice versa*. From the shear stress distribution, Figure 16(a) shows that suppressing the waves suppresses also the secondary motion.

Influence of waves upon the turbulence. We have seen that, in the vicinity of the wave region, the turbulence does not display the classical feature observed in shear layers. However, the fluctuating velocity is partly due to the wave motion. Indeed, the velocity would fluctuate under the waves even in the absence of turbulence. This suggests that the fluctuations contain two contribution:

- Onc is shear-induced turbulence that is either created by the work done by the shear stress or diffused from the region where it has been created.
- The other is wave-induced turbulence or pseudo-turbulence that comes from the penetration into the layer of the motion induced by the waves upon the liquid.

The decomposition between the two contributions has been done on the following basis. The motion induced by waves is determined within the layer from the time evolution of the interface, height: the velocity that would be observed if the flow were irrotational is then calculated (Figure 17). The resulting r.m.s. velocity is subtracted from the measured one. The result shows that the expected shear-induced turbulence distribution is now similar to single-phase flow, namely near the wave region. Thus it may reasonably anticipated that numerical simulation using LES approach would bring some interesting explanation on the physics of momentum transfer across the waves.

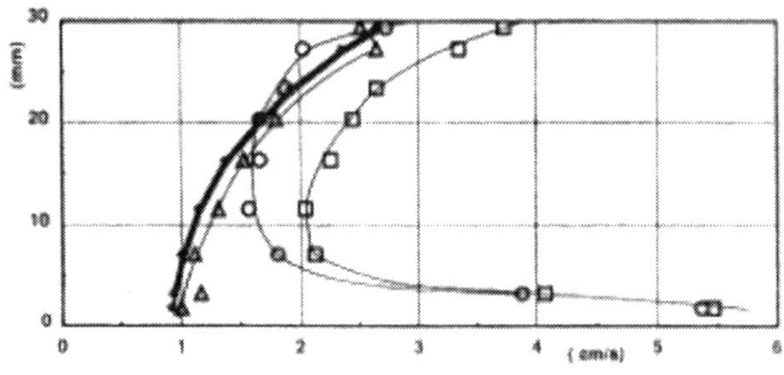

Figure 17. Vertical distribution of the rms velocity u': □: raw value, Δ: evaluated value induced by wave motion, O: corrected value.

4.2 Stratified flow in pipes

There exist few studies that describe the local structure in of stratified flow in pipes. Figure 18 shows one example of such study (Rosant, 1983). The distribution of velocity in the liquid for wavy flow has the same unexpected maximum that suggests the presence of secondary flow.

More recent data have been obtained by Strandt (1993) and Lopez (1994). They also show the same features as shown in Figure 19.

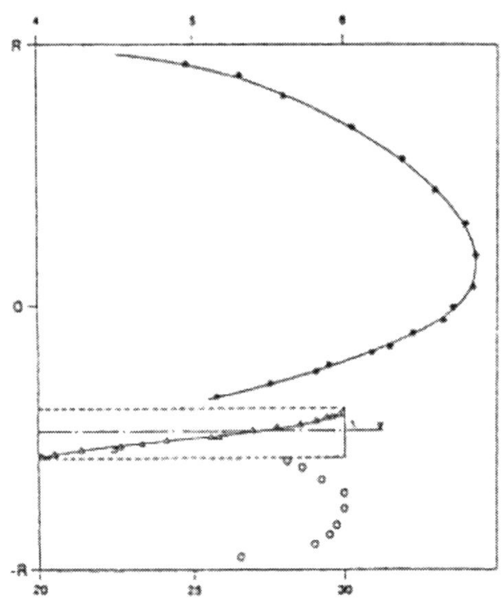

Figure 18. Vertical distribution of mean velocity and gas fraction in pipe (Rosant, 1983).

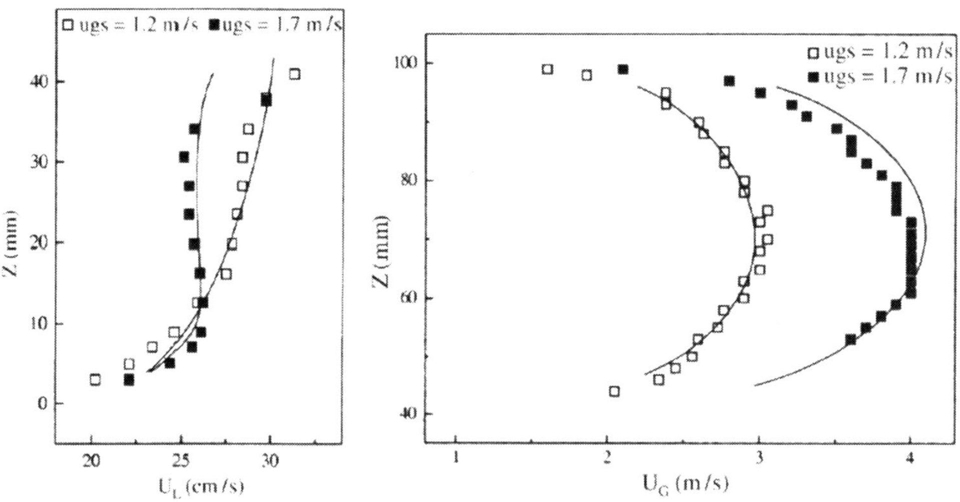

Figure 19. Vertical distribution of mean velocity in pipe (Strandt, 1993).

5 Waves: regime, amplitude, velocity

The waves are generated at the interface by complex mechanisms involving both viscous and pressure forces. As they create drag force at the liquid surface, they may increase drastically the interfacial shear stress. The specific feature of waves is their capability to propagate by carrying energy without any significant damping. Contrary to friction at the wall which results from local equilibrium (for simplicity we assume here that friction is determined by local velocity field, even if turbulence is transported), waves may be created far upstream from the location where the act. Interfacial friction is expected to be the consequence of this propagation. We will discuss in detail the wave patterns, their equilibrium amplitude and their velocity.

5.1 Wave regime

The discrimination of smooth and wavy stratified flow has been recognized in flow maps (see for example Taitel and Dukler, 1976b) although they do not characterize the type of waves. Moreover they do not predict accurately the scale effects due to change in fluid properties and pipe diameter. As the interfacial shear stress is very sensitive to the roughness caused by the waves, their detailed description and behaviour must be discussed. One may refer for this topic to the overview on separated flow modelling by Hanratty (1991).

Most of the experimental studies of wave structure have been conducted in rectangular channels. However Andritsos and Hanratty (1987) have performed experiments in pipe of 2.52 cm and 9.53 cm with liquid of various viscosities. The different regimes have been identified either by visual observation (Hanratty and Engen, 1957) or by wave spectrum or auto-correlation function (Akai, Inoue and Aoki, 1977; Fernandez-Florès, 1984). The conclusions agree qualitatively:

- At sufficiently low gas velocity, there is no noticeable disturbance at the liquid surface; the interface is smooth.
- The first waves to appear are small amplitude, short wavelength regular disturbances that extend over the channel width: they are 2-D capillary-gravity waves. The auto-correlation function exhibits a periodicity. Their energy spectrum is well represented (Lleonar and Blackman, 1980) by the function:

$$\phi(\omega) = u_i^2 \sqrt{\frac{u_i \nu_L}{\sigma}} \omega^{-3}, \tag{45}$$

where u_i is the friction velocity at the interface, and ω the wave angular frequency. From physical reasoning the capillary waves do exist for a specific frequency range. If ω is greater than $(\rho_L g^3/\sigma)^{1/4}$ the capillarity has a dominant influence provided the wave energy is not dissipated by viscosity: this condition may be fulfilled if ω is smaller than $\sigma^2 \rho_L^{-2} \nu_L^{-3}$. For viscosity ν_L greater than $(\sigma^3 \rho_L^{-3}/g)^{1/4}$ there is no frequency range corresponding to capillary waves : it may explain the absence of these waves for highly viscous liquid as pointed out by Andritsos and Hanratty. As they may be unstable to 3-D disturbances they exist over a narrow range of gas velocities.
- By increasing the gas velocity, the 2-D waves break up into 3-D pebbled structure. For small diameter pipe, this regime is not observed. The auto-correlation function does not present a clear periodicity.
- In pipe at higher gas velocity large amplitude waves with a steep front and a gradually sloping back are formed. They are not so regular than the 2-D waves observed at lower gas velocities.
- At sufficiently high gas velocities, large scale 2-D waves are superimposed to the pebbled surface: the distance between them is irregular and much greater than the liquid height. Their velocity is linked to their amplitude and in some circumstances coalescence is observed. These roll waves appear as flow surges carrying large amount of liquid. These waves are clearly seen in the time recording of the liquid height.
- At high enough gas velocity, atomisation of liquid droplets occurs.

The transitions between these different regimes are far to be clearly understood. However, Andritsos and Hanratty (1987) have confirmed the existing models of wave transition from linear stability analysis. The theory consists in perturbing a smooth interface by small amplitude sinusoidal wave of wave number *k* real, velocity *c* complex and amplitude real. The disturbance h'_L of the average liquid height is given by:

$$h'_L = \hat{h}_L \exp[ik(x - ct)] \tag{46}$$

where

$$c = c_R + ic_i$$

The fluctuations due to waves induce perturbations in pressure and shear stress at the interface:

$$p'^i = \hat{p}'^i_R \cos[ik(x - ct)] - \hat{p}'^i_I \sin[ik(x - ct)], \tag{47}$$

$$\tau''^{i} = \hat{\tau}''^{i}_{R} \cos[ik(x-ct)] - \hat{\tau}''^{i}_{I} \sin[ik(x-ct)]. \tag{48}$$

The real part of the amplitude of both pressure and shear stress perturbation is in phase with the wave height and their imaginary part is in phase with the wave slope. Two physical mechanisms may be identified: if $\hat{p}''^{i}_{R}$ is negative the perturbation is amplified because there is a suction effect at the wave crest and a compression at the through; now if $\hat{p}''^{i}_{I}$ is positive another kind of amplification occurs leading to the so-called "sheltering effect" described by Jeffreys (1925). The first mechanism is expected for long waves; the second mechanism involves wave-induced shear stress at smaller scale. The problem is solved from the linearized Navier-Stokes equations for the perturbed velocity field, in the liquid, assuming plug flow for the mean velocity. The influence of the gas is taken into account by including the boundary conditions at the interface. These conditions are chosen to illuminate one or the other mechanism. Thus for long waves pressure is hydrostatic so that plug flow is assumed for the gas. The perturbations are given from inviscid flow by:

$$\hat{p}''^{i}_{R} = -\frac{kh_L}{\tanh(kh_L)} \frac{\rho_G}{\rho_L} (U_G - c)^2 \qquad \text{and} \qquad \hat{p}''^{i}_{I} = \hat{\tau}''^{i}_{R} = \hat{\tau}''^{i}_{I} = 0 .$$

For the general case the exact theory has not yet been developed; a model of wave-induced Reynolds stress describing correctly the shear stress variation along solid wavy wall has been used.

The waves are amplified if c_I is positive. The condition c_I=0 defines neutral stability. An example of the results is shown in Figure 20 for low and high viscosities of the liquid. For low viscosity a test of sensitivity proves that the critical velocity to generate 2-D small amplitude waves is controlled by a sheltering effect (the pressure is in phase with the wave slope). On the contrary for high viscosity, the generation of waves is due to Kelvin-Helmholtz instability (the pressure is in phase with the wave height).

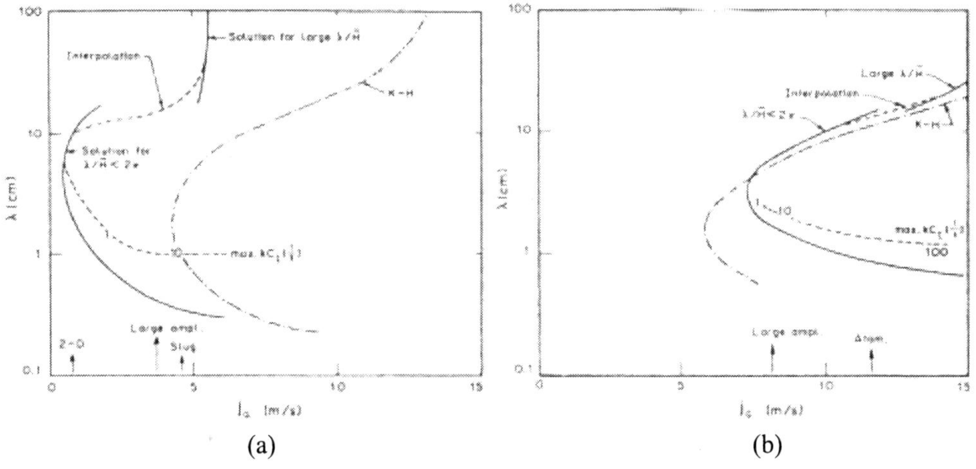

(a) (b)

Figure 20. Neutral stability curves for D = 9.5 cm: (a) μ_L = 1 cP, (b) μ_L = 80 cP.

For practical use Andritsos and Hanratty (1987) propose to consider three patterns: 2-D regular waves, large-amplitude irregular waves and atomising flows.

1. The generation of 2-D regular waves is well predicted by the sheltering criterion of Jeffreys (1925) modified by Taitel and Dukler (1976a):

$$U_{G,1} \geq \sqrt{\frac{4\nu_L \Delta\rho\, g}{s\rho_G U_L}}\,, \tag{49}$$

where s is dimensionless parameter taken equal to 0.06. However it must be kept in mind that this transition is meaningless for liquid of high viscosity. It is suggested that the capillary waves do not exist if:

$$\nu_L > K\left(\frac{\sigma^3}{\rho_L^3 g}\right)^{1/4}, \tag{50}$$

where K is a constant to be determined from experimental results.

2. For the initiation of irregular large-amplitude waves they recommend the following empirical criterion based upon the critical gas velocity corresponding to the Kelvin-Helmholtz instability:

$$U_{G,2} \geq U_L + \sqrt{\left[\frac{k\sigma}{\rho_G} + \frac{\rho_L\, g}{\rho_G k}\right]\tanh(kh_L)}\,, \tag{51}$$

The critical value corresponds to the wave number k for which the velocity is minimum:

$$k = \sqrt{\frac{\rho_L g}{\sigma}}\,. \tag{52}$$

3. For atomisation their empirical criterion is:

$$U_{G,3} \geq 1.8 U_{G,2}\,. \tag{53}$$

5.2 Wave velocity

The propagation of perturbations of finite amplitude at the surface of a liquid is a complex phenomenon, involving dispersion and non-linear effects due to pressure-velocity interactions. The problem is more intricate in the presence of gas flow. It has only been solved for asymptotic cases. First we will characterize the perturbations by the dimensionless parameter $\lambda^+ = \lambda/h$ where λ is the wavelength. The theory gives a piece of solution for small amplitude wave.

Inviscid flow. For small amplitude waves, the velocity is found from linear analysis. For an idealized horizontal unbounded problem the wave frequency ω is a function of the wave vector $\mathbf{k}$, the surface tension σ, the liquid height h_L, the layer thickness of the gas h_G, as well as the velocity field in both phases. The frequency ω takes the form:

$$\omega = \omega_0(\mathbf{k}) + \mathbf{k} \cdot \mathbf{U}_L, \tag{54}$$

where ω is the observed frequency, ω_0 is the intrinsic frequency and $\omega_0(\mathbf{k})$ the dispersion relation of the medium. The wave velocity **c** in the direction **n** is given by:

$$\mathbf{c} = \mathbf{n}\left(\frac{\omega_0}{k} + \mathbf{n} \cdot \mathbf{U}_L\right). \tag{55}$$

where k is the wave-number. The foregoing equation shows that waves are transported by the liquid velocity. They travel with respect to the fluid at the specific velocity ω_0/k. For isotropic medium, the dispersion relation $\omega_0(\mathbf{k})$ does not depend on the direction of propagation so that it reduces to $\omega_0(k)$.

Let us consider the case of two layers of fluid between two parallel plates for which the fluid will be supposed inviscid. This is of course a very idealized problem because, even weak, the viscosity has a significant influence close to the wall: the assumption corresponds to plug flow in both phases. The linearized solution of irrotational motion for each phase with interfacial mass and momentum conditions (no mass transfer or pressure jump due to surface tension) gives the well-known solution (see for example Milne-Thomson, 1955):

$$\frac{\rho_L(U_L - c)^2}{\tanh(kh_L)} + \frac{\rho_G(U_G - c)^2}{\tanh(kh_G)} = \frac{\Delta\rho\, g}{k} + \sigma k \tag{56}$$

This equation gives either two real solutions for c (neutral waves) or two imaginary solutions among which one is unstable. The limiting case corresponds to the transition due to Kelvin-Helmholtz instability plotted in Figure 20 (K-H curves):

$$U_{LG,crit} = \sqrt{\left(\Delta\rho\, g + \sigma k^2\right)\left[\frac{\tanh(kh_G)}{k\rho_G} + \frac{\tanh(kh_L)}{k\rho_L}\right]}, \tag{57}$$

where $U_{LG,crit} = U_{G,z} - U_L$.

When the second term of the l.h.s. of Eq. (56) is negligible (low pressure and low gas velocity) it becomes:

$$c = U_L \pm c_0,$$

$$c_0^2 = \left(\frac{g}{k} + \frac{\sigma k}{\rho_L}\right)\tanh(kh_L), \tag{58}$$

$$c_0^2 = gh_L \qquad \text{for shallow water } (kh_L \ll 1), \tag{59}$$

$$c_0^2 = \left(\frac{g}{k} + \frac{\sigma k}{\rho_L}\right) \qquad \text{for deep water } (kh_L \gg 1). \tag{60}$$

The waves are moving with respect to the liquid to the intrinsic velocity c_0.

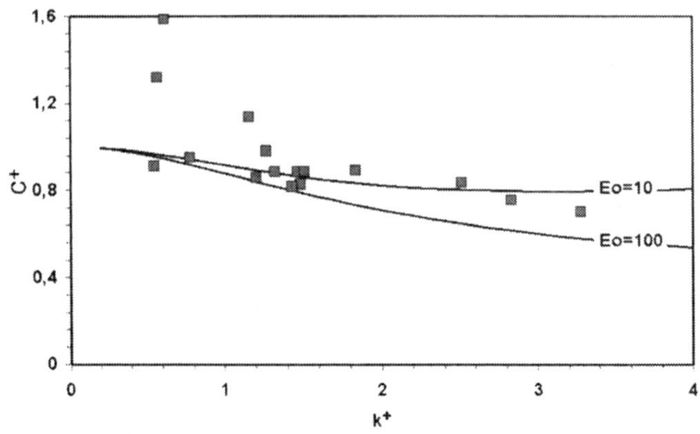

Figure 21. Wave velocity $c^+ = (gh_L)^{1/2}$ vs. wave number $k^+ = kh_L$, measurements of Fernandez (1984) compared with Eq. (58).

The velocity is plotted in Figure 21 versus the wave number in dimensionless form, for different Eötvös numbers Eo = $\Delta\rho g h_L{}^2/\sigma$. At small wave number (shallow water), the dimensionless velocity is unity: the wave motion is controlled by gravity. At large wave number (deep water), the velocity is proportional to $\sigma^{1/2}$. If U_L is less than $(gh_L)^{1/2}$ then the two sets of solution correspond to waves travelling upstream and downstream respectively. In the contrary gravity waves travel downstream.

Fernandez-Florès (1984) has measured the wave velocity as well as the dominant wavelength from the cross-correlation function of the liquid heights measured at two different locations. The experiments have been made in channel of rectangular cross-section. The results are plotted in Figure 21.

Long waves. Interfacial and wall frictions may have a significant influence on the long wave velocity. For small amplitude the solution may be found by assuming plug flow in both phases and solving the *x-t* mass and momentum equations. In the limit of the small wave numbers, the higher order terms in the momentum equations (inertia terms and surface tension term) vanishes. The wave behaves as a kinematic wave.

5.3 Wave amplitude

When the waves have been generated at the interface, they grow until they reach an equilibrium level by a complex mechanism involving gravity, dispersion friction and surface tension. There is not a complete theory that gives the equilibrium amplitude of waves even in the case of two unbounded fluids separated by an interface. Phillips (1977) has shown that the maximum height of the waves Δh_{max} depends on the wave velocity:

$$\Delta h_{\max} = \frac{(c - U_i)^2}{g}.$$

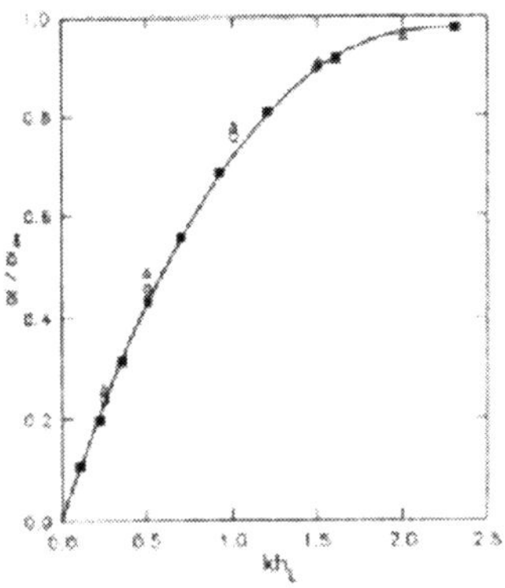

Figure 22. Wave steepness (Bontozoglou and Hanratty, 1989).

When the velocity is large enough, the velocity at the interface U_i may be neglected. For gravity waves in deep water, the maximum amplitude is expressed versus the wave number by:

$$\Delta h_{\max} = \frac{\lambda}{2\pi}. \tag{61}$$

The ratio of the wave amplitude to the wavelength is the steepness that is limited from the above relation to 0.16. This conclusion agrees with the pioneering work of Michell (1893) who gave the critical amplitude corresponding to the sharp peaked wave: he found the steepness to be 0.142. Constant steepness over a range of wave number leads also to a saturated range in the wave energy spectrum as shown by Fernandez-Florès (1984). When the steepness is high enough the local curvature increases at the crest producing capillary waves that extracts energy from large gravity waves.

The problem is even more complicated for pipe flow, due to the solid boundaries. Both the drift velocity and the liquid flow rate influence the wave energy: for high enough liquid flow rate, a saturation occurs. These results agree qualitatively with those of Cohen and Hanratty (1968). The r.m.s. wave height increases with the Froude number until it reaches the value of 0.035. The waves have been found to break into smaller waves when the Froude number reaches a value close to unity: the dominant wave never reaches the critical amplitude. An attempt has been made by Bontozoglou and Hanratty (1989) for understanding the mechanism by which 2-D waves reach an equilibrium level. In order to establish the criterion, it is assumed that the geometric limit of the wave occurs when the wave slope becomes unphysical. From a similitude analysis and physical considerations, they propose to correlate the steepness under the following form:

$$\frac{\Delta h_{\max}}{\lambda} = \alpha \frac{h_L}{\lambda} F\left(\frac{U_G - U_L}{U_{LG,crit}}\right),$$

where $U_{LG,crit}$ is the limit of the drift velocity giving rise to K-H instability, defined in Eq.(57); $\alpha(kh_L)$ is determined from numerical calculation of progressive wave of permanent form (Figure

22). The results have shown that it is almost independent of the liquid velocity and the density ratio. The above relation is in good agreement with the experimental data:

$$\frac{\Delta h_{\max}}{\lambda} = 0.079\alpha\left(kh_L\right)\frac{U_G - U_L}{U_{LG,crit}}, \tag{62}$$

where $\alpha \rightarrow \alpha_\infty kh_L$ when $kh_L \rightarrow 0$, and $\alpha \rightarrow \alpha_\infty$ when $kh_L \rightarrow \infty$, α_∞ being the value for deep water which is not specified.

6 Wall and interfacial shear stresses

6.1 Experimental determination of the mean wall shear stress

The wall and interfacial mean shear stresses that appear in Eqs. (35)-(36), cannot be determined from global measurements. Indeed these 3 equations contain 4 unknown quantities (τ_{wL}, τ_{wG}, τ_{iL}, τ_{iG}) and 6 quantities that can be measured (dP/dx, R_L) or determined (R_G, S_{wL}, S_{wG}, S_i) from the measurements. In conclusion, it is impossible to obtain trustable results on the shear stresses from pressure gradient and liquid hold-up measurements. An additional assumption must be added.

To avoid this assumption, the shear stress was measured in the channel at different locations uniformly distributed along the perimeter wetted by the gas. The detail of the method is given by Fabre *et al.* (1984) and the results by Fernandez (1984) and Fabre *et al.* (1985b). The picture of the distribution of the wall shear stress is given in Figure 23. τ_{wG} is not uniform being influenced by the secondary motion. The local values around the wall are used to determine the average value τ_G. Then the average value of the liquid shear stress τ_L is obtained from the global momentum balance (35)-(36). The results are plotted in dimensionless form in Figure 24 using the definitions (28) for the friction factors and for the Reynolds numbers definitions in agreement with (30). The comparison with the Blasius correlation used in single-phase flow shows that the non uniform distribution is not crucial for the average value and that the friction in each phase behaves as in single-phase flow. It does not seem that the presence of secondary flow has a significant influence upon the global quantities.

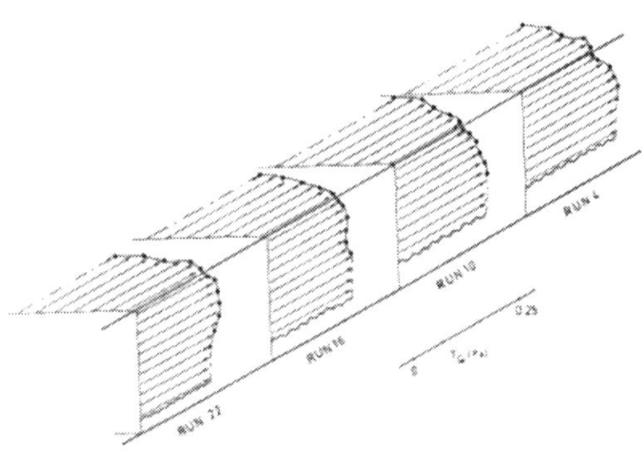

Figure 23. Distribution of the shear stress at the wall of a rectangular channel in the gas phase.

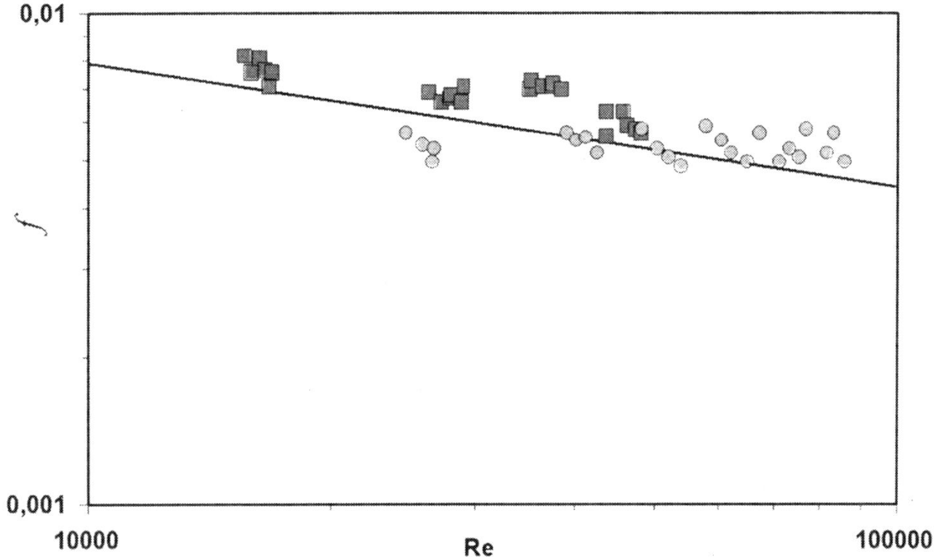

Figure 24. Wall friction factor vs. Reynolds number: liquid phase, O gas phase, the solid line corresponds to the Blasius correlation f=0.079 $Re^{-1/4}$ (Fernandez-Flores, 1984).

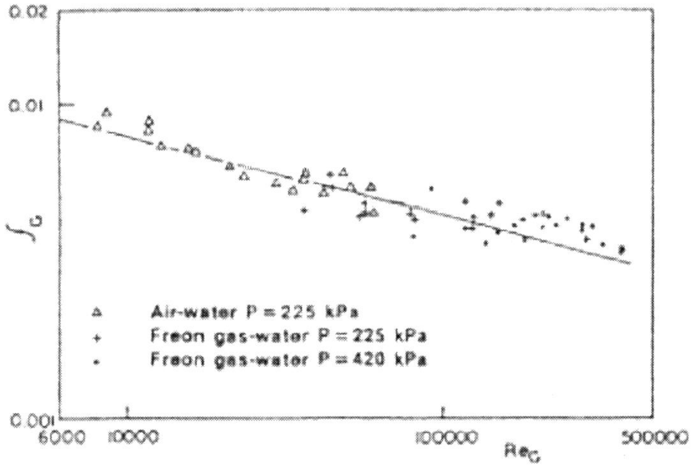

Figure 25. Friction factor at the wall in the gas phase for pipe flow (Kowalski, 1987).

To access directly to the wall shear stresses, Kowalski (1987) used a method similar to this of Fernandez-Florès (1984). His results are reported in Figure 25 for the gas phase. They indicate again that the flow behaves as in single-phase flow in the gas. We did not report his results for the liquid phase because they were obtained at liquid Reynolds number in the range 200-5000 that is clearly a transitional range.

6.2 Experimental determination of the interfacial shear stress

There are several studies that report careful determinations of the interfacial shear stress. We report here one for channel flow and one for pipe flow.

The results of channel flow experiment are given in Figure 26. They show clearly that the shear stress at the interface is much higher than at the wall. It can be 2 to 3 times greater. This shear stress is related to the presence of waves at the interface that create roughness. In these results it seems that the interfacial friction factor grows first, then saturates at some value that depends on the flow condition in the liquid phase. The experiments in pipe (Figure 27) show similar trends although it does seem to saturate as in the former case.

The evidence of the influence of wave amplitude upon the interfacial friction factor may be put in evidence in the following way. The additional contribution of waves can be quantified by calculating the equivalent sand roughness *k* (Cohen and Hanratty, 1965). An example of such results is given in Figure 28.

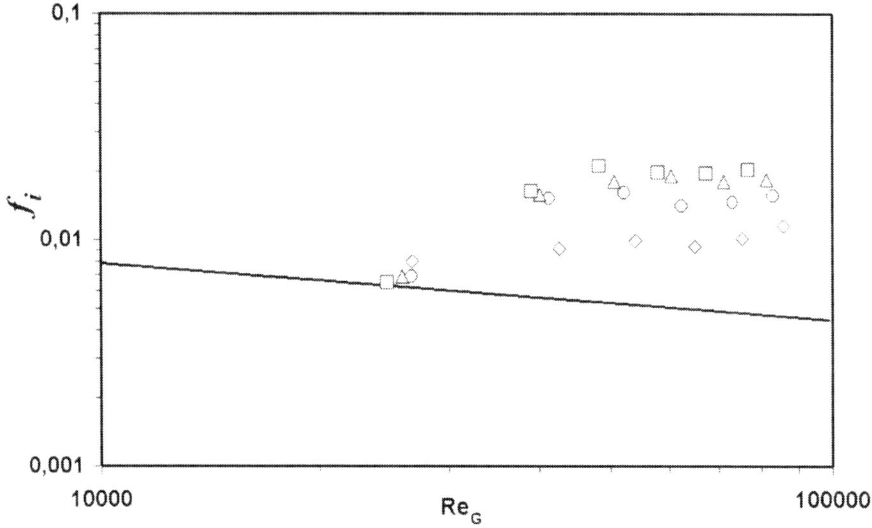

Figure 26. Interfacial friction factor vs. Reynolds number Re_G for different Re_L: ◇, 16000; O, 28000; Δ, 37000; □ 47000, — Blasius correlation f=0.079 $Re^{-1/4}$ (Fernandez-Flores, 1984).

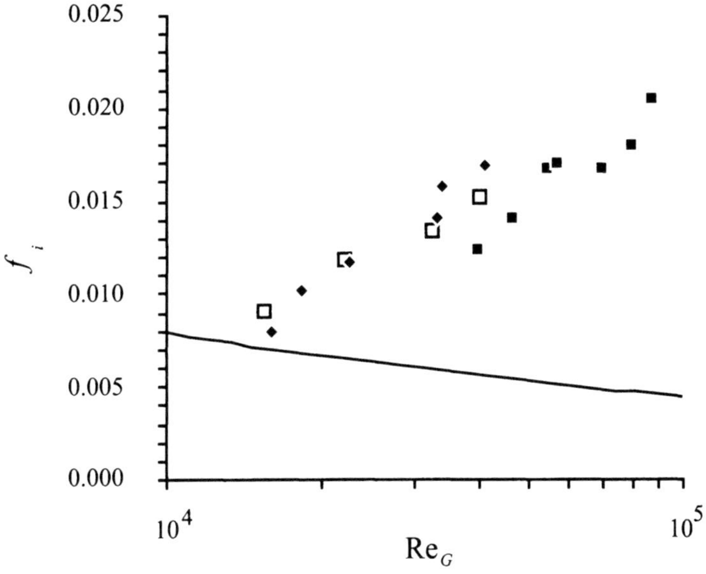

Figure 27. Interfacial friction factor vs gas Reynolds number; ■ j_L= 0.0066m/s, □ j_L= 0.02m/s, ◆ j_L= 0.05m/s, — Blasius correlation f=0.079 $Re^{-1/4}$ (Lopez, 1994).

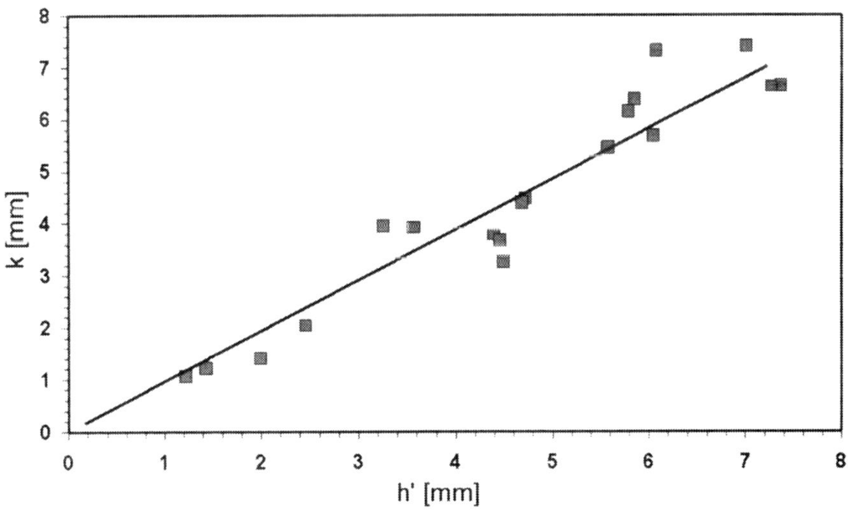

Figure 28. Interfacial sand roughness vs. modified rms height of the interface (Fernandez, 1984).

To take into account the part of the waves that emerges from the viscous sublayer, Fernandez (1984) has used a modified rms value of the interface height:

$$h' = \Delta h \sqrt{2\left[1-\left(\frac{56\nu_G}{u_i^* D_{hG}}\right)^2\right]}, \qquad (63)$$

where Δh is the rms of the height fluctuations, u_*, the friction velocity at the interface and D_h, the hydraulic diameter. In doing this, it can be seen that the modified rms fluctuation of the interface height is roughness: $h'=k$.

6.3 Useful correlations

The literature is full of various correlations whose value is limited to the range of parameters over which they were calibrated. Rather than doing one's shopping among these numerous article by choosing the most attractive closure law, one must focus on the comprehensive studies. We have made a selection of trustable solutions that have been widely used in a large variety of flow conditions. They are summarized as follows:

Wall friction factor in the gas phase. The friction factor depends on whether the flow is laminar or turbulent. Single-phase flow correlations are extended to two-phase flow as follows:

$$f_G = \frac{16}{\mathrm{Re}_G} \qquad \text{for laminar flow} \qquad (64)$$

$$\frac{1}{\sqrt{f_G}} = 3.48 - 4\log\left(\frac{2k}{D_{hG}} + \frac{9.35}{\mathrm{Re}_G\sqrt{f_G}}\right) \qquad \text{for turbulent flow} \qquad (65)$$

in which k is the sand roughness of the wall, D_h and Re the hydraulic diameter and the Reynolds number defined for the gas as:

$$D_{hG} = \frac{4R_G A}{S_{wG}+S_i}, \qquad (66)$$

$$\mathrm{Re}_G = \frac{|u_G| D_{hG}}{\nu_G} \qquad (67)$$

Note that the Poiseuille relation is recovered from Eq. (64) if $R_G=1$. It is thus an extension of this relation. Eq. (65) is the Colebrook correlation that includes both the smooth and rough regime. Rather than using the implicit correlation of Colebrook one may prefer to split it in two parts: the explicit Blasius correlation for the smooth regime and the fully rough wall correlation. On may be also interested in using the Churchill formula that covers the three different regimes. This is a matter of taste.

Wall friction factor in the liquid. Similar method may be used for the shear stress of the liquid phase at the wall with a little difference: the hydraulic diameter does not include the interfacial perimeter

$$D_{hL} = \frac{4R_L A}{S_{wL}}. \tag{68}$$

The extension of single-phase flow relation in the liquid phase leads however to some bias (Fabre et al, 1987, Rosant, 1984 and Andritsos, 1986). Cheremisinoff and Davis (1979), Rosant (1984) and Andritsos have proposed empirical correlations for f_{wL}. Nevertheless, it must be kept in mind that the crucial problem remains the closure of the interfacial friction factor f_i.

Interfacial friction factor. Among all the empirical correlations that have been developed for the interfacial friction factor, this of Andritsos and Hanratty (1987) must be recommended as it gives the best results. These authors postulated that there exists a critical gas flux $j_{G,2}$ given by Eq. (51) below which the interface is hydraulically smooth. Above this critical flux the interface is wavy and the interfacial shear stress is assumed to increase linearly with the difference $j_G - j_{G,2}$:

$$\frac{f_i}{f_{i,smooth}} - 1 = 0 \qquad \text{for } j_G \le j_{G,2} \tag{69}$$

$$\frac{f_i}{f_{i,smooth}} - 1 = 15\sqrt{\frac{h}{D}}\left(\frac{j_G}{j_{G,2}} - 1\right) \qquad \text{for } j_G \ge j_{G,2} \tag{70}$$

where $f_{i,smooth}$ is the friction factor for a smooth interface calculated from Eq. (65) with $k = 0$. The interfacial friction factor increases when the gas velocity is high enough to generate KH-waves. It must be pointed out that even if it is disturbed by J-waves, the interface is considered to behave like a smooth surface.

In the previous approach, the wavy structure is only considered through the transition between the smooth and wavy regimes. Another approach considers that the interface is seen by the gas as a surface whose roughness changes with both gas and liquid velocities. If we accept that the momentum transfer across a rough liquid surface is governed by the same mechanism as for a rough wall, it is possible to extend the single-phase flow correlation:

$$\frac{1}{\sqrt{f_i}} = 3.48 - 4\log\left(\frac{2k}{D_{hG}} + \frac{9.35}{\mathrm{Re}_G\sqrt{f_i}}\right) \tag{71}$$

The problem is completely solved provided that the roughness of the interface k may be predicted.

A solution ignoring the wave amplitude has been proposed early on by Charnock (1955) for fully developed wave field in deep water. For inviscid fluids, only gravity and pressure forces balance yielding:

$$\frac{kg}{u_i^{*2}} = \gamma, \tag{72}$$

in which

$$u_i^* = \sqrt{\frac{|\tau_{iG}|}{\rho_G}},$$

is the interfacial friction velocity and γ a constant within the range 0.1–0.5. Eq. (71) has to be solved together with Eq. (72). One obtains the following implicit relation that must be solved with an iterative procedure:

$$\frac{1}{\sqrt{f_i}} = 3.48 - 4\log\left(\gamma \,\mathrm{Fr}_G f_i + \frac{9.35}{\mathrm{Re}_G \sqrt{f_i}}\right), \quad (73)$$

where Fr_G is a Froude number defined as:

$$Fr_G = \frac{(U_G - U_L)^2}{g D_{hG}} \quad (74)$$

References

Abrams, J. (1984) *Modeling of turbulent flow over a wavy surface,* Ph. D. Thesis, Univ. of Illinois, Urbana, U.S.A.

Aggour, M.A. and Sims, G.E. (1978). A theoretical solution of pressure drop and holdup in two-phase stratified flow, *Proc. Heat Transfer Fluid Mech. Inst.*, pp. 205-217.

Agrawal, S.S., Gregory, G.A. et Govier, G.W. (1973). An Analysis of horizontal Stratified Two Phase Flow in Pipes, *Can. J. Chem. Eng.*, 51, 280.

Akai, M., Inoue, A., Aoki, S. (1977). Structure of a co-current stratified two phase flow with wavy interface, *Theo. Appl. Mech.*, vol. 25, pp. 445-456.

Akai, M., Inoue, A., Aoki, S. (1981). , The prediction of stratified two-phase flow with a two-equation model of turbulence, *Int. J. Multiphase Flow*, vol. 7, pp. 21-39.

Andritsos, N. (1986) *Effect of pipe diameter and liquid viscosity on horizontal stratified flow*, Ph. D. Thesis, Univ. of Illinois, Urbana, U. S. A.

Andritsos, N. and Hanratty, T.J. (1987) Influence of interfacial waves in stratified gas-liquid flows, *AIChE J.*, 33, p 444-454.

Andritsos, N. and Hanratty, T.J., (1987). Interfacial instabilities for horizontal gas-liquid flows in pipelines, *Int. J. Multiphase Flow*, vol. 13, no. 5, pp. 583-603.

Belcher, S. E. and Hunt, J. C. R. (1993) Turbulent shear flow over slowly moving waves, *J. Fluid Mech*, 251, p 109-148.

Benjamin, T.B. (1959). Shearing flow over a wavy boundary, *J. Fluid Mech.*, 6, 161.

Bontozoglou, V. and Hanratty, T.J. (1989). Wave height estimation in stratified gas-liquid flows, *A.I.Ch.E. Journal*, vol. 35, pp. 1346-1350.

Bruno, K. and McCready, M. J. (1989) Processes which control the interfacial wave-spectrum in separate gas-liquid flows, *Int. J. Multiphase Flow*, 15, p 531-552.

Charnock, H. (1955) Wind Stress on a Water Surface, *Q. J. Met. Soc.*, 81, p 639-640.

Cheremisinoff, N. and Davis, E.J. (1979) Stratified turbulent-turbulent gas-liquid flow, *AIChE J.*, 25, p 48-56.

Chisholm, D., (1967). A theoretical basis for the local Lockhart-Martinelli correlation for two-phase flow, *Int. J. Heat and Mass Transfer*, vol. 10, P. 1967.

Cohen, L. S. and Hanratty, T. J. (1965) Generation of waves in the concurrent flow of air and liquid, *AIChE J.*, 11, p138-144.

Cohen, L. S. and Hanratty, T. J. (1968), Effect of waves at gas-liquid interface on a turbulent air flow, *J. Fluid Mech.*, 31, p 467-.

Crouzier, O., (1978). *Ecoulements diphasiques gaz-liquide dans les conduites faiblement inclinées par rapport à l'horizontale*, Thèse, Université de Paris VI.

Davis E.J., (1969). Interfacial shear measurement for two-phase gas-liquid flow by means of Preston tubes, *Ind. Eng. Chem. Fundam.*, vol 8, pp. 153-159.

Ellis, S.R., and Gay, B. (1959). The parallel flow of two fluid streams: interfacial shear and fluid-fluid interaction, *Trans. Instn. Chem. Engrs.*, 37, 206.

Fabre, J., Masbernat, L. and Suzanne, C. (1982). New results on the structure of stratified gas-liquid flow, *Nato Advanced Research Workshop on two-phase flows and Heat Transfer*, Spitzingsee, W. Germany.

Fabre, J., Masbernat, L. and Suzanne, C. (1983). Structure cinématique de l'écoulement interne gaz-liquide à phases séparées et interface ondulée, *C.R. Acad. Sc.*, Paris, 297, 695.

Fabre, J., Masbernat, L. and Suzanne, C. (1985a). Stratified flow, Part I : local structure", International Workshop on Two-Phase Flow Fundamentals, Gaithesburg, Maryland, USA, September 22-27 1985, also in *Multiphase Science and Technology*, vol. 3, edited by G.F. Hewitt, J.M. Delhaye and N. Zuber, Hemisphere Publishing Corporation, 1987.

Fabre, J., Masbernat, L. , Fernandez-Flores, R., Suzanne, C. (1985b) Stratified flow, Part II: interfacial and wall shear stress, International Workshop on Two-Phase Flow Fundamentals, Gaithesburg, Maryland, USA, September 22-27 1985, also in *Multiphase Science and Technology*, vol. 3, edited by G.F. Hewitt, J.M. Delhaye and N. Zuber, Hemisphere Publishing Corporation, 1987.

Fernandez-Flores, R. (1984) *Etude des interactions dynamiques en écoulement diphasique stratifié*, Thèse, INP Toulouse, France.

Hanjalic, K. and Launder, B.E. (1972). Fully developed asymmetric flow in a plane channel, *J. Fluid Mech.*, vol. 51, pp. 301-335.

Hanratty, T. J. (1983) Interfacial instabilities caused by air flow over a thin liquid layer, *Waves on Fluid Interface*, Academic Press, Inc., New York.

Hanratty, T.J. (1991) Separated flow modelling and interfacial transport phenomena, *Applied Scientific Research*, vol. 48, pp. 353-390.

Hanratty, T. J. and McCready, M. J. (1992) Phenomenological understanding of gas-liquid separated flows, *Proc. Third International Workshop on Two-Phase Flow Fundamentals*, Imperial College, London, U. K., April 1992.

Hanratty, T.G. and Engen, J.M., (1957). Interactions between a turbulent air stream and a moving water surface, *J. Am. Inst. Chem. Eng.*, 3, 299.

Harris, J. A. (1993) *On the growth of water waves and the motion beneath them*, Ph. D. Thesis, Stanford Univ., U.S.A.

Henstock, W.H. and Hanratty, T.J., (1976). The interfacial drag and the height of the wall layer in annular flows, *A.I.Ch.E. Journal*, 22, 6.

Hidy, G.M. and Plate, E.J., (1966). Wind action on water standing in a laboratory channel, *J. Fluid Mech.*, 26, 651.

Hudson, J. D. (1993) *The effect of a wavy boundary on a turbulent flow*, Ph. D. Thesis, Univ. of Illinois, Urbana, U.S.A.

Jeffreys, H. (1924). On the formation of water waves by wind, *Proc. Roy. Soc.*, A., 107, pp. 189-206.

Jurman, L. A. and McCready, M. J. (1989) Study of waves on thin liquid films sheared by turbulent gas flow, *Phys. Fluids*, A1-3, p 522-535

Kordyban, E., (1974). Interfacial shear stress in two-phase wavy flow in closed horizontal channels, *J. Fluid Eng.*, 97.

Kordyban, E., (1977). Some characteristics of high waves in closed channels approaching Kelvin-Helmholtz instability, *J. Fluid Eng.*, 339.

Kowalski, J.E. (1987). Wall and interfacial shear stress in stratified flow in a horizontal pipe, *AIChE J.*, vol. 33, pp. 274-281.

Liné, A. and Lopez, D. (1995) Two-fluid model of separated two-phase flow: momentum transfer on wavy boundary, *Proc. Second International Conference on Multiphase Flow*, Kyoto, Japan

Lleonart, G.T. and Blackman, D.R., (1980). The spectral characteristics of wind-generated capillary waves, *J. Fluid Mech.*, vol. 97, pp. 455-479.

Lopez, D. (1994) *Ecoulements diphasiques à phases séparées à faible contenu de liquide*, Thèse, INP Toulouse, France

Magnaudet, J. (1989). *Interactions interfaciales en écoulement à phases séparées*. Thèse, INP Toulouse.

Michell, J.H. (1893). The highest waves in water. *Phil. Mag.* , vol. 36, pp. 430-437.

Milne-Thomson, L.M. (1955). *Theoretical hydrodynamics*, McMillan, New York.

Mouly, (1979). *Ecoulements stratifiés de gaz naturel et d'huile en conduite pétrolière*, Thèse, Paris VI.

Phillips, O.M. (1977). *The dynamic of the upper ocean*, Cambridge University Press, London.

Preston, J.H., (1954). The determination of turbulent skin friction by means of Pitot tube, *J. Roy. Aero. Soc.*, 58, 109.

Rosant, J.M. (1984) *Ecoulements diphasiques liquide-gaz en conduite circulaire*, Thèse, ENSM, Nantes, France.

Sinai, Y. L. (1985) Interfacial phenomena of fully-developed, stratified, two-phase flows, *Encyclopedia of Fluid Mechanics*, Vol 3, Gas-Liquid Flows, p 475-491, N. P. Cheremisinoff, Ed.

Smith, T.N., and Tait, R.W.F., (1966). Interfacial shear stress and momentum transfer in horizontal gas-liquid flow, *Chem. Eng. Sc.*, 21, 63.

Strand, O. (1993) *An experimental investigation of stratified two-phase flow in horizontal pipes*, Ph. D. Thesis, Univ. of Oslo, Norway.

Suzanne, C., (1985). *Structure de l'écoulement stratifié de gaz et de liquide en canal rectangulaire*. Thèse, INP Toulouse.

Taitel, Y. and Dukler, A. E. (1976) A theoretical approach to the Lockhart-Martinelli correlation for stratified flow, *Int. J. Multiphase Flow*, 2, p 591-595.

Taitel, Y. and Dukler, A. E. (1976) Model for predicting flow regime transitions in horizontal and near horizontal gas-liquid flow, *AIChE J.*, 22, p 47-55.

Thorsness, C.B., Morrisroe, P.E. and Hanratty, T.J. (1978). A comparison of linear theory with measurements of the variation of shear stress along a solid wave. *Chem. Engng. Sci.*, vol. 33, pp. 579-592.

Wallis, G.B., (1969). *One-dimensional two-phase flow*, McGraw-Hill, New York.

Wang, A.K. and Gelhar, L.W. (1974). Turbulent Couette flow, *J. Fluid Engng.*, September, pp. 265-271.

Whitham, G.B. (1974). *Linear and non-linear waves*, Wiley-Interscience.

Wu, J., (1968). Laboratory studies of wind waves interactions, *J. Fluid Mech.*, 34, 91.

Wu, J., (1975), Wind-induced drift currents, *J. Fluid Mech.*, 68, 49.

Wu, J., (1980), Wind stress coefficients over sea surface near neutral conditions : a revisit. *Journal of Physical Oceanography*, 10, 727.

Yu, H.S. and Sparrow, E.M. (1967). Stratified laminar flow in ducts of arbitrary shape, *AIChE J.*, vol. 13, pp. 10-16

Gas-Liquid Slug Flow

Jean Fabre

Institut de Mécanique des Fluides, Toulouse, France

Abstract. This chapter describes gas-liquid slug flow (also known as intermittent flow). Conservation equations are derived based on the assumptions of the unit cell model. The main results concerning the motion of long bubbles and liquid slugs are reviewed.

1 Introduction

When gas and liquid flow together in a pipe the interface between the phases may take a variety of different patterns, the most complex being probably slug flow. The primary characteristic of slug flow is its inherent intermittence. An observer looking at a fixed position along the axis would see the passage of a sequence of slugs of liquid containing dispersed bubbles, each looking somewhat like a length of bubbly pipe flow, alternating with sections of separated flow within long bubbles (Figure 1). These two states follow in a random-like manner, inducing pressure, velocity and phase fraction fluctuations: the flow is unsteady, even when the flow rates of gas and liquid, Q_L and Q_G, are kept constant at the pipe inlet.

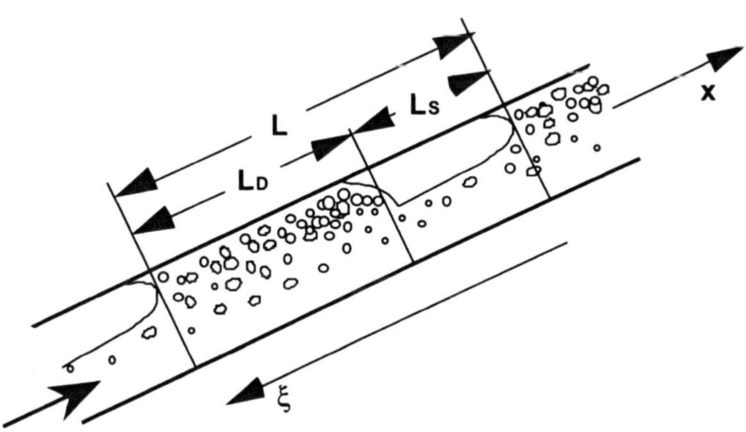

Figure 1. Gas-liquid slug flow: each cell of length L is made up of a liquid slug of length L_D and a long bubble of length L_S.

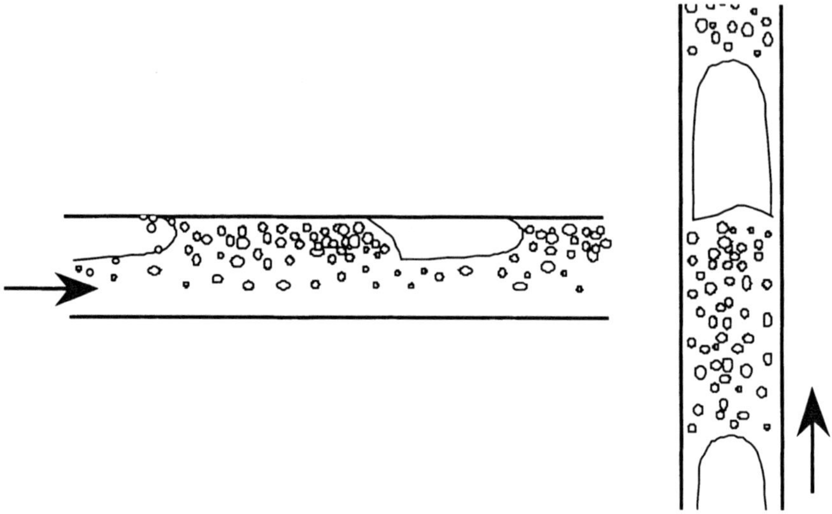

Figure 2. Horizontal (left) and vertical (right) slug flow.

The elementary part of slug flow is a cell of length L, made up of a long bubble of length L_S and of a liquid slug of length L_D. The slug of liquid containing dispersed bubbles of small diameter travels at a velocity V. It overruns a slower moving liquid in the separated film. During stable slug flow liquid is shed from the back of the slug at the same rate that liquid is picked up at the front. s a result the slug length stays more or less constant as it travels along the tube. For horizontal or near horizontal tubes, the liquid shed at the back decelerates under the influence of wall shear and forms a stratified layer. For vertical or near vertical tubes the liquid forms a falling annular film and accelerates as it moves downward. The separated section forms a set of long bubbles that carry most of the gas. In vertical flow these long bubbles are nearly cylindrical whereas in horizontal flow the interface between the gas and the liquid is almost flat (Figure 2).

From a fundamental point of view slug flow belongs to the class of intermittent flows: the flow is unsteady even if the phase flow rate remains constant with time. However unsteadiness does not mean necessarily intermittence. Strictly speaking there is no canonical definition of intermittence. Let say that a signal is intermittent if it experiences space or/and time variations that are not too frequent and of large amplitude. With this rather vague definition, slug flow is an intermittent phenomenon. At the outlet of a pipe where slug flow exists, an observer sees liquid and gas flowing alternately as if it was impossible for both phases to flow simultaneously out of the pipe. The system seems to switch from one state to another. A slug–like flow that is familiar to everybody is obtained by emptying a bottle with its neck directed downward. On the other hand, it is also possible to recognize in the flow structure a gas fraction wave that propagates along the pipe. The long bubbles are indeed formed by an expansion wave at their front, followed by a shock wave (an hydraulic jump) at the rear. The evolution of the flow structure along the pipe may be understood as wave-wave interaction.

From a practical point of view, slug flow occurs over a wide range of intermediate flow rates of both phases (see for example Figure 3 for horizontal flow and Figure 4 for vertical flow). For this reason, it presents a major interest for many industrial processes like:

- Production of oil and gas in wells and their transport in pipelines.
- Geothermal production of steam.
- Boiling and condensation processes in power generation facilities as well as in chemical plants and refineries.
- Handling and transport of cryogenic fluids.
- Emergency cooling of nuclear reactors.

It covers also a broad range of flow conditions in two-phase flow in micro-systems.

The existence of slug flow can create problems for the designer or operator. Indeed the liquid is moving in the slugs at the mixture velocity whereas it has a much smaller velocity in the long bubbles, this velocity depending upon the pipe inclination. The high momentum of the liquid slugs can create considerable force as they change direction passing through elbows, tees or other process equipment. Furthermore the low frequencies of slug flow can be in resonance with the fundamental frequency of large piping structures and severe damage can take place unless this situation is anticipated in design. In addition the intermittent nature of the flow makes it necessary to design liquid separators and their controls to accommodate the largest slug length that can exist in the system.

In contrast, there are numerous practical benefits that can result from operating in the slug flow pattern. Because of the very high liquid velocities, it is usually possible to move larger amounts of liquids in smaller lines than would otherwise be possible in two phase flow. In addition the high liquid velocities cause very high convective heat and mass transfer coefficients resulting in very efficient transport operations.

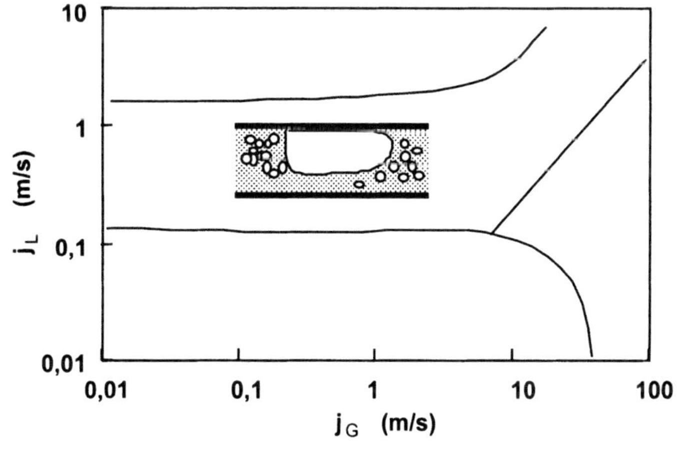

Figure 3. Occurrence of the slug flow in horizontal pipe of 5 cm diameter: the coordinates of the map are the gas and liquid fluxes, the phases are air and water.

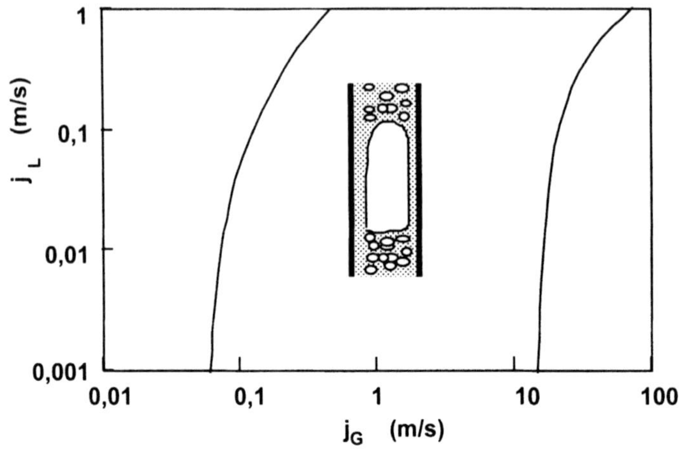

Figure 4. Occurrence of the slug flow in vertical pipe of 5 cm diameter: the coordinates of the map are the gas and liquid fluxes, the phases are air and water.

2 The Concept of Unit Cell

Modelling slug flow requires the understanding of various phenomena. A good exercise to identify these phenomena is to gain knowledge of a simple picture of an idealized slug flow made of identical cells. This concept that focuses on a unit cell (UC for short) has proved to be useful for the understanding of the physical phenomena involved in slug flow and efficient for the flow prediction as well.

Wallis was probably the first to formulate clearly the UC concept suggested by the results of Nicklin et al. Initially established for upward vertical flow, this concept was successfully applied to horizontal or slightly inclined flow. In the past twenty years the model based on this concept was improved by several investigators. It requires the two following assumptions:

- There exists a frame of a given velocity V in which the flow is steady.
- In this frame the flow in long bubbles and in liquid slugs is fully developed.

The model requires also four different closure laws on which the accuracy of the results partly relies. A review of the scientific literature reveals an abundance of these laws whose physical value is unequal (see for example the overviews of Taitel and Barnea, 1990; Fabre and Liné, 1992; Dukler and Fabre, 1994). The weaknesses of these laws often originates from the narrow range of flow conditions, fluid properties or pipe dimension that were used for their calibration. Their critical role has been discussed by Dukler and Fabre (1994).

Suppose that we have in hands the specific flow conditions such as:

- The pipe size, i.e. its diameter D or its cross-section area A, the inclination θ,

- The fluid properties, i.e. the density ρ_k (k=L,G), the kinematic viscosities ν_k and the surface tension σ,
- The volumetric flux —or superficial velocity — of each phase, $j_k = Q_k/A$.

A complete model of slug flow would produce at least the following information:
- The characteristic lengths, L_D, L_S, of the liquid slugs and long bubbles and the mean bubble size in the liquid slug;
- The form of the liquid film (stratified, annular);
- The characteristic velocities, V and the mean velocities of both gas and liquid in each part of the cell;
- The cross-sectional phase fraction in each part of the cell;
- The mean wall and interfacial shear stresses in each part of the cell;
- The pressure drop.

2.1 Assumptions of the UC Model

The main difficulty in modelling slug flow comes from its chaotic nature. This is suggested by the observation of the succession of bubbles and slugs whose length appears randomly distributed with time (Figure 5). For avoiding to account for the flow randomness, a few assumptions are needed. The initial assumption was to picture the flow as a sequence of bubbles and slugs periodic with both space and time: the UC concept was born. However by using two weaker assumptions, one arrives to the same model.

The first assumption comes from the experimental evidence illustrated in Figure 6. The probability density distribution of bubble and slug velocities shows that they are narrowly distributed about their average: in other words they are almost identical. An observer moving at the statistical mean velocity would see the whole structure almost frozen. Although this property becomes less evident at high gas or liquid flow rate, this quasi-steady behaviour in a moving frame is the key of the success of the UC model. Indeed this property leads to a great simplification since it allows transforming an unsteady problem into a steady one.

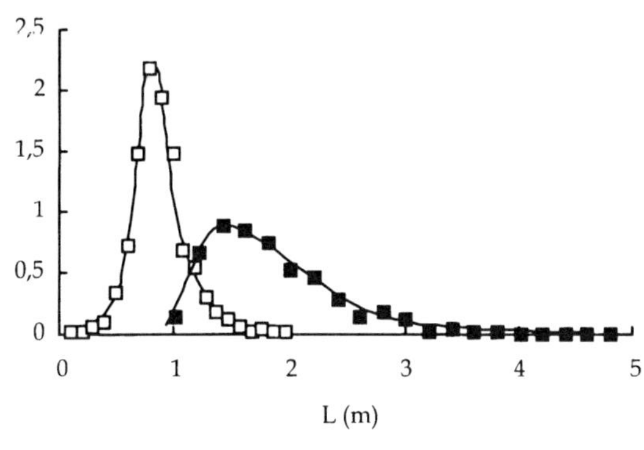

Figure 5. Probability density distribution of ■ bubble and □ slug lengths. Slug flow in pipe of 5 cm diameter at superficial velocities j_G=1.25 m/s and j_G=0.97 m/s (Fabre *et al.*, 1993).

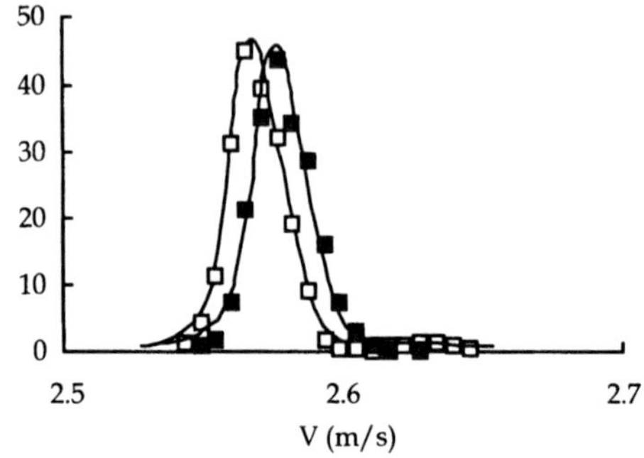

Figure 6. Probability density distribution of ■ bubble and □ slug velocities. Slug flow in pipe of 5 cm diameter at superficial velocities j_G=1.25 m/s and j_G=0.97 m/s (Fabre *et al.*, 1993).

The second assumption consists in assuming that the flow is fully developed in each part of the cell. As a consequence, the cross-sectional mean fraction and velocity of each phase do not depend on the longitudinal coordinate in the long bubbles and the liquid slugs as well. This assumption is probably stronger than the previous one. It will be revisited further.

It must be pointed out that the original concept implies that each bubble (respectively, each slug) has the same length. However this restriction is unnecessary and it must be allowed to take into account the stochastic distribution of bubble and slug lengths. We shall therefore continue to talk about the UC model since the equations are the same.

2.2 Preliminary Remarks

The inlet flow conditions will be considered constant with time t (steady conditions). They may be specified by the mass flow rates of each phase:

$$m_k = \rho_k j_k A \tag{1}$$

To simplify the discussion the density of both phases will be considered constant so that the steadiness $dm_k/dt = 0$ is equivalent to $dj_k/dt = 0$.

For a comprehensive study, it is sufficient to consider physical quantities average over a pipe cross-section. As the flow is unsteady in a fixed frame, these quantities are functions of axis-coordinate x and time t. We define the following local quantities:

- χ, a Boolean function that denotes the presence of the separated flow (presence of long bubble).
- U_k, the mean velocity of phase-k.
- P_k, the mean pressure in phase-k.
- R_k, the fraction of phase-k that satisfies the condition

$$\sum_{k=L,G} R_k = 1 \tag{2}$$

2.3 Balance Equations

Mass conservation. Following the first assumption, there exists a frame moving at the velocity V in which the flow is steady. Thus by using the new coordinate (see Figure 1):

$$\xi = Vt - x \tag{3}$$

the physical quantities does not depend on time. In the new coordinate system all the quantities remain unchanged except the velocity ($U_k \rightarrow V - U_k$).

The conservation of mass is expressed as:

$$\frac{d}{d\xi}\left[R_k\left(V - U_k\right)\right] = 0, \tag{4}$$

that leads to

$$\varphi_k = R_k\left(V - U_k\right), \tag{5}$$

where φ_k is constant. Like j_k for the standing frame, φ_k represents the volumetric flux of phase-k entering the long bubble region, and shed from the liquid slug.

Momentum conservation. The momentum equation for phase-k is:

$$\frac{d}{d\xi}\left[\rho_k R_k\left(V - U_k\right)^2 + R_k P_k\right] = -\frac{\tau_{wk} S_{wk} + \tau_{ik} S_i}{A} + \rho_k R_k g \sin\theta \tag{6}$$

where τ is the x-component of the shear stress exerted upon the phase-k by the wall (subscript w) or the interface (subscript i), S is the wetted perimeter and g the gravity; θ is the pipe inclination (angle of x-axis with the horizontal). The above equation simplifies in using Eq. (5):

$$\rho_k \varphi_k^2 \frac{dR_k^{-1}}{d\xi} + \frac{dR_k P_k}{d\xi} = -\frac{\tau_{wk} S_{wk} + \tau_{ik} S_i}{A} + \rho_k R_k g \sin\theta \tag{7}$$

with the jump condition

$$\sum_{k=L,G} \tau_{ik} = 0. \tag{8}$$

2.4 Space Average Balance Equations

Space average. In what follows, space average will be used rather than the local quantities. We note the space-average over a distance $\Delta\xi$ that is large enough compared to the length of the largest cell.

$$\overline{F} = \frac{1}{\Delta\xi} \int_{\xi}^{\xi+\Delta\xi} F d\xi. \tag{9}$$

As the flow is steady this average may be also interpreted as the time average that would be obtained at some given location in the standing frame.

Rate of intermittence. The rate of intermittence is noted β. It is defined as the space average of χ.

$$\beta = \overline{\chi} \tag{10}$$

It may be viewed also as the ratio of the sums of bubble to cell lengths that are contained within the interval Δξ. For an observer located at a fixed point *x*, it be viewed as the time rate during which long bubbles are observed.

Average of phase fraction, velocity... in bubbles, in slugs. We will use also the conditional average over the long bubbles (resp. the liquid slugs). For a quantity, let say, *F*, it is defined as:

$$F_S = \frac{\overline{\chi F}}{\beta}, \qquad F_D = \frac{\overline{(1-\chi)F}}{(1-\beta)}. \tag{11}$$

It defines an average weighted by the presence of phase. Applied to the phase fraction, the pressure, the shear stress, the flux, these definitions yields:

$$\begin{aligned} R_S &= \frac{\overline{\chi R}}{\beta}, \quad R_D = \frac{\overline{(1-\chi)R}}{(1-\beta)}, \\ P_S &= \frac{\overline{\chi P}}{\beta}, \quad P_D = \frac{\overline{(1-\chi)P}}{(1-\beta)}, \\ \tau_S &= \frac{\overline{\chi \tau}}{\beta}, \quad \tau_D = \frac{\overline{(1-\chi)\tau}}{(1-\beta)}, \\ j_S &= \frac{\overline{\chi j}}{\beta}, \quad j_D = \frac{\overline{(1-\chi)j}}{(1-\beta)}. \end{aligned} \tag{12}$$

It must be pointed out that, under the assumption that the flow is fully developed in the long bubble (resp. the liquid slug), the physical quantities *R*, *dP/dξ*... changes with ξ only when χ changes. In this case R_S is equal to the value taken by *R* in the long bubbles, R_D, the value of *R* in the liquid slugs, etc. Although they are trivial, these averages will be used further in a less restrictive assumption framework.

For the velocity, it is more convenient to define it from the flux. By using (12), it yields:

$$U_S = \frac{j_S}{R_S}, \qquad U_D = \frac{j_D}{R_D}. \tag{13}$$

Average of phase fraction. With these conditional averages, the averages of a given quantity becomes:

$$\overline{F} = \overline{\chi F} + \overline{(1-\chi)F} = \beta F_S + (1-\beta)F_D. \tag{14}$$

Using this definition, one may express the mean phase fraction of phase-k and the mean flux as:

$$R_k = \beta R_{kS} + (1-\beta) R_{kD}\,, \tag{15}$$

$$j_k = \beta j_{kS} + (1-\beta) j_{kD}\,. \tag{16}$$

It must be noted that the intermittence factor may be deduced from the definition (15):

$$\beta = \frac{R_G - R_{GD}}{R_{GS} - R_{GD}}\,. \tag{17}$$

Space average of governing equations. The phase fractions follow the general geometrical rule:

$$\sum_{k=L,G} R_{kS} = 1\,, \qquad \sum_{k=L,G} R_{kD} = 1\,, \qquad \sum_{k=L,G} R_k = 1\,. \tag{18}$$

The space average of mass balance equation (Eq. 5) over the separated region and the dispersed region leads to:

$$\varphi_k = R_k\left(V - U_k\right) = R_{kS}\left(V - U_{kS}\right) = R_{kD}\left(V - U_{kD}\right). \tag{19}$$

This equation expresses that the flux of phase k entering the long bubble is equal to the one entering the liquid slug. Because φ_k may be written as the sum $\beta\varphi_k + (1-\beta)\varphi_k$, Eqs. (19) leads to:

$$\varphi_k = R_k\left(V - U_k\right) = R_{kS}\left(V - U_{kS}\right) = R_{kD}\left(V - U_{kD}\right).$$

The bracketed term may be expressed using Eq. (16) so that the flux in the moving frame expresses versus the flux in the standing frame:

$$\varphi_k = R_k V - j_k\,. \tag{20}$$

The averaging of the momentum balance (Eq. 7) requires the second assumption. As the flow is fully developed in each part of the cell, the space derivatives cancel except for pressure. Eq. (7) simplifies and may be averaged over the separated and dispersed region respectively. Coming back to the expression in the standing frame it yields:

$$R_{kS}\frac{dP_S}{dx} = \frac{\tau_{kwS} S_{kwS} + \tau_{kiS} S_{iS}}{A} - \rho_k R_{kS}\, g \sin\theta\,, \tag{21}$$

$$R_{kD}\frac{dP_D}{dx} = \frac{\tau_{kwD} S_{kwD} + \tau_{kiD} S_{iD}}{A} - \rho_k R_{kD}\, g \sin\theta\,. \tag{22}$$

Since the flow is fully developed, the pressure gradient is the same in both phases and must not be distinguished. The mean pressure gradient over the cell results from the mean pressure gradient over each part of the cell weighted by their rate of occurrence:

$$\frac{dP}{dx} = \frac{\beta\left(\tau_{wLS} S_{wLS} + \tau_{wGS} S_{wGS}\right) + (1-\beta)\left(\tau_{wLD} S_{wLD}\right)}{A} - \left(\rho_L R_L + \rho_G R_G\right) g \sin\theta\,. \tag{23}$$

The pressure gradient involves two contributions: the weight and the wall friction.

2.5 Closure Problem

It is worth noting that the fully developed flow assumption makes the equation independent of the cell length. Only the intermittence factor β appears. This point will be discussed further.

The pressure gradient appears only in Eq. (23). Therefore, once the phase fractions and the velocities are determined, the wall friction and the weight of the phases may be calculated. This remarks suggests splitting the solution of the problem into two steps. In a first step we discuss how to determine the phase fractions, R_L, R_{LS}, R_{LD}. These physical quantities are not coupled with the pressure gradient that will be determined in a second step.

To figure out the closure problem for the determination of the phase fractions, the 5 independent algebraic equations have been grouped in Table 1. They are nonlinear and they contain 9 unknown quantities: R_L, R_{LS}, R_{LD}, U_{LS}, U_{LD}, U_{GS}, U_{GD}, V, β. This is the role of the 4 closure equations to restore the missing information. There exist different strategies to close the set of equations. We shall however limit the discussion to the most classical method that requires to model:

- The velocity V of the long bubble.
- The gas fraction R_{GD} in the liquid slug that results from a complex mechanism of gas shedding and re-coalescence at the rear of the long bubbles.
- The drift velocity of the gas in the liquid slug U_{GS}–U_{LS}.
- The liquid hold-up R_{GD} in the long bubbles that requires a model for either stratified flow or annular flow.

Table 1. Equations for liquid hold-up.

$$R_L = \beta R_{LS} + (1-\beta) R_{LD}$$

$$j_L = \beta R_{LS} U_{LS} + (1-\beta) R_{LD} U_{LD}$$

$$j_G = \beta(1-R_{LS})U_{GS} + (1-\beta)(1-R_{LD})U_{GD}$$

$$R_{LS}(V-U_{LS}) = R_{LD}(V-U_{LD})$$

$$(1-R_{LS})(V-U_{GS}) = (1-R_{LD})(V-U_{GD})$$

Whereas the phase fractions are not coupled to the pressure gradient, the pressure gradient does depend on the phase distribution as Eq. (23) shows. Even for horizontal flow in which the weight vanishes, they still have a great influence on the pressure gradient through the intermittence factor β. For the pressure gradient to be calculated, it requires the models for:

- The shear stress at the wall in the separated region τ_{wkS}.
- The shear stress at the wall in the dispersed region τ_{wkD} including the contribution of the hydraulic jump at the rear of the bubble.

Different models have been published in the scientific literature. Most of them fall into the type of the UC model, i.e. they have in common the set of equations presented in the previous section. What makes the difference is the choice of the closure laws.

3 Long Bubbles: Motion, Shape

As most of the gas is conveyed by the large bubbles the accurate prediction of their motion and their shape is essential. If the models proposed in the literature or used in computer codes are successful, this is mainly due to the reliability in predicting their velocity V.

It is indeed possible to get a crude estimate of the gas fraction by assuming that the gas is conveyed at velocity V in these large bubbles and that the liquid slug does not contain dispersed bubbles:

$$R_G \cong \frac{j_G}{V}. \tag{24}$$

This relation, obtained by assuming $\varphi_G = 0$ in Eq. (20) is frequently used as a first guess in the iterative solution of slug flow equations. It does a fairly good job in some simplified cases. This shows that the phase fractions are primarily sensitive to the long bubble velocity.

Our present knowledge of the motion of long bubbles in tubes comes from both the theory and the considerable amount of data for various flow conditions, fluid properties and pipe diameters (see the review of Dukler and Fabre, 1994).

In what follows, we limit our attention to the case of long bubbles. These bubbles are sometimes called Taylor or Dumitrescu bubbles for cylindrical bubbles in vertical pipe or Benjamin bubbles for bubbles in horizontal pipe. So how long should be a long bubble? From a descriptive point of view, one should probably answer: several diameters. From a modelling point of view, the answer should be: long enough for their motion to be controlled by the size of the channel or the pipe in which they move. This will be our favourite definition even if it does fit the intuitive perception.

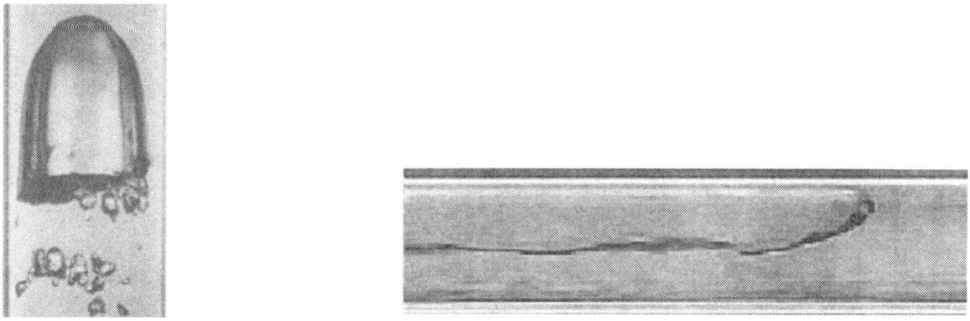

Figure 7. Long bubbles moving in liquid in tube.

Figure 7 shows two examples: a bubble rising in a vertical tube and a bubble pushed by the liquid motion in a horizontal tube. The rising bubble looks short whereas the horizontal one is long enough for the picture to show only the front part. However in both case, their motion does not depend on their length. In fact the motion of an isolated bubble is controlled by the flow close to its tip. It will be seen further that, at a distance of about one diameter from its nose, the liquid flow becomes supercritical so that none of the perturbations created downstream can influence the flow upstream.

3.1 Motion of Long Bubbles in Still Liquid

The motion of a long bubble in a channel or a pipe filled of a liquid is driven by the motion of the liquid itself and/or by the effects of gravity, i.e. buoyancy and weight. If the liquid is at rest the only force that move the bubble is the gravity. However other forces may have some secondary effects: this is the case of viscous and surface tension that slow down the motion. Neglecting the gas viscosity, the bubble velocity V_∞ in still liquid may be expressed under the general form:

$$V_\infty = f(g, D, \theta, \rho_L, \rho_G, \nu_L, \sigma),$$

where g is the acceleration of gravity, D the diameter of the pipe, θ the pipe inclination, ρ_L and ρ_G the densities, ν_L the kinematic viscosity, σ the surface tension. The foregoing relation may be written in dimensionless form. A simplification arises from the fact that the gravity is necessarily associated with the difference $\Delta\rho=\rho_L-\rho_G$ so that g is replaced by a modified gravity $g^*=g\Delta\rho/\rho_L$. The foregoing relation may be written under the form:

$$V_\infty = C_\infty \sqrt{g^* D}\,, \tag{25}$$

to underline the dominant role of gravity in driving the motion of the bubble. A simple similarity analysis shows that C_∞ is a function of three independent dimensionless numbers:

$$C_\infty = C_\infty(\mathrm{N_f}, \mathrm{Eo}, \theta). \tag{26}$$

These numbers have been chosen so that they do not include the bubble velocity in their definition (note that the star will be removed for clarity on g):

- $N_f=D^{3/2}g^{1/2}/\nu_L$ is the dimensionless inverse viscosity. It is the ratio between the root of the Froude number and the Reynolds number.
- $Eo=\rho_L g D^2/\sigma$ the Eötvös –or Bond– number. It quantifies the ratio between the gravity and surface tension forces.

Rise velocity in vertical pipe: the theory of Dumitrescu. For vertical pipes the theory has been very successful, in reducing the problem to the determination of the inviscid flow in the liquid near the bubble nose. This assumption applies only when viscosity has a negligible influence: this is the case of the *inertia controlled regime* occurring in vertical flow when $N_f>300$ (e.g. Wallis, 1969), i.e. when $D > 50(\nu_L^2/g^*)^{1/3}$.

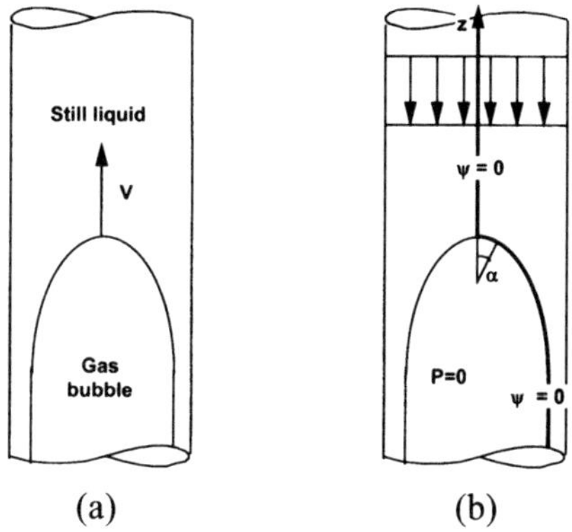

Figure 8. Bubble rising in a vertical tube: (a) in the standing frame, (b) in the moving frame.

Historically the theory was first developed for the case of negligible surface tension when Eo>100, i.e. when the pipe diameter satisfies the condition $D>10(\sigma/\rho_L g^*)^{1/2}$. For an air bubble rising in water at atmospheric pressure this is realized if D > 2.7 cm.

In inviscid fluid the pressure and velocity at the bubble surface are related through the Bernoulli equation. If both gas motion and surface tension are ignored, the pressure inside the bubble is assumed constant. Neat the tip where the bubble radius *R* is nearly constant it yields (Figure 8):

$$\frac{u^2}{2} - gR(1-\cos\alpha) = 0\,. \qquad (27)$$

As *R* is unknown, one must find the solution of the velocity field that fit the condition (27). As the velocity is uniform at infinity, the flow is irrotational in the whole domain. The general solution of the Stokes's stream function for a potential flow in a cylinder can be put under the form of a series of cylindrical harmonics:

$$\psi = -\frac{1}{2}Vr^2 + r\sum_i A_n \exp\left(\frac{-k_n z}{a}\right)\mathrm{J}_1\left(\frac{k_n r}{a}\right),$$

where *a* is the radius of the cylinder and k_n a root of J_0=0, J_0 and J_1 being Bessel functions. The predicted rise velocity depends upon the number of terms retained in the series and the method for selecting the correct solution.

Dumitrescu (1943) was the first to give the solution for σ=0 (Eo$\to\infty$), by retaining three terms in the series expansion of the boundary condition near the tip of a prescribed spherical front, leading to the well-known solution:

$$C_\infty(\infty,\infty,90°) = 0.351\,. \qquad (28)$$

His result agrees closely with the widely accepted value of 0.345 from the experiments of Harmathy (1960), White & Beardmore (1962), Nicklin *et al* (1962), and Zukoski (1966), as well as with the numerical simulations of Mao & Dukler (1990). Davies and Taylor (1950), published after Dumitrescu a less accurate solution. They retained only the first term of the series and the lowest root k_n=3.832. As the Bernoulli condition cannot be satisfied everywhere they arbitrarily chose to fix the condition (27) at r/a=1/2 leading to C_∞=0.328.

These results must be considered as asymptotic values limited to bubbles that are long enough: indeed their volume must be greater than $(0.4\ D)^3$. The ratio R/a is equal to 0.71 as confirmed by the numerical simulations of Mao & Dukler.

Drift motion in horizontal pipe: the theory of Benjamin. For a horizontal pipe the motion of long bubbles has been less studied theoretically for the obvious reason that the symmetry with respect to the axis is lost. It has for a long time been a matter of controversy [as pointed out by Weber (1981), some investigators believed that the bubbles should be stationary while others did not]. This problem has been discussed theoretically by Benjamin for the *inertia-controlled regime* in a nice paper published in 1968. He considered the case of a horizontal cavity filled with liquid and open at one end. As the tube is emptying, a bubble front propagates towards the closed end. Let us consider the control volume of Figure 9. The conservation of mass and *x*-momentum reads:

$$VH = Uh\,, \tag{29}$$

$$\rho VH(U-V) = H\left(p_a - \frac{1}{2}\rho gH\right) - h\left(p_c - \frac{1}{2}\rho gh\right), \tag{30}$$

where the surface tension is ignored and the zero reference of pressure is taken in the gas. Applying the Bernoulli relation between *a* and *O* and between *O* and *c* leads to:

$$p_a + \frac{1}{2}\rho V^2 = 0\,, \tag{31}$$

$$p_c + \frac{1}{2}\rho U^2 - \rho g(H-h) = 0\,. \tag{32}$$

The non trivial solution is:

$$h = \frac{H}{2} \quad \text{and} \quad V = 0.5\sqrt{gH}\,. \tag{33}$$

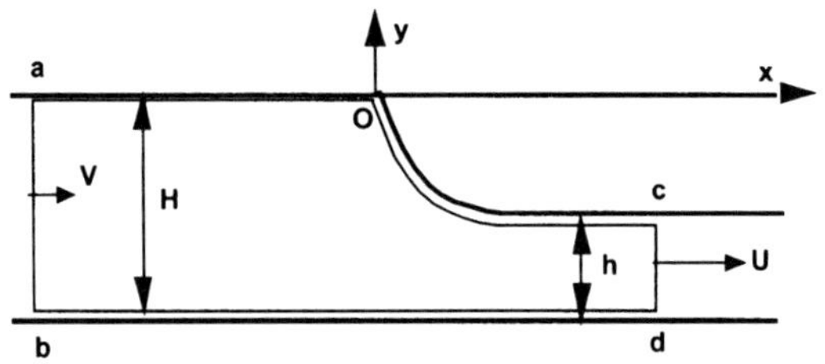

Figure 9. Front of a long bubble moving in horizontal channel.

The form of the solution is similar to that of vertical motion. For the case of a tube, the solution may be obtained with the same method. It gives $V = 0.54\sqrt{gH}$:

$$C_\infty(\infty,\infty,90°) = 0.54 \tag{34}$$

The value is in agreement with the experimental values of Zukoski (1966) corresponding to the highest Eötvös numbers. A drift velocity greater in horizontal than in vertical situations is not, intuitively, what one would expect.

Effect of surface tension. The physical influence of surface tension may be understood as follows. If we add the pressure jump due to surface tension into the Bernoulli equation, one obtains:

$$u\frac{du}{ds} = -g\frac{dz}{ds} - \frac{2\sigma}{\rho R^2}\frac{dR}{ds}, \tag{35}$$

where s is the curvilinear coordinate taken at the bubble surface. The forces appear in the r.h.s. Whatever the case -vertical or horizontal- the gravity effect is positive since at the surface z decreases when s increases: this is a driving force. However the mean radius of curvature increases with s: for cylindrical bubble it goes roughly from D/4 at the tip to D/2 far from it and for horizontal flow it goes from a finite positive value at the stagnation point to infinity far from it. Thus the second term of the r.h.s. is negative indicating that the contribution of surface tension is to resist to the motion. It may even cancel the gravity effect when surface tension is large enough. Indeed, in very small tubes, one can observe that long bubbles don't move even when the tube is vertical. The explanation was given by Bretherton (1961): he demonstrated that if the Eötvös number is less than some critical value (Eo<3.37), a bubble takes a form that prevents its motion.

In vertical flow, the analysis of Dumitrescu has been extended to the case where surface tension is not negligible (not too large Eötvös number). The influence has been analysed theoretically by Bendiksen (1985) who found that surface tension monotonically reduces the rise velocity.

$$C_\infty(\infty,\mathrm{Eo},90°) = 0.344\frac{1-0.9e^{-0.0165\,\mathrm{Eo}}}{\left(1-0.52e^{-0.0165\,\mathrm{Eo}}\right)^{3/2}}\sqrt{1+\frac{20}{\mathrm{Eo}}\left(1-\frac{6.8}{\mathrm{Eo}}\right)} \tag{36}$$

Figure 10 compares the above theories to experimental results and to the correlation of Wallis (1969):

$$C_\infty(\infty,\mathrm{Eo},90°) = 0.369\sqrt{1-\frac{6.94}{\mathrm{Eo}}}\,. \tag{37}$$

The theory has some unexplainable behaviour at large Eötvös number. It seems also to be less accurate than the correlation that is preferable for practical purpose.

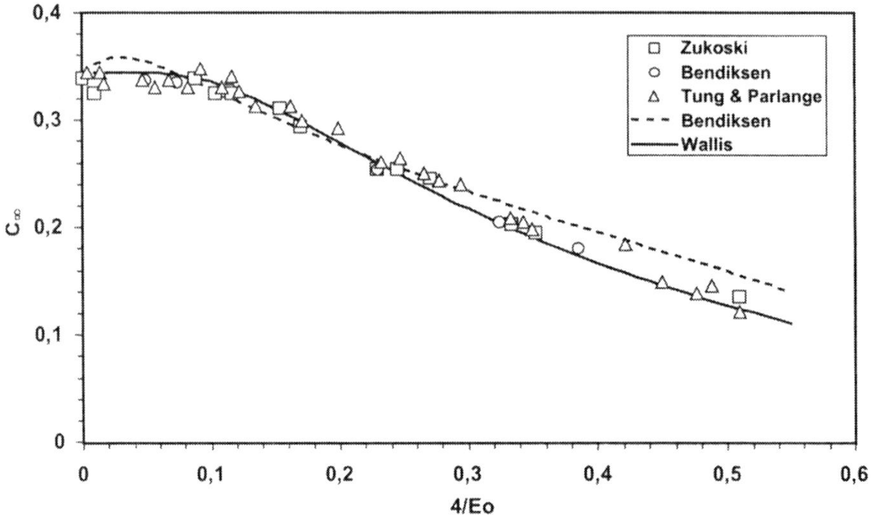

Figure 10. Influence of surface tension on the dimensionless rise velocity of long bubbles.

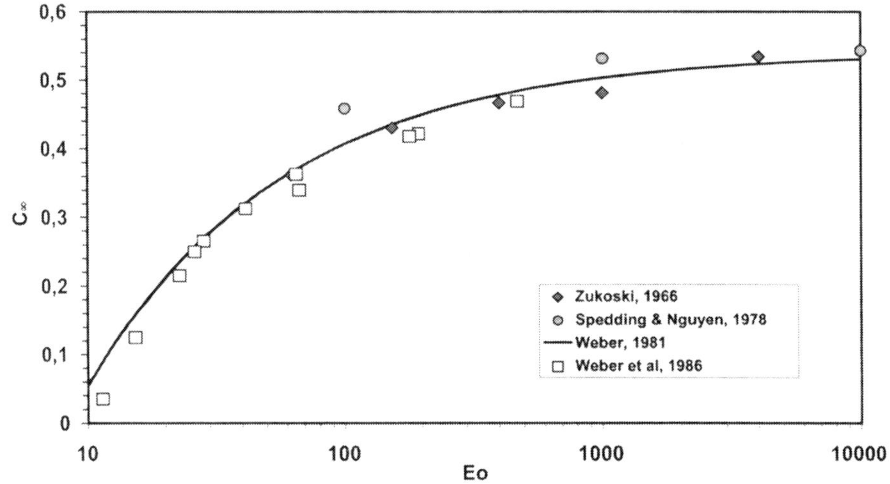

Figure 11. Effect of surface tension on bubble drift in horizontal pipe.

In horizontal situation, the drift velocity decreases with surface tension more strongly than in vertical one: the tendency is shown in Figure 11. The experimental results are well predicted by the correlation mentioned by Weber (1981):

$$C_\infty(\infty, \mathrm{Eo}, 0°) = 0.54 - 1.76\,\mathrm{Eo}^{-0.56}\,. \tag{38}$$

The conclusion is that one must reach very high Eötvös number for the drift to be independent of surface tension, in contrast to the case of vertical flow.

Effect of viscosity. The effect of viscosity can be seen on the map proposed by White and Beardmore (1962) from their experiments (Figure 12). There exists a purely *viscous regime* when N_f is less than 2 (Wallis 1969). In this regime, the inertia has no effect. For the condition to be fulfilled the pipe diameter must be less than $1.6(\nu_L^2/g^*)^{1/3}$, which may arise only with highly viscous liquids. The dimensional analysis leads to C_∞ being expressed as:

$$C_\infty(N_f < 2, \infty, 90°) = 0.01 N_f\,, \tag{39}$$

where the coefficient was determined experimentally by White & Beardmore (1962).

For the mixed regime Wallis (1969) proposed a general correlation that fits their experimental data reasonably well:

$$C_\infty(N_f, \infty, 90°) = 0.345\left(1 - e^{-0.01 N_f}\right). \tag{40}$$

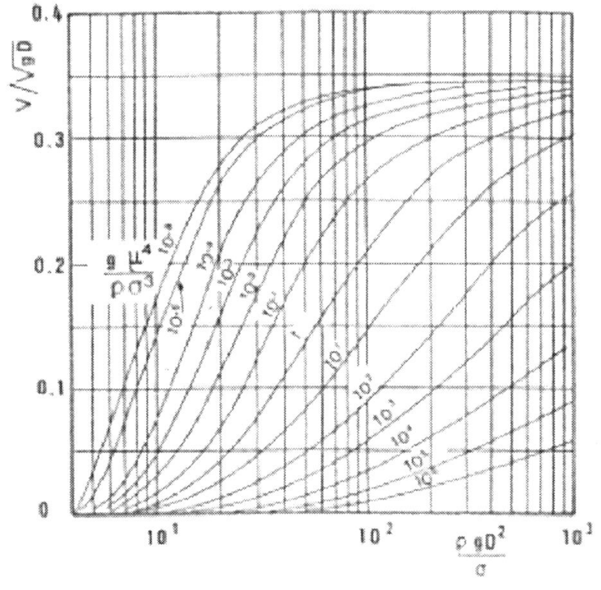

Figure 12. Rise velocity of long bubble in vertical tube (after White and Beardmore, 1962).

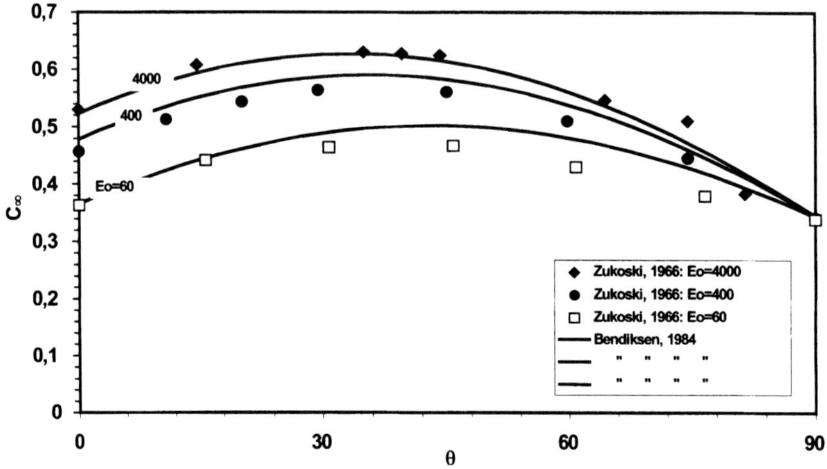

Figure 13. Rise velocity of long bubble for different pipe inclinations.

Influence of pipe inclination. The influence of pipe inclination has been investigated experimentally by Zukoski (1966), Spedding & Nguyen (1978), and Weber et al (1986) for pipe inclinations ranging from 0 to 90°. The effect of inclination (Figure 13) is complex because of the change in bubble geometry. Below 30° the tube is wetted by the gas, the contact angle of the bubble at the wall being acute; beyond 40° this angle is obtuse. At high Eötvös numbers the velocity is a maximum for an inclination in the range 35 – 45°, roughly corresponding to contact at right angles with the wall. A general theory for an inclined pipe is lacking at present. The empirical correlation of Bendiksen (1984)

$$C_\infty(\theta) = C(0°)\cos\theta + C(90°)\sin\theta \ . \tag{41}$$

may be used for inertia-controlled regime.

3.2 Motion of a Bubble With the Liquid Moving Ahead of It

We discuss now the influence of the motion of the liquid that modifies the bubble behaviour in pushing it in the tube. This motion may contribute to increase the effect of gravity (up-flow) or to decrease it (down-flow).

Vertical motion: flow regime transition. Let us focus at first upon the case of vertical flow that has been early on investigated by Nicklin *et al* (1962) for up-flow and down-flow of liquid. Their study was a major contribution because they proposed a law that makes the slug flow models robust and predictive:

$$V = C_0 j_L + C_\infty \sqrt{gD} \ . \tag{42}$$

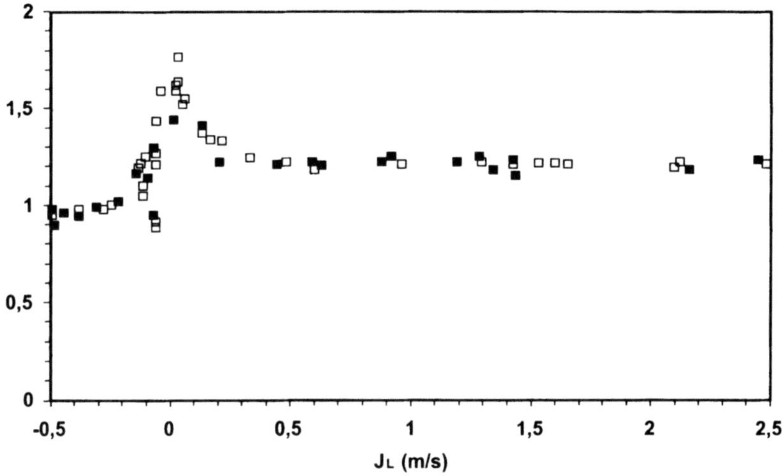

Figure 14. Coefficient C_0 replotted from the experiments of Nicklin *et al.* (1964).

The main assumption contained in this law is to postulate that the gravity effect and the liquid motion are uncoupled. From their experimental results they found that over a large range of liquid flux the bubble velocity is linear with respect to j_L.

However, if one has the curiosity to replot their results in a different way, it shows that the coefficient C_0 is still a function of j_L (Figure 14): it increases from 0.9 for negative values of liquid velocity to a maximum of 1.8 near $j_L = 0$ and then decreases towards an asymptotic value of 1.2, which is reached when j_L is greater than 0.5. From this figure it is clear that there exists some kind of transition in the bubble motion near $j_L = 0$ that will be discussed further.

As the value of C_0 at high Reynolds number is close to the ratio of the maximum to the mean velocity, they said "the bubble velocity is very nearly the sum of the velocity on the centre-line above the bubble plus the characteristic velocity in still liquid". While crude, this explanation predicts the rise velocity with sufficient accuracy for most purposes. In particular, any physical mechanism having an effect on the velocity distribution of the liquid upstream the nose is expected to affect the rise velocity. This happens if the flow is laminar ahead of the bubble: one recovers a coefficient that is close to 2, very near from the maximum value measured by Nicklin *et al.*

A theoretical analysis of this problem has been carried out by Collins et al (1978) and has been extended by Bendiksen (1985) in order to take into account surface tension effects under the restrictive assumption of an inviscid fluid. The results are valid only for the *inertia-controlled regime*. As previously mentioned, since the velocity distribution must have an important effect on the bubble motion the rise velocity is expected to depend on whether the flow is laminar or turbulent upstream the nose. The law must take the form C_0=f(Re) where Re is the Reynolds number characterizing the flow regime within the tube. It might seem surprising to discuss this effect of liquid viscosity in the framework of inviscid theory as Collins *et al.* and Bendiksen did. However, viscosity acts essentially to develop the liquid velocity profile far ahead of the bubble —but it has no influence near the bubble front if inertia still dominates: this condition is satisfied provided N_f>300. For inviscid axis-symmetric rotational flow, Stokes's stream function satisfies a Poisson equation that is solved by applying the boundary conditions at the bubble surface. Collins *et al.* (1978) used this approach to obtain an approximate solution for both laminar and turbulent flow with prescribed upstream vorticity. Using two different methods for flow approximation, they found two different solutions for laminar flow of the form

$$V = u_m + \Phi\left(\frac{u_m}{\sqrt{gD}}\right)\sqrt{gD}\ , \tag{43}$$

where u_m is the velocity at the tube axis. An asymptotic behaviour of the equations is given for the small values of the argument of Φ leading to:

$$V = 2.27\, j_L + 0.361\sqrt{gD}\ , \qquad \text{or} \qquad V = 2.16\, j_L + 0.347\sqrt{gD}\ . \tag{44}$$

Collins et al extended their method to the case of turbulent flows. However, in contrast to laminar flow, the function that defines the vorticity distribution is not constant, so that an approximate solution of the Poisson equation must be found. Restricting their analysis to the case of smooth wall and using the velocity profile of Reichard (1951) to describe the vorticity far ahead the bubble nose, they arrived at the conclusion that the coefficient C_0 must be given in function of the Reynolds number.

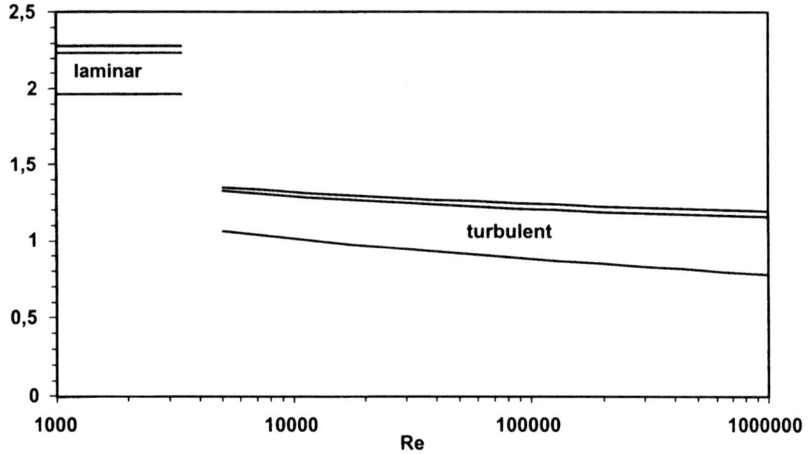

Figure 15. Evolution of C_0 with the Reynolds number for Eo=100, 1000, 10000 (the coefficient increases with Eo).

We summarize below the theoretical laws for the inertia-controlled regime in laminar and turbulent flow, by giving the solution extended by Bendiksen to the case where surface tension has some effect:

$$C_0 = 2.29\left[1 - \frac{20}{\mathrm{Eo}}\left(1 - \mathrm{e}^{-0.0125\mathrm{Eo}}\right)\right] \qquad \text{laminar flow} \qquad (45)$$

$$C_0 = \frac{\log\mathrm{Re} + 0.309}{\log\mathrm{Re} - 0.743}\left[1 - \frac{2}{\mathrm{Eo}}\left(3 - \mathrm{e}^{-0.025\mathrm{Eo}}\right)\log\mathrm{Re}\right] \qquad \text{turbulent flow} \qquad (46)$$

The result is plotted in Figure 15. From the theory we know little about the transition between the two regimes. It is interesting to note that both C_0 and C_∞ decrease when surface tension increases whereas C_0 increases and C_∞ decreases when viscosity increases. They have not the same behaviour and it is not very intuitive that the bubble velocity of the bubble can increases when the viscosity increases.

Transition between up-flow and down-flow. We know little about the existence of a transition between up-flow and down-flow. Griffith & Wallis (1961) were probably the first to report the unstable motion of cylindrical bubbles in downward liquid flow, eccentrically located towards the pipe wall. By looking at the flow seen by the bubble ahead of it (Figure 16), we note that there is a major change between the situations of up- and down-flow. A qualitative explanation of the difference of behaviour could be as follows. The bubble tip that controls the motion tends to move up under the gravity force. In such motion, it follows the path where there is the smallest resistance to its displacement, i.e. in the region where the momentum of the liquid phase is the smallest. For up-flow, it is not surprising that the tip of the bubble be located on the axis where the liquid velocity is the smallest. For down-flow it is expected that the tip tends to migrate near the wall. However the surface tension prevents the radius of curvature of the bubble to be too small: it may be possible only if the tip is not too close to the wall. For large tubes, the effect must be more pronounced than for small ones.

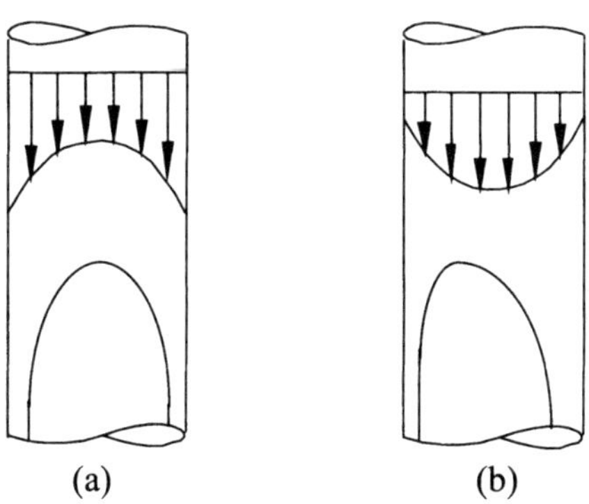

Figure 16. Scheme of the bubble shape in up-flow (a) and down-flow (b).

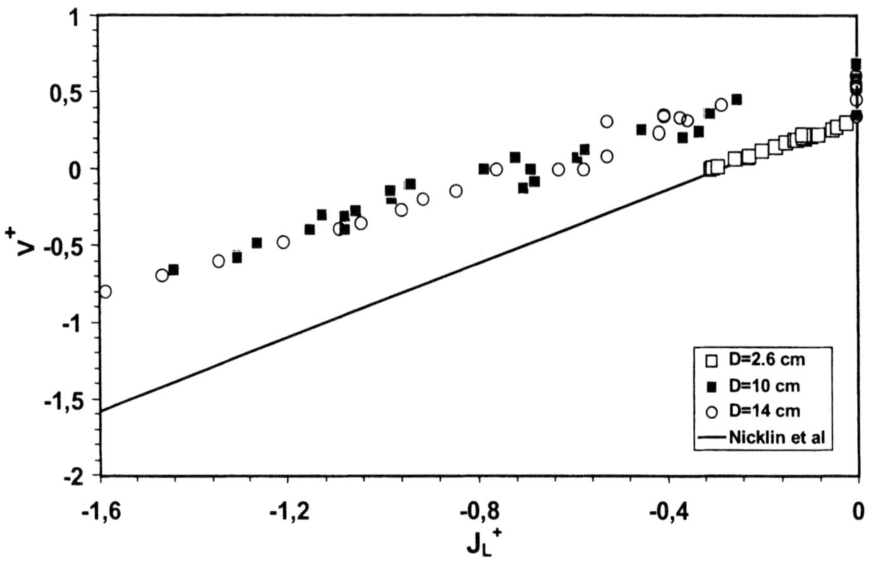

Figure 17. Bubble velocity in down-flow from Martin (1976): $V^{+}=V/(gD)^{1/2}$, $j_L^{+}=j_L/(gD)^{1/2}$.

Martin (1976) carried out specific experiments of downward flow in pipes of different diameters (Figure 17). For the smallest diameter pipe (Eo≈100), the results do not display any difference with the Nicklin *et al* correlation. For the largest diameter pipes (Eo≈1400 and 2700), they do: C_0=0.90 (resp. 0.86) for D=10 cm (resp. 14 cm).

Horizontal and inclined motion: shape transition and bubble turning transition. The motion of bubbles in horizontal and inclined pipe was investigated by Bendiksen (1984). He carried out experiments in pipe for inclinations ranging from –30° to 90°. They show that the law (42) of Nicklin *et al* proposed for vertical motion still applies for horizontal and inclined motion.

However the experimental data put in evidence two transitions. The first one concerns the change of shape that the bubble experiences when the inertia becomes greater than the stratifying effect of gravity. The other one is the so-called "bubble turning" transition that happens in downward motion.

The shape transition was first identified in slug flow. In horizontal or slightly inclined flow, the shape of the bubble at the nose changes when the liquid velocity increases (Figure 18). At low velocity, the bubble has the characteristic shape of the bubbie described by Benjamin (1968): the tip is located close to the upper wall. At increasing velocity the tip of the bubble moves away from the wall. At very large velocity the bubble is nearly centred in the tube. The gravity force that stratifies the liquid in the film is in competition with the liquid inertia that tends to centre the bubble. This competition is quantified by the Froude number $Fr=j/(gD)^{1/2}$.

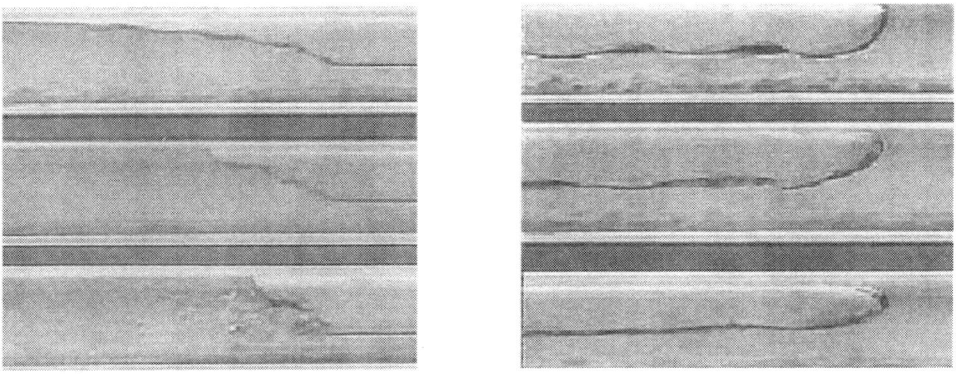

Figure 18. Evolution of bubble shape in horizontal flow when the liquid velocity increases (from top to bottom).

From his experiments in small diameter pipe Bendiksen (1984) suggested that the transition might occur at $Fr_C = 3.5$. At this critical value the location of the bubble tip with respect to the axis was observed to change from 0.75·D for the lowest velocity to 0 for the highest. This shape modification leads to a change in the values of the coefficients of the Nicklin's correlation. In particular the coefficient C_0 changes from 1 to values close to the one it takes in vertical flow when the bubble is centred (Figure 19). In addition, C_∞ also changes. When the bubble becomes more centred there is no more the driving effect of gravity and C_∞ cancels. We summarize the behaviour as follows:

- For $Fr<Fr_c$ $C_0\approx 1$ and $C_\infty\approx$ value for $j_L=0$
- For $Fr>Fr_c$ $C_0\approx 1.2$ in turbulent flow and $C_\infty\approx 0$.

The bubble turning phenomenon happens for negative slopes (Figure 20). For velocity below some critical value depending upon the slope the bubble tip points against the liquid flow as it does in counter-current flow. Nevertheless, the bubble does not necessarily move up, it may be pushed downward. At increased velocity, the bubble tip points in the same direction than the liquid flow: towards the bottom. The bubble moves as in co-current flow.
The consequences can be seen in Figure 20: the bubble moves faster when it behaves as in co-current flow. The results can be summarized as follows:

- When the liquid velocity is smaller than a critical value, $C_0<1$ and $V_\infty<0$
- When it is greater $C_0>1$ and $V_\infty>0$: for the highest liquid velocity $C_0\rightarrow 1.19$ and $V_\infty\rightarrow 0$ similarly to horizontal flow.

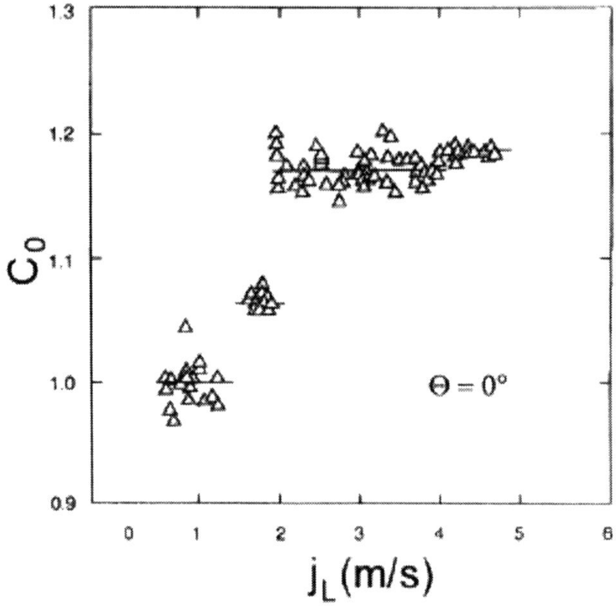

Figure 19. Experimental data on C_0 after Bendiksen (1984).

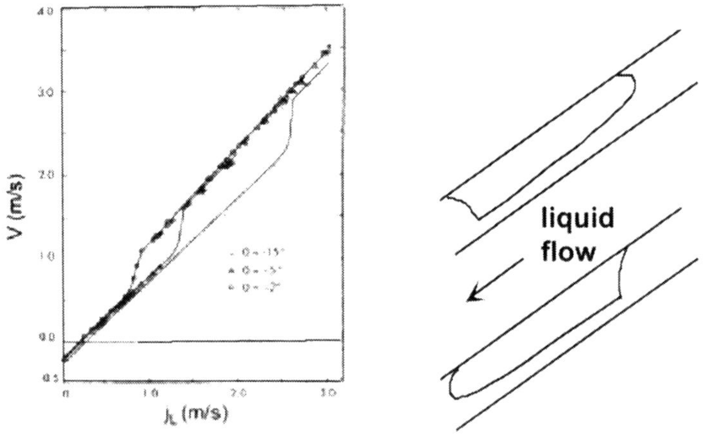

Figure 20. Bubble turning for liquid down-flow.

Motion in viscous regime. Now let us examine the case when gravity is negligible, i.e. for Eo<<1. Then the pipe inclination is no longer a relevant parameter and we expect to find roughly the same results for vertical and horizontal flows. Although this case corresponds to very extreme conditions it may be of interest for two-phase flow in very small tubes and for 0-gravity flow. We refer here to the experimental investigations of Fairbrother and Stubbs (1935) and Taylor (1961) and to the theoretical work of Bretherton (1961). Taylor and Bretherton introduces a dimensionless drift coefficient m defined by:

$$m = \frac{V - j_L}{V}. \tag{47}$$

If we report this definition in a Nicklin-like relation by ignoring the term due to gravity, it may be found that:

$$m = 1 - \frac{1}{C_0}. \tag{48}$$

In the absence of driving force ($g = 0$) m must be a function of the dimensionless number:

$$\mathrm{Ca} = \frac{\mu V}{\sigma} \qquad (\text{or } \mathrm{Ca} = \frac{\mu j}{\sigma}) \tag{49}$$

that is a capillary number.

The conclusions of Taylor, Fairbrother and Stubbs, and Bretherton are summarized as follows:

- When the capillary number is small enough, the velocity is obtained by retaining the viscous force and the surface tension in the momentum equation Bretherton (1961). Then m is shown theoretically to vary as:

$$m = 1.29(3\mathrm{Ca})^{2/3} \qquad \text{for Ca} < 0.003. \tag{50}$$

- For higher values of Ca the experimental results (Fairbrother and Stubbs, 1935) are well fitted by:

$$m = 1.0\sqrt{\mathrm{Ca}} \qquad \text{for Ca} < 0.09. \tag{51}$$

- When the capillary number is high enough we do not expect any influence from surface tension. In other words m must tend asymptotically towards a constant. It happens at a rather low value of Ca of 1.7:

$$m = 0.56 \qquad \text{for Ca} > 1.7. \tag{52}$$

Eq. (52) may be written using C_0 rather m. In this case it happens that $C_0 = 2.27$, in total agreement with the first solution of Collins et al (1979) for laminar flow when the gravity term is discarded (Eq. 44).

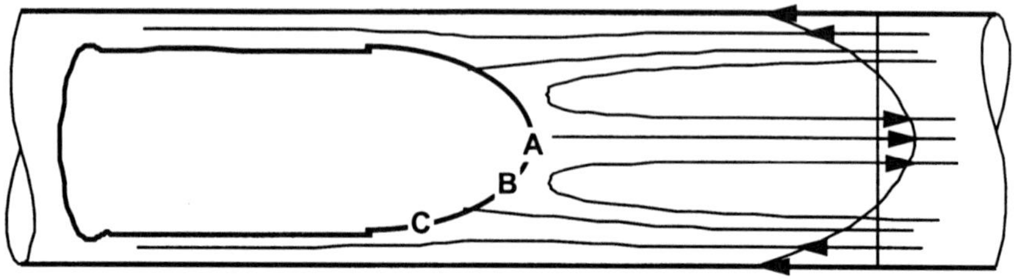

Figure 21. Picture of the flow upstream a moving bubble in viscous regime for m < 0.5.

An interesting point about the solution at low capillary number was suggested by Taylor. The velocity profile in the moving frame far upstream the bubble is:

$$\frac{u-V}{V}=\left[2-\left(\frac{2r}{D}\right)^2\right]-C_0 .$$

When $C_0 < 0$, i.e. for m < 0.5, the velocity is positive on the axis while negative at the wall. This gives the probable picture of Figure 21 for the streamlines, with one stagnation point on the axis and a stagnation circumference. This solution was visualized by Cox (1964).

3.3 Bubble Shape

Bubbles rising in vertical tubes have the shape of a prolate spheroid independent of their length. The nose appears smooth except for high liquid velocity for which it fluctuates, probably under the effect of large turbulent eddies that could modify the shape equilibrium. It has been also pointed out that in counter-current flow, the nose is displaced towards the tube wall where it has not a stable position: it turns or oscillates in search of a stable position. The shape at the rear depends on whether or not the viscous force is negligible. When negligible, the bubble has a flat back indicating that flow separation and vortex shedding occur (see for example the picture of (Figure 7). When it is not, the rear of the bubble may take the form depicted in Figure 21.

The shape of the bubble depends upon the pipe inclination. Indeed the experiments of Zukoski in still liquid (1966) show clearly that the eccentricity increases when the pipe is deviated from the vertical position. As a consequence, when the inclination decreases from 90° to the horizontal, the cross-sectional area of the film far from the nose departs from a centered annulus to an eccentric annulus, then to a segment of the circle indicating that stratified flow is reached in the liquid film at some distance behind the nose which varies with inclination angle. According to Spedding and Nguyen (1978) this change in shape occurs between 30° and 40°.

In horizontal flow, we have already mentioned the characteristic of the bubble nose as well as the shape transition that occurs for certain critical Froude number. Fangundes *et al* (1999) carried out experiments in which they released isolated bubbles in a horizontal pipe. From these experiments it can be seen (picture of Figure 18 and recording of Figure 22) that:

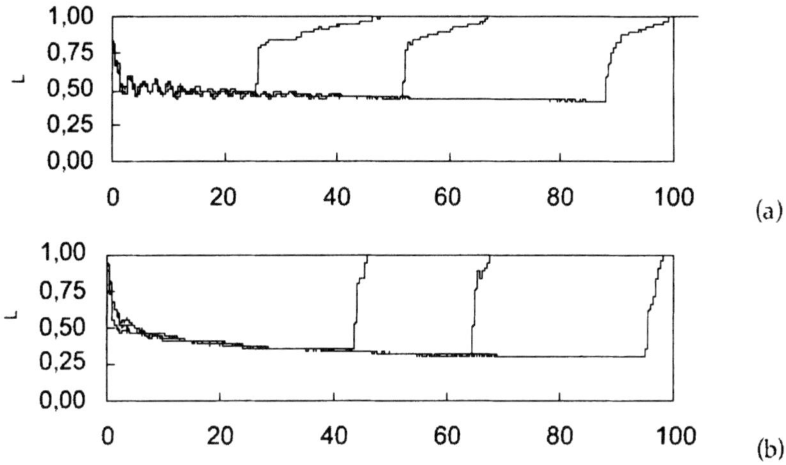

Figure 22. Influence of the bubble volume: $j_L = 0.6$ m/s (a) 1.2 m/s (b), the x-coordinate is the non dimensionless distance to the bubble nose, the y-coordinate is the local liquid hold-up.

- The shape of the bubble is independent of its length.
- Like the nose, the rear of the bubble experiences a shape evolution when the liquid velocity that pushes the bubble increases. At low velocity, the bubble presents at the rear a smooth evolution whereas at high velocity the rear looks like a hydraulic jump.

The bubble is in fact composed of four different parts (Figure 23): a nose controlled by inertia and gravity whose length is about 1 D, a body controlled by friction and gravity that may extend over several diameters, a hydraulic jump controlled by inertia and gravity with about 1 D length and finally a tail of a few diameters.

The body controls the length of the bubble and the height of the film upstream the jump. Its shape may be predicted using a shallow water approximation. Knowing the shape and the bubble volume it is possible to determine the flow conditions at the end of the liquid film, just before the jump. Then the intensity (i.e. the jump of height) across the jump may be calculated.

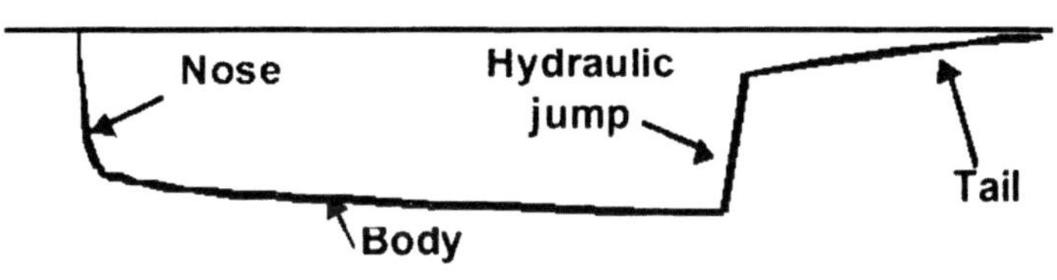

Figure 23. The various regions of a long bubble.

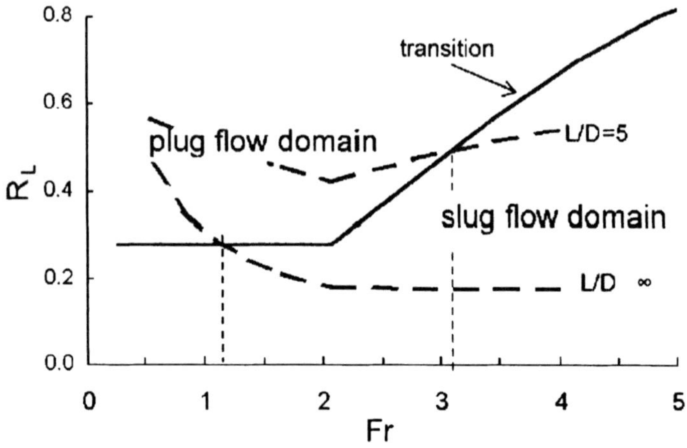

Figure 24. Map of plug to slug transition after Fagundes *et al.* (1999).

The presence of the tail depends on the jump intensity:

- If the jump intensity is small enough so that the height downstream does not equal D, the tail exists and we are in the so-called plug flow domain. This happens if the momentum $A\varphi^2/R_{LS}$ of the liquid film that enters into the jump is small enough, i.e. at low bubble velocity V and large enough film thickness, i.e. long enough bubble.
- If these conditions are not fulfilled, the interface after the jump reaches the upper part of the pipe, the tail disappears and we are now in the slug flow domain. To satisfy the conservation of momentum across the hydraulic jump, bubble shedding must occurs at the rear part of the long bubbles.

Figure 24 shows the result of the model of Fagundes *et al.* It is seen that for long bubbles one reaches the transition at a smaller Froude number than for short ones.

3.4 Developing Length of a Bubble

The practical models used for predicting slug flow usually contain the assumption that the flow is full developed in the long bubbles. To see how good is this assumption it is possible to determine the evolution of the liquid hold-up along the film. To illustrate this question we discuss the example of horizontal motion that was treated by Fagundes *et al.* (1999). It may be easily extended to vertical motion.

In the shallow water approximation (1D two fluids model), the momentum equation in the moving frame reads:

$$\frac{dR_L}{d\xi} - g\frac{R_L^3}{\varphi_L^2}\frac{dh_L}{d\xi} = -2f_L\frac{S_{wL}}{\pi D^2}\left(\frac{R_L - R_{L\infty}}{R_{L\infty}}\right)^2, \tag{53}$$

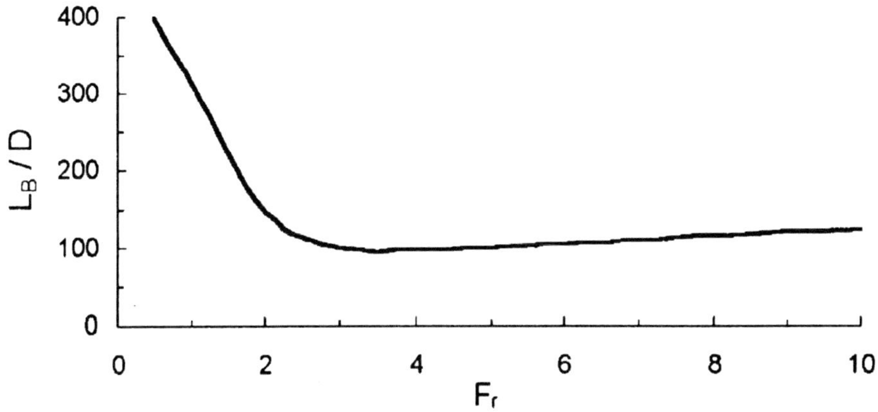

Figure 25. Developing length of a bubble in horizontal pipe after Fagundes *et al.* (1999).

where R_L is the local liquid hold-up, $h_L(R_L)$ the height of the liquid film, $R_{L\infty}$ the liquid hold-up for a bubble of infinite length, f_L the friction factor at the wall and $\varphi_L = V - j_L$ the liquid flux (Eq. 19). $R_{L\infty}$ can be determined for horizontal flow by the following approximate expression given by Fagundes *et al.* (1999):

$$R_{L\infty} = \frac{\varphi_L}{V}\left[1 + 1.7\left(0.5 + \frac{\varphi_L}{V}\right)\left(\frac{\nu_G \rho_G^4}{\nu_L \rho_L^4}\right)^{1/7}\right], \tag{54}$$

provided its value falls between 0.1 and 0.5.

The order of magnitude of the bubble length L_S can be found by putting Eq. (53) in dimensionless form. It follows that:

$$\frac{L_S}{D} = \mathcal{G}\left[\frac{gD}{f_L (V - j_L)^2}\right]. \tag{55}$$

The calculations have been carried out and plotted in Figure 25. They show that the length of a long bubble such as $R_L = 0.95\ R_{L\infty}$ is greater than 100 D. Thus, for bubbles shorter than L_S the assumption of fully developed flow in the film is questionable.

3.5 Motion of Train of Bubbles in Slug Flow

Measured bubble velocities are shown in Figure 26 for vertical flow, and in Figure 27 for horizontal flow. They are plotted in dimensionless scale versus the mixture velocity defined as:

$$j = j_L + j_G\,. \tag{56}$$

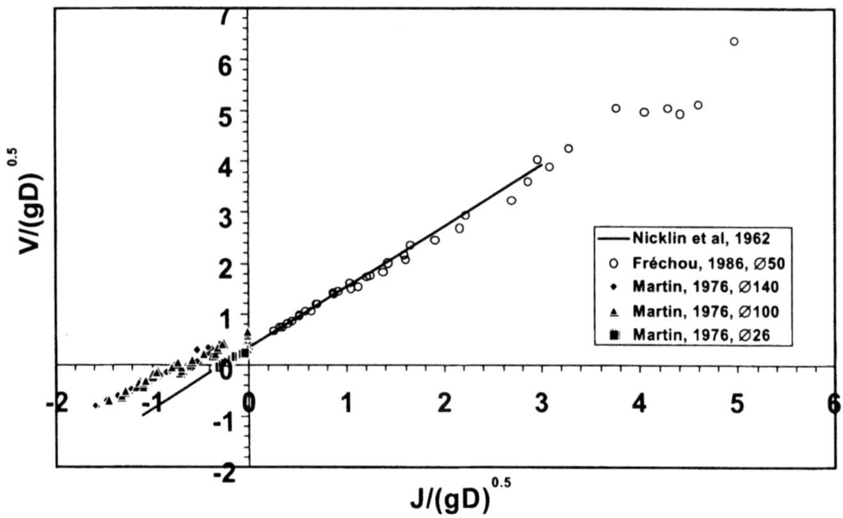

Figure 26. Velocity of long bubbles vs. mixture velocity, $\theta = 90°$

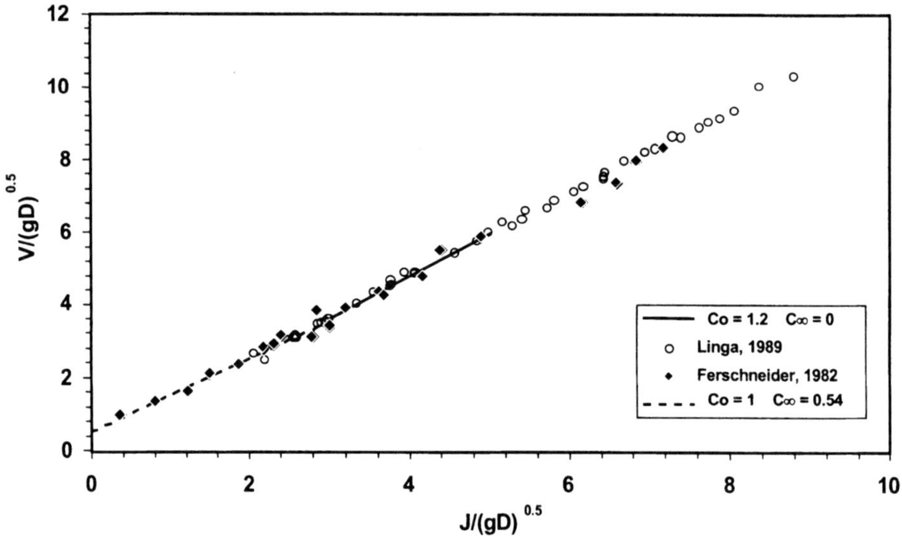

Figure 27. Velocity of long bubbles vs. mixture velocity, $\theta = 0°$

It must be noted that at high velocity the data are scattered. These figures illustrate some general trends that will be briefly discussed. For a more extensive analysis, see the review of Fabre and Liné (1992) or Dukler and Fabre (1994).

The V(j) relation is linear over certain ranges of mixture velocity j thus supporting the assumption of Nicklin *et al.* for single bubble motion. The velocity is thus given by:

$$V = C_0 j + C_\infty \sqrt{gD}\ . \tag{57}$$

where the law is similar to Eq. (42) except that j_L is replaced by j. C_0 and C_∞ remains constant for some range of mixture velocity and fluid properties.

The **flow regime transition** put in evidence in vertical flow suggests that the bubble move faster when the liquid flow is laminar upstream the bubble nose than when it is turbulent. Fréchou carried out experiments in slug flow (1986) with fluids of different viscosity to vary the Reynolds number of the mixture, $Re = jD/\nu_L$, over a wide range. The transition was found near a critical Reynolds number $Re_C = 1000$; as shown in figure 28, the data are reasonably fitted by:

$$C_0 = \frac{2.27}{1+(Re/Re_c)^2} + \frac{1.2}{1+(Re_c/Re)^2} \tag{58}$$

The **up-flow/down-flow transition** in vertical flow is clearly visible in the vicinity of j=0 in Figure 26. However much has to be done for down-flow condition to understand the mechanism that controls the bubble motion.

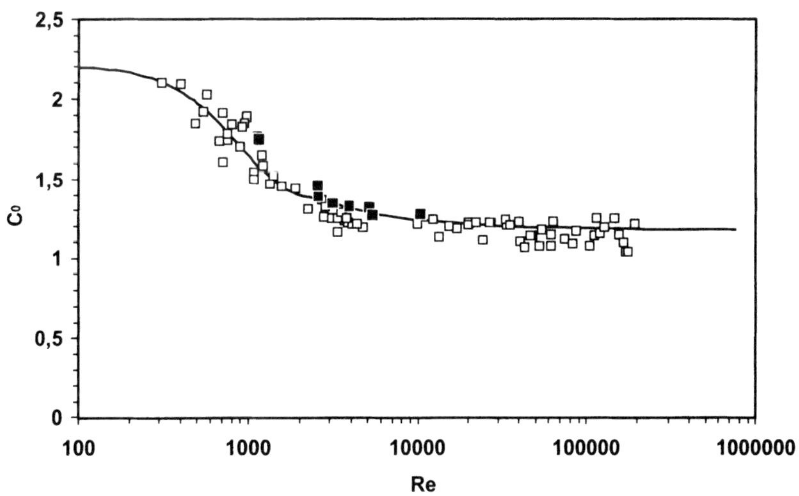

Figure 28. Influence of the flow regime on bubble motion □: Fréchou; ■: Mao & Dukler; ▬: Eq. (58).

The **shape transition** in horizontal flow has for a long time been a matter of controversy. It is shown in Figure 27 less clearly than the previous one. In the various experiments (Ferré, 1979; Théron, 1989; Ferschneider, 1982; Linga, 1989) the authors do not agree on the value of the critical Froude number at which the transition occurs. However these experiments were carried out with different pipe diameters and different fluids leading to different values of the Eötvös number. As surface tension is expected to have some influence, the critical Froude number should be a function of the Eötvös number: $Fr_C = f(Eo)$. It may be suggested that this function be chosen so as to the two laws shown in Figure 27 intersect at $Fr = Fr_C$ to ensure the continuity of the bubble velocity, i.e.:

$$V + C_\infty \sqrt{gD} = 1.2V ,$$

that gives, by using Eq. (38) for expressing C_∞:

$$Fr_c = 2.7 - 8.8Eo^{-0.56} \tag{59}$$

The **bubble-turning transition** is also observed in slug flow. The question regarding the direction of the bubble and thus its motion can be solved from the following consideration. Slugging occurs when stratified flow is unstable. Lets us now consider the two stratified flow patterns that can be observed in descending flow (see chapter on stratified flow).

- If the gravity force is high enough compared to the pressure force due to gas friction, the liquid is moved independently from the gas: it controls the hold-up and the cross-section offered to the gas flow. If the gas velocity is so small that the velocity difference $U_G - U_L$ can induce a K-H instability, then slug flow will form with bubble pointing upward (Figure 29a).
- If the pressure force due to gas friction is greater than the gravity force in the liquid, the gas velocity is greater than that of liquid. If the condition required for a K-H instability is fulfilled, slug flow will form with bubble pointing downward (Figure 29b). Note that in both cases, the bubble moves downward.

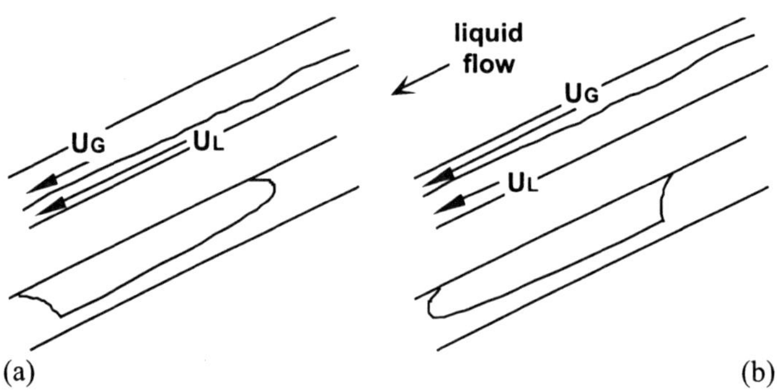

Figure 29. Upper pictures: stratified flow; lower pictures: resulting slug flow.

3.6 Liquid Hold-Up in Long Bubbles

The general method generally used to determine the hold-up in large bubble starts from the assumption that the separated flow region between the nose and the tail is fully developed. The liquid hold-up may be known by eliminating the pressure gradient between Eqs. (21) for k=L,G:

$$\frac{\tau_{wGS} S_{wGS}}{R_{GS}} + \frac{\tau_{iGS} S_{iS}}{R_{GS} R_{LS}} - \frac{\tau_{wLS} S_{wLS}}{R_{LS}} + A\Delta\rho g \sin\theta = 0 . \qquad (60)$$

This is the hold-up equation already put in evidence in stratified flow. In the foregoing equation the shear stresses at both wall and interface are expressed as follows:

$$\tau_{kwS} = -f_{kwS} \rho_k \frac{U_{kS} |U_{kS}|}{2} , \qquad (61)$$

$$\tau_{GiS} = -f_{GiS} \rho_G \frac{(U_{GS} - U_{LS}) |U_{GS} - U_{LS}|}{2} , \qquad (62)$$

in which the friction factors f have to be closed following the method indicated in the chapter "Stratified flow". Solving Eq. (60) addresses two important issues:

- The pattern of the interface within the bubble must be known. For vertical flow the interface forms an annulus, whereas it is flat in horizontal flow. A transition thus occurs which must be modelled. As already said, very little is known on this problem.
- The assumption of fully developed flow is rather strong. It has to be revisited by considering the evolution of the thickness of the liquid film as a function of the length of the bubble.

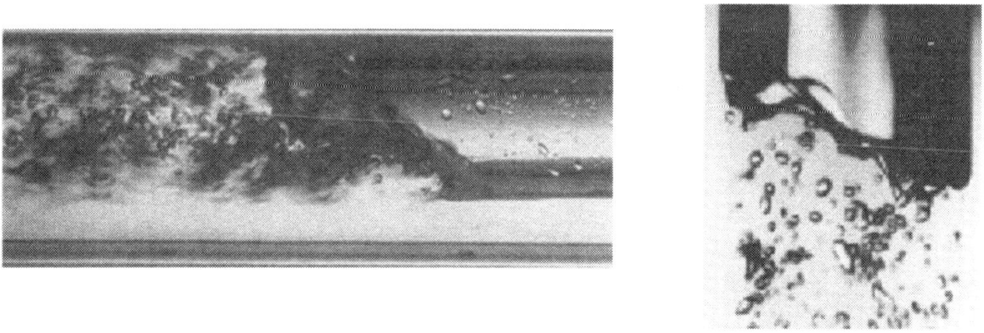

Figure 30. Pictures of the entrainment of small bubbles at the rear of the long ones.

4 Liquid Slugs

One of the most specific features of slug flow is the entrainment of small bubbles at the rear of the gas slugs, generating a bubbly mixture that flows from the rear of a long bubble to the front of the next one.

This phenomenon, pictured in Figure 30 for both horizontal and vertical motions, addresses several issues:

- The generation of small bubbles from the gas slugs.
- Their motion in the liquid slugs.
- The development of the bubbly mixture in the liquid slugs.

4.1 Entrainment of Small Bubbles

In the recent decade, some experimental data of gas fraction in the liquid slugs have been published. Some of these results obtained with similar flow conditions but different pipe inclinations are illustrated in Figure 31 - the data are replotted here versus dimensionless mixture. This presentation shows that the evolution of the gas fraction with the mixture velocity has the same trend in horizontal and in vertical pipe. This suggests that the same physical process take place and that the same modelling can be used for both cases.

The mechanism of entrainment may be explained as indicated in Figure 32. The liquid shed from the rear of a liquid slug, flows around the nose of the long bubble to form a stratified or annular film flowing downward. This film enters at a relatively high velocity into the front of the next slug at high relative velocity. As the liquid film enters the slug it entrains some gas. In the mixing zone at the front of the next slug there is a local region of high void fraction that is clearly observable. In this region of high turbulence level, the mixing process carries some of the bubbles to the front of the slug where they coalesce back into the long bubble.

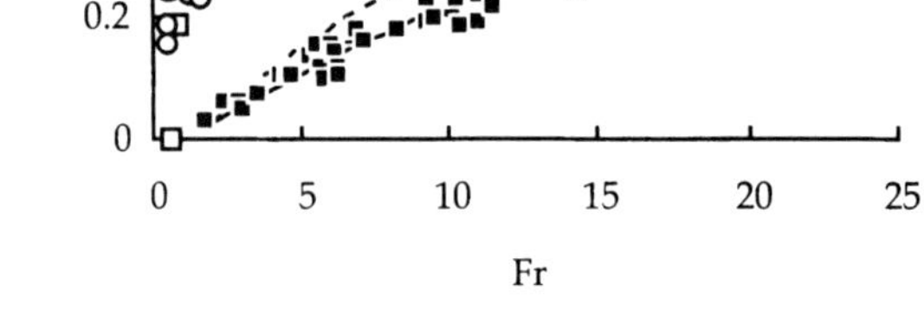

Figure 31. Gas fraction in liquid slugs: Vertical flow: □ Barnea & Shemer, ○ Mao & Dukler. Horizontal flow: ■ Andreussi & Bendiksen; —— horizontal flow; - - - vertical flow.

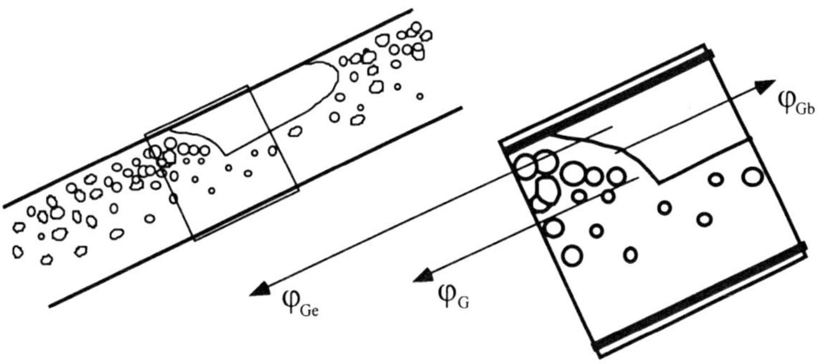

Figure 32. Entrainment at the rear of a long bubble.

Material balance considerations require that:

$$\varphi_G = \varphi_{Ge} - \varphi_{Gb} \tag{63}$$

where φ_G, φ_{Ge}, φ_{Gb}, are the net flux, the flux entrained from the tail and the flux back to the long bubble respectively.

What is the basic difference between horizontal and vertical flow? In experiments of Figure 31 the fluid properties and the pipe diameter were the same and it appears that the gas content in the liquid slugs is higher in vertical flow than in horizontal flow. This does not tell us however whether the gas flux is different between both cases. This flux is given versus the gas fraction by:

$$\varphi_G = R_{GD}\left(V - U_{GD}\right). \tag{64}$$

Even if the gas fraction is higher in vertical than in horizontal flow, the net gas flux entrained could be the same provided that the relative bubble velocity is smaller. Since the bubble drift is higher in vertical than in horizontal flow, this could be true. However $V >> U_{GD}$ and we can firmly state that the gas flux is higher in vertical than in horizontal flow.

The gas entrainment raises another question. Figure 31 shows that below some mixture velocity there is no bubble in the slugs. There exists some critical velocity difference above which gas is entrained: this is the onset of bubble entrainment. In vertical flow the velocity difference is always sufficient to generate small bubbles at the tail of the long ones.

There are a few models in the literature that were developed for predicting the gas fraction in the liquid slugs. We shall not make room for those that are less than satisfactory. These models were developed on purpose either for horizontal flow or vertical flow and the result is quite disappointing when one try to apply each to the other case. Keeping in mind that the mechanism of entrainment is basically the same whatever the pipe slope, a reliable model should do a good job in both cases.

Andreussi and Bendiksen (1989) proposed a model that applies satisfactorily to horizontal or slightly inclined flows. They postulated that the flux of gas entrained at the tail of the long bubble is proportional to the flux of liquid entering the front of the slug once this film velocity exceeds a critical threshold value. Part of this gas is returned to the long bubble at a rate proportional to the void fraction in the liquid slugs. Using a simplified expression of the long bubble velocity, it may be demonstrated that the gas fraction is expressed as:

$$R_{GD} = \frac{j - j_f}{(j + j_0)^n}. \tag{65}$$

In this equation, the critical mixture velocity j_f and the velocity scale j_0 are expressed by:

$$j_f = 2.6\left[1 - 2\left(\frac{d_0}{D}\right)^2\right]\sqrt{gD}, \tag{66}$$

$$j_0 = 2400\left(1 - \frac{\sin\theta}{3}\right)Eo^{-3/4}\sqrt{gD}, \tag{67}$$

with d_0=25 mm and n an exponent depending on the density ratio of both phases. The values of the numerical coefficients are chosen for the best fit with experimental data. According to Eq. (66) the onset of entrainment must not be sensitive to the pipe diameter provided it is large enough leading to a critical mixture velocity of about $2.6\cdot(gD)^{1/2}$. This is probably not fortuitous that this critical mixture velocity is close to the velocity at the shape transition predicted by Eq. (59). Indeed, as previously mentioned, we believe that the entrainment takes place to balance the momentum condition across the hydraulic jump.

4.2 Gas Drift in Liquid Slugs

For vertical or inclined flow it is generally assumed that buoyancy causes the bubbles in the slug to move upward relative to the liquid at a velocity identical to that in bubbly flow. This is consistent with the idea that the liquid slug is equivalent to a section of a pipe which carries distributed bubbly flow. Because the Harmathy equation modified for the presence of a swarm of bubbles has been successfully used for this drift velocity in bubbly flow it has been assumed that it would describe the process equally well in slug flow.

$$U_{GD} - U_{LD} = 1.54(1 - R_{GD})^{0.5}(\sigma g)^{0.25}. \tag{68}$$

We recommend however to use this law with care since it is not expected to work properly when the viscosity of the liquid is too high.

Another choice is to use a drift flux model for the bubbly region. This model has been proposed from theoretical grounds by Kowe *et al.* It leads to the following expression for the gas velocity:

$$U_{GD} = C_1 j + (1 - C_m R_{GD})(1 - R_{GD})V_B, \tag{69}$$

in which C_1 accounts for the velocity and gas fraction distribution, C_n is the entrained mass coefficient whose value is 0.5 for spherical bubbles, and V_B is the rise velocity of bubble in still liquid. This velocity may be calculated for vertical flow by using the set of relations given by Wallis. For inclined pipe the question has not yet been resolved.

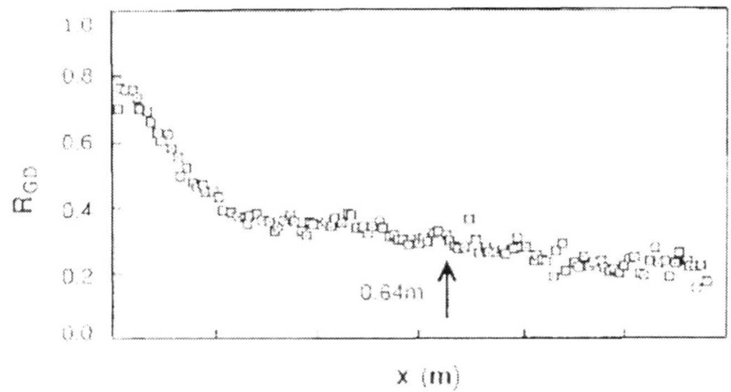

Figure 33. Evolution of the gas fraction along the liquid slug. The arrow indicates the mean slug length.

As the bubble diameter is needed it may be postulated that their size results mainly from turbulence breakup. In this case the model of Hinze is well accepted: it suppose that at the critical diameter the pressure fluctuations that tend to break the bubble is balanced by the interfacial force. It turns out that the bubble diameter may be expressed as a function of the wall shear stress in the slugs:

$$d_B = 1.15\left(\frac{R_{LD}^2\sigma^3}{\rho_L}\right)^{1/5}\left|\frac{\tau_{LwD}S_{LwD}U_{LD}}{A}\right|^{-2/5}. \tag{70}$$

Because of the lack of experimental data on the gas drift, it is hard to conclude.

4.3 Development of the Flow Structure in the Liquid Slug

As the flow in the long bubbles evolves with the distance to its front, the flow in the liquid slugs does the same. The few measurements that are available show that the region just behind the rear of the long bubble is highly aerated by numerous small bubbles. These bubbles are entrapped in the bubble wake from where they escape to flow towards the next bubble. Figure 33 gives an example of such evolution: one can see that at a distance equal to the mean slug length, the flow is not yet fully developed. The assumption that is usually taken in the slug flow models is still more questionable for the slug region than for the bubble region.

5 Slug Structure

It has been shown in previous sections that neither the characteristic length scale L of the cells not their frequency f are needed to calculate void fraction and pressure gradient.

5.1 Mean Length and Frequency

However there is a practical need for knowing the time or length scales of slug flow. For example in hydrocarbon two-phase transportation the maximum slug size is important for the design of slug catchers.

From the times of passage T_{Di}, T_{Si} and the velocities V_{Di}, V_{Si}, of each slug and bubble the mean lengths L_D, L_S may be determined from the statistical average of the products $T_{Di}V_{Di}$, $T_{Si}V_{Si}$. However since the probability distribution of the velocities are narrowly distributed about their average, the assumption $V_{Di} = V_{Si} = V$ leads to:

$$\overline{L}_D = V\overline{T}_D \quad \text{and} \quad \overline{L}_S = V\overline{T}_S \tag{71}$$

where T_D, T_S are the mean times of residence of slugs and long bubbles. The mean slug length $\overline{L}_D$ is one of the characteristic length scale. A characteristic time scale is the mean time of passage of the cell $\overline{T} = \overline{T}_D + \overline{T}_S$: to this scale one substitutes $n = 1/\overline{T}$, generally referred to as the slug frequency. Note that n is the number of cells per unit time seen by a fixed observer with no implication as to periodicity. It can be shown that the mean slug length and the slug frequency are related by:

$$\overline{L}_D = (1-\beta)\frac{V}{n}. \tag{72}$$

Let us discuss the case of horizontal flow. When the superficial gas velocity increases the mean length of the liquid slugs increases and then reaches an asymptotic value lying between 30 to 40D. Concerning the slug frequency, the experiments show that when the mixture velocity increases, it goes through a minimum. Gregory & Scott proposed a correlation based on their data and those of Hubbard:

$$n = 0.0157\left[\frac{j_L}{gD}\left(\frac{V_m^2}{V} + V\right)\right]^{1.2} \quad \text{(where } n \text{ is in s}^{-1}\text{)} \tag{73}$$

where V_m is the slug velocity at the minimum frequency. Unfortunately this relation suffers from two weaknesses: it is not dimensionless and it requires the velocity V_m to fit the data.

A theoretical method has been proposed by Tronconi. He assumed that the number of slugs formed by unit time is inversely proportional to the period of the finite amplitude wave prior to the pipe bridging:

$$n = 0.61\frac{\rho_G}{\rho_L}\frac{U_G}{h_G} \tag{74}$$

where h_G and U_G are the thickness and the velocity of the gas layer prior to slugging. The method requires however to determine the equivalent stratified flow.

References

Andreussi, P., Bendiksen, K. (1989). An investigation of void fraction in liquid slugs for horizontal and inclined gas-liquid pipe flow. *Int. J. Multiphase Flow.* 15, 937–946.

Barnea, D., Shemer, L. (1989). Void fraction measurements in vertical slug flow: applications to slug characteristics and transition. *Int. J. Multiphase Flow* 15, 495–504.

Bendiksen, K. H. (1984). An experimental investigation of the motion of the long bubbles in inclined tubes. *Int. J. Multiphase Flow* 10, 467–83.

Benjamin, T. B. (1968). Gravity currents and related phenomena. *J. Fluid Mech.* 31, 209–48.

Collins, R., de Moraes, F. F., Davidson, J. F., Harrison, D. (1978). The motion of large bubbles rising through liquid flowing in a tube. *J. Fluid Mech.* 89, 497–514.

Delfos, R., Wisse, C.J. Oliemans, R.V.A. (2001) Measurement of air entrainment from a stationary Taylor bubble in a vertical tube. *Int. J. Multiphase Flow.* 27, p. 1769-1787.

Dukler, A.E. and J. Fabre, (1994) Chapter 7: Gas liquid slug flow: knots and loose ends, in *Multiphase science and technology. Two Phase flow fundamentals*, G.F. Hewitt, J.H. Kim, R.T. Lahey, J.M. Delhaye, and N. Zuber, Editors, Begell House: Wallinford, UK. pp. 355-470.

Dumitrescu, D. T. (1943). Strömung an einer Luftblase im senkrechten Rohr, *Z. Angew. Math. Mech.* 23, 139–49.

Fabre, J., Grenier, P., Gadoin, E. (1993). Evolution of slug flow in long pipe, *6th International Conference on Multi-Phase Production*, Cannes, France, June 1993,in Multi Phase Production, Ed. A. Wilson, pp. 165-177, MEP, London.

Fabre, J., Liné, A. (1992). Modelling of two phase slug flow. *Annu. Rev. Fluid Mech.* 24, 21-46.

Fagundes Netto, J.R., J. Fabre, and L. Péresson (1999) Shape of long bubbles in horizontal slug flow. *Int. J. Multiphase Flow.* 25(6-7): p. 1129-1160.

Fagundes Netto, J.R., J. Fabre, and L. Péresson (submitted), Behaviour of long bubbles in horizontal tubes: transient motion and overtaking mechanism. *Int. J. Multiphase Flow.*

Ferschneider, G. (1982). Ecoulements gaz-liquide à poches et à bouchons en conduite. *Rev. Inst. Fr. Pét.* 38, 153–82.

Fréchou, D. (1986). *Etude de l'écoulement ascendant à trois fluides en conduite verticale.* Thèse, Inst. Natl. Polytech. de Toulouse, France.

Gregory, G. A., Scott, D. S. (1969). Correlation of liquid slug velocity and frequency in horizontal cocurrent gas-liquid slug flow. *AIChE J.* 15, 833–35.

Griffith, P., Wallis, G. B. (1961). Two-phase slug flow. *J. Heat Transfer.* 83, 307–20.

Harmathy, T. Z. (1960). Velocity of large drops and bubbles in media of infinite or restricted extent, *AIChE J.* 6, 281–88.

Hinze, J. O. (1955) Fundamentals of the hydrodynamic mechanism of splitting in dispersion processes. *AIChE J.* 1, 289–95.

Hubbard, M.G. (1965). *An analysis of horizontal gas-liquid slug flow*, Ph.D. Thesis, University of Houston.

Kowe R., Hunt J.C.R., Hunt A., Couet B., Bradbury L.J.S. (1988) *Int. J. Mult. Flow.* 14, 587–606.

Linga, H. (1991). Flow pattern evolution; some experimental results obtained at the SINTEF Multiphase Flow Laboratory, *5th International Conference on Multi Phase Production*, Cannes, France, June 1991, in Multi Phase Production, Ed. A.P. Burns, pp. 51-67, Elsevier.

Mao Z., Dukler, A. E. (1989). An experimental study of gas-liquid slug flow. *Exp. Fluids.* 8, 169–82.

Mao Z., Dukler, A. E. (1991). The motion of Taylor bubbles in vertical tubes. II. Experimental data and simulations for laminar and turbulent flow. *Chem. Eng. Sci.* 46, 2055-64.

Martin, C. S. (1976). Vertically downward two-phase slug flow. *J. Fluids Eng.* 98, 715–22.

Nicklin, D. J., Wilkes, J. O., Davidson, J. F. (1962). Two phase flow in vertical tubes. *Trans. Inst. Chem. Engs.* 40, 61–68.

Spedding, P. L., Nguyen, V. T. (1978). Bubble rise and liquid content in horizontal and inclined tubes. *Chem. Eng. Sci.* 33, 987–94.

Taitel, Y., Barnea, D. (1990). Two-phase slug flow. *Adv. Heat Transfer.* 20, 83–132.

Tronconi, E. (1990). Prediction of slug frequency in horizontal two-phase slug flow. *AIChE Journal* 36, 701–709.

Wallis, G. B. (1969). *One-Dimensional Two-Phase Flow.* New-York, McGraw-Hill.
Zukoski, E. E. (1966). Influence of viscosity, surface tension and inclination angle on motion of long bubbles in closed tubes. *J. Fluid Mech.* 25, 821–37.

One Dimensional Models for Pressure Drop, Empirical Equations for Void Fraction and Frictional Pressure Drop and Pressure Drop and other Effects in Fittings

Barry Azzopardi and John Hills

School of Chemical, Environmental and Mining Engineering,, University of Nottingham, U.K.

Abstract. This chapter considers overall methods for pressure drop in gas liquid two-phase flow in pipes. The various elements of pressure drop are identified. Empirical methods are presented. The performance of these equations against available data are reported. Pressure drop in two-phase flow on the shell side of tube bundles is then considered. Geometries other than pipes are then considered. Firstly, pressure drop across fittings such and expansions and contractions are examined. Secondly, other aspects of two-phase flow at fittings are reviewed. Finally, the division of phase split at T-junctions is studied.

1 One Dimesional Models

Though some progress might be made in analysing single-phase flow through integration of the velocity profiles etc., the complexities of two-phase flow are too large for this approach to be used. Instead we can progress using a simple, time-averaged one-dimensional model. This chapter considers this approach and describes the empirical equations employed to complete the method. The accuracy of this has been tested against large banks of data. The results of this are presented.

2 Separated Flow Concept

To describe a two-phase flow, especially when mass is transferred between the phases as in boiling and condensation, it is usual to specify the total mass flow rate, $\dot{M}$, and the fraction of that flow which is travelling as vapour, which is called the *quality*, x_g. The individual mass flow rates of the liquid and gas, $\dot{M}_l$ and $\dot{M}_g$, are related to $\dot{M}$ by:

$$\dot{M} = \dot{M}_g + \dot{M}_l \tag{1}$$

whilst the quality

$$x_g = \frac{\dot{M}_g}{\dot{M}_g + \dot{M}_l} \tag{2}$$

Flow conditions are often best represented by the mass flux, $\dot{m}$:

$$\dot{m} = \frac{\dot{M}}{S} = \frac{\dot{M}_g + \dot{M}_l}{S} \tag{3}$$

or, for each individual phase

$$\dot{m}_g = \frac{\dot{M}_g}{S}; \quad \dot{m}_l = \frac{\dot{M}_l}{S} \tag{4}$$

Here S is the cross-sectional area of the channel. One can also use volume fluxes; for an individual phase this is called the superficial velocity, u_{is}, and equals the mean velocity, which the phase would travel at if it occupied the whole of the channel cross section.

$$u_{gs} = \frac{\dot{m}\, x_g}{\rho_g}; \quad u_{ls} = \frac{\dot{m}(1 - x_g)}{\rho_l} \tag{5}$$

A complete description of a two-phase flow would involve specifying the phase present, and its velocity, at each point in the flow and at all times. This, or a close approximation to it, is the approach used in Computational Fluid Dynamics (CFD), which is still at an early stage of development for two-phase flows. Drastic simplifications are needed for an analytical approach, and one such, often employed, assumes that the phases each occupy part of the cross section, and that each phase flows at its own velocity, as shown in Figure 1.

The fraction of cross-sectional area occupied by the gas is called the void fraction, ε_g, and the phase velocities, u_l and u_g are related to the superficial velocities u_{ls} and u_{gs} by:

$$u_l = \frac{u_{ls}}{1 - \varepsilon_g} \quad \text{and} \quad u_g = \frac{u_{gs}}{\varepsilon_g} \tag{6}$$

These two velocities can be used to calculate the mass flow rates. Combining the individual phase equations gives

$$\frac{(1 - x_g)\dot{M}}{x_g \dot{M}} = \frac{\rho_l u_l (1 - \varepsilon_g) S}{\rho_g u_g \varepsilon_g S} \tag{7}$$

Rearranging yields

$$\varepsilon_g = \frac{1}{1 + U_R \dfrac{(1 - x_g)}{x_g} \dfrac{\rho_g}{\rho_l}} \tag{8}$$

where U_R ($= u_g/u_l$) is known as the slip velocity ratio or slip ratio.

The case where $u_l = u_g$, so that $U_R = 1$, is known as homogeneous flow. In this case the equation for void fraction reduces to

$$\varepsilon_{gH} = \frac{1}{1 + \dfrac{(1 - x_g)}{x_g} \dfrac{\rho_g}{\rho_l}} \tag{9}$$

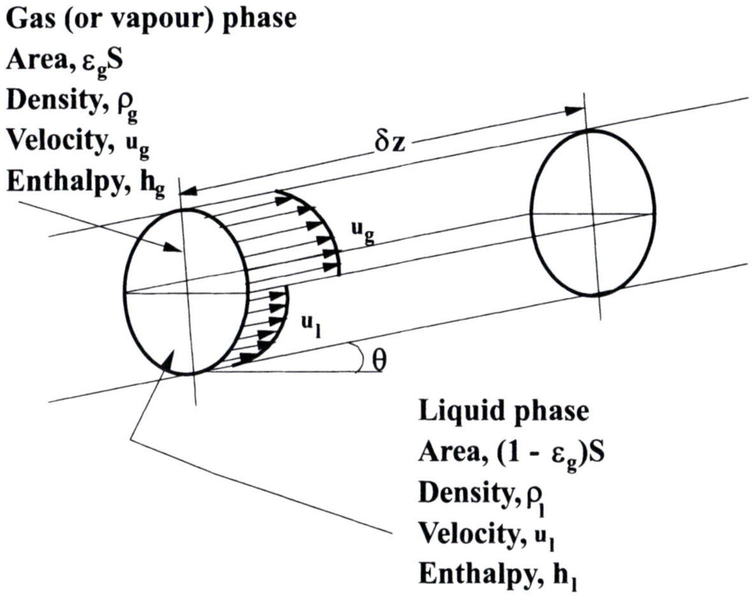

Figure 1. Element of duct with separated flow.

3 Momentum Equation

Two basic approaches can be used to relate the average pressure gradient in a two-phase flow to the parameters describing flow discussed above. The one adopted in this lecture uses mass and momentum balances, the other mass and energy balances. Although both approaches would result in two equally valid expressions from which pressure drop could be determined, a review of published material indicates that most workers have used the momentum approach. One possible reason for the lack of popularity of the energy approach could be the fact that one of the terms (the irreversible loss term) does not necessarily reduce to zero as the flow rate reduces to zero because of the possibility of energy losses in the interaction between the phases.

3.1 Basic Equations

For the small element of channel illustrated in Figure 1, a force balance can be written. This is

$$\left\{\begin{array}{l}\text{the net applied force in}\\\text{the direction of flow}\\\text{over the element}\end{array}\right\} = \left\{\begin{array}{l}\text{rate of increase of}\\\text{momentum in the direction}\\\text{of flow over the element}\end{array}\right\} \qquad (10)$$

The effects of static pressure, wall shear stress and gravity provide the forces. In integral form, the net applied force in the direction of flow can be written as

$$net\ force = \int_S p\, dS - \int_S (p + \frac{dp}{dz}\delta z)\, dS - \int_S \rho\, g \sin\theta\, \delta z\, dS - \int_P \tau\, \delta z\, dP \tag{11}$$

In order to evaluate the integrals for a one-dimensional representation of the flow, the following assumptions are made:

(a) the static pressure is constant over any cross section;
(b) the fluid density is ρ_g over the fraction ε_g of the cross section area and ρ_l over the remainder;
(c) the wall shear stress can be represented by a uniform value τ

The resulting one-dimensional equation is therefore

$$net\ force = -\frac{dp}{dz}\,\delta z\ S - [\varepsilon_g \rho_g + (1-\varepsilon_g)\rho_l]\, g \sin\beta\, \delta z\, S - \bar{\tau}\, \delta z\, P \tag{12}$$

The rate of increase of momentum in the direction of flow over the element is evaluated from the difference between the rates of momentum transport out of and into the element. The rate of momentum transport, γ, across any cross section is given by

$$\gamma = \int_S u^2 \rho\, dS \tag{13}$$

The rate of increase of momentum transport across the element is therefore given by

$$[\gamma_i + \frac{d\gamma}{dz}\delta z] - \gamma_i = \left(\frac{d}{dz}\int_S u^2 \rho\, dS\right)\delta z \tag{14}$$

Which, using the above assumptions, can be written as

$$\frac{d\gamma}{dz}\,\delta z = \frac{d}{dz}[(\bar{u}_g^2 \rho_g \varepsilon_g + \bar{u}_l^2 \rho_l (1-\varepsilon_g))\, S]\,\delta z = \frac{d}{dz}\left[\dot{m}^2\left(\frac{x_g^2}{\varepsilon_g \rho_g} + \frac{(1-x_g)^2}{(1-\varepsilon_g)\rho_l}\right) S\right]\delta z \tag{15}$$

Substituting Equations (12) and (15) into Equation (10), dividing through by S δz and simplifying yields

$$-\frac{dp}{dz} = \bar{\tau}\frac{P}{S} + [\varepsilon_g \rho_g + (1-\varepsilon_g)\rho_l]\, g \sin\beta + \frac{d}{dz}\left[\dot{m}^2\left(\frac{x_g^2}{\varepsilon_g \rho_g} + \frac{(1-x_g)^2}{(1-\varepsilon_g)\rho_l}\right)\right] \tag{16}$$

Equation (16) can be considered to be the sum of three components

$$-\frac{dp}{dz} = -\left(\frac{dp}{dz}\right)_{friction} - \left(\frac{dp}{dz}\right)_{gravitation} - \left(\frac{dp}{dz}\right)_{acceleraton} \tag{17}$$

where

$$-\left(\frac{dp}{dz}\right)_{friction} = \bar{\tau}\frac{P}{S} \tag{18}$$

$$-\left(\frac{dp}{dz}\right)_{gravitation} = [\varepsilon_g \rho_g + (1 - \varepsilon_g)\rho_l]\, g \sin\beta \tag{19}$$

$$-\left(\frac{dp}{dz}\right)_{acceleration} = \frac{d}{dz}\left[\dot{m}^2\left(\frac{x_g^2}{\varepsilon_g \rho_g} + \frac{(1 - x_g)^2}{(1 - \varepsilon_g)\rho_l}\right)\right] \tag{20}$$

Correlation and prediction of two-phase pressure drops then becomes a matter of treating each of the three terms, Equations (18) - (20), separately.

3.2 Frictional Component.

For single-phase flow, Equation (18) is normally transformed into an expression involving a friction factor and the dynamic head, i.e.,

$$\tau\frac{-P}{S} = \frac{4\,f}{D_t}\frac{\rho \bar{u}^2}{2} = \frac{4\,f}{D_t}\frac{\dot{m}^2}{2\,\rho} \tag{21}$$

A similar expression can be written for two-phase flow

$$\tau\frac{-P}{S} = \frac{4\,f_{TP}}{D_t}\frac{\rho_{TP}\,\bar{u}_{TP}^2}{2} = \frac{4\,f_{TP}}{D_t}\frac{\dot{m}_{TP}^2}{2\,\rho_{TP}} \tag{22}$$

which may also be written as the product of a single-phase pressure gradient and a multiplying factor, traditionally written ϕ^2 and called the *two-phase multiplier.*

$$\tau\frac{-P}{S} = \frac{4\,f_{SP}}{D_t}\frac{\dot{m}_{SP}^2}{2\,\rho_{SP}}\,\phi^2 \tag{23}$$

where

$$\phi^2 = \left[\frac{f_{TP}}{f_{SP}}\frac{\dot{m}_{TP}^2}{\dot{m}_{SP}^2}\frac{\rho_{TP}}{\rho_{SP}}\right] \tag{24}$$

Correlations for two-phase pressure drop have been developed through this multiplier.

There are in fact four possible single phase pressure gradients, depending on what single phase flow is considered as corresponding to the given two-phase flow, and each results in a corresponding multiplier. In using any published correlation, it is essential to establish which has been used. Table 1 gives the four possibilities.

For boilers, "all flow as liquid" is the natural choice, and will be used in this and the following chapters. However, there are a number of important correlations which are based on the "liquid fraction only". The frictional component of pressure drop will always result in a loss of pressure.

Table 1. Definition of different two-phase multipliers

Title	Mass flux	Density	Reynolds number to obtain friction factor	Subscript
All flow as liquid	$(\dot{m}_g + \dot{m}_l)$	ρ_l	$(\dot{m}_g + \dot{m}_l)D_t / \eta_l$	lo
Liquid fraction only	$\dot{m}_l$	ρ_l	$\dot{m}_l D_t / \eta_l$	l
Gas fraction only	$\dot{m}_g$	ρ_g	$\dot{m}_g D_t / \eta_g$	g
All flow as gas	$(\dot{m}_g + \dot{m}_l)$	ρ_g	$(\dot{m}_g + \dot{m}_l)D_t / \eta_g$	go

3.3 Gravitational Component

As can be seen from Equation (19), this depends on the mean void fraction ε_g, and it is calculated from correlations or measurements of void fraction, as will be shown in the subsequent sections. This component of pressure loss gives a loss of pressure for up flow and a gain or recovery of pressure for down flow. An exception occurs for fully annular (all liquid as a film) or stratified down flow where there is minimal recovery of pressure, i.e., the gravitational component can be ignored. This can be best illustrated by considering falling film flow, down flow of liquid along the walls without any gas flow. If there were any gain in pressure because of the increased head of liquid, the pressure in the gas would also rise. However, with a continuous gas path the pressure cannot rise. This apparent anomaly can be explained by the possible pressure increase being balanced by the shear of the liquid on the wall. For a two-phase flow, there will be an overall pressure loss - that of gas friction.

3.4 Accelerational Component

Equation (20), integrated along the flow path from z_1 to z_2 gives:

$$- \Delta p_{accelerational} = \left[\dot{m}^2 \left(\frac{x_g^2}{\varepsilon'_g \rho_g} + \frac{(1 - x_g)^2}{(1 - \varepsilon'_g) \rho_l} \right) \right]_{z_1}^{z_2} \quad (25)$$

This equation represents the difference in momentum flux into and out of the section of duct.

Little work has been done on the prediction of momentum flux for moderate velocity flows because the accelerational component of pressure drop is small even if evaporation or condensation are occurring. Momentum flux can be measured by allowing the flow to impact on a device, which alters its flow direction by 90°. The impact force on the device is then measured. Rose (1964), Andeen and Griffiths (1968) and Wiafe (1970) performed such measurements.

These experiments yield a momentum flux larger than that predicted by Equation (25) if the true void fraction ε_g is used. This discrepancy arises because velocities and void fractions have been averaged across the duct; to avoid confusion the void fraction in the accelerational term will be written ε'_g. In fact, the momentum flux is generally more closely predicted by evaluating Equation (25) using the homogeneous void fraction calculated from Equation (9). This agreement must be fortuitous, since the homogeneous model does not even represent a theoretical maximum for momentum flux since velocity profiles or fluctuating components of velocity will lead to higher values than for uniform steady flow.

Assuming that in any engineering calculation, the purpose is to predict the maximum likely pressure drop, the following recommendations can be made:

(a) For adiabatic and evaporating flows where changes in momentum flux will increase the pressure gradient, the pressure drop can be calculated using the homogeneous model (see section 3.1) as:

$$- \Delta p_{accelerational} = \left[\frac{\dot{m}^2}{\rho_H} \right]_{z1}^{z2} = \left[\dot{m}^2 \left(\frac{x_g}{\rho_g} + \frac{(1 - x_g)}{\rho_l} \right) \right]_{z1}^{z2} \tag{26}$$

(b) For condensing flows where changes in momentum flux will lead to a pressure recovery, the pressure drop can be calculated using Equation (25) with $\varepsilon'_g = \varepsilon_g$.

Designers wishing to obtain the best prediction of pressure drop may, however, wish to consider the approach suggested by Morris (1984) which predicts momentum fluxes between the two extremes and reasonable agreement with much of the experimental data:

$$- \Delta p_{accelerational} = \left[\frac{\dot{m}^2}{\bar{\rho}} \right]_{z1}^{z2} \tag{27}$$

where

$$\frac{1}{\bar{\rho}} = \left(\frac{x_g}{\rho_g} + U_R \frac{(1 - x_g)}{\rho_l} \right) \left(x_g + \frac{(1 - x_g)}{U_R} (1 + (U_R - 1)^2 \beta) \right) \tag{28}$$

$$U_R = \left[(1 - x_g) + x_g \frac{\rho_l}{\rho_g} \right]^{1/2} \tag{29}$$

$$\beta = \left[\left(\frac{\rho_l}{\rho_g} \right)^{1/2} - 1 \right]^{-1} \tag{30}$$

3.5 Combined Equation.

Substituting from above into the integrated version of Equation (17) using the "all flow as liquid" definition for frictional pressure gradient gives

$$- \Delta p = \int_{z_1}^{z_2} \frac{4 f_{lo}}{D_t} \frac{\dot{m}^2}{2 \rho_l} \phi_{lo}^2 \, dz \; + = \int_{z_1}^{z_2} [\varepsilon_g \rho_g + (1 - \varepsilon_g) \rho_l] \, g \sin \beta \, dz \; + \left[\dot{m}^2 \left(\frac{x_g^2}{\varepsilon'_g \rho_g} + \frac{(1 - x_g)^2}{(1 - \varepsilon'_g) \rho_l} \right) \right]_{z1}^{z2} \tag{31}$$

In principle, Equation (31) can be used for any flow in any duct, pipeline or heat exchanger with or without heat transfer. In practice, special versions are used for calculating pressure changes across pipe fittings or devices with sudden changes of flow area and for high speed or critical flows.

Equation (31) is therefore mostly used for calculating the pressure changes along straight lengths of pipe. Assuming that the duct has constant cross sectional area and hence constant mass velocity, $\dot{m}$, the equation may be rewritten as

$$-\Delta p = \frac{2\,\dot{m}^2}{D_t}\frac{f_{lo}}{\rho_l}\int_{z_1}^{z_2}\phi_{lo}^2\,dz + g\sin\beta\int_{z_1}^{z_2}[\varepsilon_g\rho_g + (1-\varepsilon_g)\rho_l]\,dz + \dot{m}^2\left[\left(\frac{x_g^2}{\varepsilon'_g\rho_g} + \frac{(1-x_g)^2}{(1-\varepsilon'_g)\rho_l}\right)\right]_{z_1}^{z_2} \tag{32}$$

As explained above, the best predictions of accelerational pressure drop are not obtained using the true void fraction ε_g, and hence ε_g' is used in the acceleration term in Equation (32). There are thus 3 unknowns to be calculated, namely ϕ^2_{lo}, ε_g and ε'_g. The next section discusses the methods, which have been suggested to specify these parameters.

A further development would be to allow that the gas density varies according to a gas law. The pressure gradient equation can then be written as

$$-\frac{dp}{dz} = \frac{g\ \sin(\theta)\ [\rho_g\,\varepsilon_g + \rho_l\,(1-\varepsilon_g)] + \dfrac{f_{TP}\ \dot{m}\ (u_{gs}+u_{ls})}{2\ D_t}}{1-[\rho_g\,\varepsilon_g + \rho_l\,(1-\varepsilon_g)]\dfrac{(u_{gs}+u_{ls})\ u_{gs}}{p}} \tag{33}$$

One of the major requirements of a two phase flow model is the prediction of pressure drop, since this is involved in the calculation of flow rates under a given head loss, and of pumping requirements for a given flow rate. In the section above the one-dimensional model was developed in which all variables - pressure, mass and volume flow rates of each phase, void fraction, etc - are averaged across the flow cross-section, so that all have unique values at any longitudinal position along the pipe.

Use of a momentum balance or using the energy equation broke down the local static pressure gradient into three components, either, and because the vast majority of published work has used the momentum approach, this was adopted. In this approach, the breakdown is:

$$\text{Pressure drop} = \text{wall frictional term} + \text{gravitational term} + \text{accelerational term} \tag{34}$$

The acceleration term has been covered earlier: this lecture concentrates on the gravitational and wall friction terms, and looks at empirical ways of predicting the unknown parameters. The gravitational (or static head) term is given by:

$$\left(\frac{dp}{dz}\right)_{\text{grav}} = [\varepsilon_g\,\rho_g + (1-\varepsilon_g)\,\rho_l\,]\ g\ \sin\beta \tag{35}$$

which contains the unknown parameter ε_g, the void fraction or gas holdup, and the prediction of this parameter will be discussed below.

The wall friction term was developed in the previous chapter by analogy with single-phase flow as:

$$\left(\frac{dp}{dz}\right)_f = \frac{\bar{\tau} P}{S} = \frac{4 f_{SP}}{D_t} \frac{\dot{m}_{SP}^2}{2 \rho_{SP}} \phi^2 \tag{36}$$

which contains the unknown parameter ϕ^2, called the two-phase multiplier. (f_{SP}, the single-phase friction factor, is assumed to be known, as it is calculable from the appropriate phase flow rate and duct geometry by well-known correlations). The prediction of this parameter will be covered in Section 6.

4 Data Base

There have been a very large number of experimental studies of void fraction in two-phase flow in tubes. Hewitt (1978) reviews the available techniques, which include (a) quick-closing valves, (b) capacitance probes, (c) X-ray and γ-ray attenuation techniques and (d) local impedance and optical reflection probes. Method (a) measures a true average over a restricted length of pipe at one instant in time. Method (b) measures a complex spatial average over a short length of pipe. Method (c) measures an average along a chord and method (d) measures "point" values which can be averaged numerically over as long a time as the experimenter chooses. For these, and other reasons, the various techniques do not agree among themselves, even if used in the same flow; this inherent scatter in the data bank has repercussions on the attempt to fit empirical correlations to it.

The published literature contains over 9000 experimental measurements of void fraction in adiabatic two-phase flow, and while the majority are for air-water or steam-water, other systems have been studied, such as air-oil (higher viscosity), nitrogen-mercury, liquid oxygen-gaseous oxygen, liquid potassium-gaseous potassium, and argon-ethanol. The data cover horizontal, vertical and inclined flows.

Experimentally, it is not possible to measure frictional pressure drop directly, so this term in the total pressure drop equation is obtained by subtracting the measured acceleration and gravity terms from the measured total pressure drop.

Table 2. Parameter ranges of variable in data base

	Void fraction	**Pressure drop**
Mass flux (kg/m^{2}s)	1.1-9087	1.7-24990
Quality (-)	0.0-1.0	0.0-1.0
Tube diameter (m)	0.0074-0.216	0.0074-0.305
Pressure (Pa)	6520-2.3 10^8	8800-1.976 10^7
Void fraction (-)	0.006-0.999	
Pressure gradient (N/m^3)		1.96 10^5-3.76 10^7
Liquid density (kg/m^3)	467-13540	498.3-13550
Gas density (kg/m^3)	0.07-268	0.103-164
Liquid viscosity (Ns/m^2)	5.4 10^{-5}-0.097	5.85 10^{-5}-0.128
Gas viscosity (Ns/m^2)	7.4 10^{-6}-3.0 10^{-5}	1.09 10^{-5}-1.8 10^{-4}
Surface tension (N/m)	0.0007-0.47	0.00132-0.456

Because of the fluctuating nature of two-phase flow, even an apparently simple measurement like that of static pressure difference between two tapping points poses significant problems associated with the definition and measurement of a mean value. To this must be added the considerable uncertainty in the void fraction, and the large scatter found in the published frictional pressure drop measurements is not surprising.

A number of extensive databases of void fraction and pressure drop have been created. That assembled by HTFS at Harwell consists of 9152 void fraction data points and 19000-pressure drop data points. These banks cover horizontal, vertical and inclined flows and cover the parameter ranges in Table 2:

Though they cover wide ranges of parameters, the data banks do not have points for every combination. For example, for the small dimensions and mass fluxes typical of compact heat exchangers, data have only recently been provided by studies such as those of Bao *et al.* (1994), Holt *et al.* (1999) and Mishima and Hibiki (1996). An idea of where data is available is given in Figure 2.

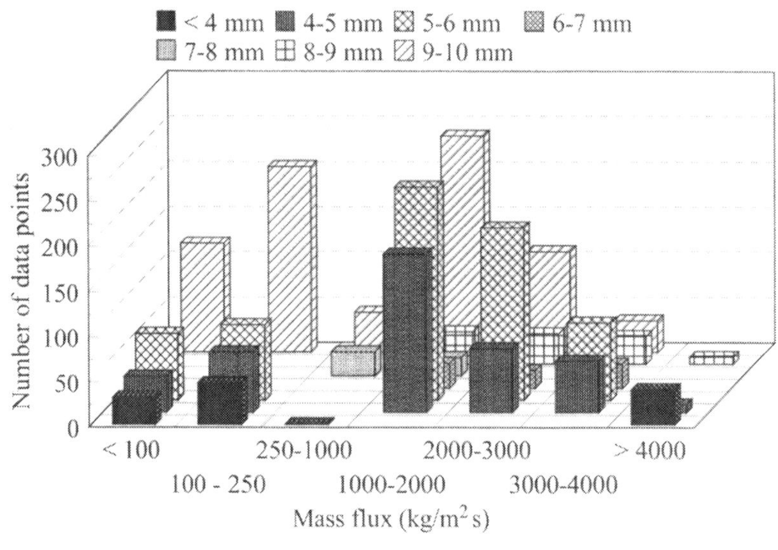

Figure 2. Conditions relevant to compact exchangers where two-phase pressure drop data are available

5 Void Fraction Equations

Based on the void fraction data-bank, or sub-sets of it, a very large number of correlations have been suggested. While many of them give excellent fits to restricted data sets (often of the authors' own gathering!). A correlation that is going to be of use to the practising engineer should have wider application, and, ideally, should still give reasonable answers if extrapolated to extreme conditions. Such asymptotically correct behaviour should include:

$$\begin{array}{llll} \text{a)} & \varepsilon_g \to 0 \quad \text{as} \quad x_g \to 0 & \text{(single phase liquid)} \\ \text{b)} & \varepsilon_g \to 1 \quad \text{as} \quad x_g \to 1 & \text{(single phase gas)} \\ \text{c)} & \varepsilon_g \to \varepsilon_{gH} \quad \text{as} \quad p \to p_{crit} & \text{(gas \& liquid identical)} \end{array} \qquad (37)$$

where ε_{gH} is the voidage as given by the homogeneous model described earlier and p_{crit} is the critical pressure of the system.

Apart from some direct, totally empirical correlations described below, the majority of published correlations are based on developments of the homogeneous flow model. Because of its simplicity the relevant equations can often be manipulated to particularly useful forms. This one-dimensional model assumes that gas and liquid have equal velocity, and since the mean gas and liquid velocities are related to the known superficial velocities:

$$\bar{u}_g = \frac{u_{gs}}{\varepsilon_g} \quad \text{and} \quad \bar{u}_l = \frac{u_{ls}}{1 - \varepsilon_g} \qquad (38)$$

Equating $\bar{u}_g$ and $\bar{u}_l$ leads to the result:

$$\varepsilon_{gH} = \frac{1}{1 + \dfrac{u_{ls}}{u_{gs}}} = \frac{1}{1 + \dfrac{(1 - x_g)}{x_g} \dfrac{\rho_g}{\rho_l}} \qquad (39)$$

This homogeneous model needs correction on two counts: first, because of non uniform velocity and voidage profiles which affect the averaging process implied in a one-dimensional model, and secondly, because of "slip" - the fact that, in general, the mean gas and liquid velocities are not equal. These corrections can be applied in various ways: either by applying an empirical multiplier to Equation (39), or by allowing for the slip by including the a factor called the "slip ratio" in the denominator:

$$U_R = \frac{\bar{u}_g}{\bar{u}_l} \qquad (40)$$

or by attempting to correct for the two effects separately in the so-called "drift flux" model. These three approaches will be discussed below.

5.1 Empirical Multiplier on the Homogeneous Equation

Here, the void fraction is the homogeneous value, Equation (39), multiplied by a correction factor A. Expressions for A are given by Armand (1946), Massena (1960), Bankoff (1960), Armand and Treschev (1947), Hughmark (1962) and Rooney (1962).

5.2 Correlations for Slip Ratio

In this case the void fraction is given by

$$\varepsilon_g = \frac{1}{\left(1 + U_R \dfrac{(1 - x_g)}{x_g} \dfrac{\rho_g}{\rho_l}\right)} \qquad (41)$$

and all correlations are through U_R.

Chisholm (1972) published a simple algebraic correlation:

$$U_R = \left[1 - x_g \left(1 - \frac{\rho_l}{\rho_g} \right) \right]^{0.5} \tag{42}$$

Smith (1971) proposed a simple model for annular two-phase flow with a fraction e of the liquid entrained in the gas core. He postulated equal momentum flux in film and core, which led to slip ratio given by:

$$U_R = e + (1 - e) \left(\frac{\dfrac{\rho_l}{\rho_g} + \dfrac{e(1 - x_g)}{x_g}}{1 + \dfrac{e(1 - x_g)}{x_g}} \right)^{\frac{1}{2}} \tag{43}$$

Though more complicated, this equation is not, in general, any more accurate than the simpler Equation (42).

Zivi (1964), Levy (1960), Thom (1964), Marchaterre and Hoglaund (1962) have published other equations or graphical methods.

Three much more general, but much more complicated correlations have been suggested. Each is based on analysis of a large data bank. The first was produced by Premoli *et al.* (1970) at CISE, and it is as follows:

$$U_R = 1 + E_1 \left(\frac{j}{1 + E_2 j} - jE_2 \right)^{0.5} \quad for \frac{1}{1 + jE_2} > E_2 \tag{44}$$

$$U_R = 1 \qquad \text{otherwise}$$

Where

$$j = \frac{\varepsilon_{gH}}{1 - \varepsilon_{gH}} \tag{45}$$

$$E_l = 1.578 \;\; Re^{-0.19} \left(\frac{\rho_l}{\rho_g} \right)^{0.22} \tag{46}$$

$$E_2 = 0.0273 \; We \; Re^{-0.51} \left(\frac{\rho_l}{\rho_g} \right)^{-0.08} \tag{47}$$

and

$$Re = \frac{(\dot{m}_l + \dot{m}_g) D_t}{\eta_l} ; We = \frac{(\dot{m}_l + \dot{m}_g)^2 D_t}{\sigma \; \rho_l} \tag{48}$$

HTFS at Harwell have produced another correlation of UR in terms of three empirically chosen parameters:

$$U_R = f\left((\dot{m}_g + \dot{m}_l), \frac{\varepsilon_{gH}}{1 - \varepsilon_{gH}}, \frac{\rho_g}{\rho_l}\left(\frac{\eta_l}{\eta_g}\right)^{0.2} \right) \tag{49}$$

details are proprietary and only available to subscribers.

5.3 Drift Flux Correlations

The drift flux model was introduced by Wallis (1969), and is particularly applicable to flows where the liquid phase is continuous (bubble, slug/plug and churn). It introduces a parameter called the drift velocity, which is the velocity of the gas relative to the gas-liquid mixture (cf slip velocity which is gas relative to liquid). Zuber and Findlay (1965) developed a series of correlations based on this idea. The mean gas velocity is given by (u_{gs} / ε_g) and they equated this to the mixture velocity plus the drift velocity, v_{gd}. They allowed for non-uniformities in the distribution of velocity and voidage by means of a constant C_0 multiplying the mixture velocity:

$$\frac{u_{gs}}{\varepsilon_g} = C_0 \, (u_{gs} + u_{ls}) + v_{gd} \tag{50}$$

This can be rearranged as an equation in ε_g:

$$\varepsilon_g = \frac{u_{gs}}{C_0 \, (u_{gs} + u_{ls}) + v_{gd}} \tag{51}$$

where v_{gd} usually has the form

$$v_{gd} = K \left(\frac{\sigma \, g \, (\rho_l - \rho_g)}{\rho_l^2} \right)^{0.25} \tag{52}$$

Zuber and Findlay (1965) suggested a value of 1.13 for C_0 and 1.4 for K. Harmathy (1960) proposed K = 1.53. For a bubble rising in a swarm of bubbles, Zuber and Hench (1962) suggested that the correct value would be $K = 1.53\,(1 - \varepsilon_g)$. Recent work related to pressure relief systems, Fruendt et al. (1997), indicates that for liquids with viscosities much higher than water, a much lower value of K might be more appropriate.

One possible advantage of the drift flux model is its ability to provide a means of correlating the effect of flow directions, with v_{gd} being positive for upflow, negative for downflow and zero for horizontal flow. Clark and Flemmer (1985), Bhaga and Weber (1972) and Oshinowa and Charles (1974) have all examined this possibility, but it seems that C_0 is also dependent on flow direction and tube diameter (Clark and Flemmer, 1985), and no simple relationship could be derived. Despite this drawback, the drift flux approach is extensively used to correlate data in liquid phase continuous regimes.

The relation between the drift flux model, Equation (51), and the modified homogeneous flow model, Equation (39) multiplied by A, can be seen by dividing numerator and denominator of Equation (51) by $C_0 u_{gs}$, which gives:

$$\varepsilon_g = \frac{1/C_0}{\left(1+\frac{u_{ls}}{u_{gs}}\right)+\frac{v_{gd}}{C_0\, u_{gs}}} = \frac{1/C_0}{\left(1+\frac{(1-x_g)}{x_g}\frac{\rho_g}{\rho_l}\right)+\frac{v_{gd}}{C_0\, u_{gs}}} \tag{53}$$

Thus the factor A becomes $1/C_0$, which has a value of 0.885 from the work of Zuber and Findlay (1965); this can be compared to the 0.833 suggested by Armand (1946). The fact that Equation (53) does not have the correct asymptotic behaviour as $x_g \to 1$ serves to underline the fact that the drift flux model is only suited to liquid-phase continuous flows, for which x_g does not reach high values.

5.4 Direct Correlations

Lockhart and Martinelli (1949) presented data for low-pressure air-water flow which they were able to present as a single curve by plotting against a parameter X, given by

$$X = \sqrt{\frac{\left(\frac{dp}{dz}\right)_l}{\left(\frac{dp}{dz}\right)_g}} \tag{54}$$

Although many people have taken their work as a universal prediction of void fraction, this was not their intention, and the paper is more important for its introduction of the parameter X, used in frictional pressure drop correlations, as shown in Section 6.

Other correlations were given by Martinelli and Nelson (1948) (for steam-water), Baroczy (1963) and Beggs and Brill (1973). This last correlation is of particular significance as it is very popular in the oil industry having been derived from a database involving tests from inclined pipes. The Beggs and Brill (1973) correlation for void fraction (or more correctly for hold up = 1- void fraction) starts from the homogeneous void fraction, Equation (39) and a Froude number defined by

$$Fr = \frac{(u_{gs}+u_{ls})^2}{g\, D_t} \tag{55}$$

These two parameter are combined to give the equation for void fraction which has the general form

$$\varepsilon_l(0) = \frac{C_{B1}\,(1-\varepsilon_{gH})^{C_{B2}}}{Fr^{C_{B3}}}$$
$$\varepsilon_g = 1 - \frac{C_{B1}\,(1-\varepsilon_{gH})^{C_{B2}}}{Fr^{C_{B3}}}\left[1 - C_{B4}\sin(1.8\,\beta) - 0.333\sin^3(1.8\,\beta)\right] \tag{56}$$

where the constants C_{B1} to C_{B4} take the values given in Table 3 and $\varepsilon_l(0) \geq (1-\varepsilon_{gH})$; $0 > \varepsilon_g > 1$.

Table 3. Constants for Beggs and Brills equation for void fraction

Constant	Fr<L$_1$	Fr>L$_1$ and >L$_2$	L$_1$<Fr<L$_2$
C_{B1}	0.98	1.065	0.845
C_{B2}	0.4846	0.5824	0.5351
C_{B3}	0.0868	0.0609	0.0172
C_{B4} for up flow	eqn(57)	0	eqn(59)
C_{B4} for down flow	eqn (58)	eqn(58)	eqn(58)

with

$$C_{B4} = \varepsilon_{gH} \ln\left[\frac{0.011\ N_{lV}^{3.539}}{(1 - \varepsilon_{gH})^{3.768}\ Fr^{1.614}}\right] \tag{57}$$

$$C_{B4} = \varepsilon_{gH} \ln\left[\frac{4.7\ N_{lV}^{0.1244}}{(1 - \varepsilon_{gH})^{0.3692}\ Fr^{0.5056}}\right] \tag{58}$$

$$C_{B4} = \varepsilon_{gH} \ln\left[\frac{2.96\ (1 - \varepsilon_{gH})^{0.305}\ N_{lV}^{0.0978}}{Fr^{0.4472}}\right] \tag{59}$$

$$N_{lV} = u_{ls}\left(\frac{\rho_l}{g\ \sigma}\right)^{0.25} \tag{60}$$

and $L_1 = \exp(-4.62 - 3.757X - 0.481X^2 - 0.0207X^3)$, $L_2 = \exp(1.060 - 4.602X - 1.609X^2 - 0.179X^3 + 0.000635X^5)$, where $X = ln\ (1 - \varepsilon_{gH})$. The three columns of the Table were identified by Beggs and Brill as segregated (stratified, wavy and annular), distributed (bubbly and mist) and intermittent (plug and slug). However, these boundaries are empirical.

1.1 Test of Equations

The equations for void fraction are required for the estimation of the gravitational pressure drop component as given by Equation (35):

$$\left(\frac{dp}{dz}\right)_{\mathrm{grav}} = [\varepsilon_g \rho_g + (1 - \varepsilon_g)\rho_l]\, g \sin\beta \tag{35}$$

Therefore, they are tested in terms of their prediction of the mean density given by:

$$\bar{\rho} = \varepsilon_g \rho_g + (1 - \varepsilon_g)\rho_l \tag{61}$$

which is the form in which ε_g appears in Equation (35).

The database discussed in section 4 has been used to test a number of correlations by dividing it into ranges of mass flux, density ratio, pressure, tube diameter, and a number of other variables. Here we shall only present overall data for a particular pipe inclination (vertical upflow or horizontal), together with an overall result independent of inclination. Results, given in Table 3, give the following information:

N The number of data points in the sample;

F The "correction factor": this is the average number by which the predicted value must be multiplied to give the experimental value. Values of F greater than unity mean that the given correlation under-predicts the mean density.

R The "range factor": this gives the factor by which the corrected result must be multiplied and divided to give the true mean density with 99% probability and 95% confidence.

From Table 4, it can be seen that the CISE correlation is the best of those available in the open literature for vertical upflow and overall, but that the much simpler Chisholm correlation runs it close overall and is actually significantly better in horizontal flow. It may be noted that, with the large sample sizes used here, R = 1.5 corresponds to about 20% standard deviation and R = 2.1 to about 40%. Table 5 shows the accuracy of the homogeneous description for different ranges of mass flux. As can be seen it does particularly badly at low mass fluxes but moderately well at the higher fluxes when the phases might be expected to be well mixed. Similar trends are found for the fit as a function of pressure, where predictions improve as the pressure approaches the critical value.

Table 4. Test of void fraction equations

Method	Horizontal (N = 2312)		Vertical up (N = 3954)		All data (N = 6266)	
	F	R	F	R	F	R
Homogeneous	2.26	9.4	2.44	9.1	2.37	9.0
Zivi	0.87	2.8	0.95	3.5	0.92	3.3
Zuber	1.13	2.4	0.87	3.4	0.96	3.1
Smith	1.15	2.3	1.19	2.5	1.18	2.4
Chisholm	1.15	2.2	1.16	2.3	1.16	2.3
CISE	1.22	2.9	1.02	1.8	1.09	2.2
HTFS	0.98	2.1	1.00	1.7	0.99	1.8
Bryce (steam-water only)	1.03	1.4	1.02	1.5	1.03	1.5
HTFS (steam-water only)	0.97	1.4	0.98	1.6	0.98	1.5

Table 5. Test of Homogeneous void fraction equation

m (kg/m^{2}s)	<100	100-250	250-500	750-1000	2000-3000	>4000
N	558	811	471	317	230	183
F	7.9	4.25	2.64	1.66	1.28	1.12
R	9.4	7.9	7.0	3.8	2.1	1.8

6 Frictional Pressure Drop Equations

The correlations used for frictional pressure drop are normally based on the two-phase multiplier introduced in at the beginning of this lecture, Equation (23):

$$\left(\frac{dp}{dz}\right)_{frict} = \frac{\bar{\tau}\,P}{S} = \frac{4\,f_{SP}}{D_t}\,\frac{\dot{m}_{SP}^2}{2\,\rho_{SP}}\,\phi^2 \tag{23}$$

and, as explained above, workers seem to have settled on the choice of "all flow as liquid" as the single phase flow with which the two-phase data are compared, so that they correlate the multiplier ϕ_{lo}^2. Thus ρ_l is used for ρ_{SP} and the term $\dot{m}_{SP}$ is interpreted as being equal to $\dot{m}_{TP}$, i.e.:

$$\dot{m}_{SP} = \dot{m}_{TP} = \dot{m}_g + \dot{m}_l \tag{62}$$

Almost all the experimental work has been based on smooth pipes, so that the term f_{SP} was set equal to the value it would have in a smooth pipe at the Reynolds number:

$$Re_{SP} = \frac{\dot{m}_{SP}\,D_t}{\eta_l} \tag{63}$$

The problem then arises as to what value of f_{SP} should be used when applying the value of ϕ_{lo}^2 calculated from the chosen correlation to a real, and possibly rough pipe. There is some evidence that pressure drop is less affected by roughness in two-phase flow than it is in single-phase flow, in which case the smooth pipe correlation would be the one to use: however, the use of smooth pipe f_{SP} could cause an under-prediction in very rough tubes, and would also lead to a discontinuity at the point where the flow becomes two-phase in evaporating flows (boiler tubes, etc), and hence it is recommended that the f_{SP} term should be calculated from a single-phase correlation allowing for pipe roughness, such as that of Haaland (1983).

As for the void fraction correlations discussed in Section 5, it is possible to suggest asymptotic forms of any proposed correlation for ϕ_{lo}^2. These include:

$$\begin{aligned}
&\phi_{lo} \to 1 \quad \text{as} \quad x_g \to 0\\
&\phi^2_{lo} \to 1 \quad \text{as} \quad p \to p_{crit}\\
&\phi^2_{lo} \ \text{increases as} \ x_g \ \text{increases}\\
&\phi^2_{lo} \ \text{increases as} \ \rho_l/\rho_g \ \text{increases}\\
&\phi^2_{lo} \ \text{decreases as} \ \dot{m} \ \text{increases}
\end{aligned} \tag{64}$$

In the sections below, we present some of the correlations suggested for ϕ_{lo}^2.

6.1 Homogeneous Model

In the homogeneous model, the two-phase flow behaves as a single fluid, and hence the suffixes $_{SP}$ in Equation (23) can be replaced by $_{TP}$, and ϕ^2 set equal to unity. $\dot{m}_{TP}$ is given by Equation (62), while ρ_{TP} is the homogeneous two-phase density, given by:

$$\frac{1}{\rho_{TPH}} = \frac{x_g}{\rho_g} + \frac{(1 - x_l)}{\rho_l} \tag{65}$$

The friction factor will be replaced by a two-phase homogeneous factor:

$$f_{TPH} = f\,(Re_{TPH}, e/D_t) \tag{66}$$

where the appropriate single-phase function is used and e/D_t is the roughness factor and

$$Re_{TPH} = \frac{\dot{m}_{TP}\ D_t}{\eta_{TPH}} \tag{67}$$

The only remaining unknown is the homogeneous viscosity, η_{TPH}, and five possible ways of calculating it have been suggested.

$$\eta_{TPH} = \eta_l \qquad \text{(Owens, 1961)} \tag{68}$$

$$\frac{1}{\eta_{TPH}} = \frac{x_g}{\eta_g} + \frac{1 - x_g}{\eta_l} \qquad \text{(Isbin \textit{et al.}, 1958)} \tag{69}$$

$$\eta_{TPH} = \eta_g\, x_g + \eta_l\,(1 - x_g) \qquad \text{(Cicchitti \textit{et al.}, 1960)} \tag{70}$$

$$\frac{\eta_{TPH}}{\rho_{TPH}} = \frac{\eta_g}{\rho_g}\, x_G + \frac{\eta_l}{\rho_l}\,(1 - x_g) \qquad \text{(Dukler \textit{et al.}, 1965)} \tag{71}$$

$$\eta_{TPH} = \eta_l\,(1 - \varepsilon_{gH})\,(1 + 2.5\,\varepsilon_{gH}) + \eta_g\,\varepsilon_{gH} \qquad \text{(Beattie and Whalley, 1982)} \tag{72}$$

6.2 Graphical Correlations

Lockhart and Martinelli (1949) suggested that the multipliers ϕ_l^2 and ϕ_g^2 should be correlated with the function X^2, where:

$$\phi_l^2 = \frac{\Delta\, p_{TP}}{\Delta\, p_l}; \quad \phi_g^2 = \frac{\Delta\, p_{TP}}{\Delta\, p_g}; \quad X^2 = \frac{\Delta\, p_l}{\Delta\, p_g} \tag{73}$$

and Δp_l and Δp_g are the pressure drops which would result from the liquid or gas phases flowing alone.

Δp_l and Δp_g are calculated from the Friction factor - Reynolds number correlations for single-phase flow, and it is possible to identify four groupings for the calculation of X^2, depending on whether the liquid and vapour flows are laminar or turbulent. Lockhart and Martinelli provided four separate correlations for these four cases, which are labelled on Figure 3 as vv, vt, tv and tt where v = viscous and t = turbulent and the liquid is quoted first.

Chisholm (1967) suggested an analytical form for this correlation, by writing:

$$\phi_l^2 = 1 + \frac{C}{X} + \frac{1}{X^2} \tag{74}$$

where C is a constant. The four curves can be represented by:

$$C_{vv} = 5; \quad C_{tv} = 10; \quad C_{vt} = 12; \quad C_{tt} = 20 \tag{75}$$

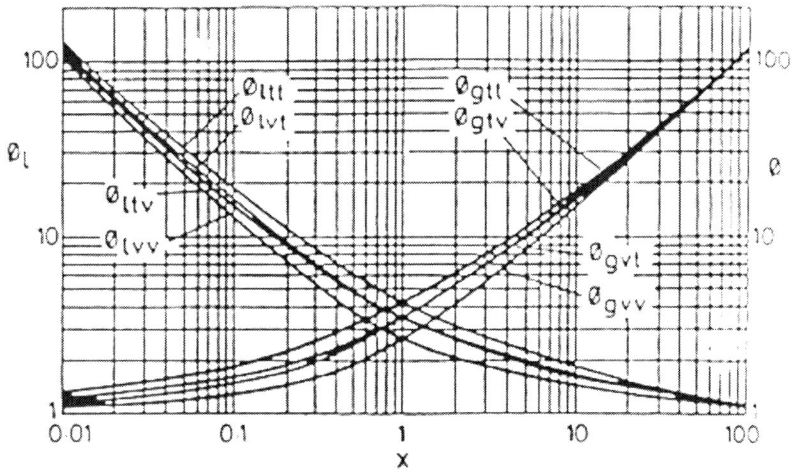

Figure 3. Lockhart & Martinelli Correlation for Wall-Friction

Recently, this form of equation has been suggested as a suitable correlation for gas/liquid flows in small diameter tubes, Holt *et al.* (1997), Mishima and Hibiki (1996). In this case, the constant, *C*, was described as a function of the hydraulic diameter of the channel.

Martinelli and Nelson (1948) used a limited set of steam-water data to derive the correlation of ϕ^2_{lo} as a function of x_g, with pressure as a parameter. Other suggestions were made by Thom (1964), Kirillov *et al.* (1982) and Baroczy (1966).

6.3 Algebraic Correlations

These are much more suited to computers than the graphical correlations, and hence are gaining in popularity. Occasionally, simple curve fitting of the graphs is possible, as for example Chisholm's equations for the Lockhart-Martinelli curves (Equations 74 and 75), but in general this procedure is too complex, and new empirical algebraic correlations are preferred.
Friedel (1979) proposed:

$$\phi^2_{lo} = A_1 + A_2 \tag{76}$$

where

$$A_l = (1 - x_g)^2 + x_g^2 \frac{\rho_l}{\rho_g} \frac{f_{lo}}{f_{go}} \tag{77}$$

and

$$A_2 = \frac{3.24\, x_g^{0.78} (1 - x_g)^{0.224} \left(\frac{\rho_l}{\rho_g}\right)^{0.91} \left(\frac{\eta_g}{\eta_l}\right)^{0.19} \left(1 - \frac{\eta_g}{\eta_l}\right)^{0.7}}{F_r^{0.045}\, W_e^{0.035}} \tag{78}$$

for horizontal and vertically upward flow, while

$$A_2 = \frac{48.6\, x_g^{0.8} (1 - x_g)^{0.29} \left(\frac{\rho_l}{\rho_g}\right)^{0.9} \left(\frac{\eta_g}{\eta_l}\right)^{0.73} \left(1 - \frac{\eta_g}{\eta_l}\right)^{7.4} Fr^{0.03}}{W_e^{0.12}} \tag{79}$$

for vertically downward flow. In these equations, f_{lo} and f_{go} are the single phase friction factors if all the flow were liquid or gas respectively, and

$$We = \frac{\dot{m}_{TP}^2\, D_t}{\rho_{TPH}\, \sigma} \tag{80}$$

where ρ_{TPH} is the homogeneous two-phase density, defined in Equation (81).

$$\rho_{TPH} = \left(\frac{x_g}{\rho_g} + \frac{(1 - x_g)}{\rho_l}\right)^{-1} \tag{81}$$

The HTFS staff at Harwell have produced correlations for ϕ^2_{lo} based on their large data bank. They are of the form:

$$\phi^2_{lo} = f(\dot{m}_{TP}, \Lambda, X) \tag{82}$$

where Λ is the Baroczy property index

$$\Lambda = \frac{\rho_g}{\rho_l} \left(\frac{\eta_l}{\eta_g}\right)^{0.2} \tag{83}$$

and X the Lockhart-Martinelli parameter. The equation is proprietary and only available to subscribers to HTFS.

Muller-Steinhagen and Heck (1986) based their correlation on the observation that for a given total mass flux the two-phase frictional pressure drop at a quality of 100 % is usually the same as the frictional pressure drop at a quality of 50 %. This can be generalised into the two-phase multiplier

$$\phi_{lo}^2 = \left[1 + (\Gamma^2 - 1)\frac{x_g}{x_o}\right](1 - x_g)^{\frac{1}{3}} + \Gamma^2\, x_g^3 \tag{84}$$

where

$$x_o = 1 \ \ if \ \ Re_{lo}\, Fr_{lo} < 0.1 \tag{85}$$

for mass flux < 25 kg/m^{2}s, otherwise:

$$x_o = \min\left[1.6\left(\frac{\eta_g}{\eta_l}\right)^{0.15}[1+\log_{10}(Re_{lo}\ Fr_{lo}]^{-0.33}\ ,\ 1\right] \tag{86}$$

Here

$$\Gamma^2 = \frac{f_{go}}{f_{lo}}\frac{\rho_l}{\rho_g} \tag{87}$$

Wooley and Muller-Steinhagen (1989) indicated that for an overall performance this and the Lockhart/Martinelli correlations gave the best results for flow in horizontal pipes. Beggs and Brill (1973) used a direct correlation method. The frictional pressure gradient was

$$\left(\frac{dp}{dz}\right)_f = \frac{0.5\ f_{TP}\ \dot{m}\ (u_{gs} + u_{ls})}{D_t} \tag{88}$$

where

$$f_{TP} = f_{SP}\exp(S) \tag{89}$$

and

$$S = \frac{\ln(y)}{(-0.0523 + 3.182\ln(y) - 0.8725[\ln(y)]^2 + 0.01853[\ln(y)]^4)} \tag{90}$$

outside the interval $1.0 < y < 1.2$ and by

$$S = \ln(2.2\ y - 1.2) \tag{91}$$

within it. In this

$$y = \frac{(1 - \varepsilon_{gH})}{(1 - \varepsilon_g)^2} \tag{92}$$

with ε_g calculated from Equation (56). The total pressure drop is calculated from Equation (33).

6.4 Test of Overall Pressure Drop Prediction

This section deals with the overall accuracy of prediction of two-phase pressure drop, ie, the comparison of the measured pressure drop with that calculated by addition of the acceleration, gravity, and friction terms. It is this overall accuracy, rather than that of the three individual terms, which is ultimately of importance to the designer.

As in the case of mean density, the data are presented in terms of the three numbers N (number of observations), F (correction factor) and R (range factor) defined above. The accelerational term is normally calculated with sufficient accuracy from the homogeneous model, and this has been done here. Where authors have presented correlations for both friction and gravity, it might be fairest to make a comparison using their combined methods, but since other authors have only calculated one or the other, the method adopted here is to use the modified

HTFS correlation for gravity (shown in Table 6 to be the most generally accurate), combined with the different authors' predictions for wall friction. The only exception is where the frictional correlation is based on the homogeneous model: since the gravity prediction for this model contains no adjustable constants, it is used in preference to the modified HTFS for consistency.

Three tables of comparison are given: Table 6 is for overall predictions (all fluids, all pipe orientations), Table 7 illustrates the accuracy of the homogeneous equation at various mass fluxes while Table 8 shows the data broken down into ranges of different parameters (mass flux, pipe diameter and fluid density ratio) and compared to the HTFS correlation for both wall friction and gravity. From Table 6, it would appear that the Friedel correlation (1979) is probably the best one available in the open literature. However, recent testing by Holt *et al.* (1995) has illustrated that it looses accuracy at low gas to liquid density ratios.

One source of difference between experiments, which accounts for some of the scatter in the data from ostensibly the same experimental conditions, concerns the question as to whether a given flow is fully developed. Experiments in a vertical pipe have indicated that as much as 200 pipe diameters are needed to establish fully developed flow, and such a length is rarely provided in experimental rigs. While it might be possible to establish a set of data in fully developed flow, the practical value would be small since very few industrial two-phase flow systems have more than 200 diameters of straight, uninterrupted pipe run without bends, T-junctions or other fittings.

Table 6. Test of overall pressure drop.

Method	N	F	R
Homogeneous (Owens)	14320	0.77	7.0
Homogeneous (Dukler *et al.*)	14320	1.26	3.2
Lockhart-Martinelli	14320	0.79	3.5
Martinelli-Nelson*	5141	0.71	2.4
Baroczy**	9202	0.94	2.2
Chisholm	14320	0.81	3.4
Friedel***	13223	0.88	2.9
HTFS	14320	0.99	2.2
Muller-Steinhagen & Heck	11586	1.08	2.6

* steam water only ** $339 < m,\_/ \text{kgm}^2\text{s} < 4068$ *** $\eta_l/\eta_g < 1000$

Table 7. Test of homogeneous pressure drop equation.

m (kg/m^{2}s)	<100	100-250	250-500	750-1000	2000-3000	>4000
N	2123	1917	2233	1486	1036	323
F	1.56	1.3	1.31	1.26	1.0	0.89
R	6.4	3.6	2.9	2.5	2.0	2.9

Table 8. Breakdown of pressure drop comparisons.

M (kg/m²s)	<100	100 – 250	250-500	750-1000	2000-3000	>4000
N	2123	1917	2233	1486	1036	323
F	1.0	0.98	0.96	1.02	0.93	0.86
R	2.9	2.6	2.3	2.1	1.7	2.5

D (m)	<0.0032	0.0032-0.01	0.01-0.032	0.032-0.1	>0.1
N	39	1437	7403	4848	593
F	0.42	0.96	1.02	0.99	0.87
R	4.7	1.7	2.2	2.4	1.9

ρ_l/ρ_g	<100	100-178	178-316	316-562	562-1000	>1000
N	4718	450	498	2423	5280	951
F	0.99	0.98	1.08	1.15	0.95	0.87
R	1.9	2.0	2.6	2.1	2.6	1.9

Perhaps in theory, one could include length of pipe run among the parameters in the data bank, but it is probable that flow development is affected by the nature of the fitting at the start of the straight run, and possibly the nature of the flow upstream of that; in practice, it seems that we must live with a considerable degree of inaccuracy in our predictions for the time being.

7 Shell Side Void Fraction and Pressure Drop

Compared with the wealth of data for flow in tubes, data for shell-side flow is very sparse indeed, and consequently there is far less testing of correlations. Major works in this area are Schrage *et al.* (1987), Dowlati *et al.* (1988, 1990) and Ulbrich and Mewes (1995). All used upflow of air and water through a rectangular section fitted with horizontal rods (to simulate tubes) with tube diameter pitch to diameter ratios and tube layouts typical of shell and tube heat exchangers. They found that the homogeneous model gave an overprediction. Ulbrich and Mewes tested a number of correlations against a bank containing most previously published data and concluded that a drift flux model gave the best results. For this they used the same value as C_o as quoted above for flow in pipes but used the expression for v_{gd} suggested by Lahey and Moody (1977),

$$v_{gd} = 2.9\left[\frac{g\,\sigma\,(\rho_l - \rho_g)}{\rho_l^2}\right]^{0.25} \tag{93}$$

As for void fractions, there is not much data on frictional pressure drop on the shell side of an exchanger. Chisholm (1985) suggested:

$$\phi_{lo}^2 = 1 + (\Gamma^2 - 1)\left[B\ f(n, x_g) + x_g^{2-n}\right] \tag{94}$$

where Γ^2 is similar to the Lockhart-Martinelli parameter X^2, except that Δp_l and Δp_g are replaced by Δp_{lo} and Δp_{go}, the pressure drops which would occur if all the flow were liquid or all the flow were gas. Calculation of *B, n* and the function *f* are very complex and depend on flow patterns within the heat exchanger.

Schrage *et al.* (1987) and Dowlati *et al.* (1990) attempted to correlate their data by plotting the liquid only two-phase multiplier ϕ_l^2 as used by Lockhart and Martinelli against the turbulent-turbulent parameter X_{tt} given by

$$X_{tt}^2 = \left(\frac{\dot{m}_l}{\dot{m}_g}\right)^{1.8} \frac{\rho_g}{\rho_l} \left(\frac{\eta_l}{\eta_g}\right)^{0.2} \tag{95}$$

The results for the data of Dowlati *et al.* are shown in Figure 4 for the two pitch/diameter ratios used. It can be seen that there is a significant mass flow effect un accounted for in this simple plot. They attempted to fit Chisholm's equation to the data:

$$\phi_l^2 = 1 + \frac{C}{X} + \frac{1}{X^2} \tag{96}$$

by selecting different values of the parameter *C* for different ranges of mass flow, but their suggestions are unlikely to be translatable to other exchanger designs. Again it seems that extreme caution is needed until more data are available.

7 Two-phase Flow through Fittings

In pipework, fittings, or singularities as they are sometimes known, can have as important a role in the pressure losses as the straight sections of pipe. It is only in long distance transmission pipe lines that pressure drop from singularities can be trivial. In heat exchangers, there are losses through the inlet and outlet nozzles and at the start and end of tubes. In addition, for vertical and horizontal reboilers, the pressure drop in the pipework linking these exchangers to the column bottom are most influential in the recirculation rates through these units. This lecture commences by illustrating the importance of pipe and fitting pressure loss. The important feature of the pressure profile through singularities are considered next. Single-phase flow is then considered with basic principles being illustrated on sudden enlargements and contractions. Other fittings are then considered. In the section on two-phase flow, we review the range of methods that have been employed; some important ones are illustrated in application to a sudden enlargement. Other fittings are then considered. Next, effects other than pressure drop on fittings are considered. Finally, the division of phases at T-junctions are looked at.

Figure 4 shows two typical sections of pipework. The first is from a chemical process plant. We can see that in this small portion there are 11 fittings (mainly 90° bends). An estimate has been made of the distribution of pressure losses about the circuit. This shows that 29% of the pressure drop is due to bends, 13% to valves and only 58% to the pipes. Obviously fittings are very important. A second example, taken from Nostrebo (1985) shows similar behaviour with 49% of the pressure drop being due to valves, 23.8% to bends, 14.5% to tees and only 12.7% to the pipes.

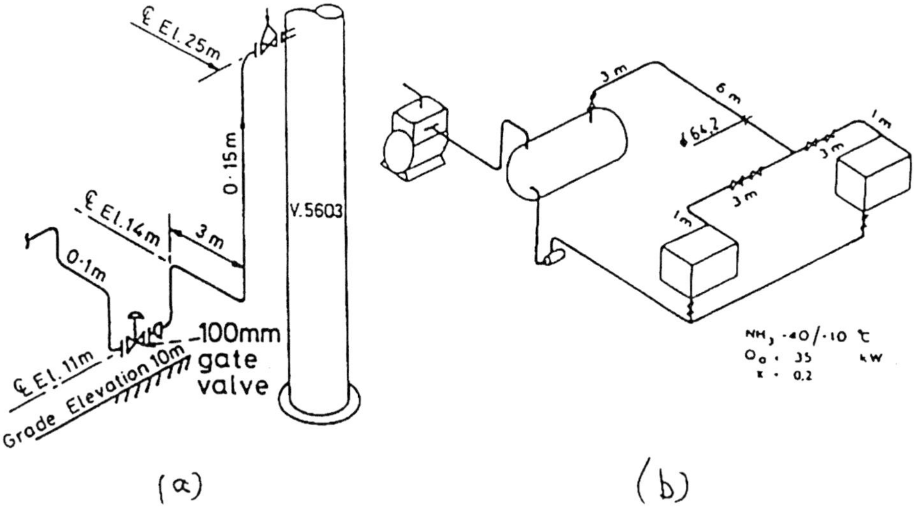

Figure 4. Typical pipework geometry.

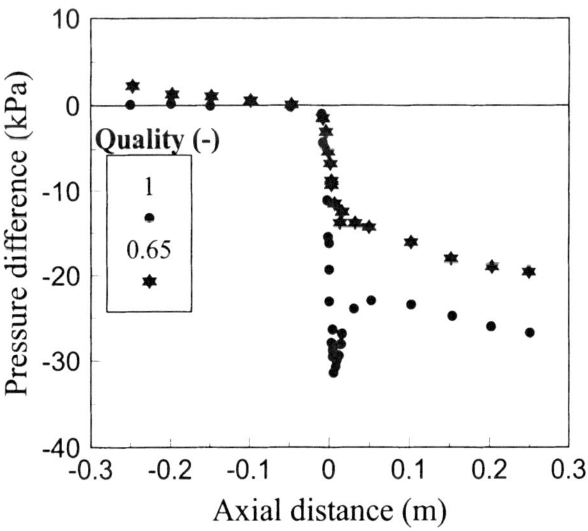

Figure 5. Typical pressure profile.

A typical pressure profile across a singularity is shown in Figure 5. This is for a sudden contraction. From this we can see several important features. Firstly the singularity causes a perturbation, recovery from which can take a significant distance (> 200 diameters has been noted for bends). The flow is deemed to have recovered when the pressure gradient achieves a constant value. The pressure drop is usually determined by extrapolating these constant gradient lines from upstream and downstream. It is noted that the upstream and downstream gradient can be different, e.g. when there is a change in cross sectional areas. The data for gas only flow shows a significant difference from the two-phase data. The latter does not show the minimum found in the gas flow case and which is a symptom of the vena contracta which occurs at certain fittings.

8 Single Phase Flow

The analysis of pressure drop across fittings is usually considered to be under steady stat conditions. Frictional and gravitational terms are usually considered negligible. From mass, momentum and energy balances, for a singularity extending from z_i to z_o, the equations can be manipulated and integrated to give

$$\Delta p = p_o - p_i = -\frac{\dot{M}^2 I_M}{\rho} \tag{97}$$

and

$$\Delta p = -\rho \Delta\mu - \frac{\dot{M}^2 I_E}{2\rho} \tag{98}$$

where $\Delta\mu$ is the change in internal energy of the fluid as it passes through the singularity and the integrals I_M, I_E are given by

$$I_M = \int_{z_i}^{z_o} \frac{1}{S} + \frac{d}{dz}\left(\frac{1}{S}\right) dz \qquad I_E = \int_{z_i}^{z_o} \frac{d}{dz}\left(\frac{1}{S^2}\right) dz \tag{99}$$

The two terms on the right hand side of Equation (98) can be regarded as the irreversible and reversible components of pressure drop. We can then write

$$\Delta p_I = \frac{\dot{M}^2}{2p}[2 I_M - I_E] \tag{100}$$

and

$$\Delta p_R = \dot{M}^2 \frac{I_E}{2\rho} \tag{101}$$

Now for most singularities or fittings the geometry is too complex to permit evaluation of the integrals, I_M, I_E. In addition, even if the inlet and outlet cross sectional areas are the same additional losses can occur due to wall shear and changes in velocity distribution. The following two sections consider cases where analysis is possible and other cases.

8.1 Enlargements and Contractions

For enlargements, Delhaye (1981) has written complete equations. He has examined each term and considered the approximations made and found that the simplified approach shown here is valid and reasonable. In this case we can write a simple momentum balance between planes 1 and 2 where the flow is parallel, Figure 3.

Assuming that the pressure p_1 acts over the area S_2 at plane 1 (this is found to be a good approximation) then a simple momentum balance can be written

$$p_1 S_2 + \dot{M} u_1 = p_2 S_2 + \dot{M} u_2 \tag{102}$$

and as

$$u_1 S_1 = u_2 S_2 \tag{103}$$

Equation (102) can be rewritten as

$$p_2 - p_1 = \frac{\dot{m}_1^2}{\rho} S(1-S) \tag{104}$$

where $S = S_1/S_2$. Comparison with Equation (1) shows that

$$I_M = \frac{S(1-S)}{S_1^2} \tag{105}$$

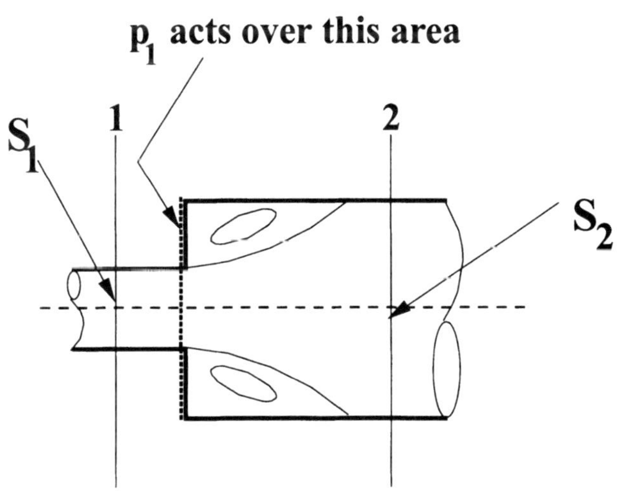

Figure 6. Sudden enlargement

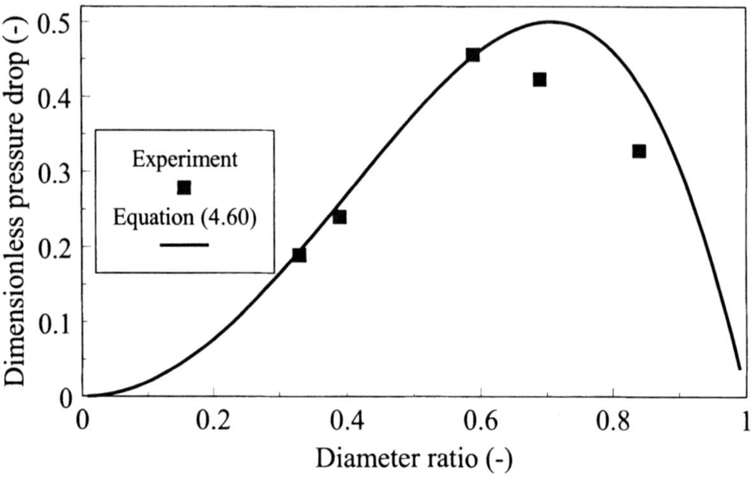

Figure 7. Test of predictions for sudden enlargement.

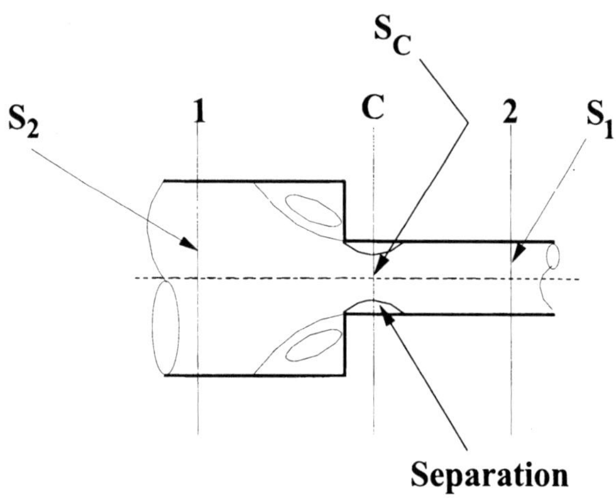

Figure 8. Sudden contraction.

Hewitt (1984) gives detailed derivations of I_M and I_E, the latter being

$$I_E = \frac{S^2 - 1}{S_1^2} \tag{106}$$

Thence the irreversible and reversible pressure drops are given by

$$\Delta p_I = -\frac{\dot{m}_1^2}{2\rho}(1 - S)^2 \tag{107}$$

and

$$\Delta p_R = -\frac{\dot{m}_1^2}{2\rho}(1 - S^2) \tag{108}$$

The predictions of total pressure drop, Equation (104), have been compared against the data of Archer (1913), Figure 7. As seen, good agreement is achieved.

In the case of a sudden contraction, Figure 8, there is a vena contracta formed just downstream of the contraction. Contraction of the fluid to plane C is found to be reversible with the main irreversible loss occurring beyond this point. Equations for total, irreversible and reversible pressure drops are

$$\Delta p = -\frac{\dot{m}^2}{2\rho}\left[S^2\left(\frac{1}{C_c} - 1\right)^2 + (S^2 - 1)\right] \tag{109}$$

$$\Delta p_I = -\frac{\dot{m}_1^2}{2\rho} S^2\left(\frac{1}{C_c} - 1\right)^2 \tag{110}$$

and

$$\Delta p_R = -\frac{\dot{m}^2}{2\rho}(S^2 - 1) \tag{111}$$

Chisholm (1983) and Benedict (1980) give equations for the contraction coefficient, C_c.

8.2 Other Fittings

Fittings other than enlargements and contractions include bends, valves and orifices. As these are not amenable to analytical solution it is the usual engineering practice to describe the pressure drop through a loss coefficient, k. This is defined through

$$\Delta p = k\frac{\dot{m}^2}{2\rho} \tag{112}$$

For bends, Ward Smith (1980) has shown k to depend on the radius ratio and the inlet Reynolds number.

For valves, loss coefficient for valves are often taken from manufacturers data, e.g. Crane (1983). For junctions this problem is complicated by there being two pressure drops, Figure 9. T-junctions are considered in more detail below.

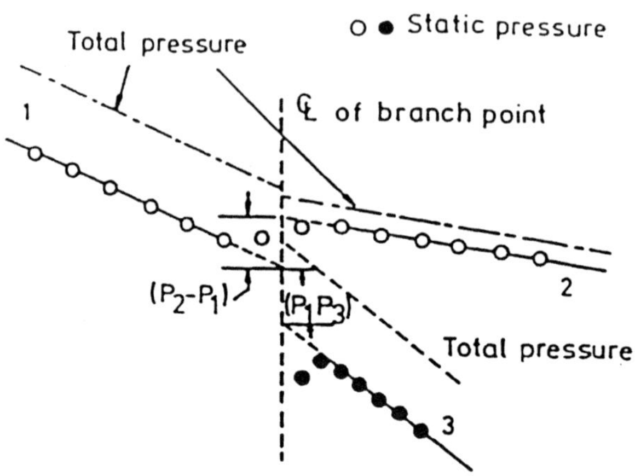

Figure 9. Pressure profiles at Ts.

9 Two-Phase Pressure Drop

9.1 Basics

The complexity of flow through a fitting is further complicated if the fluid is a gas liquid mixture. In spite of the importance of fittings, the body of knowledge for two-phase flow is limited. The review by Azzopardi and Sudlow (1993) gives an idea of the data available. Subsequently, there have been publications by Attou *et al.* (1997), Attou and Bolle (1997a, b) Guglielmini *et al.* (1997) and Schmidt and Friedel (1997) which present new data. They also and give details of models more complex than the approach developed here of using a single phase pressure drop times a two-phase multiplier. Some the results presented by these new papers call into question some of the basic assumptions that would normally be employed. For example, Schmidt and Friedel (1997) have made a careful study of the flow through a sudden contraction. They propose that the vena contracta does not form downstream of the contraction citing as evidence for this the fact that they did not measure the dip and recovery in the axial pressure profile that is shown in Figure 5 and which is associated with the vena contracta. In contrast, Guglielmini *et al.* (1986) and Arosio *et al.* (1990) who also studied a sudden contaction did find the minimum and by implication the vena contracta. This apparent contradiction might be explained by reference to the flow pattern occurring. Examination of the conditions, at which Schmidt and Friedel obtained their data, indicates that this was mainly taken in the annular flow pattern. In contrast, the data of Guglielmini et al., for vertical upflow, were in the bubble and slug patterns and Arosio *et al.*, for horizontal flow, were in the stratified and slug flow patterns. In annular type flows atomisation of liquid from the wall film can occur at the contraction. The resulting drops deposit on the opposite wall and cause suppression of the recirculation of fluid, which characterises the vena contracta. Such behaviour would not be expected in bubble, slug or stratified flows.

Another problem that has been raised is that of pressure recovery. In single-phase flow, it is not unreasonable to expect that some of the pressure drop, which occurs at a contaction, could be recovered at the next expansion. In many of the analyses, which have been presented for two-phase flow, it is assumed that the same occurs. However, as there is so much energy dissipation occurring because of the highly mobile gas/liquid interface, it is possible that potential for a reversible pressure drop is not justified in two-phase flow.

This section presents the conventional approach, which assumes that the single phase pressure drop times a two-phase multiplier can be used. The latest model of Schmidt and Friedel will also be presented. Because, in two-phase gas-liquid flows, the possibility exists of each phase travelling at a different velocity a number of descriptions which take account of the variation in velocities and phases are examined.

Separated flow considers that the phases travel in two distinct parts of the channel and can be characterised by a velocity for each phase.

Homogeneous flow assumes that the phases travel at the same velocity. It is thus a special case of separated flow.

Annular flow (with entrainment) In the previous three descriptions it was assumed that all of each phase behaves in the same way. However observation of specific flow patterns, e.g. annular flow, shows to be a simplistic description. In this case three fields must be considered, liquid film, entrained drops and gas core. An example is that suggested by Morris (1985).

$$\phi_{lo}^2 = \left[x\frac{\rho_l}{\rho_g} + K(1-x) \right] \left[x + \frac{(1-x)}{K} \left(1 + \frac{(K-1)^2}{(\rho_l/\rho_g)-1} \right) \right] \tag{113}$$

where $K = \left[1 + x\left(\frac{\rho_l}{\rho_g} - 1 \right) \right]^{1/2}$.

Chisholm (1983) and Simpson et al. (1985) have suggested alternative approaches based on empirical equations. Chisholm puts forward two approaches (B and C equations) which, it will be show below are equivalent to the separated and homogeneous models when particular values of the parameters B and C are used. Chisholm's equations are

$$\frac{\Delta p}{\Delta p_l} = \phi_l^2 = 1 + \frac{C}{X} + \frac{1}{X^2} \tag{114}$$

where Δp_l is the pressure drop for only the liquid flowing and X is the Lockhart-Martinelli parameter and

$$\frac{\Delta p}{\Delta p_{lo}} = \phi_{lo}^2 = [1 + \left(\frac{\rho_l}{\rho_g} - 1 \right)(B\,x(1-x) + x^2)] \tag{115}$$

where Δp_{lo} is the pressure drop for all of the flow travelling as liquid. Note Δp_l is equal to $(1-x)^2 \Delta p_{lo}$. Simpson *et al.* (1985) proposed a two-phase multiplier of the form

$$\phi_{lo}^2 = \left(1 + x\left[\left(\frac{\rho_l}{\rho_g}\right)^{1/6} - 1\right]\right)\left(1 + x\left[\left(\frac{\rho_l}{\rho_g}\right)^{5/6} - 1\right]\right) \quad (116)$$

9.2 Enlargements and Contractions.

As in single-phase flow simple geometries such as enlargements and contractions are amenable to analysis. Figure 10 shows separated flow through a sudden enlargement.

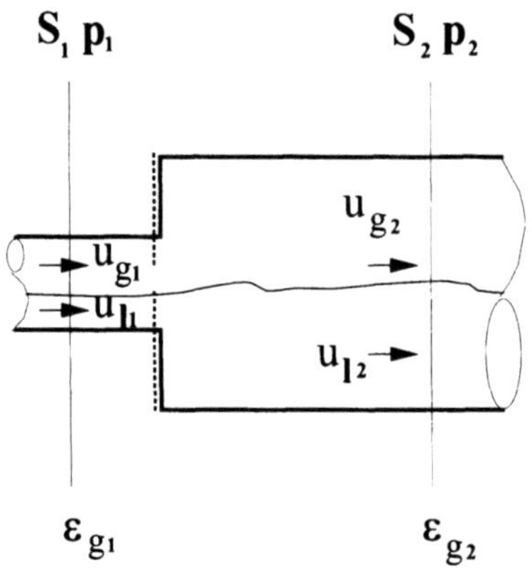

Figure 10. Sudden enlargement - two-phase flow.

As in single-phase flow the pressure p_1 is assumed to apply over the area S_2 just after the expansion. A momentum balance gives

$$p_2 S_2 + \dot{M}_g u_{g2} + \dot{M}_l u_{l2} = p_1 S_2 + \dot{M}_g u_{g1} + \dot{M}_l u_{l1} \quad (117)$$

which on rearranging becomes

$$p_2 - p_1 = \frac{\dot{M}x}{S_2}(u_{g1} - u_{g2}) + \frac{\dot{M}(1-x)}{S_2}(u_{l1} - u_{l2}) \quad (118)$$

Now

$$u_g = \frac{\dot{m}x}{\varepsilon_g \rho_g} \quad (119)$$

$$u_l = \frac{\dot{m}(1-x)}{(1-\varepsilon_g)\rho_l} \quad (120)$$

Substituting into Equation (22) yields

$$p_2 - p_1 = \frac{\dot{m}_1^2}{\rho_l} S \left[\left\{ \frac{(1-x_g)^2}{(1-\varepsilon_{g1})} + \frac{\rho_l}{\rho_g} \frac{x_g^2}{\varepsilon_{g1}} \right\} - S \left\{ \frac{(1-x_g)^2}{(1-\varepsilon_{g2})} + \frac{\rho_l}{\rho_g} \frac{x_g^2}{\varepsilon_{g2}} \right\} \right] \tag{121}$$

where $S = S_1/S_2$ and $\dot{m}_1 = \dot{M}_1 / S_1$. This is often known as Romie's (1958) equation. If we assume that there is no flashing, i.e. $x_1 = x_2$ and that the voidage does not change over the expansion, Equation (121) can be written

$$p_2 - p_1 = \frac{\dot{m}_1^2}{2\rho_l} [2S(1-S)] \left[\frac{(1-x)^2}{(1-\varepsilon_g)} + \frac{\rho_l}{\rho_g} \frac{x^2}{\varepsilon_g} \right] \tag{122}$$

This is the equation for total pressure drop for separated flow.
These equations are of the form

$$\Delta p = \frac{\dot{m}_1^2}{2\rho_l} k \, \phi_{lo}^2 \tag{123}$$

where k is a loss coefficient and ϕ_{lo}^2 is a two-phase multiplier. Use the definition for void fraction

$$\varepsilon_g = \frac{1}{1 + U_R \dfrac{1-x_g}{x_g} \dfrac{\rho_g}{\rho_l}} \tag{124}$$

where U_R - u_g/u_l is the slip ratio, in Equation (122) yields

$$p_2 - p_1 = \frac{k\,\dot{m}^2}{2\rho_l} (1-x_g)^2 \phi_l^2 = \frac{k\,\dot{m}^2}{2\rho_l} (1-x_g)^2 \left[1 + \left(\frac{1}{U_R} \sqrt{\frac{\rho_l}{\rho_g}} + U_R \sqrt{\frac{\rho_g}{\rho_l}} \right) \frac{1}{X} + \frac{1}{X^2} \right] \tag{125}$$

where

$$X^2 = \frac{\Delta p_l}{\Delta p_g} = \frac{(1-x)^2}{x^2} \frac{\rho_g}{\rho_l} \tag{126}$$

This is Chisholm's C Equation (114) with

$$C = \frac{1}{U_R} \sqrt{\frac{\rho_l}{\rho_g}} + U_R \sqrt{\frac{\rho_g}{\rho_l}} \tag{127}$$

Alternatively we can write

$$C = C^* \left(\sqrt{\frac{\rho_l}{\rho_g}} + \sqrt{\frac{\rho_g}{\rho_l}} \right) \tag{128}$$

A value of $C^* = 1$ reduces Equation (125) above to the homogeneous equation.

Following on from single-phase flow we can write the separated flow equations for the irreversible and reversible components of pressure drop

$$\Delta p_I = -\frac{\dot{m}_l^2}{2}\left[-\frac{2S(1-S)}{\rho_M} - \frac{(S^2-1)\rho_H}{\rho_E^2}\right] \quad (129)$$

and

$$\Delta p_R = -\frac{\dot{m}_l^2 (1-S^2)\rho_H}{2\rho_E^2} \quad (130)$$

These equations have been tested against experimental data by a number of workers. Patrick and Swanson (1959) and Richardson (1959) note that the approximation $\varepsilon_{g1} = \varepsilon_{g2} = \varepsilon_g$, which leads to the simplification of Equation (121) to Equation (122) is only partially correct. The data of Ferrell and McGee (1966) (mainly $\dot{m}_l < 2000\ kg/m^2 s$) shows values below the homogeneous prediction but reasonable agreement when the known void fraction was used. Janssen and Kervinen (1964) show that equation (122) with void fraction given by an appropriate correlation works well. At higher mass fluxes $(\dot{m}_l > 2000\ kg/m^2 s)$. The data of Fitzsimmons (1964) and Weisman (1974) is well predicted by the homogeneous model. Harshe *et al.* (1976) have produced an extensive comparison. ESDU (1989), who carried out an independent comparison between the different equations and available data, recommend the equation of Simpson *et al.* (1985) for the two-phase multiplier.

For sudden contractions, the separated flow equations for total, irreversible and reversible pressure drops obtained by the above approach are

$$\Delta p = -\frac{\dot{m}_l^2}{2\rho_M}\left[\frac{S^2}{C_c^2} - 1 - 2S^2\left(\frac{1}{C_c} - 1\right)\right] \quad (131)$$

$$\Delta P_I = -\frac{\dot{m}_l^2}{2}\left(\frac{\left[\frac{S^2}{C_c^2} - 1 - 2S^2(\frac{1}{C_c} - 1)\right]}{\rho_M} + \frac{(S^2-1)\rho_H}{\rho_E^2}\right) \quad (132)$$

and

$$\Delta p_R = -\frac{\dot{m}_l^2 (S^2-1)\rho_H}{2\rho_E^2} \quad (133)$$

Comparison with experiment has shown that data is well represented by the homogeneous model, Cemak *et al.* (1963), Geiger and Rohrer (1966), Harshe *et al.* (1976) and ESDU (1989). The applicability of the homogeneous model should not be surprising as the jetting of fluids into the contraction probably acts as a homogeniser.

The model developed by Schmidt and Friedel (1997) for a sudden contraction differs from those discussed above in considering that there are three fluid streams: gas, drops and film. In addition, they also took into account the pressure on the annular upstream facing wall. Unlike earlier analyses, they did not assume that this pressure was uniform or equal to the upstream pressure. The momentum balance can thus be written as

$$p_1 S_1 - \int_{D_{t2}}^{D_{t1}} \int_0^{2\pi} p_{base} p_2 S_2 - \left(F_{fric} + F_{grav}\right) = \frac{\dot{m}_2^2}{\rho_{eff,2}} K_{M,2} S_2 - \frac{\dot{m}_1^2}{\rho_{eff,1}} K_{M,1} S_1 \tag{134}$$

where ρ_{eff} is the density based on the sum of the momentum of the three fluid streams and calculated from

$$\rho_{eff} = \left[\left(\frac{x_g}{\rho_g} + U_R \frac{(1 - x_g)}{\rho_l} \right) \left(x + \frac{(1 - x_g)}{U_R} \left[1 + (U_R - 1)^2 \frac{e}{1 - e} \right] \right) \right] \tag{135}$$

with the slip ratio

$$U_R = \frac{x_g}{1 - x_g} \frac{1 - \varepsilon}{\varepsilon} \frac{\rho_l}{\rho_g} \tag{136}$$

the fraction of liquid entrained as drops (in volume terms)

$$e = \frac{\rho_g}{\rho_l} \varepsilon_E U_R \tag{137}$$

$$\varepsilon_E = \frac{1}{U_R} \left[1 - \frac{(1 - x_g)}{1 - x_g (1 - 0.18\, We^{0.18}\, Re^{0.5})} \right] \tag{138}$$

$$Re = \frac{\dot{m}\,(1 - x_g)\, D_t}{\eta_l} \tag{139}$$

$$We = \dot{m}^2 x_g^2 \frac{D_t}{\rho_g \sigma} \frac{(\rho_l - \rho_g)}{\rho_g} \tag{140}$$

and a void fraction

$$\varepsilon = 1 - \frac{2\,(1 - x_g)^2}{1 - 2x_g + \sqrt{1 + 4x_g (1 - x_g)\left(\frac{\rho_l}{\rho_g} - 1\right)}} \tag{141}$$

This results in the following equation for pressure drop

$$\Delta p_{con} = \frac{\dot{m}_2^2 \left[\frac{1}{\rho_{eff,2}} - \frac{S}{\rho_{eff,1}} + f_{con}\, \rho_{eff,2} \left(\frac{x_g}{\rho_g\, \varepsilon_2} - \frac{1 - x_g}{\rho_l\,(1 - \varepsilon)} \right)^2 (1 - \sqrt{S}\,)^2 \right]}{1 + \Gamma_{con}\left(\frac{1}{S} - 1 \right)} \tag{142}$$

where the dimensionless base pressure coefficient is given by

$$\Gamma_{con} = 0.77\ S\ (\ 1\ -\ S^{0.306}\ ) \tag{143}$$

and the relevant friction factor is

$$f_{con} = 5.2\,10^{-3}\ x^{0.1}\ (\ 1\ -\ x\ ) \left(S\ \frac{\eta_l}{\eta_{g,3}} \right)^{0.8} \tag{144}$$

Schmidt and Friedel (1997) tested this approach against a database of more than 1200 data points with a mean error of 1% and a standard deviation of 34%.

9.3 Valves.

For valves Chisholm has recommended appropriate B values. However, later work by Simpson *et al.* (1983), Fairhurst (1983) and Simpson *et al.* (1985) show that these B values can give poor predictions. Better agreement was found with the equation of Simpson *et al.* (1985), equation (116). ESDU (1989) recommend the equation of Simpson *et al.* (1985), Equation (116), for gate and ball valves and the equation of Morris (1985), Equation (113), for globe and diaphragm valves.

Simpson *et al.* (1985) discusses the problem of flashing of liquid to vapour in passing through the valve. They found that a correct necessary to account for this but that at mass fluxes of 500 kg/m^{2}s this only amounted to 2%. At mass fluxes of 1360 kg/m^{2}s it had risen to 10%. Obviously account must be taken of this for high flow cases. Friedel and Kissner (1985) discuss relief valves.

9.4 Other fittings

Here we will examine pressure drops in trends and valves and consider briefly other items such as orifice plates and venturis. Data for bends is sparce. It can be difficult to obtain as long recovery lengths are required downstream of the bends. Chisholm (1983) recommends use of his *B* equation with

$$B = 1 + \frac{2.2}{k_{lo}\,(2 + R_c / D_t)} \tag{145}$$

for 90° bends k_{lo} is the loss coefficient for all of the flow travelling as liquid and R_c is the radius of curvature of the bend. For bend angles < 90°, Chisholm recommends the value for 90°. For angles> 90°

$$B = 1 + [B_{90} - 1]\frac{k_{90}}{k} \tag{146}$$

where $k_{9.0}$ is the loss coefficient for the 90° bend and k is the value for the actual bend.
Orifice plates can be used for metering purposes. As with contractions vena contracta is formed and it is the pressure drop to this point that is used. Lin (1985) has given a comprehensive review of equations published for this geometry and puts forward a well-tested method. Venturi meters have been used in two-phase flow to measure the flow quality, x, if the mass flow is known, e.g. Harris and Shires (1972). It is noted that Harris (1967) shows these units sensitive to details of upstream pipework.

10 Other Effects of Fittings

Pressure drop is not the only effect of singularities that is of interest. In some cases it is the change in spatial distribution of the phases which is required. This might be motivated by a wish to minimise disturbance to flows upstream of phase separators so not to produce fine drops which would cause operational difficulties to the separator. Azzopardi and Sudlow (1993) produced a thorough review. Here we review briefly the information for three cases, bends, the effect of inserts on film flow in annular flow and the effect of orifices on void fraction.

The amount of information on the behaviour of two-phase flow in bends is very limited. For example, Gardner and Neller (1969) have studied bubble/slug flow in a bend from a vertical to a horizontal pipe. They found that gas can flow either on the outside or the inside of the bend depending on the balance between centrifugal forces (tending to push the water to the outside) and gravity (tending to pull it to the inside). They proposed a critical Froude number, $Fr_\theta = V^2/gR_c sin\theta = 1$, where V is the mixture mean velocity, R_c the radius of curvature of the bend and θ the angle of the band, to distinguish the two types of behaviour.

Kooijman and Lacey (1968), Anderson and Hills (1974) and Maddock *et al.* (1974) have studied annular in 90° bends between vertical and horizontal pipes. They observed that the film flows predominantly on the inside of the bend. They suggested that this indicates that the radial pressure gradients and gravity are stronger than the centrifugal forces. Only Chakrabati (1976) has studied bends in a horizontal plane through Banerjee *et al.* (1967), Hewitt and Dell (1968) and Whalley (1980) have made observations and measurements on tubs coiled about a vertical axis. However, it is believed that this information can be used to assess the flow behaviour in bends as Lacey and Kooijman have observed that the flow is very similar at 90° and 360° into a coil. These studies present clear evidence that, whereas one would expect centrifugal forces to cause a separation of the phases with the denser phase flowing around the outer part of the duct perimeter, under certain conditions it may flow mainly along the inner part.

Banerjee *et al.* showed that this behaviour, which they called film inversion, could be quantitatively accounted for solely by differences in radial pressure gradients in the gas moving gas and the slower liquid film. They presented a criterion based on a balance of centrifugal and gravitational forces. The subsequent studies of Hewitt and Dell, and Whalley show qualitative but not clear quantitative confirmation of this analysis. Under certain other conditions, particularly higher liquid flow rates, the liquid in the film flows on the outside of the bend. However, no theoretical models exist which can predict the circumferential variation of film flow rate.

The fractions of drops of different sizes, which deposit on the bend have been studied by James *et al.* (2000). Simple models for the deposition of droplets onto the walls of a bend have been developed by a number of workers, notably Burkholz (1989), who studied gas-liquid separators of the wave-plate or chevron type. In his (two-dimensional) analysis, he considered flow around a bend of angle α in a channel of width W. By assuming equilibrium between centrifugal and radial forces, and that the droplet mass flow and concentration are uniform over the pipe cross-section, he obtained the following expression for η, the droplet collection efficiency for the bend:

$$\eta = \frac{\rho_d V_B d_d^2}{18 \mu_F W} \alpha \tag{147}$$

Here ρ_d is the droplet density, V_B is the continuous phase bulk velocity, d_d is the droplet diameter, and μ_F is the continuous phase dynamic viscosity. James *et al.* (2000) suggested a modified version of this equation which is more appropriate for pipe bends. This applies the same argument to a circular pipe cross-section rather than to a two-dimensional channel. The result is

$$\eta = 1 - \left(\frac{2\theta - sin(2\theta)}{\pi} \right) \tag{148}$$

$$\begin{aligned} cos(\theta) &= \frac{L}{D_t} & 0 \le L \le D_t \\ cos(\theta) &= 1 & L > D_t \end{aligned} \tag{149}$$

Here L is the radial displacement of a circular disk of droplets at the exit to the bend and D_t is the pipe diameter. Following Burkholz's analysis it can be shown that

$$L = \frac{\pi V_B d_d^2 \rho_d}{36 \mu_F} \tag{150}$$

Although this expression for efficiency cannot be verified directly, as there are mechanisms such as droplet coalescence and re-entrainment which will affect any measurements, it will be interesting to compare the efficiencies it predicts with results from the numerical simulations. These simple calculations have been confirmed by James et al. (2000) using Computational Fluid Dynamics (CFD) to obtain the gas velocity and turbulence fields and a drop tracking/eddy interaction model to follow the behaviour of drops. Figure 11 shows good agreement between the approaches in determining the efficiency of deposition of drops. This implies that most of the drops will deposit. However, recent studies indicate that though there is less entrained liquid after the bend, there is still some. This is probably due to re-entrainment of liquid. Indirect confirmation of this is provided by measurements of drop sizes by Ribeiro *et al.* (2001) who found that the mean drop size after the bend was almost twice that upstream, Figure 12 This is probably due to the downstream drops having been created from the totally deposited liquid which having more liquid in the film probably had thicker films and thus larger drops.

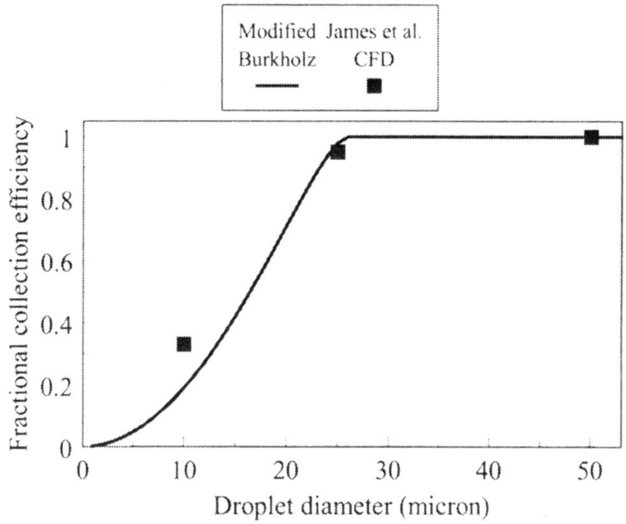

Figure 11. Comparison of deposition efficiency from droplet trajectory calculations with equations of Burkholz (1989) - air/water.

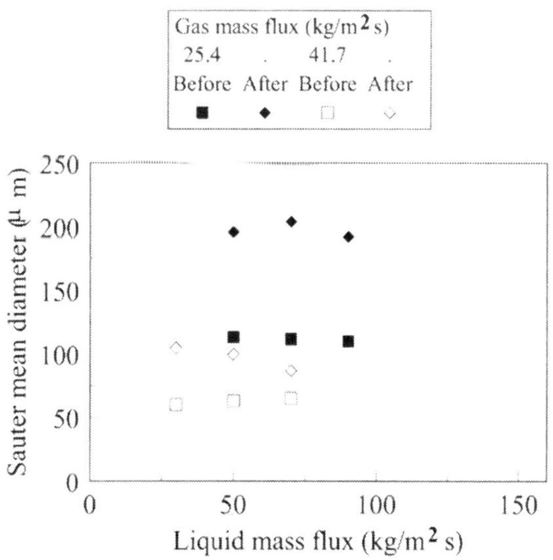

Figure 12. Drop sizes upstream and downstream of a bend - after Ribeiro et al. (2001).

McQuillan and Whalley (1984) and Fryer and Whalley (1982) have studied the effect on inserts on the film flow rate in vertical annular flow. The former considered an orifice plate whilst the latter study considered the effect of a twisted tape. As expected the orifice plate caused extra atomisation of the film, Figure 13. The liquid returns to the film downstream of the orifice plate. Similarly the swirl induced by the twisted tape increases the film flow rate though as the swirl decays downstream of the insert the film flow rate reverts to its original value. If we consider the effects of the two inserts for the same inlet conditions, we see that the return to equilibrium is much slower for the swirl tape than the orifice plate. In fact for the film flow rate to recover by 50% requires 5/6 times the axial distance for the swirl tape relative to the orifice plate.

The effect of the end of the twisted tape can affect the amount of liquid, which is forced onto the wall. Hills *et al.* (1996) showed much higher film flow rates when the end is cut to a fish tail than when it is flat. This is probably due to the fish tail being better at directing liquid, which lands on the twisted tape onto the wall.

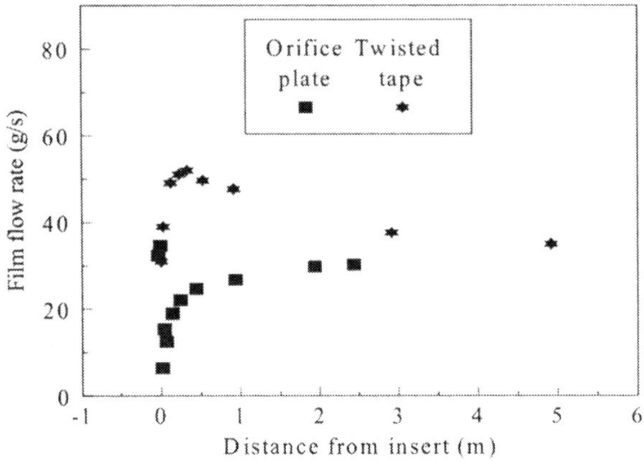

Figure 13. Variation of water film flow rate with distance downstream of an orifice plate or a twisted tape - gas flow rate = liquid flow rate = 0.063 g/s. Tube diameter = 32 mm, orifice diameter = 12 mm. Pressure = 1.5 bar.

Salcudean *et al.* (1983) have studied the effect of flow obstructions on void fraction in horizontal tubes. For bubbly flow about a central obstruction the voidage increases, bubbles are slowed down and slip ratios of less than 1 are observed. In contrast, slug flow about a peripheral obstruction shows a decrease in void fraction.

11 Phase Separation at T Junction

In the division of two-phase flow at junctions, it has been noted that there is an almost inevitable maldistribution of the phases between the outlets. This can constitute a major problem when it occurs in chemical process and oil and gas production and refinery plant as the maldistribution can have significant effect on the behaviour of equipment downstream of the junction. For example, when steam injection is being used to effect enhanced recovery of viscous oils, the steam is usually generated at a central point and distributed to a number of wells. This can involve several junctions. In this process it is important to know where the water (either that coming from the boiler because of incomplete evaporation or that due to condensation of steam along the transmission lines) goes to. Water having a much lower enthalpy than steam is much less effective at lowering the viscosity of the oil.

An example of the problem that the phase maldistribution can cause is found in the case of a bank of air-cooled heat exchangers operating as condensers. Because of the distance from the column there could be partial condensation in the pipe to the condensers. The flow is divided between the individual exchangers by a series of T-junctions. It was found that one of the condensers under performed. Laboratory trials revealed that at the first junction thirty per cent of the vapour and ten per cent of the liquid was taken off. The same occurred at the next two junctions. This left seventy per cent of the liquid and ten per cent of the vapour for the last condenser, which, being over loaded with liquid, had very poor heat transfer.

As the maldistribution is so common, it is reasonable to ask if the effect could be utilised as part of phase separation equipment. It will be shown below that the separation is usually not perfect. Therefore, any application must be as a partial separator. Such an approach could lessen the load on the main separator, possibly leading to smaller units. Such a reduction in size is an important consideration for high-pressure systems because of the difficulties in manufacturing large, thick-walled vessels. For applications in offshore oil/gas production, there is an added saving as any weight saving results in economies in the supportive structural steelwork.

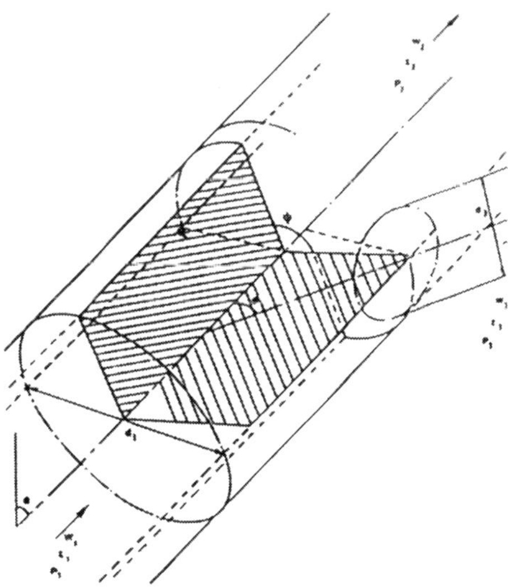

Figure 14. Junction showing relevant variables.

The geometry of the junction strongly influences the flow split and associated pressure drops. Gardel (1957) has shown that for single-phase flow the pressure drops across the junction are affected by the angles between the inlet and outlet pipe, ϕ, the ratio of side arm to main tube diameter, D_3/D_1, and the degree of rounding of the corner, r/D_1, see Figure 14. In the case of two-phase flow, when gravity can cause stratification, it is also necessary to specify the angle between the inlet pipe and the vertical, θ, and the angle, ψ, between the side arm and the top of the main tube. There are a great many possible combinations of values of these parameters. In practice, however, only certain values of these parameters are used, usually $\theta = 0°$ or $90°$ $\phi = 45°, 90°, 135°$, and $\psi = 0°, 90°$ or $180°$, and some combinations are not possible (e.g., ψ is not relevant if $\theta = 0°$). The problem thus reduces to one of potentially manageable proportions. Those cases where the two pipes meet at right angles are termed T junctions; in other cases the term Y junction is used.

The division of fluids at junctions depends on the pressures in the two downstream pipework systems. Obviously, the lower the pressure (i.e., the more powerful the suction), the greater the proportion of fluid passing down that outlet. However, in addition to pressure drops due to friction and pipe fittings, there are extra pressure drops (rises) across the junction as illustrated in Figure 6. The problem can be illustrated if we imagine a simple experiment in which two sources of fluid, gas and liquid provide a two-phase flow with constant and given flow rate The settings of valves on the two downstream legs varies the flow split between 0% (no flow to the side arm) and 100% (all flow to the side arm). All other variables of interest result from these settings. It is noted that in determining the dependent variables, Saba and Lahey (1984) and Lahey *et al.* (1985) have suggested that, apart from the geometry, there are eight variables, which define the flow split. These are the total flow rate and quality for each of the three legs of the junction together with the two pressure drops associated with the junction (Δp_{12}, Δp_{13}). As three of the variables might normally be specified, there would be five unknowns requiring five equations to solve the problem. Overall and component mass balances could provide two equations. Momentum or energy balances might provide two more. The remaining relationship could be identified with the phase partition at the junction. Lemonnier and Hervieu (1992) argue that though the approach is fundamentally correct it will be invalidated unless appropriate forms of the individual equation are used.

It is interesting to note that most of the information on flow split at junctions has appeared in the last ten years. Even with this growing database, there are hardly any cases with replication of data. This should not be surprising because of the large number of variables that are relevant. Only a limited amount of data analysis has been carried out to ascertain the applicability of combinations of parameters. One obvious combination is the gas momentum. Intuitively one might expect the processes in the split to be inertia dominated. In one systematic experiment, Azzopardi *et al.* (1988) set both gas flow rate and system pressure to values, which gave the same superficial gas momentum. The flow split was very similar in the two cases. However, in other cases where the phase momentum was the same, there were differences in the phase split. It is believed that this is caused by differences in the flow pattern approaching the junction.

Data has been obtained on the flow split with all the major flow patterns approaching the junction. Before considering the trends in the available data, it is probably instructive to review the main features of flow pattern and how they might influence phase split at T-junctions.

In vertical annular flow there is cylindrical symmetry. The gas occupies the majority of the channel with part of the liquid being dragged along the channel walls as a film, the rest being carried as drops in the gas phase. This flow can exist in dynamic equilibrium, with constant interchange of liquid between the film and the drops. As noted above, the momentum of the fluids are probably important factors in the partition of the phases. The measurements of Azzopardi and Teixeira (1994) and the calculations of Schadel *et al.* (1990) for annular flow show that the drops travel at approximately the gas superficial velocity. The velocity of the film can be deduced, from measurements of film flow rate and thickness, to be much lower than that of the gas. Relative to the momentum of the gas, the momentum of the film could vary from 0.1 to 20 whilst that for the drops might range from 300 to 1250.

The description of vertical annular flow given above is also valid for horizontal pipes at very high gas flows. However, at more moderate gas velocities, the effect of gravity becomes increasingly important. This manifests itself in a strong asymmetry in the flow rate and thickness of the liquid film. In fact recent observations have indicated that the boundary between annular and stratified flows might be less clear cut than hitherto expected. Atomisation can occur from the stratified liquid layer at velocities below those at which the liquid forms a film around the entire circumference of the pipe. Moreover, the description of stratified flow, which postulates a liquid layer with a straight horizontal interface, is only valid at higher liquid flow rates. At lower flow rates the liquid occupies a crescent shaped area. In addition, recent observations indicate that the can be a small (but significant) amount of liquid travelling as entrained drops at the higher gas velocity end of these conditions. For crescent stratified flow the simple description of gas and film might be sufficient. However, at higher liquid flow rates, where the interface can be considered as horizontal, there could be significant velocity effects in the liquid flow which would render the simplification of considering the liquid as having low momentum inadequate. Horizontal slug flow can be considered as having high momentum liquid slugs interspersed with the equivalent of stratified flow. The all-liquid part of the slug will travel at approximately the mixture velocity. The liquid in the stratified layer is much slower.

Available data have been collected in a number of review articles, the most recent and comprehensive of which is that of Azzopardi (1999).

11.1 Parametric trends

Figure 15 illustrates the way data on phase split are presented. This is a plot of fraction of incoming liquid emerging through the side arm versus the corresponding gas fraction. For an infinite resistance in the side arm leg the data lies at 0, 0. If the main pipe is blocked it lies at 1, 1. The actual split will lie on the locus between these two points and depend on the resistances in the two legs. The data in Figure 15 are for vertical annular flow approaching the junction. Here, a decrease in the gas flow rate results in a higher fraction of liquid taken off. Figure 16 shows that a decrease in the liquid flow rate gives an increase in the fraction of liquid taken off. Data for a horizontal junction show similar trends. In vertical churn/slug/bubble flows, the take off ranges from liquid dominated, i.e., above the $x = y$ line to gas dominated, below $x = y$, Figure 17. The occurrence of each of these types of splits is systematic as shown on the map, Figure 18. However, the boundary does not correspond to any specific transition. The given curve is empirical.

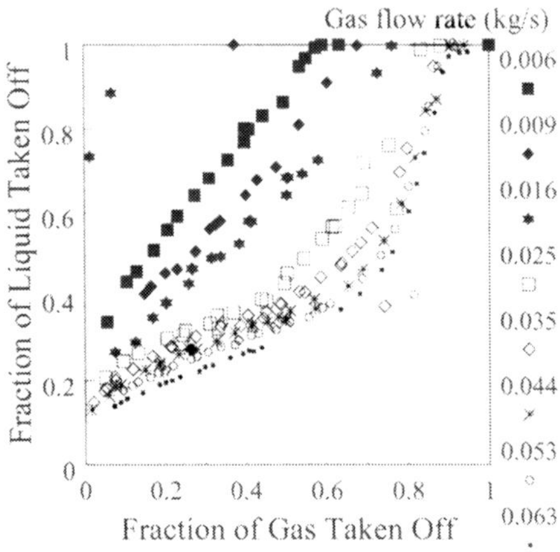

Figure 15. Effect of gas flow rate on phase split with a vertical main pipe - liquid superficial velocity = 0.08 m/s; Main pipe and side arm diameter = 0.032 m; pressure = 1.5 bar. Reprinted from International Journal of Multiphase Flow, vol. 14, Azzopardi, B.J., Measurements and observations of the split of annular flow at a vertical T junction, pp 701-710, Copyright (1988), with permission from Elsevier.

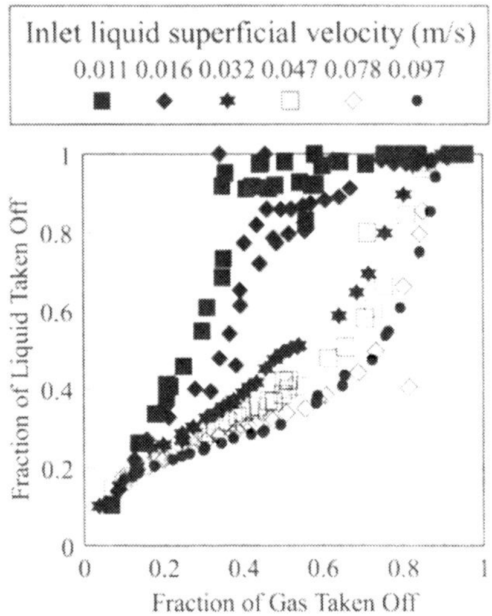

Figure 16. Effect of liquid flow rate on phase split with a vertical main pipe - gas superficial velocity = 37 m/s; Main pipe and side arm diameter = 0.032 m; pressure = 1.5 bar. Reprinted from International Journal of Multiphase Flow, vol. 14, Azzopardi, B.J., Measurements and observations of the split of annular flow at a vertical T junction, pp 701-710, Copyright (1988), with permission from Elsevier.

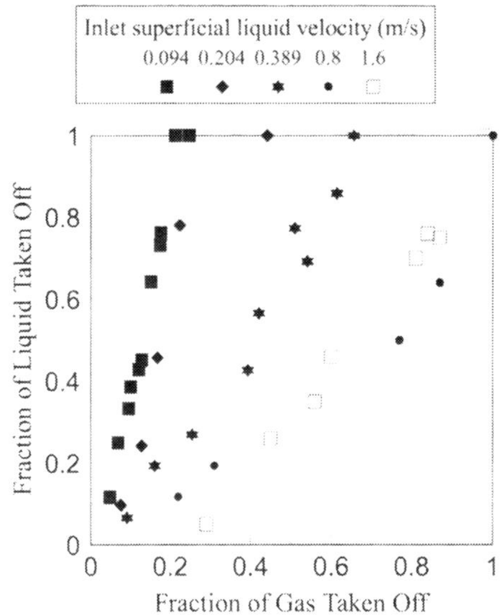

Figure 17. Effect of liquid flow rate on phase split with a vertical main pipe - gas superficial velocity = 0.63 m/s; Main pipe and side arm diameter = 0.076 m; pressure = 1.0 bar.

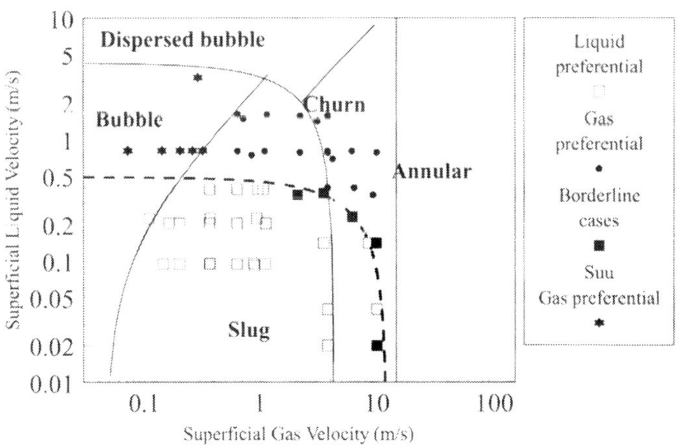

Figure 18. Flow pattern map showing where liquid dominated and gas dominated take off occurs. Main pipe and side arm diameter = 0.076 m; pressure = 1.0 bar. —— Taitel et al. (1980) transitions; ---- Empirical phase dominance boundary.

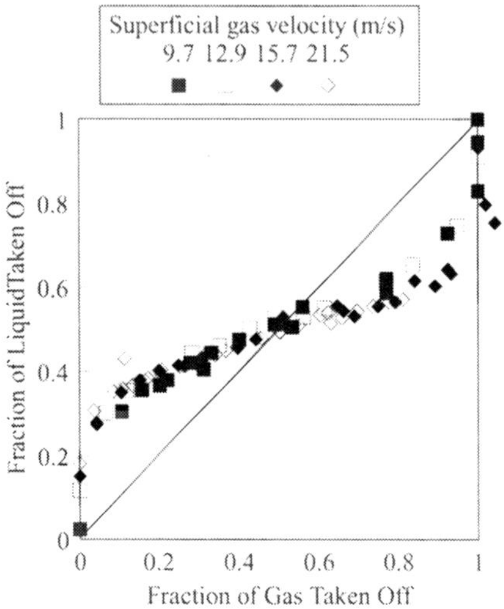

Figure 19. Phase split at an impacting T-junction. Air/water; Liquid superficial velocity = 0.08 m/s; Main pipe and side arm diameter = 0.032 m; pressure = 1.75 bar.

Note, even when the division occurs at an impacting junction – both outlets at right angles to the inlet – there can still be maldistribution unless the downstream resistances are exactly equal. Figure 19 shows data from such a geometry. In this case the inlet pipe was vertical and the two-phase mixture approaching the junction was in annular flow, Azzopardi et al. (1989). A review of two-phase division at impacting junctions has been published by El-Shaboury *et al.* (2001).

There has been no systematic study of the effect of scale (i.e., diameter) on the flow split. For vertical annular flow, comparison has been made between data from a 0.032 m diameter pipe and one of 0.125 m diameter. The momentum for the gas (in terms of superficial velocities) was 1715 and 1924 kg/ms^2 respectively while values for the liquid were 0.25 and 0.42. No apparent trend with diameter can be seen. In contrast, for horizontal flows, there can be significant difference even when the phase momentum (based on superficial velocities) is the same for smaller and larger diameters. This is probably due to the proximity to the stratified/annular boundary and the degree of circumferential maldistribution of the liquid.

Observation of the flow split indicates that, though part of the liquid is taken off through the side arm directly, there are other phenomena, which cause the liquid to be taken off. In vertical flow, there is evidence that once the gas velocity in the main pipe above the junction falls below a critical value, liquid cannot be carried up and it returns to the junction and is easily taken off. This phenomenon is illustrated in Figure 20. Observations have shown that the behaviour is complicated further if there is a bend beyond the junction and there is not sufficient distance between them. In that case liquid could be caught up at the junction and not so much would fall back. There is also evidence that not all of the liquid falling back is taken off, some can be caught up again by the gas and emerge through the main pipe, Charron and Whalley (1995). At low liquid rates, for gas take off beyond a critical value, the film can be seen to be brought to a stop at the junction. This enables this liquid to be easily extracted through the side arm. The phenomenon can occur with both vertical and horizontal main pipes. Figure 21 shows the occurrence of such film stopping for the case of horizontal main pipe and vertically upward side arm, Azzopardi and Smith (1992). Azzopardi (1989) explains the phenomenon by considering the pressure changes in the main pipe. Measurement has shown that the pressure rises, a not very surprising result as the gas velocity has decreased. The liquid film, particularly that not directly taken off, reacts to this pressure change by decreasing its momentum, i.e., it slows down which leads to an increase in film thickness. Eventually a critical value of gas take off is reached where the film is brought to a complete halt, reacts to lateral pressure gradients and is taken off through the side arm, Figure 22.

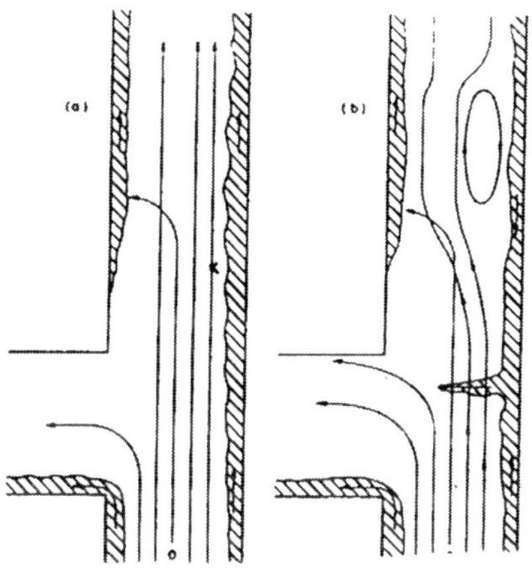

Figure 20. Sketches illustrating flooding affecting flow split at junction.

Figure 21. Example of liquid at junction as part of "film stop" mechanisms. Reprinted from International Journal of Multiphase Flow, vol. 18, Azzopardi, B.J. and Smith, P.A., Two-phase flow split at a T junction: Effect of side arm orientation and downstream geometry, pp 861-875, Copyright (1992), with permission from Elsevier.

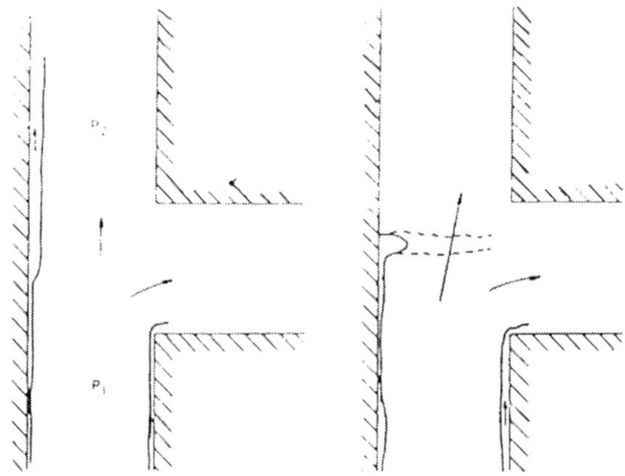

Figure 22. Sketch illustrating principal features of film stop mechanism.

Observations of the liquid layer in stratified flow experiments have shown that there can be significant differences in the layer height before and after the junction. This was noticed in experiments where both the main pipe and side arm were horizontal. However, it is much more clearly seen in Figure 23 taken from experiments with a vertically upward side arm, Azzopardi and Smith (1992). For gas take off of up to 55%, there is no liquid take off and the only visible change is a lengthening of the wavelength of waves on the interface. However, at a slightly higher gas take off (75%), a significant increase in layer height manifests itself. This facilitates the inception of liquid take off. Increasing the gas take off further both enhances the height difference and greatly increases the liquid take off. This phenomenon appears to be very similar to the classic hydraulic jump well know in open channel flows.

The trends with gas and liquid flow rates described above can probably be explained if we consider the split of liquid between film and drops and the asymmetry of the film in horizontal flow because of the action of gravity. The fraction of liquid traveling as drops increases with increase of both gas and liquid flow rates. The difference in momentum of the film and drops probably means that the film but not the drops will be easily taken off. As in vertical flow the film is axisymmetric, the decrease in liquid taken off with increasing gas flow rate can probably be explained by there being less liquid in the film. In the case of horizontal flows, the asymmetry in the film flow rate means that, as the gas flow rate decreases, there is less liquid near the junction and hence lower liquid take off. The trend with liquid flow rate is probably related to the film momentum, the lower the momentum the easier it is to divert the liquid.

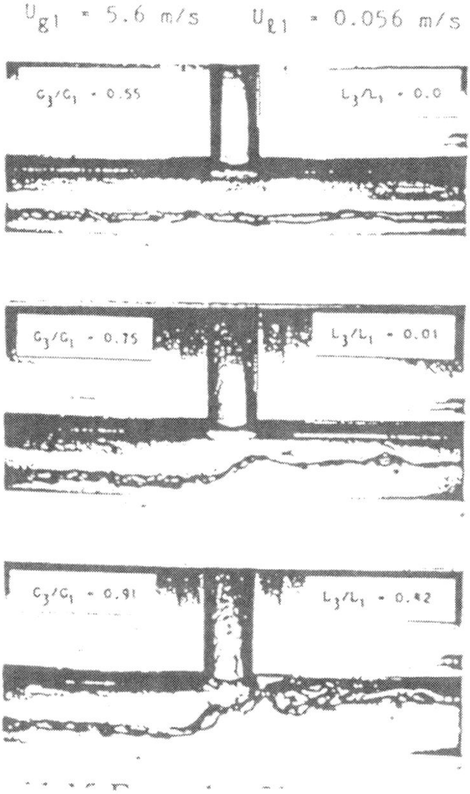

Figure 23. Example of increase of liquid level at the junction during stratified flow. Reprinted from International Journal of Multiphase Flow, vol. 18, Azzopardi, B.J. and Smith, P.A., Two-phase flow split at a T junction: Effect of side arm orientation and downstream geometry, pp 861-875, Copyright (1992), with permission from Elsevier.

11.2 Models of Phase Separation

Evidence from single-phase flow, McNown (1954), Charron and Whalley (1995), indiciates that the fluid taken off through the side arm comes from the segment of the main pipe nearest the side arm. For side arm diameters smaller than that of the main pipe, the segment assumption is reasonable. However, for equal diameter junctions, the boundary line is curved with the edges being further from the side arm than the central part. If both phases taken off were assumed to come from the segment, the relationship between liquid and gas taken off would be as shown in Figure 24 for annular and stratified flow. These curves would be independent of gas and liquid flow rates. However, published experimental data indicates that this description is inadequate. Here, two models based on the phenomena or mechanisms, which affect the phase spilt, are considered in detail. Other published models are considered briefly.

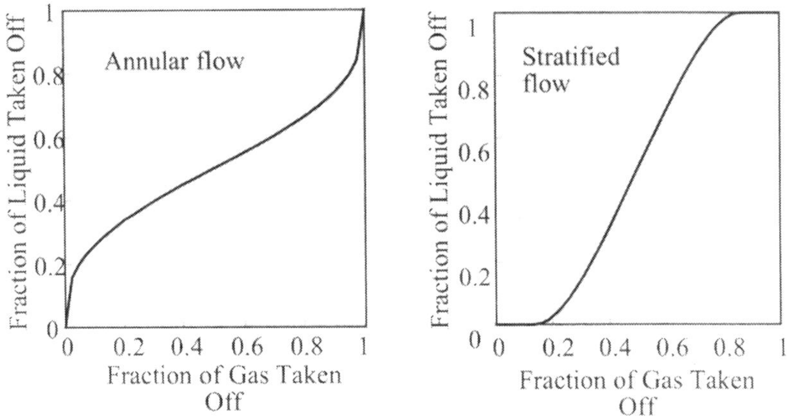

Figure 24. Separation according to fluids in segment - annular and stratified flow.

For vertical annular flow, Azzopardi and Whalley (1982) suggested that the film and gas both came from the same segment of main tube the side arm. The drops, because of their higher momentum, were assumed to carry straight on. This results in the following equations for the fraction of gas and liquid taken off through the side arm.

$$L'_{seg} = \frac{\dot{M}_{l3}}{\dot{M}_{l1}} = K \frac{\theta}{2\pi}(1 - E) \qquad G' = \frac{\dot{M}_{g3}}{\dot{M}_{g1}} = \frac{1}{2\pi}(\theta - \sin\theta) \tag{151}$$

K is an empirical correction factor for the effect of diameter ration suggested by Azzopardi (1984) which is given by $K = 1.2\left(\frac{D_{t3}}{D_{t1}}\right)^{0.4}$ and probably reflects the difference between equal diameter and reduced diameter junctions. At higher take off in vertical flow, Azzopardi (1988, 1989) suggested that flow reversal and flooding would reverse the direction of the liquid in the main pipe above the junction. This liquid falling back down to the junction would have low momentum and be easily taken off. The equation of Wallis (1961) was used to determine the conditions at which the liquid starts falling back, the amount of liquid fall back and the conditions at which no liquid was carried up. This equation is

$$\sqrt{u_g^*} + \sqrt{u_l^*} = 0.88 \tag{152}$$

where $u^*_i = u_{is2}(\rho_i/([\rho_l-\rho_g]gD_{t1})^{0.5}$ is a non-dimensionalised velocity and i refer to liquid or gas. For the start of flow reversal, $u_{ls2} = 0$ and $u_{gs2} = u_{g1}(1-G'_R)$ which can be determined explicitly from Equation (80). In other cases we also have to specify u_{ls2}. When stopping of the film does not take place the method suggested by Azzopardi (1988) that all the entrained liquid deposits in the pipe above the junction can be used and

$$u_{ls2} = \frac{\dot{M}_{l1} - \dot{M}_{lseg}}{\rho_l \pi D_t 1^2 /4} \tag{153}$$

where M_{lseg} is the liquid take off value calculated from Equation (57). If the film which does not enter the junction directly is brought to a halt by the film stopping mechanism, then only that liquid, which was originally entrained and which deposits in the pipe downstream of the junction, is available to fall back down to the junction. Then

$$u_{ls2} = \frac{\dot{M}_{lE}}{\rho_l \pi D_t 1^2 /4} \tag{154}$$

In the first case of flow reversal and flooding we must solve the combination of Equations (58-60). In the latter case of film stop, equation (154) should be substituted into (152) to give an explicit expression for G'_f, the fraction of gas take off beyond which there is no liquid carried up. In order to analyse the film stopping behaviour, the pressure drop across the junction must be specified. The extended Bernoulli equation can be written as

$$p_1 + \frac{\rho u_1^2}{2} = p_2 + \frac{\rho u_2^2}{2} + k_{12} \frac{\rho u_1^2}{2} \tag{155}$$

where k_{12} is a loss coefficient. For two-phase flow appropriate definitions of ρ and u are required. However, as film stopping occurs at very low liquid flow rates, the equation could be reasonably well approximated using the gas density and velocity. k_{12} can be described by an equation of Gardel (1957) who fitted data from single-phase flow experiments

$$k_{12} = 0.03 - 0.26G' + 0.58\, G'^2 \tag{156}$$

where G' is the mass fraction of gas going into the side arm. Combining Equations (62) and (63), together with using $u_{g2} = (1\text{-}G')u_{g1}$ and rearranging gives

$$p_1 - p_2 = \frac{\rho u_{g1}}{2} [0.03 - 2.26G' + 1.58\, G'^2] \tag{157}$$

For the liquid we can write

$$p_1 + \frac{\rho u_{f1}^2}{2} = p_2 + \frac{\rho_l u_{f2}}{2} + D \tag{158}$$

where u_{f1}, u_{f2} are the mean velocities of the liquid film before and after the junction respectively and D is a term to account for the dissipation of energy. As there is no information available to help specify D it has been assumed to be equal to zero in subsequent calculations. However, this is a limiting case and so D has been retained in the equations so that an indication can be obtained of the effect of a finite positive value of D on any results. Rearranging Equation (64) and combining with (63) yields

$$\rho_l u_{f2}^2 = \rho_l u_{f1}^2 + \rho_g u_{g1}^2 [0.03 - 2.26G' + 1.58\, G'^2] - 2D \tag{159}$$

Calculations show that u_{f2} can decrease at certain values of G'. Similarly, the film thickness δ_2, calculated from

$$\delta_2 = \frac{\Gamma}{u_{f2}} \tag{160}$$

shows a sudden increase. Obviously, a limiting case occurs when u_{f2} drops to zero. At this condition Equation (64) can be rearranged to give the critical value of G'

$$G'_C = 0.715 - \sqrt{0.493 - 0.633\frac{\rho_l u_{f1}^2}{\rho_g u_{g1}^2} + \frac{1.266D}{\rho_g u_{g1}^2}} \tag{161}$$

When D is set equal to zero, G'_C depends only on the film and gas momentum upstream of the junction. Inspection of Equation (161) shows that positive values of D would give lower values of G'_C. The liquid that is stopped in this way is assumed to be split between the side arm and the main pipe according to the shear imposed on it by the dividing gas flow. Allowing that the liquid already extracted or flowing as drops is not relevant to this division, an expression for the extra fraction of liquid of liquid taken off following film stop is

$$L'_s = (1 - L'_{seg} - E)\left(\frac{G'^2}{1 - 2G' + 2G'^2}\right) \tag{162}$$

L'_{seg} is the fraction of liquid taken off directly and obtained from equation (151). The fraction of liquid travelling as drops, E, was calculated using the method of Hewitt and Govan (1990).

There is no reason why the above approach should not be applied to horizontal annular flow and horizontal stratified flow of the crescent type. Obviously, flooding is irrelevant for these cases. However, the observations of Azzopardi and Smith (1992) indicate that the film stop phenomenon still occurs in horizontal pipes. Another difference from the vertical case is the variation of film flow rate around the main pipe. In this case an integral version of Equation (151) must be used. Available models of horizontal annular flow such as Fukano and Ousaka (1989) could provide the variation of film flow rate. For entrained fraction, Williams *et al.* (1996) provides an alternative correlation to that of Hewitt and Govan (1990). For the crescent stratified flow the film flow rate variation and the fraction of pipe covered by the liquid are required. If it is assumed that the film flow is uniform over the wetted section of pipe, then the correlations of Hart *et al.* (1989) can be used.

11.3 Comparison of Model with Data

Though there has not, as yet, been a systematic comparison between the published models and available data, some workers have compared their predictions with data other than their own. Here we consider some of those comparisons. For vertical annular flow approaching a junction, Azzopardi (1989) has shown that the prediction of his three phenomena model successfully predicts the data of Azzopardi (1988). Figure 25 illustrates how the inclusion of the different effects improves the predictive capability of the model. The effect of liquid flow rate is correctly handled. In most cases only direct take off of the liquid film and flooding are relevant, the film stopping is only relevant in the lowest two liquid flow rates. It is interesting to note that the models of Shoham *et al.* (1987) and Hart *et al.* (1991) also give reasonable predictions over most of the range. Of course, none of these models can cope with effect of flooding. The model of Ballyk and Shoukri (1990) is also close to the data.

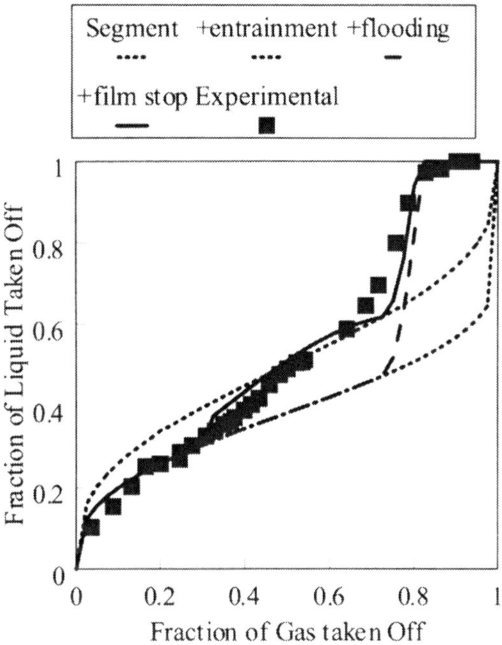

Figure 25. Cumulative effect of corrections to segment approach.

For horizontal annular flow, the models of Azzopardi (1989) and Ballyk and Shoukri (1990) successfully handle the effect of side arm orientations. Comparison of the predictions of the former with the data of Azzopardi and Memory (1989) is shown in Figure 26. The model of Fukano and Ousaka. (1989) has been used to predict the circumferential variation of film flow rate. These figures illustrate the importance of allowing for the non-uniformity of the film flow rate around the pipe circumference. Most of the available comparisons ignore the circumferential variation. Nevertheless, reasonable agreement can be achieved for horizontal side arms. It is interesting that the models of Shoham *et al.* (1987) and Hart *et al.* (1991), though giving the correct general trends, are not as close as that of Azzopardi (1989) which invokes two mechanisms as opposed to the one of the other models.

11.4 Pressure Drop

As seen in Figure 9, there are specific pressure drops associated with the junction. These can be important in some cases. However, it is probable that the pressure drops in the two downstream lines are far more important. Azzopardi and Hervieu (1994) reviewed available information on pressure drops.

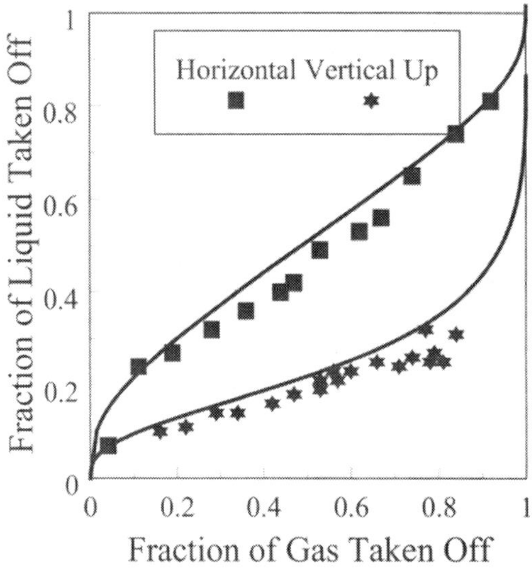

Figure 26. Horizontal annular flow.

11.5 Use of Computational Fluid Dynamics

A quite different approach has been presented by Lahey (1990) and by Issa and Oliveira (1994), who present the results of three-dimensional Computational Fluid Dynamic (CFD) predictions of phase separation in T-junctions. Their method is based on two different multi-dimensional two-fluid models, which involve momentum and mass balance equations for each phase. Two different turbulence models have also been implemented. In the turbulence model proposed by Lahey (1990) standard liquid phase k-e transport equations are used, assuming that the total liquid phase turbulence is the superposition of that due to bubble-induced effects and the shear-induced turbulence given by the k-ε model. The turbulence model of Issa and Oliveira (1994) accounts for instantaneous void fraction fluctuations, thus introducing additional correlation terms in the equations. The authors give details of the closure laws used in these models. The set of equations was solved numerically through a control-volume discretization, using a non-orthogonal mesh in the region of the junction.

Lahey (1990) showed that this approach gave good agreement with the KfK stratified flow data for various side arm orientations. Issa and Oliveira (1994) have also applied this approach to two-phase flows though a T-junction connecting rectangular cross-sectional channels. Comparisons were then possible with the local measurements of Popp and Sallet (1983). The first comparison concerns single-phase liquid flow, when 81% of the inlet flow rate is deflected into the side branch. The authors report good agreement between the measured and calculated velocity profiles in the side arm and in the main channel far downstream of the junction, but poor agreement in the plane of symmetry of the junction itself. This can be explained by the presence of the recirculation zone in the run, inducing strong 3-D secondary flows, as discussed by Hervieu (1988).

Popp and Sallet (1983) provide only a few profiles for the water velocity in two-phase bubbly flow, measured far downstream of the junction in the branch and in the run. These data are well predicted by the model of Issa and Oliveira (1994). Predictions of the phase split for bubbly flow also gave reasonable agreement with the data of Hewitt et al. (1990), Figure 27. Issa and Oliveira (1994) did not present pressure loss predictions. However, the predictions of Lahey (1990) showed good agreement with the KfK two-phase pressure drop data.

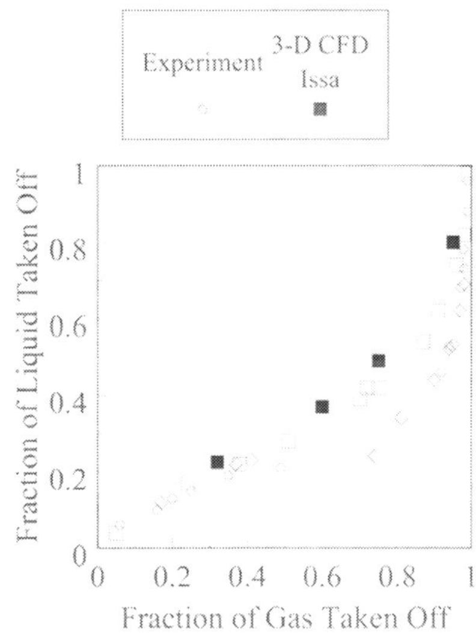

Figure 27. Comparison of prediction of Issa and Oliveira (1994) with data of Hewitt *et al.* (1990).

An interesting feature of the paper of Issa and Oliveira (1994) is the discussion concerning how the predictions of the model are affected by the expression of the drag coefficient. Hervieu (1988), who showed that the drag correlation developed for rigid spheres drastically underpredicts the interfacial friction effects, pointed out this trend. Hervieu recommended not only the use of a drag correlation accounting for the dynamic oscillations of the interface, but also to include the added mass effects when expressing the momentum exchange between continuous and dispersed phases.

To tackle horizontal annular flow dividing at a junction with all pipes horizontal, Issa and Adechy (2002) used CFD for the gas/drop flow and an integral method for the liquid film. The accuracy of the method in predicting film thickness variations at inlet and outlet of the junction is described in lecture 4. It gives good predictions of the phase split.

12 Use of a T-junction as partial phase separator

This case study is taken from the paper by Azzopardi *et al.* (2002). The output of a reactor is fed to a distillation column. Originally liquid, a significant part of it flashes as the pressure is dropped across a valve and along several metres of pipeline. In the column the liquid is being carried upwards with the vapour and causing the column to loose efficiency. This is a standard problem whose solution is to introduce a vessel to act as a phase separator into the line. The vapour and liquid are then introduced into separate points in the column. This approach has economic and safety implications. It has been suggested that the phases could be separated sufficiently using a T-junction so as to feed the vapour and liquid in a different points and thus restore efficiency. The current pipe work is 0.3 m internal diameter and the flow rates and physical properties of the vapour and liquid are typical of low boiling hydrocarbons. In order to assess the potential of a T-junction as a phase separator for this case it is important to understand the way in which the phases are distributed about the cross section of the feed pipe. The first step is to calculate the flow pattern. Here the method of Taitel and Dukler (1976) is used. It is deduced that the flow is annular. The next stage is to determine the fraction of liquid travelling as drops. The method described in lecture 4 is for vertical upflow not horizontal flow. Because of this, calculations were also made with the empirical equations of Hoogerndoorn and Welling (1965) and Williams *et al.* (1996). These methods all give values very close to 1, i.e., the liquid is travelling as drops. For a simple T-junction with all pipes in the horizontal plane, the model of Azzopardi (1989) described in above indicates that the liquid will carry on along the main pipe and the gas will exit from the side arm. However, because of the low surface tension, checks were made of the size of drops as it was feared that these might be small enough to follow the gas into the side arm. A value of 50 μm was obtained. If this and the Upper Limit Log Normal equation are used to obtain the distribution of sizes, will result in a significant part of the drops being in small sizes. More than 50% will be <50μm. Consequently, because of the small size and the high gas velocity, the assumption of drops carrying on along the main pipe might not be tenable in this case. An alternative approach will therefore be necessary. When an annular or stratified flow is forced to flow around a bend, the gas will follow the direction of the pipe. Drops will tend to impinge on the outside of the bend. Only the smallest drops will be able to follow the gas round the bend. The film will be thickened at specific places around the pipe wall depending on the orientation of the bend. The limited available evidence which is discussed in above shows that the entrained fraction after a bend is less than that before the bend for upstream entrained fractions greater than 0.1 and that the ratio of upstream entrained fraction to that downstream decreases as the upstream entrained fraction increases. Drop sizes, were found to be larger after the bend than before it. To calculate the effectiveness of deposition of drops at a bend, the method of Burkholz (1989) can be used, Equations (147-150). Calculations have been made of the deposition, which might occur at a bend using the methods described above. The equation of Azzopardi (1985) was used to calculate the Sauter mean diameter. This was employed in the Upper Limit Log Normal equation to give the distribution of sizes. Equations (147-150) were used, calculating the fraction depositing at one-micron intervals and summing the effects of all drop sizes. From this it was expected that 96% of the liquid should have deposited. As this liquid would initially be on the outside of the bend, the design illustrated in Figure 28 was proposed. The U bend was included to provide a liquid trap and prevent vapour exiting with the liquid.

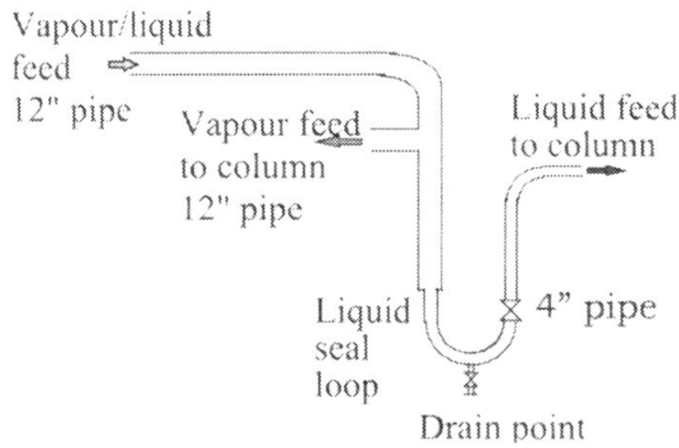

Figure 28. Proposed design. Reprinted from Chemical Engineering Research and Design, vol. 80, Azzopardi, B.J., Colman, D.A. and Nicholson, D., Plant application of a T-junction as a partial phase separator, pp 87-96, Copyright (2002), with permission from the Institution of Chemical Engineers.

As further evaluation of the T-junction as a separator a CFD model was set up using the Fluent code. This modelled the 3D gas flow in the pipe section shown in Figure 28 including flow around the bend and through the T-junction. A structured grid of 56000 cells was used and the turbulent flow of the single-phase gas was modelled using the Reynolds Stress Model. The gas inlet was just before the bend and all of the gas exited through the side arm. The run or nominal liquid leg was treated as a dead end. Trajectories of a range of drop sizes (33-70 μm) were calculated using stochastic tracking and the fate of these drops recorded. If a drop strikes a wall before it enters the side arm then it is considered 'captured'. If it enters the side arm then it is considered lost to the vapour stream. Figure 29 shows the trajectories of 50 μm drops (the mean size predicted above for a surface tension of 0.015 N/m).

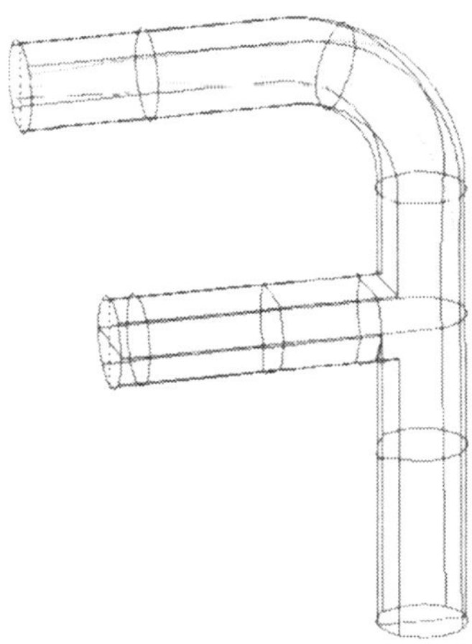

Figure 29. Trajectories of 50 μm drops. Reprinted from Chemical Engineering Research and Design, vol. 80, Azzopardi, B.J., Colman, D.A. and Nicholson, D., Plant application of a T-junction as a partial phase separator, pp 87-96, Copyright (2002), with permission from the Institution of Chemical Engineers.

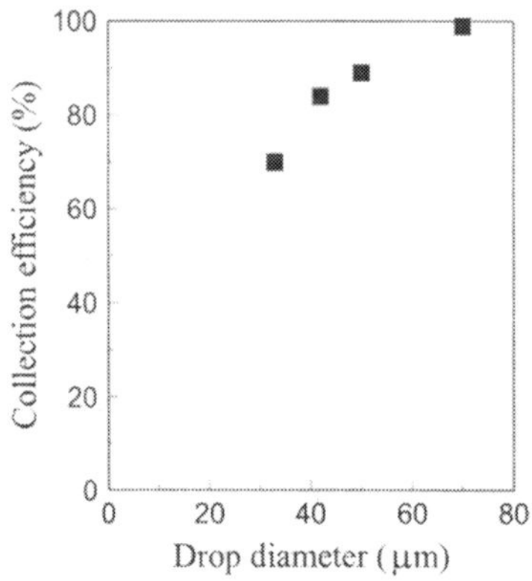

Figure 30. Collection efficiency predicted from drop trajectory computations.

From the calculation presented above, the mean drop size of the multi-phase flow in the pipe is expected to be 50 - 70 micron so the capture rate of the T-junction for all the liquid may be expected to be of the order 80 % for these physical conditions, the calculated grade efficiency is shown in Figure 30. This gas confidence in design and it was installed on plant. No direct measurements of the flows into and out of the Tee are possible, however non-intrusive methods have been used to infer the separation performance by comparison with flash calculations and the overall mass balance.

When the T-junction was initially started a nucleonics density measurement was made of the flows out of the T. This confirmed that the liquid leg contained liquid and hence was providing a liquid trap to force all of the vapour out of the side arm. It was noted that instead of a well-defined liquid level the density fell over a distance of about 1 m above the 'solid' liquid indicating the presence of a foam or dense mist here. It was not believed to be falling drops. If a foam or mist is present then this could allow some slippage of liquid into the vapour phase. The nucleonics method was not sufficiently sensitive to detect liquid in the vapour phase.

An ultrasonic mass flowmeter has been used to measure the liquid flow from the liquid leg. This gave a value of 22.2 tonnes/hr. While this was being measured a mass balance over the downstream column was used to calculate the total mass flow feed into the T-junction. In order to obtain a separation efficiency of the T-junction the theoretical liquid flow in the feed was calculated using an ASPEN flash calculation on the feed stream and compared with the measured liquid flow. The results for a calculated total mass flow of 57.1 t/hr gave 19.6 tonnes/hr vapour and 37.5 tonnes/hr liquid. Assuming no errors in the ASPEN flash calculation, but allowing for errors in the flow measurement of: 10% for the ultrasonic flow-meter 5% for the plant data. Then the efficiency = 51 - 69 %. If errors in the ASPEN flash of ±20% are considered then the efficiency range is 42 to 85%.

The plant management are very happy with the unit.

References

Andeen, G.B., and Griffiths, P. (1968) Momentum flux in two-phase flow. *Journal of Heat Transfer* 90:211-222.

Anderson, G.H., and Hills, P.D. (1974) Two-phase annular flow in tube bends. *Symposium on. Multiphase Flow Systems*, University of Strathclyde, paper J1, published in *Instution of Chemical Engineers Symposium Series* No. 38.

Archer, W.H. (1913) Experimental determination of loss of head due to sudden enlargement in circular pipes. *Transactions of the Americaan Society of Civil Engineers* 76:999-1026.

Armand, A.A. (1946) The resistance during the movement of a two-phase system in horizontal pipes *Izv. Vsesoyuznogo Tepl. Inst.* 1:16-23.

Armand, A.A., and Treschev, G.G. (1947) Investigation of resistance during the movement of steam-water mixtures in heated boiler pipes at high pressures. *Izv. Vsesoyuznogo Tepl. Inst* 4:1-5.

Arosio, S., Guglielmini, G., Lorenzi, A., Muzzio, A., and Sotgia, G. (1990) Two-phase pressure drop through sudden area contractions in horizontal flow. *Heat Transfer 1990* (Proceedings of the 9th International Heat Transfer Conference, Jerusalem, 19-24 Aug. 1990), Hemisphere Publishing Corperation 6:59-64.

Attou, A., and Bolle, L. (1997a) Integral formulation of balance equations for two-phase flow through a sudden enlargement - part 1: basic approach. *Proceedings of the Institution of Mechanical Engineers* 211C:387-397.

Attou, A., and Bolle, L. (1997b) Integral formulation of balance equations for two-phase flow through a sudden enlargement - part 2: a new interlocked volumes semi-empirical model. *Proceedings of the Institution of Mechanical Engineers* 211C:399-408.

Attou, A., Giot, M., and Seynhaeve, J.M. (1997) Modelling of steady-state two-phase bubbly flow though a sudden enlargement. *International Journal of Heat and Mass Transfer* 40:3375-3385.

Azzopardi, B.J. (1984) The effect of side arm diameter on two phase flow split at a T junction., *International Journal of Multiphase Flow* 10:509-512.

Azzopardi, B.J. (1985) Drop sizes in annular two-phase flow. *Experiments in Fluids* 3:53-59.

Azzopardi, B.J. (1988) Measurements and observations of the split of annular flow at a vertical T junction. *International Journal of Multiphase Flow* 14:701-710.

Azzopardi, B.J. (1989) The split of annular-mist flows at vertical and horizontal Ts. *Proceedings of the Eighth International Conference on Offshore Mechanics and Arctic Engineering*, The Hague, Netherlands, 19-23 March, ASME

Azzopardi, B.J. (1999) Phase split at T-junctions. *Multiphase Science and Technology* 11:223-329.

Azzopardi, B.J. and Whalley, P.B. (1982) The effect of flow pattern on two phase flow in a T junction. *International Journal of Multiphase Flow* 8:481-507.

Azzopardi, B.J., Purvis, A., Govan, A.H. (1987) Annular two-phase flow split at an impacting T. *International Journal of Multiphase Flow* 13:605-614.

Azzopardi, B.J., Wagstaff, D., Patrick, L., Memory, S.B., and Dowling, J. (1988) The split of two-phase flow at a horizontal T - annular and stratified flow. UKAEA Report AERE R13059.

Azzopardi B J and Memory, S.B. (1989) The split of two-phase flow at a horizontal T - annular and stratified flow. *4th International Conference on Multi-phase Flow*, Nice, France, 19-21 June (Pub. BHRA)

Azzopardi, B.J., and Smith, P.A. (1992) Flow split at a T junction: effect of side arm orientation and downstream geometry. *International Journal of Multiphase Flow* 18:861-875.

Azzopardi, B.J., and Sudlow, C.A. (1993) The effect of pipe fittings on the structure of two-phase flow. *XIth UIT National Heat Transfer Conference, Milan, Italy, 24-26 June.*

Azzopardi, B.J., and Hervieu, E. (1994) Phase separation at junctions. *Multiphase Science and Technology* 8:645-714.

Azzopardi, B.J., and Teixeira, J.C.F. (1994) Detailed measurements of vertical annular two phase flow - Part I: drop velocities and sizes. *Journal of Fluids Engineering* 116:792-795.

Azzopardi, B.J., Colman, D.A., and Nicholson, D. (2002) Plant application of a T-junction as a partial phase separator. *Transactions of the Institution of Chemical Engineers* 80A:87-96.

Ballyk, J.D., and Shoukri, M. (1990) On the development of a model for predicting phase separation phenomena in dividing two-phase flow. *Nuclear Engineering and Design* 123:67-75.

Banerjee, S., Rhodes, E., and Scott, D.S. (1967) Film inversion of co-current two-phase flow in helical coils. *American Institute of Chemical Engineers* 13:189-191.

Bankoff, S.G. (1960) A variable density single-fluid model for two-phase flow with particular reference to steam-water flow. *Journal of Heat Transfer* 82:265-272.

Bao, Z.Y., Bosnich, M.G., and Haynes, B.S. (1994) Estimation of void fraction and pressure drop for two-phase flow in fine passages. *Transactions of the Institution of Chemical Engineers* 72A:625-632.

Baroczy, C.J. (1963) Correlation of liquid fraction in two-phase flow with application to liquid metals. *6th National Heat Transfer Conference American Institute of Chemical Engineers*, Preprint No 26.

Baroczy, C.J. (1966) A systematic correlation for two-phase pressure drop. *Chemical Engineering Progress, Symposium Series*, 62:232-249.

Beattie, D.R.H., and Whalley, P.B. (1982) A simple two-phase frictional pressure drop calculation method. *International Journal of Multiphase Flow* 8:83-87

Beggs, H.D., and Brill, J.P. (1973) A study of two-phase flow in inclined pipes. *Journal of Petroleum Technology*,25:607-617.

Benedict, R.P. (1980) *Fundamentals of pipeflow*. Wiley-Interscience, New York.Bhaga,D. and Weber, M.E., (1972) Holdup in vertical two- and three-phase flow. *Canadian Journal of Chemical Engineering* 50:323-328.

Burkholz, A. (1989) *Droplet Separation*. VCH, Weinheim, Germany.

Cemak, J.O., Jicha, J.J., and Lightner, R.G. (1963) Two-phase pressure drop across vertically mounted thick plate restrictions. *ASME paper* 63-HT-11.

Chakbratai, P. (1976) Some aspects of annular two-phase flow in a horizontal tube PhD Thesis, Imperial College, London.

Charron, Y., and Whalley, P.B. (1995) Gas-liquid annular flow at a vertical tee junction - part I. Flow separation. *International Journal of Multiphase Flow* 21:569-589.

Chisholm, D. (1967) A theoretical basis for the Lockhart-Martinelli correlation for two-phase flow. *International Journal of Heat and Mass Transfer* 10:1767-1778.

Chisholm, D. (1972) An equation for velocity ratio in two-phase flow. *N.E.L.*, Report No 535.

Chisholm, D. (1983) *Two-phase flow in pipelines and heat exchangers* Pitman Press Ltd., Bath, England.

Chisholm, D. (1985) Two-phase flow in heat exchangers and pipelines. *Heat Transfer Engineering* 6:48-57.

Cicchitti, A., Lombardi, C., Silvestri, M., Soldani, G., and Zavatarelli, R. (1960) Two-phase cooling experiments - pressure drop, heat transfer and burnout experiments. *Energia Nucleare* 7:407-425.

Clark, N.N., and Flemmer, R.L. (1985) Predicting the holdup in two-phase bubble upflow and downflow using the Zuber and Findlay drift-flux model. *American Institute of Chemical Engineers Journal* 31:500-503.

Crane (1983) *Flow of fluids through valves, fittings and pipe*. Crane Ltd., London.

Delhaye, J.M. (1981) Singular pressure drops. In *Two-phase flow and heat transfer in the power and process industries*. A.E. Bergles (Ed), Hemisphere Pub. Corp.

Dowlati, R., Kawaji, M., and Chan A.M.C. (1988) Void fraction and friction pressure drop in two-phase flow across a horizontal tube bundle. *American Institute of Chemical Engineers Symposium Series* 84:126-132.

Dowlati, R., Kawaji, M., and Chan A.M.C. (1990) Pitch-to diameter effect on two-phase flow across an in-line tube bundle. *American Institute of Chemical Engineers Journal* 36:765-772

Dukler, A.E., Moye Wicks III, and Cleveland, R.G. (1965) Frictional pressure drop in two-phase flow - B: an approach through similarity analysis. *American Institute of Chemical Engineers Journal* 10:44-51.

El-Shaboury, A.M.F., Soliman, H,M., and Sims, G.E. (2001) Current state of knowledge on two-phase flow in horizontal impacting tee junctions. *Multiphase Science and Technology* 13:139-178.

Engineering Science Data Unit (ESDU) (1989) Two-phase flow pressure losses in pipeline fittings. ESDU Item No. 89012.

Fairhurst, C.P. (1983) Component pressure loss during two-phase flow. *International Conference on the Physical Modelling of Multiphase Flow*, Coventry, England, Paper A1, 1-24.

Ferrell, J.K., and McGee, J.W. (1964) Two-phase flow through abrupt expansions and contractions. TID 23394.

Fitzsimmons, D.E. (1964) Two-phase pressure drop in piping components. HW 80970 Rev. 1

Friedel, L. (1979) Improved friction pressure drop calculations for horizontal and vertical two-phase pipe flow. *European Two-phase Flow Group Meeting.*

Friedel, L., and Kissner, H.M. (1985) Pressure loss in safety valves during two-phase gas/vapour-liquid flow. *Proceedings of the 2nd International Conference on Multiphase Flow*, London, June (BHRA), 39-66.

Fruendt, J., Steiff, A., and Weinspach, P.-M. (1997) Pressure relief with highly viscous fluids. *Process Safety Progress* 16:57-59.

Fryer, P.J., and Whalley, P.B. (1982) The effect of swirl on the liquid distribution in annular two-phase flow. *International Journal of Multiphase Flow* 8:285-289.

Fukano, T., and Ousaka, A. (1989) Prediction of the circumferential distribution of film thickness in horizontal and near-horizontal gas-liquid annular flow. *International Journal of Multiphase Flow* 15:403-419.

Gardel, A. (1957) Les pertes de charge dans les ecoulementes au travers de branchements en te. Bulletin Technique de la Suisse Romande 9:122-130 and 10:143-148.

Gardner, G.C., and Neller, P.H. (1969) Phase distributions in flow of an air-water mixture round bends and past obstructions at the wall of a 76 mm boil tube. *Proceedings of the Institution of Mechanical Engineers* 184:36.

Geiger, G.E., and Rohrer, W.M. (1966) Sudden contraction losses in two-phase flow. *Journal of Heat Transfer* :1-9.

Guglielmini, G., Lorenzi, A., Muzzio, A., and Sotgia, G. (1986) Two-phase pressure drops across sudden area contractions - pressure and void fraction profiles. *Heat Transfer 86, (Proceedings of the 8th International Heat Transfer Conference*, San Francisco, 17-22 Aug., C.L. Tien et al. Ed.), Hemisphere 5:2361-2366.

Guglielmini, G., Muzzio, A., and Sotgia, G. (1997) The structure of two-phase flow in ducts with sudden contractions and its effect on the pressure drop. *Proceedings of the Conference on Experimental Heat Transfer, Fluid Mechanics and Thermodynamics* :1023-1036..

Haaland, S.E. (1983) Simple and explicit formulas for the friction factor in turbulent pipe flow. *Journal of Fluids Engineering* 105:89-

Harmathy, T.Z. (1960) Velocity of large drops and bubbles in media of infinite or restricted extent. *American Institute of Chemical Engineers Journal* 6:281-288.

Harris, D.M. (1967) Calibration of a steam-quality-meter for channel power measurement in the prototype SGHW Reactor. *European Two-Phase Flow Group Meeting*, Bournemouth.

Harris, D.M., and Shires, G.L. (1972) Two-phase pressure drop in a Venturi. *National Engineering Laboratory UK*, Report No. 549, pp. 18-33.

Harshe, B., Hussain, A., and Weisman, J. (1976) Two-phase pressure drop across restrictions and other abrupt area changes. Cincinnati University Ohio, Report NUREG 0062.

Hart, J., Hamersma, P.J., and Fortuin, J.M.H. (1989) Correlations predicting frictional pressure drop and liquid holdup during horizontal gas-liquid pipe flow with a small liquid holdup, *International Journal of Multiphase Flow* 15:947-964.

Hart, J., Hamersma, P.J., and Fortuin, J.M.H. (1991) A model for predicting liquid route preference during gas-liquid flow through horizontal branched pipelines. *Chemical Engineering Science* 46:1609-1622.

Hervieu, E. (1988) Ecoulement monophasique et diphasique a bulles dans un branchement en T: étude théorique et expérimentale. Thèse de Doctorat de l'Institut National Polytechnique de Grenoble, France.

Hewitt, G.F., and Dell, F.R. (1968) Two-phase flow of air-water mixtures in horizontal coiled tubes. Private communication.

Hewitt, G.F. (1978), *Measurement of two -phase flow parameters.* Academic Press, London

Hewitt, G.F. (1984) Two-phase flow through orifices, valves, bends and other singularities. *8th Lecture Series on Two-Phase Flow*, Trondheim.

Hewitt, G.F., Gill, L.E., Roberts, D.N., and Azzopardi, B.J. (1990) The split of low inlet quality gas/liquid flow at a vertical T - Experimental data. UKAEA Report AERE M3801. Report AERE M2459.

Hewitt, G.F., and Govan, A.H. (1990) Phenomenological modelling of non-equilibrium flow with phase change. *International Journal of Heat and Mass Transfer* 32:229-242.

Hills, J.H., Azzopardi, B.J., and Barhey, A.S. (1996) Spatial unsteadiness - a way towards intensive gas-liquid reactors *Transactions of the Institution of Chemical Engineers* 74A:567-574.

Holt, A.J., Azzopardi, B.J., and Biddulph, M.W. (1995) The effect of density ratio on two-phase frictional pressure drop. *International Symposium on Two-Phase Flow Modelling and Experimentation*, Rome, 9-11 October.

Holt, A.J., Azzopardi, B.J., and Biddulph, M.W. (1997) Two-phase pressure drop and void fraction in narrow channels. 5th U.K. National Heat Transfer Conference, London.

Holt, A.J., Azzopardi, B.J., and Biddulph, M.W. (1999) Calculation of two-phase pressure drop for vertical upflow in narrow passages by means of a flow pattern specific model. *Transactions of the Institution of Chemical Engineers* 77A:7-15.

Honan, T.J., and Lahey, R.T. (1981) Measurement of phase separation in wyes and tees. *Nuclear Engineering and Design* 64:93-102.

Hoogerndoorn, C.J., and Welling, W.A. (1965) Experimental studies on the character of annular-mist flow in horizontal pipes. *Symposium on Two-phase Flow*, Exeter, 21-23 June, Paper C3.

Hughmark, G.A. (1962) Hold-up in gas-liquid flow. *Chemical Engineering Progress* 58:62-65.

Isbin, H.S., Moen, R.H., Wickey, R.O., Mosher, D.R., and Larson, H.C. (1958) Two-phase steam-water pressure drops. *Nuclear Science and Engineering Conference, Chicago.*

Issa, R.I., and Oliveira, P.J. (1994) Numerical prediction of phase separation in two-phase flow through T-junctions. *Computers Fluids* 23:347-372.

Issa, R.I, and Adechy, D. (2002) Modelling of annular flow through pipes and t-junctions. Submitted.

James, P.W. Azzopardi, B.J. Graham, D.I., and Sudlow, C.A. (2000) The effect of a bend on droplet size distribution in two-phase flow. *7th International Conference on Multiphase Flow in Industrial Plants,* Bologna, 13-15 September.

Janssen, E., and Kervinen, J.A. (1964) Two-phase pressure losses - final report. US Atomic Energy Comm., Report No. GEAP 4634.

Kirillov, P.R., Smogalev, I.P., Doroshenko, V.A. (1982) A graphical method of predicting the losses of pressure due to friction with a rising steam-water flow in round tubes. *Thermal Engineering* 29:171-172

Kooijman, J.M., and Lacey, P.M.C. (1968) Unpublished experiments. University of Exeter.

Lahey, R.T., and Moody, F.J. (1977) *The Thermal Hydraulics of a Boiling Water Nuclear Reactor, American Nuclear* Society.

Lahey, R.T., Azzopardi, B.J., and Cox, M. (1985) Modelling two-phase flow division at T-junctions. *2nd International Conference on Multiphase Flows*, London, 19-21 June (ed. BHRA).

Lahey, R.T. (1990) The analysis of phase separation and phase distribution phenomena using two-fluid models. *Nuclear Engineering and Design* 122:17-40.

Lemonnier, H., and Hervieu, E. (1991) Theoretical modelling and experimental investigation of single-phase and two-phase division at a tee junction. *Nuclear Engineering and Design* 125:201-213.

Levy, S. (1960) Steam-slip - theoretical prediction from momentum model. *Journal of Heat Transfer* 82:113-124.

Lin, Z.H. (1985) Two-phase flow measurements with orifices. In *Encyclopedia of Fluid Mechanics* (Ed. N. Cherenmisinoff) Gulf Publishing Co. Houston.

Lockhart, R.W., and Martinelli, R.C. (1949) Proposed correlation of data for isothermal, two-phase, two-component flow in pipes. *Chemical Engineering Progress* 45:39-48.

McNown, J.S. (1954) Mechanics of manifold flow. *ASCE Transactions* 119:1103-1142.

McQuillan, K.W., and Whalley, P.B. (1984) The effect of orifices on the liquid distribution in annular two-phase flow. *International Journal of Multiphase Flow* 10:721-73.

Maddock, C., Lacey, P.M.C., and Patrick, M.A. (1974) The structure of two-phase flow in a curved pipe. *Symposium on. Multiphase Flow Systems*, University of Strathclyde, paper J2, published in *Instution of Chemical Engineers Symposium Series* No. 38.

Marchaterre, J.F., and Hoglund, B.M. (1962) Correlation for two-phase flow. *Nucleonics* 8:142-

Martinelli, R.C., and Nelson, D.B. (1948) Prediction of pressure drop during forced circulation boiling of water", *Transaction of the American Society of Mechanical Engineers* 70:695-702.

Massena, W.A. (1960) Steam-water pressure drop. *Hanford report* H.W. 65706.

Mishima, K., And Hibiki, T. (1996) Some characteristics of air-water two-phase flow in small diameter vertical tubes. *International Journal of Multiphase Flow* 22:703-712.

Morris, S.D. (1984) A simple model for estimating two-phase momentum flux. *Institution of Chemical Engineers Symposium Series* No. 86, 2:773-784.

Morris, S.D (1985) Two-phase pressure drop across valves and orifice plates. European Two-phase Flow Group *Meeting*, Marchwood Engineering Laboratory, Southampton, England.

Muller-Steinhagen, H., and Heck, K. (1986), A Simple Friction Pressure Drop Correlation for Two-Phase Flow in Pipes, *Chemical Engineering Process* 20:297-308.

Norstebo, A. (1985) Pressure drop in bends and valves in two-phase refrigerant flows. *2nd International Conference on Multiphase Flows*, London, 19-21 June (ed. BHRA).

Oshinowo, T., and Charles, M.E. (1974) Vertical two-phase flow. *Canadian Journal of Chemical Engineering* 52:438-448.

Owens, W.L. (1961) Two-phase pressure gradient. *International Heat Transfer Conference, Boulder, Colorado.*

Patrick, M., and Swanson, B.S. (1959) Expansion and contraction of an air-water mixture in vertical flow. *American Institute of Chemical Engineers* 5:440-445.

Popp, M., and Sallet, D.W. (1983) Experimental investigation of one and two-phase flow through a tee-junction. Paper B3, *International Conference on the Physical Modelling of Multiphase Flows,* Coventry, England, April 19-21.

Premoli, A., Francesco, D., and Prina, A. (1970) An empirical correlation for evaluating two-phase mixture density under adiabatic conditions. *European Two-Phase Flow Group Meeting.*

Ribeiro, A.M., Bott, T.R., and Jepson, D.M. (2001) The influence of a bend on drop sizes in horizontal annular two-phase flow. *International Journal of Multiphase Flow* 27:721-728.

Richardson, B.E. (1959) Some problems in horizontal two-phase two component flows. ANL 5949.

Romie, F. (1958) Unpublished information.

Rooney, D.H. (1968) Void fraction prediction under saturated conditions. *N.E.L.* Report No 386.

Rose, S. (1964) Some hydrodynamic Characteristics of bubbly mixtures flowing vertically upwards in tubes. ScD Thesis, Massachusetts Institute of Technology.

Saba, N., and Lahey, R.T. (1984) The analysis of phase separation phenomena in branched conduits. *International Journal of Multiphase Flow* 10:1-20.

Salcudean, M., Chun, C.H., and Groenveld, D.C. (1983) Effect of flow obstructions on void distribution in horizontal air-water flow. *International Journal of Multiphase Flow* 9:91-96.

Schadel, S.A., Leman, G.W., Binder, J.L., and Hanratty, T.J. (1990) Rates of atomisation in vertical annular flow. *International Journal of Multiphase Flow* 16:363-374.

Schmidt, J., and Friedel, L. (1997) Two-phase pressure drop across sudden contractions in duct area. *International Journal of Multiphase Flow* 23:283-299.

Schrage, D.S., Hau, J.T., and Jensen, M.K. (1987) Void fractions and two-phase multipliers in a horizontal tube bundle. *American Institute of Chemical Engineers Symposium Series* **No 257**.

Shoham, O., Brill, J.P., and Taitel,Y. (1987) Two-phase flow splitting in a Tee junction - experiment and modelling. *Chemical Engineering Science* 42:2667-2676.

Simpson, H.C., Rooney, D.H., and Gratton, E. (1983) Two-phase flow through gate valves and orifice plates. *International Conference on the Physical Modelling of Multiphase Flow*, Coventry, England, Paper A2, :25-40.

Simpson, H.C., Rooney, D.H., and Callender, T.M. (1985) Pressure loss through gate valves with liquid-vapour flows. *2nd International Conference on Multiphase Flows*, London, 19-21 June (ed. BHRA) :67-80.

Smith, S.L. (1971) Void fraction in two-phase flow - a correlation based upon equal velocity head model. *Heat and Fluid Flow* 1:22-39

Taitel, Y., and Dukler, A.E. (1976) A model for predicting flow regime transitions in horizontal and near-horizontal gas-liquid flow. *American Institute of Chemical Engineers Journal* 22:47-55.

Thom, J.R.S. (1964) Prediction of pressure drop during forced circulation boiling of water. *International Journal of Heat and Mass Transfer* 7:709-624.

Ulbrich, R., and Mewes, D. (1995) Experiemtnal study of gas void fraction for two-phase flow across tube bundles. *Proceedings of the International Symposium on Two-Phase Flow Modelling and Experimentation*, Rome, 9-11 October, 1211-1218.

Wallis, G.B. (1961) Flooding velocities for air and water in vertical tubes. UKAEA Report AEEW R123.

Wallis, G.B. (1969) *One-dimensional Two-phase Flow.*, McGraw-Hill.

Ward Smith, A.J. (1971) *Pressure losses in ducted flows.* Butterworth, London.

Weisman, J. (1974) Two-phase pressure drop studies. Second Water Reactor Information Meeting, Washington, October.

Whalley, P.B. (1980) Air-water two-phase flow in a helically coiled tube. *International Journal of Multiphase Flow* 6:345-356.

Wiafe, F.K. (1970) Two-phase flow through rough tubes. PhD Thesis, University of Strathclyde.

Williams, L.R., Dykhno, L.A., and Hanratty, T.J., (1996) Droplet flux distributions and entrainment in horizontal gas-liquid flows. *International Journal of Multiphase Flow* 22:1-18.

Wooley, D.M., and Muller-Steinhagen, H. (1989) Prediction of frictional pressure drop for two phase flow in horizontal pipes. *Proceedings of the Seventeenth Australian Chemical Engineering Conference*, 184-190.

Zivi, S.M. (1964) Estimation of steady state steam void fraction by means of the principle of minimum entropy production. *Journal of Heat Transfer* 86:247-252.

Zuber, N., and Findlay, J.A. (1965) Average volumetric concentration in two-phase flow systems. *Journal of Heat Transfer* 87:453-468.

Zuber, N., and Hench, J. (1962) Steady state and transient void fraction of bubbling systems and their operating limits, Part I: Steady state operation. General Electric Report 62GL100.

Liquid-Liquid Two-Phase Flow Systems

Neima Brauner

School of Engineering,
Tel-Aviv University
Tel-Aviv 69978, Israel

1 General Description of Liquid-Liquid Flows: Flow Patterns

Flows of two immiscible liquids are encountered in a diverse range of processes and equipments. In particular in the petroleum industry, where mixtures of oil and water are transported in pipes over long distances. Accurate prediction of oil-water flow characteristics, such as flow pattern, water holdup and pressure gradient is important in many engineering applications. However, despite of their importance, liquid-liquid flows have not been explored to the same extent as gas-liquid flows. In fact, gas-liquid systems represent a very particular extreme of two-fluid systems characterized by low-density ratio and low viscosity ratio. In liquid-liquid systems the density difference between the phases is relatively low. However, the viscosity ratio encountered extends over a range of many orders of magnitude. Table 1.1 summarizes experimental studies reported in the literature on horizontal oil-water pipe flows, while studies on inclined and vertical systems are summarized in Table 1.2 and 1.3. (The tables can be found at the end of the end of this article before the bibliography). These tables reflect the wide range of physical properties encountered. Moreover, oils and oil-water emulsions may show a Newtonian or non-Newtonian rheological behavior. Therefore, the various concepts and results related to gas-liquid two-phase flows cannot be readily applied to liquid-liquid systems.

Diverse flow patterns were observed in liquid-liquid systems. In most of the reported studies the identification of the flow pattern is based on visual observations, photographic/video techniques, or on abrupt changes in the average system pressure drop. In some recent studies, the visual observation and pressure drop measurements are backed-up by conductivity measurements, high frequency impedance probes or Gamma densitometers for local holdup sampling, or local pressure fluctuations and average holdup measurements (see Tables 1.1 to 1.3). The flow patterns can be classified into four basic prototypes: Stratified layers with either smooth or wavy interface; Large slugs, elongated or spherical, of one liquid in the other; A dispersion of relatively fine drops of one liquid in the other; Annular flow, where one of the liquids forms the core and the other liquid flows in the annulus. In many cases, however, the flow pattern consists of a combination of these basic prototypes.

Sketches of various possible flow patterns observed in horizontal systems are given in Figure 1.1. Stratified flow with a complete separation of the liquids may prevail for some limited range of relatively low flow rates where the stabilizing gravity force due to a finite density difference is dominant (Figure 1.1a). It is possible that one of the layers is discontinuous, and the flow structure is stratified layers of a free liquid and a dispersion of the other liquid (Figure 1.1c-d). With increasing the flow rates, the interface displays a wavy character with possible entrainment of drops at one side or both sides of the interface (Figure 1.1b, 1.1e-g). The entrainment process increases with increasing the flow rates. When the lighter and heavier phases are still continuous at the top and bottom of the pipe, but there is a concentrated layer of drops at the interface, a three layer structure is formed (Figure 1.1h). Eventually, for sufficiently high water flow rate, the entire oil phase becomes discontinuous in a continuous water phase resulting in an oil-in-water dispersion or emulsion (Figure 1.1i). An emulsion is a stable dispersion. Vice versa, for sufficiently high oil flow rate, the water phase may be completely dispersed in oil phase resulting in a water-in-oil dispersion or emulsion (Figure 1.1j). It is also possible for oil-in-water and water-in-oil dispersions to coexist. Impurities and high mixture velocities may yield a foam like structure of intensively intermixed oil and water, possibly with occasional appearance of clusters of one of the liquids. There are operating conditions under which

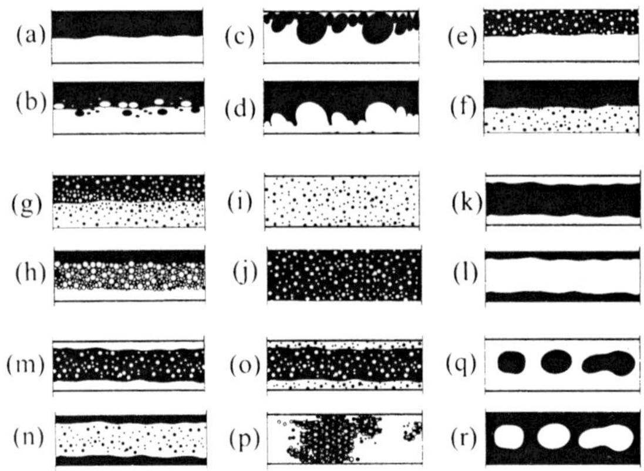

Figure 1.1 **a/b** Stratified flow of two separated layers (S, possibly with mixing at the interface, SM). **c/d** Stratified layers of a free-liquid and a dispersion of the other liquid (e.g. oil-in-water dispersion above a water layer, Do/w&w). **e/f** Stratified layers of a free liquid and a dispersion in the other liquid (e.g. oil and oil-in-water dispersion, Do/w&o; water and water-in-oil dispersion, Dw/o&w). **g/h** Layers of dispersions (e.g. water-in-oil dispersion above oil in water dispersion Dw/o&o/w, possibly with pure oil at the top and/or water at the bottom). **i/j** Fully dispersion or emulsion of one liquid in the other liquid (e.g. water-in-oil or oil-in-water dispersion or emulsion, Dw/o or Do/w). **k/l** Core-annular flow – a core of one liquid within the other liquid (e.g. a core of viscous oil and water in the annulus, ANw. Oil in the annulus, ANo). **m/n** Annular flow of a liquid with a dispersion in the core (water in the annulus DANw, oil in the annulus DANo). **o** Core-annular flow of two dispersions (CADw or CADo). **p** Intermittent flow (one liquid alternately occupying the pipe as a free liquid or as a dispersion, Io or Iw). **q/r** Large elongated or spherical bubbles of one liquid in the other (SLo,Bo or SLw,Bw).

an oil-in-water dispersion will change to water-in-oil dispersion. This phenomena is referred in the literature as phase inversion and is associated with an abrupt change in the frictional pressure drop (see Figure 15 in Brauner, 1998). [1]

Under certain conditions, the oil and water may stabilize in annular-core configuration (Figures 1.1k-ℓ). The flow of a viscous oil in a core, which is lubricated by a water film in the annulus (core flow), is most attractive from the viewpoint of pressure drop reduction in transportation of highly viscous oils. With increasing the water rate, the viscous core breaks up to either large slugs and bubbles, (Figure 1.1q) or into oil dispersion flowing in a continuous water phase (Figure 1.1i or q). It is possible to have also "inverted" annular flow with the oil flowing in the annulus (Figures 1.1ℓ or n). Comparison of experimental flow pattern maps reported in the literature for horizontal oil-water systems of relatively low viscosity ratio, $\mu_o/\mu_w < 100$ and $\Delta\rho/\rho_w \geq 0.1$ (shown in Brauner, 1998, Figure 2) indicated a general similarity between the sequence of the observed flow patterns and the stratified flow boundaries, but differences in the classification of the various dispersed flow regimes and the associated transitional boundaries. However, some of the reported transitional boundaries actually represent gradual changes in the dispersions structure and the associated pressure drop, and are therefore susceptible to subjective judgment

[1] A copy of this reference can be downloaded from http://www.eng.tau.ac.il/~brauner/LL-Flow

and variations. When the water is the continuous phase, oil viscosity seems to have a minor effect on the flow patterns. However, the oil viscosity affects the location of the phase inversion from Dw/o to Do/w. The input water-cut, U_{ws}/U_m required to invert the dispersion decreases with increasing the oil viscosity. Core flow (water annulus) is usually not obtained in oil-water systems of relatively low oil viscosity and relatively high $\Delta\rho$.

As in gas-liquid systems, the flow pattern depends on the liquids flow rates and physical properties, tube diameter and inclination. However, due to the relatively low density differential between the two-fluids, the role of gravity in liquid-liquid systems diminishes. Therefore, wall-wetting properties of the liquids and surface tension forces become important and may have a significant effect on the flow pattern. For instance, in stratified flow the interface between the liquid phase is not necessarily planar. The common assumption of a plane interface (Fig. 1.2a) is appropriate for horizontal gas-liquid systems, which are dominated by gravity. In fact, systems of low density differential as oil-water systems, resemble reduced gravity systems and capillary systems, where surface forces become important. The wetting liquid tends to climb over the tube wall resulting in a curved (concave or convex) interface (Fig. 1.2b or h). Stratified flows with curved interfaces in liquid-liquid systems have been obtained both experimentally (Valle and Kvandel, 1995, Angeli et al., 2002, Gat, 2002) and in numerical simulations (Ong et al., 1994). The possible stratified flow configurations extend from fully eccentric core of the upper phase (Fig. 1.2c) to fully eccentric core of the lower phase (Fig. 1.2g). Hydrodynamic forces may also cause the core phase to detach from the wall surface to form an eccentric core-annular configuration. However, due to a density differential between the core phase and the annular phase, the core usually stabilizes in an eccentric position (Fig. 1.2d or f) rather than in a concentric position (Fig. 1.2c). Break-up of the top (or bottom) wall film due to the float-up tendency of light (or heavier) core phase results in stratification of the fluids.

The occurrence of annular flow in liquid-liquid systems is therefore more frequently encountered in oil-water systems of low density differential, $\Delta\rho$ and small diameter tubes. These systems are characterized by a small non-dimensional Eotvös number, $Eo_D = \frac{\Delta\rho g D^2}{8\sigma} \ll 1$ and resemble micro-gravity systems. In such systems, an annulus of the wetting phase (surrounding a core of the non-wetting phase) is a natural configuration which complies with surface tension forces and wall-adhesion forces. However, for specified operational conditions, different flow patterns may result by changing the tube material (hydrophobic or hydrophilic). The start-up procedure (oil flowing in the pipe and then introducing water or vice versa), which affects the effective liquids-wall adhesion, or entry conditions (type of nozzle used to introduce the two-liquids) are also important factors in controlling the flow pattern.

In vertical upward flow and low oil viscosities, the observed flow patterns typically include oil drops, bubbles or slugs in water, transitional flow (TF, churn), water drops in oil and oil-in-water or water-in-oil emulsions (see Figure 5.3, Section 5). The physical interpretation of flow patterns transitions is similar to that described for vertical gas-liquid systems. However, the clearly defined bullet-shaped bubbles that characterize slug flow in gas-liquid slug flow are normally not observed in oil-water systems. The churn flow is characterized as intermittent flow of complex and irregular structures of continuous oil phase (oil-dominated) and continuous water phase (water-dominated). The drops size decreases with increasing the mixture velocity, and for high velocity the liquids, either homogeneous Dw/o or Do/w of fine droplets are formed.

The organized flow pattern data on inclined liquid-liquid systems in the literature is rather limited (see Table 1.2). In large Eo_D systems, a considerable drift between the lighter oil phase and water phase exists at low mixture velocities. Under such conditions, even a moderate inclination from the vertical affects intermittency in the flow with regions of back-flow of the heavier phase (Vigneau et al., 1988 and Flores et al., 1997, Figure 7 shown in Brauner, 1998). It is to be noted that the stratified pattern typically vanishes for steeper upward inclinations than $\approx 30^o$, compared to gas-liquid systems where the stratified flow vanishes already for shallow upward inclinations.

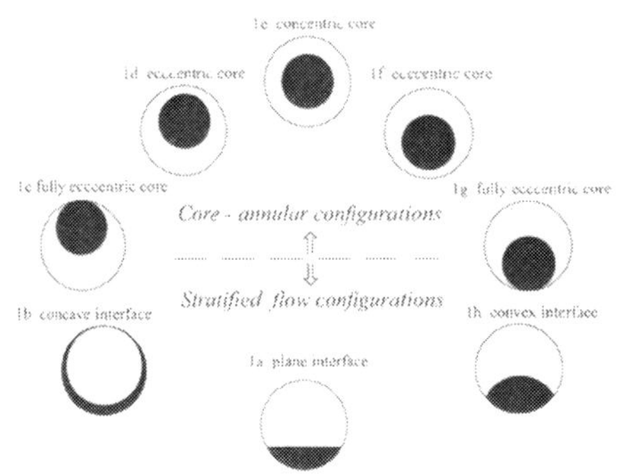

Figure 1.2 Schematic description of various configurations of separated flows.

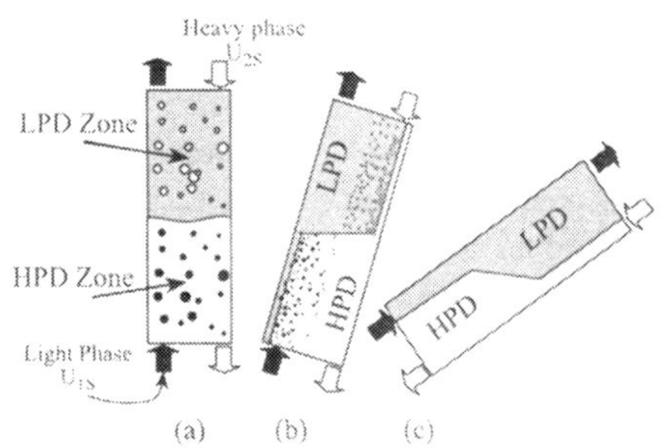

Figure 1.3 Flow patterns in countercurrent flow, a. vertical, b. off-vertical, c. inclined column.

Counter-current liquid-liquid flow is frequently encountered in the process industry. Figure 1.3 shows a schematic description of the flow patterns obtained in a column in counter-current flow of relatively low flow rates (Ullmann, et al., 2001). In a vertical column (Figure 1.3a), the basic flow pattern is dispersed flow, with either the heavy phase dispersed in the light phase (light phase dominated, LPD), or the light phase dispersed in the heavy phase (heavy phase dominated, HPD). These two configurations of dispersed flow can be simultaneously obtained in the column, separated by an interface. The latter can be placed at any position along the column by manipulating the resistance at the heavy phase outlet. With a sufficiently low (high) resistance, the flow pattern in the entire column is LPD (HPD), respectively. In systems of $Eo_D \gg 1$, the phases tend to segregate with a slight off-vertical positioning of the column (Figure 1.3b). The two configurations obtained in this case correspond to stratified-dispersed flow both in the HPD (Do/w&o) and in the LPD (Dw/o&w) zones. Further inclining the tube results in a complete segregation of the phases. In an inclined tube (Figure 1.3c), the basic flow pattern in both zones is stratified flow with either a wavy or smooth interface. The flow in the HPD (LPD) zone corresponds to a thick (thin) layer of the heavy phase flowing counter-currently to a thin (thick) layer of the light phase. Similarly to the operation of a vertical column, the location of the interface between these two zones can be controlled by adjusting the resistance at the heavy phase outlet. Thereby, the entire column can be occupied by either one of these two flow configurations, or by both of them.

The various flow patterns are associated with different pressure drop, in situ holdup, heat transfer coefficient and other related phenomena, such as fouling and corrosion of the pipe. Therefore, generalized models which attempt to cover the whole range of different liquid properties and different flow patterns (e.g. Charles and Lillelcht, 1966, Theissing, 1980) can only be approximate. The accepted approach today consists of predicting the flow pattern under specified operational conditions (see Section 5) and applying an appropriate model (see Sections 2,3,4).

2 Stratified Flow

Stratified flow is considered a basic flow pattern in horizontal or slightly inclined liquid-liquid systems of a finite density differential, since for some range of sufficiently low flow rates, the two liquids phases tend to segregate. The modeling of liquid-liquid stratified flows requires the consideration of additional aspects in comparison to gas-liquid stratified flows. Due to the variety of physical properties that may be encountered, it is not a priori evident which of the phases is the faster (for specified operational conditions). Therefore, the ambiguity concerning the appropriate closure law for representing the interfacial shear is even greater than in the case of gas-liquid flows. Multiple solutions can be obtained for specified operation conditions in co-current and counter-current inclined flows, which are relevant in practical applications. Moreover, as a result of the relatively low density difference, surface tension and wetting effects become important, and the interface shape (convex, concave, plane) is an additional field that has to be solved.

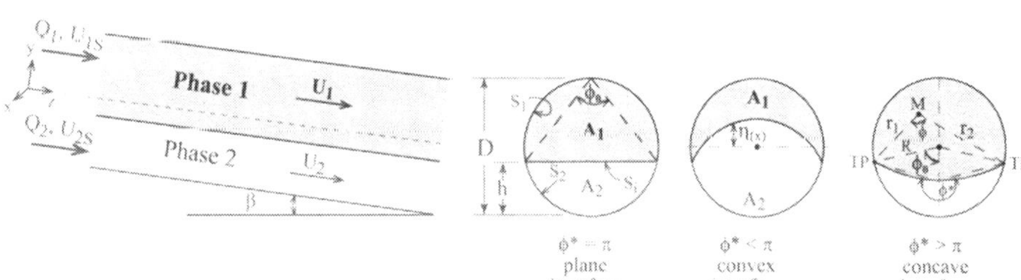

Figure 2.1 **Schematic description of stratified flow configuration and parameters.**

The stratified flow configuration and coordinates are illustrated in Figure 2.1. A configuration of a curved interface is associated with a different location of the triple point (TP) and thus, with a variation in the contact area between the two fluids and between the fluids and the pipe wall. Depending on the physical system involved, these variations can have prominent effects on the pressure drop and transport phenomena. On the other hand, the feasibility of obtaining exact solutions for stratified flows is restricted to laminar-laminar flows, which are of limited relevance to practical applications of gas-liquid two phase flows. However, laminar flow in both phases is frequently encountered in liquid-liquid systems.

Given the location of the fluids interface, the 2-D velocity profiles in steady and fully developed axial laminar flow of stratified layers, $u_1(x,y), u_2(x,y)$ are obtained via analytical or numerical solutions of the following Stokes equations (in the z direction, see Figure 2.1):

$$\mu_1\left(\frac{\partial^2 u_1}{\partial x^2}+\frac{\partial^2 u_1}{\partial y^2}\right)=\frac{\partial P_1}{\partial z}-\rho_1 g\sin\beta\ ;\ \ \mu_2\left(\frac{\partial^2 u_2}{\partial x^2}+\frac{\partial^2 u_2}{\partial y^2}\right)=\frac{\partial P_2}{\partial z}-\rho_2 g\sin\beta \quad (2.1)$$

The required boundary conditions follow from the no-slip condition at the pipe wall and continuity of the velocities and tangential shear stresses across the fluids' interface. For a given axial pressure drop, the solution for u_1 and u_2 can be integrated over the fluids flow cross sections to yield the corresponding volumetric flow rates Q_1 and Q_2. From the practical point of view, we are interested in a solution for the pressure drop and flow geometry (interface location) for given flow rates. However, the inverse problem is much more complicated, since the shape of fluids interface is, in fact, unknown.

2.1 The interface shape

The location of the interface can be obtained by considering the Navier-Stokes equations in the y and x directions:

$$\frac{\partial P_j}{\partial y}+\rho_j\ g\cos\beta=0\quad ;\quad \frac{\partial P_j}{\partial x}=0\quad ;\quad j=1,2 \quad (2.2)$$

Note that equations (2.2) yield $\frac{\partial}{\partial y}(\partial P_j/\partial z)=0$ and $\frac{\partial}{\partial x}(\partial P_j/\partial z)=0$. Thus, the pressure gradient in the axial direction is the same for the two fluids ($\partial P_1/\partial z=\partial P_2/\partial z=\partial P/\partial z$). Integration of (2.2) in the y direction yields a linear variation of the pressure in this direction due to the hydrostatic pressure:

$$P_1=P_{1i}-\rho_1(y-\eta)g\cos\beta\quad ;\ P_2=P_{2i}-\rho_2(y-\eta)g\cos\beta \quad (2.3)$$

where P_{1i}, P_{2i} are the local pressures at either side of the fluids interface, at $y=\eta(x)$. For an axial, fully developed flow, the hydrodynamic stresses normal to the fluids interface vanish. In this case, the equation for the interface location evolves from the condition of equilibrium between the pressure jump across the interface and the surface tension force:

$$P_{1i}-P_{2i}=\frac{\sigma}{R_i} \quad (2.4)$$

where σ is the surface tension (assumed constant) between the two fluids and R_i is the local radius of the interface curvature:

$$R_i=\left\{\frac{d}{dx}\frac{d\eta/dx}{[1+(d\eta/dx)^2]^{1/2}}\right\}^{-1}=-\left\{\frac{d}{d\eta}\frac{dx/d\eta}{[1+(dx/d\eta)^2]^{\frac{1}{2}}}\right\}^{-1} \quad (2.5)$$

The interfacial curvature in the axial direction is infinite. Equation (2.4) is the well-known Laplace (1806) formula that can be put in the following form:

$$\sigma\frac{d}{dx}\left\{\frac{d\eta/dx}{[1+(d\eta/dx)^2]^{\frac{1}{2}}}\right\}-(\rho_2-\rho_1)\eta g\cos\beta=\text{const}=\lambda \quad (2.6)$$

Equation (2.6) is a non-linear differential equation for $\eta(x)$. Thus, for the flow field under consideration, the position of the fluids interface can be obtained by solving the quasi-static situation. The solution for $\eta(x)$ should comply with the wettability condition at the pipe wall and symmetry with respect to the y axis. It is also constrained by the fluids in-situ holdup available in the flow.

The same differential equation (2.6) can be also obtained from the variational problem of minimizing the total system free energy (Bentwich, 1976, Gorelik and Brauner, 1999). Given the fluids holdup, the components of the free energy, that are subject to variation with changes in the interface shape, are the potential energy in the gravity field and the surface energy (due to the liquids contact with the pipe wall and the liquid-liquid interface). The fact that the same differential equation evolves suggests that the formulation of a variational problem that minimizes the system potential and surface energies is consistent with the hydrodynamic equations for unidirectional and fully developed axial flow. Hence, no other energies (such as the fluids kinetic energies) should be included in the analysis. Equation (2.6) was solved numerically by Bentwich (1976) and analytically by Gorelik and Brauner (1999) and Ng et al., (2001) in terms of elliptical integrals. The analytical solution includes the shape of the interface, $\eta(x)$ and the dimensionless capillary pressure, $\Lambda = 4\lambda/(\Delta\rho g \cos\beta D^2)$, in terms of three dimensionless parameters: the Eotvös number, Eo_D, the fluid/wall wettability angle, α and the fluids holdup. The function $\eta(x)$ determines the geometry of the fluids distribution in the pipe cross section and contact with the pipe wall, whereas Λ is required for calculation of the pressure distribution.

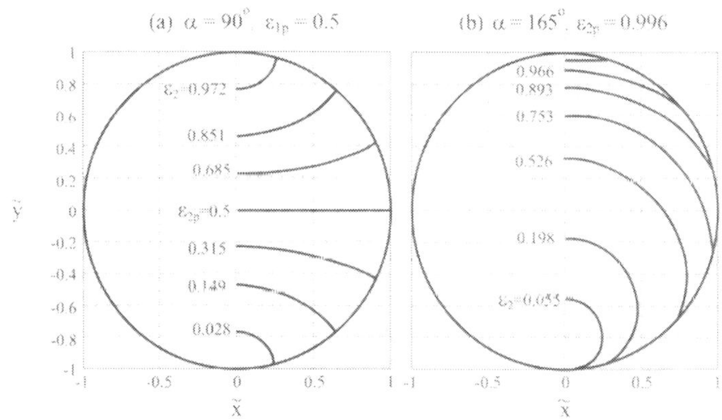

Figure 2.2 Interface configuration for $Eo_D = 1$: effect of holdup for $\alpha = 90^0$ and $\alpha = 165^0$.

An important point to realize is, that in a pipe, the interface shape varies with the fluids holdup. This is demonstrated in Figure 2.2 where the solutions for $\eta(x)$ are given for a constant Eotvös number and different fluids holdup. The case of $\alpha = 90^o$ (Figure 2.2a) corresponds to equal wettability of the two fluids. In this case, the interface is convex for relatively low holdup of the lower phase, $\epsilon_2 < 0.5$ and concave for $\epsilon_2 > 0.5$. For the particular case of $\epsilon_2 = \epsilon_{2p} = 0.5$, the interface is plane, since this configuration satisfies the wettability condition at the solid wall. When the upper phase is the more wetting phase ($\alpha > 90^o$), ϵ_{2p} increases, as shown in Figure 2.2b (for $\alpha = 165°$, $\epsilon_{2p} \simeq 0.996$). The value of $\epsilon_{1p} = 1 - \epsilon_{2p}$ approaches zero as $\alpha \to \pi$ (ideal wettability of the lighter phase) and the interface is convex independently of the fluids holdup. Similarly, $\epsilon_{2p} \to 0$ as $\alpha \to 0$ and the interface is always concave. However, for partial wettability ($\alpha \neq 0, \pi$), there is a particular value of holdup, ϵ_{2p}, where adhesion forces to the wall are just balanced at the triple point and the system behaves as pseudo gravitational - the interface is plane independently of the Eotvös number. For $\epsilon_2 \neq \epsilon_{2p}$ the interface curvature increases with reducing Eo_D. The dependence of the interface shape on the fluids holdup is a basic difference between pipe flow and channel flow. In a rectangular cross section, the interfacial shape is invariant with the fluids holdup, except for extremely low holdup of one of the phases, where the interface shape may be constrained by the contact with either the upper, or lower wall.

The variation of the TP point location with the holdup and the Eotvös number can be studied in view of Figure 2.3. The location of the TP point corresponding to a plane interface is given by the curve for $Eo_D \to \infty$, when surface forces vanish. The other extreme of no gravity force is described by the curve of $Eo_D = 0$. The figure shows that the location of the TP (represented by ϕ_0) deviates from that predicted by a plane interface (ϕ_0^P) already for $Eo_D = 200$. Figure 2.3a is for almost ideal wettability of the upper phase ($\alpha = 175^0$). For this case $\epsilon_{1p} \to 0$ and the interface is practically convex for any non-vanishing value of the capillary number, $\phi_0 < \phi_0^P$. The effect of Eo_D becomes less pronounced as $\alpha \to 90^0$ (Figure 2.3b).

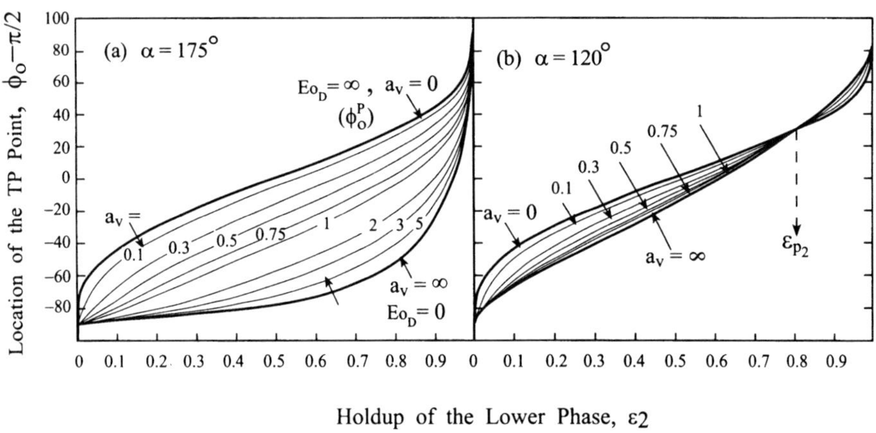

Figure 2.3 The fluids contact with the wall: variation of ϕ_o with the holdup for various Eotvos numbers ($Eo_D = 2/a_v^2$) and wettability anlgles.

2.2 Constant curvature approximation for the interface shape

Exact analytical solutions for the velocity profiles $u_1(x,y), u_2(x,y)$ in laminar flows can be obtained when the fluids interface can be described by a constant curvature curve. In this case, the bipolar coordinate system can be applied to obtain a complete analytical solution for the velocity profiles, distribution of shear stresses along the pipe wall and fluids interface, axial pressure drop and in-situ holdup, in terms of prescribed flow rates and fluids viscosities (Bentwich, 1964, Brauner et al.,1995, 1996a, Moalem Maron et al., 1995). Otherwise, given the location of the interface $\eta(x)$, numerical schemes must be used for solving the Stokes equations (2.1)(see Ng et al., 2002). The assumption of a constant curvature is trivially satisfied for a zero interfacial surface tension, where $R_i \to \infty$ in eq. (2.4). In this case, the interface is plane with a zero pressure difference across the interface, and the flow geometry can be described by the thickness of the (lower) fluid layer, h (Figure 2.1). Analytical solutions for flow with a plane interface are given in several publications (Semenov and Tochigin 1962, Bentwich 1964, Ranger and Davis 1979, Brauner et al., 1996a). However, in view of eq. (2.6), the assumption of a constant interfacial curvature is evidently also valid when the effect of the gravitational field is negligible, as under microgravity conditions or when $\rho_2 \simeq \rho_1$, whereby $Eo_D \to 0$.

In an attempt to bridge the gap between large and small Eotvös numbers, Brauner et al., (1996b) modelled the shape of the interface by a constant characteristic interfacial curvature. The appropriate characteristic interfacial curvature was derived by formulating the variational problem of minimizing the sum of the system potential (E_p) and surface energies (E_s) with approximate configurations that are described by a priori unknown constant curvature. The curvature and the location of the TP are subject to variations, which are constrained by a prescribed holdup.

In case the interface is of constant curvature, the flow configuration can be described in terms of two variables: ϕ_0 and ϕ^* (Fig. 2.1). The view angle of the interface from a point on the upper wall, ϕ_0 determines the distribution of the two phases over the tube wall. The interface curvature is determined by ϕ^*, which is the view angle of the two triple points (TP) from a point situated on the phases interface. Given ϕ_0 and ϕ^*, geometrical relationships yield the phases flow areas (A_1 and A_2) contact lengths with the tube wall (S_1 and S_2) and interface length, S_i (see Table 2.1). A plan interface is described by $\phi^* = \pi$. Convex interfaces are described by ϕ^* less than π, up to the limit of $\phi^* = 0, \phi_0 = 0$, which corresponds to a fully eccentric core of the lower phase touching the tube bottom. Concave interfaces are described by $\phi^* > \pi$, up to the limit of $\phi^* = \pi \;\; \phi_0 = 2\pi$, which corresponds to a fully eccentric core of the upper phase. It is to be noted that ϕ^* is always bounded between ϕ_0 and $\phi_0 + \pi$.

Taking a configuration of plane interface as a reference, the expression obtained for the system free energy reads:

$$\frac{\Delta \widetilde{E}}{L} = \frac{\Delta Es + \Delta Ep}{LR^3(\rho_2 - \rho_1)g\cos\beta} = \left[\frac{\sin^3\phi_0}{\sin^2\phi^*}(ctg\phi^* - ctg\phi_0)\left(\pi - \phi^* + \frac{1}{2}\sin(2\phi^*)\right)\right.$$

$$\left. + \frac{2}{3}\sin^3\phi_0^P\right] + Eo_D^{-1}\left[\sin\phi_0\frac{(\pi - \phi^*)}{\sin\phi^*} - \sin\phi_0^P + \cos\alpha(\phi_0^P - \phi_0)\right] \quad (2.7)$$

Table 2.1: Geometrical relationships for curved and plane interfaces

	Curved interface, $\phi^* \neq \pi$	Plane interface, $\phi^* = \pi$
$\widetilde{A} = \frac{A}{D^2}$	$\pi/4$	$\pi/4$
$\widetilde{A}_1 = \frac{A_1}{D^2}$	$\frac{1}{4}\left\{\pi - \phi + \frac{1}{2}\sin(2\phi) - \left(\frac{\sin\phi_0}{\sin\phi^*}\right)^2\left[\pi - \phi^* + \frac{1}{2}\sin(2\phi^*)\right]\right\}$	$\frac{1}{2}\left[\pi - \phi_0^P + \frac{1}{2}\sin(2\phi_0^P)\right]$
$\widetilde{A}_2 = \frac{A_2}{D^2}$	$\frac{1}{4}\left\{\phi_0 - \frac{1}{2}\sin(2\phi_0) - \frac{\sin^2\phi_0}{\sin^2\phi^*}\left[\phi^* - \pi - \frac{1}{2}\sin(2\phi^*)\right]\right\}$	$\frac{1}{4}\left[\phi_0^P - \frac{1}{2}\sin(2\phi_0^P)\right]$
$\widetilde{S}_1 = \frac{S_1}{D}$	$\pi - \phi_0$	$\pi - \phi_0^P$
$\widetilde{S}_2 = \frac{S_2}{D}$	ϕ_0	ϕ_0^P
$\widetilde{S}_i = \frac{S_i}{D}$	$(\pi - \phi^*)\sin(\phi_0)/\sin(\phi^*)$	$\sin(\phi_0^P)$
$\widetilde{U}_1 = \frac{U_1}{U_{1S}}$	$\pi/\left\{\pi - \phi_0 + \frac{1}{2}\sin(2\phi_0) - \left(\frac{\sin\phi_0}{\sin\phi^*}\right)^2\left[\pi - \phi^* + \frac{1}{2}\sin(2\phi^*)\right]\right\}$	$\pi/\left[\pi - \phi_0^P + \frac{1}{2}\sin(2\phi_0^P)\right]$
$\widetilde{U}_2 = \frac{U_2}{U_{2S}}$	$\pi/\left\{\phi_0 - \frac{1}{2}\sin(2\phi_0) + \left(\frac{\sin\phi_0}{\sin\phi^*}\right)^2\left[\pi - \phi^* + \frac{1}{2}\sin(2\phi^*)\right]\right\}$	$\pi/\left[\phi_0^P - \frac{1}{2}\sin(2\phi_0^P)\right]$

Given Eo_D, α and $\epsilon_2 = A_2/A$, the equilibrium interface shape is determined by ϕ_0 and ϕ^* which correspond to a minimum of $\Delta\widetilde{E}$ subject to the constraints:

$$\epsilon_2 = \frac{1}{\pi}\left\{\phi_0 - \frac{1}{2}\sin(2\phi_0) + \left(\frac{\sin\phi_0}{\sin\phi^*}\right)^2\left[\pi - \phi^* + \frac{1}{2}\sin(2\phi^*)\right]\right\} \;;\phi^* \neq \pi$$
$$\epsilon_2 = \frac{1}{\pi}\left[\phi_0^P - \frac{1}{2}\sin(2\phi_0^P)\right] \;;\phi^* = \pi \tag{2.8}$$

The particular ϕ^* which corresponds to minimal energy is thus a function of the three non-dimensional parameters $\phi^* = \phi^*(Eo_D, \alpha, \text{ holdup}) = \phi^*(Eo_D, \alpha, \phi_0)$. Note that in the approximate solution, the ϕ^* only approximately satisfies the wettability condition. However, in the extremes of $Eo_D = 0$ or $Eo_D \rightarrow \infty$, (where also the exact interfacial shapes correspond to a constant curvature) the approximate and exact solutions coincide.

$$\phi^*(\phi_0, \alpha) = \pi \Rightarrow \text{ plane interface; } Eo_D \rightarrow \infty$$
$$\phi^*(\phi_0, \alpha) = (\pi - \alpha) + \phi_0; \quad Eo_D \rightarrow 0 \tag{2.9}$$

The largest deviations of the approximate solution were obtained for $Eo_D \simeq 0.5 \div 1$ and ideal wettability of either of the phases. However, Figure 2.4a shows that even in this range of parameters, the approximate solution closely follows the exact solution in describing the effect of the holdup on the curving of the interface. Figure 2.4b summarizes results for ϕ_o obtained with various wettability angles and shows that the comparison improve as $\alpha \rightarrow \pi/2$. The largest deviations are for $\alpha \rightarrow 180°$(or $\alpha \rightarrow 0°$). However, for $\alpha = 135°$ (or 45°) the differences between the two solutions are already un-noticeable.

The results obtained for the interface shape, which corresponds to minimum energy of eq. (2.7), have been used to construct the so-called 'interface monograms' (Fig. 9 in Brauner, 1998). Given the Eotvös number and the wall/phases adhesion properties as reflected by the wettability angle, a curve relating the interfacial curvature, ϕ^*, to the

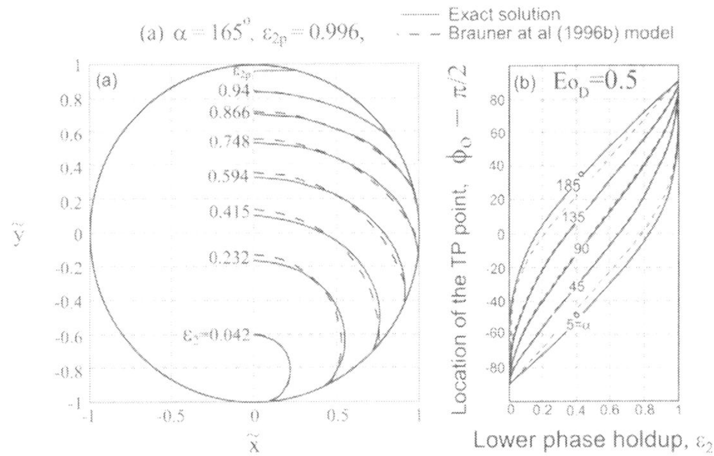

Figure 2.4 **Comparison of the characteristic curvature model with the exact solution $Eo_D = 0.5$:**
a) curves of $\tilde{\eta}(x)$ for $\alpha = 165^0$, b) value of ϕ_0 vs. ϵ_2 for various α.

phases distribution angle ϕ_0, is obtained. Each point along an interface monogram is associated with a different holdup. That form of the interface monogram can be conveniently combined with the solution of the flow problem, where the phases in situ holdup is obtained via the solution of the flow equations (see Figure 2.5 below).

2.3 Exact solutions of two-phase laminar pipe flow

The appropriate coordinate system for solving the flow problem, for stratified flow with a curved interface is the well-known bipolar coordinate system. Coordinate ϕ represents the view angle of the two triple points (TP) from an arbitrary point in the flow domain (Figure 2.1). Coordinate ξ relates to the ratio of the radius vectors r_1, r_2 ($\xi = ln(r_1/r_2)$). The pipe perimeter and the interface between the fluids are isolines of coordinates ϕ, so that the upper section of the tube wall bounding the lighter phase is represented by ϕ_o, while the bottom of the tube, bounding the denser phase, is represented by $\phi = \phi_0 + \pi$. The interface coincides with the curve of $\phi = \phi^*$. Thus, the two-phase domains map into two infinite strips in the (ϕ, ξ) domain and are defined by:

$$\begin{aligned}&\text{Upper phase}: -\infty < \xi < \infty\,; \quad \phi_0 < \phi < \phi^* \\ &\text{Lower phase}: -\infty < \xi < \infty\,; \quad \phi^* < \phi < \phi_0 + \pi\end{aligned} \tag{2.10}$$

Analytical solutions of the Stokes equations for horizontal stratified flow with an interface of an arbitrary curvature were explored by Brauner et al., (1995, 1996a). In these studies, analytical expressions in terms of Fourier integrals in the bipolar coordinate system were provided for the velocity profiles ($\tilde{u}_{1,2} = u_{1,2}/U_R$ and the distribution of shear stresses over the tube wall (τ_1, τ_2) and free interface (τ_i) (see also Moalem Maron et al, 1995):

$$\widetilde{u}_1(\phi,\xi) = 2\sin\phi_0\{\frac{\sin(\phi-\phi_0)}{\cosh\xi-\cos\phi}+2(1-\widetilde{\mu})\frac{\sin(\phi^*-\phi_0)}{\sin(\phi^*)}\int_0^\infty W_{1v}(\omega,\phi)\cos(\omega\xi)d\omega\} \tag{2.11}$$

$$\widetilde{u}_2(\phi,\xi) = 2\widetilde{\mu}\sin\phi_0\{\frac{\sin(\phi-\phi_0)}{\cosh\xi-\cos\phi}+2(1-\widetilde{\mu})\frac{\sin(\phi^*-\phi_0)}{\sin(\phi^*)}\int_0^\infty W_{2v}(\omega,\phi)\cos(\omega\xi)d\omega\} \tag{2.12}$$

where: $U_R = \frac{D^2}{16\mu_1}\left(\frac{-\partial P}{\partial z}\right)$, $\widetilde{\mu} = \mu_1/\mu_2$, and the spectral functions are given by:

$$W_{1v}(\omega,\phi) = \frac{\sinh[\omega(\phi^*-\pi)]}{\psi(\omega)\sinh(\pi\omega)}\,\frac{\sinh[\omega(\phi-\phi_0)]}{\cosh[\omega(\phi^*-\phi_0)]} \tag{2.13}$$

$$W_{2v}(\omega,\phi) = \frac{\sinh[\omega(\phi^*-\pi)]}{\psi(\omega)\sinh(\pi\omega)}\,\frac{\sinh[\omega(\phi-\pi-\phi_0)]}{\cosh[\omega(\phi^*-\pi-\phi_0)]} \tag{2.14}$$

$$\psi(\omega) = \tanh[\omega(\phi^*-\phi_0)] + \widetilde{\mu}\tanh[\omega(\pi+\phi_0-\phi^*)] \tag{2.15}$$

Thus:

$$\widetilde{u}_{1,2} = \frac{u_{1,2}}{U_R} = \widetilde{u}(\phi_0,\phi^*,\widetilde{\mu});\ \widetilde{\boldsymbol{\tau}}_1,\widetilde{\boldsymbol{\tau}}_2,\widetilde{\boldsymbol{\tau}}_i = \widetilde{\boldsymbol{\tau}}(\phi_0,\phi^*,\widetilde{\mu}) \tag{2.16}$$

where $\widetilde{\tau} = \frac{\tau}{\tau_R}$; $\tau_R = \frac{R}{2}\left(-\frac{\partial P}{\partial z}\right)$. Note that the velocity and shear stress scales used for normalization include the unknown pressure drop. The phases flow rates are obtained by integrating the phases velocities over the corresponding flow areas A_1 , A_2 (see Table 2.1). For a given pressure drop and a viscosity ratio, the integration yields Q_1 and Q_2 as functions of (ϕ_0, ϕ^*, $\widetilde{\mu}$, $\partial P/\partial z$). The ratio of the two fluids flow rates, however, is independent of the system pressure drop; $\widetilde{Q} = \frac{Q_1}{Q_2} = \widetilde{Q}(\phi_0,\ \phi^*,\ \widetilde{\mu})$. The corresponding pressure drop (normalized with respect to the superficial pressure drop of the upper fluid) is $\left(\frac{d\widetilde{P}_1}{dZ}\right) = \frac{(-\partial P/\partial z)}{(\partial P/\partial z)_{1S}} = \frac{d\widetilde{P}_1}{dZ}(\phi_0,\ \phi^*,\ \widetilde{\mu})$. Therefore, once the fluids viscosities and flow rates are known, the solution of the flow equations provide a relationship between ϕ_0 and ϕ^*:

$$\phi_0 = \phi_0(\phi^*) \quad \rightarrow \quad \text{Flow Monogram for specified } \widetilde{\mu} \text{ and } \widetilde{Q} \tag{2.17}$$

Once the interface curvature is also specified, for instance, a plane interface ($\phi^* = \pi$), the corresponding ϕ_0 can be obtained, and then the system pressure drop, dimensional velocity profiles and shear stress profiles can be computed. However, the interfacial curvature should comply with the continuity of normal stresses (pressure and surface tension forces) across the interface and with solid/fluids adhesion forces (wettability angle). Hence, the closure relationship needed for the interfacial curvature is provided by the system 'interface monogram' $\phi^* = \phi^*(Eo_D,\alpha,\phi_0)$ as described in Section 2.2.

A convenient frame for obtaining a complete solution (which includes the interface curvature) is via the construction of the system 'operational monograms' (Fig. 2.5). These monograms combine the system 'interface monogram' (dashed curves) with the system

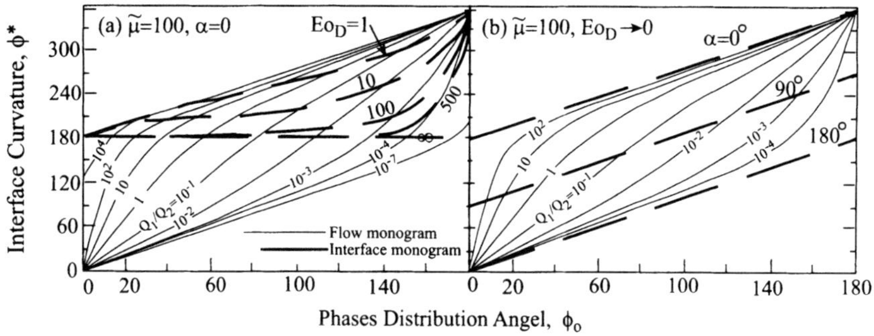

Figure 2.5 **Operations monogram - effect of Eotvös number and wettability angle on the stratified flow geometry (ϕ_0 and ϕ^*).**

'flow monograms'. The intersection points of the 'interface' and 'flow' monograms represent all stratified flows solutions obtained for various Q_1/Q_2 ratios. Fig. 2.5 indicates that for a given physical system parameters $(\tilde{\mu}, \alpha, Eo_D)$ and operational conditions Q_1/Q_2, there exists a single solution (ϕ^*, ϕ_0) which determines the resulting flow characteristics. Fig. 2.5a shows that as Eo_D decreases, the solutions for the flow configuration correspond to stratified flow with curved interfaces ($\phi^* \neq 180^0$). The interfacial curvature increases with decreasing Eo_D. Stratified configurations with curved interfaces may also be obtained in systems of low Eotvös number with partial wettability of the fluids (Fig. 2.5b for $0 < \alpha < 180^o$). But, for $Eo_D \to 0$ and ideal wettability of either one of the phases ($\alpha = 0$ or $\alpha = 180^0$ in Fig. 2.5) the solutions obtained correspond to fully eccentric core-annular configuration, irrespective of the phases flow rates (and viscosities). When the upper phase is the wetting phase, $\alpha = 0$, the solution is $\phi_0 = 180^0$, $\phi^* = 360^0$, which corresponds to a fully eccentric core of the upper non-wetting phase touching the upper tube and surrounded by an annulus of the wetting phase. For $\alpha = 180^0$, the solution is $\phi_0 = 0$ and $\phi^* = 0$, in which case the lower phase forms a fully eccentric core at the tube bottom, which is surrounded by the upper wetting phase. Indeed, the occurrence of annular flow in liquid-liquid system is more frequently encountered in oil-water systems of low density differential and small diameter tubes, which are characterized by small Eotvös number.

Exact solutions of the Stokes equations (2.1) for inclined flows assuming a plane interface between the fluids ($\phi^* = \pi$) were obtained by Masliyah and Shook, 1978, Biberg and Halvodsen, 2000 and Goldstein, 2002. These solutions are valid only for large Eo_D systems.

2.4 The two-fluid model (TFM)

For laminar stratified flows, exact solutions of the Stokes equations can be obtained which include the characteristic interface curvature and all the details of the local and integral flow characteristics. But, these analytical solutions still involve extensive computations. In many practical situations, one of the phases (or both) is turbulent. Therefore, for practical applications, there is a need for a model which can also handle turbulent flows and mixed flow regimes in horizontal and inclined systems. To this end, the two-fluid model can be used (Brauner and Moalem Maron, 1989, Brauner et al., 1998). The model equations presented here and in a unified form that is applicable both to co-current and counter current stratified flows (Ullmann et al., 2002a). Assuming a fully developed flow, the integral forms of the momentum equations for the two fluids are (see Figure 2.1):

$$-A_1\left(\frac{\mathrm{d}P}{\mathrm{d}z}\right)+\tau_1 S_1+\tau_i S_i+\rho_1 A_1 g\sin\beta=0$$

$$-A_2\left(\frac{\mathrm{d}P}{\mathrm{d}z}\right)+\tau_2 S_2-\tau_i S_i+\rho_2 A_2 g\sin\beta=0 \tag{2.18}$$

Eliminating the pressure drop yields:

$$-\tau_1\frac{S_1}{A_1}+\tau_2\frac{S_2}{A_2}-\tau_i S_i\left(\frac{1}{A_1}+\frac{1}{A_2}\right)+(\rho_2-\rho_1)\ g\ \sin\beta=0. \tag{2.19}$$

The Blasius equation can be used to provide the closure laws required for the wall and interfacial shear stresses ($\tau_1,\ \tau_2,\ \tau_i$) in terms of the average velocities, $U_1,\ U_2$ and the friction factors f_1, f_2 and f_i:

$$\tau_1=-\frac{1}{2}f_1\rho_1 U_1|U_1|;\quad f_1=C_1\left(\frac{\rho_1 D_1|U_1|}{\mu_1}\right)^{-n_1}$$

$$\tau_2=-\frac{1}{2}f_2\rho_2 U_2|U_2|;\quad f_2=C_2\left(\frac{\rho_2 D_2|U_2|}{\mu_2}\right)^{-n_2}$$

$$\tau_1=-\frac{1}{2}f_i\rho(U_1-U_2)|U_1-U_2|\ . \tag{2.20}$$

where U_1, U_2 are positive (negative) for downward (upward) flow ($\beta>0$ in both cases). The Reynolds numbers for the two fluids in Eq. (2.20) are based on the equivalent hydraulic diameters, which are defined according to the relative velocity of the phases. In co-current flow, the interface is considered as "free" for the slower phase and as a "wall" for the faster phase. When the velocities are of the same order, the interface is considered "free" with respect to both phases:

$$D_1=\frac{4A_1}{(S_1+S_i)};\ D_2=\frac{4A_2}{S_2};\ \rho=\rho_1 \text{ and } f_i=F_i f_1 \quad \text{for } |\mathbf{U_1}|>|\mathbf{U_2}|.$$

$$D_1=\frac{4A_1}{S_1};\ D_2=\frac{4A_2}{(S_2+S_i)};\ \rho=\rho_2 \text{ and } f_i=F_i f_2 \quad \text{for } |\mathbf{U_2}|>|\mathbf{U_1}|.$$

$$D_1=\frac{4A_1}{S_1};\ D_2=\frac{4A_2}{S_2};\ \tau_i\simeq 0 \text{ for } \mathbf{U_1}\simeq\mathbf{U_2}\ . \tag{2.21}$$

In counter-current flow, each of the layers is dragged by the other one opposite to its flow direction, therefore:

$$D_2=\frac{4A_2}{S_2+S_i}\quad;\quad D_1=\frac{4A_1}{S_1+S_i} \tag{2.22}$$

A value of $F_i > 1$ can be introduced in eqs. (2.21) to account for a possible augmentation of f_i due to irregularities at the free interface. However, due to the lower density (hence velocity) difference and lower surface tension encountered in liquid-liquid systems, the interface appears less roughened compared to gas-liquid systems. The main issue here concerns the decision as to which of the liquids actually dominates the interfacial interactions. In case of a perturbed interface, the effects of drop entrainment and the consequential mixing at the interface, rather than the wave phenomenon, have to be considered.

Introducing non-dimensional variables (length normalized by D, area by D^2 and velocities by superficial velocities U_{1s}, U_{2s}, see Table 2.1), the various geometric parameters and the non-dimensional velocities $\widetilde{U}_1, \widetilde{U}_2$ are all functions of the phases distribution angle over the tube wall, ϕ_0 and the interface curvature ϕ^*. Given the flow regime in the two-layers ($C_{1,2}$ and $n_{1,2}$ in eq. (2.20) are prescribed), the general relation stated by the dimensionless form of the combined momentum equation (2.19) is:

$$f(\chi^2, \widetilde{Q}, Y, \phi*, \phi_0) = 0\ ; \quad \text{Flow monogram} \tag{2.23}$$

The three non-dimensional parameters of the solution χ^2, Y and $\widetilde{Q}$, which evolve through the normalization of the combined momentum equation, are given by:

$$\chi^2 = \frac{2C_2/D|\mathbf{U}_{2s}D/\nu_2|^{-n_2}\rho_2|\mathbf{U}_{2s}|\,U_{2s}}{2C_1/D|\mathbf{U}_{1s}D/\nu_1|^{-n_1}\rho_1|\mathbf{U}_{1s}|\,U_{1s}} = \frac{(-dP/dz)_{2s}}{(-dP/dz)_{1s}} \tag{2.24}$$

$$Y = \frac{(\rho_2-\rho_1)g\sin\beta}{2C_1/D|\mathbf{U}_{1s}D/\nu_1)|^{-n_1}\rho_1|\mathbf{U}_{1s}|U_{1s}} = \frac{(\rho_2-\rho_1)g\sin\beta}{(-dP/dz)_{1s}}\ ;\ \widetilde{Q} = \frac{Q_1}{Q_2} = \frac{U_{1s}}{U_{2s}}$$

It is worth noting that for co-current flow U_{1s}, U_{2s} are positive in case of downward flow and are negative both for upward flow, whereas, for counter-current flow U_{1s} is negative, (the light phase flows upward). Therefore, concurrent flows correspond to positive X^2 with $Y > 0$ for down-flow and $Y < 0$ for up-flow. Countercurrent flows correspond to negative X^2 with $Y < 0$. The number of non-dimensional parameters, which eventually define the flow monogram, depend on the flow regime in both phases. In particular, for horizontal laminar (L-L) flows, $Y = O, \chi^2 = (\widetilde{\mu}\widetilde{Q})^{-1}$, and as in the exact solution, the two-fluid flow monogram yields: $\phi_0 = \phi_0(\widetilde{Q}, \widetilde{\mu}, \phi*)$; while for horizontal turbulent (T-T) flows, the solution is also dependent on fluids density ratio, whereby: $\phi_0 = \phi_0(\widetilde{Q}, \widetilde{\mu}, \widetilde{\rho}, \phi*)$. For mixed flow regime in the two layers, more information is needed, which includes the superficial Reynolds number of either one of the phases. In all cases, the closure relation for the interfacial curvature introduces two-additional non-dimensional parameters, the Eotvös number and the wettability angle.

Results of the two-fluid model are demonstrated in Figure (2.6) (see also Brauner et al, 1998). The height of water climbing on the wall, $h_f = 0.5(1-\cos\phi_0)$, and the location of the oil water interface on the tube centerline, $h_0 = 0.5[1 - \cos\phi_0 + \sin\phi_0 \mathrm{ctg}(\phi^*/2)]$, can be computed once ϕ_0 and ϕ^* are known. The latter can be determined by combining the solution of the two-fluid flow equations, as represented by the flow monogram for specified oil and water flow rates, with the interface monogram for $Eo_D = 10, \alpha = 0$. Fig. 2.6 (a and b) shows a comparison between the experimental data for h_f and h_0 and the predicted values. The gap between h_f and h_o indicates the extent of water climbing over the wall surface. The height of the water film increases with increasing the water rate or reducing the oil rate, whereby the flow configuration gradually approaches a fully eccentric core-annular configuration. Given the flow geometry, the pressure drop can be computed by either of eqs. (2.18). Figure 2.6c demonstrates that the values predicted for the pressure drop are also in a reasonable agreement with the experimental data indicating a water lubrication effect.

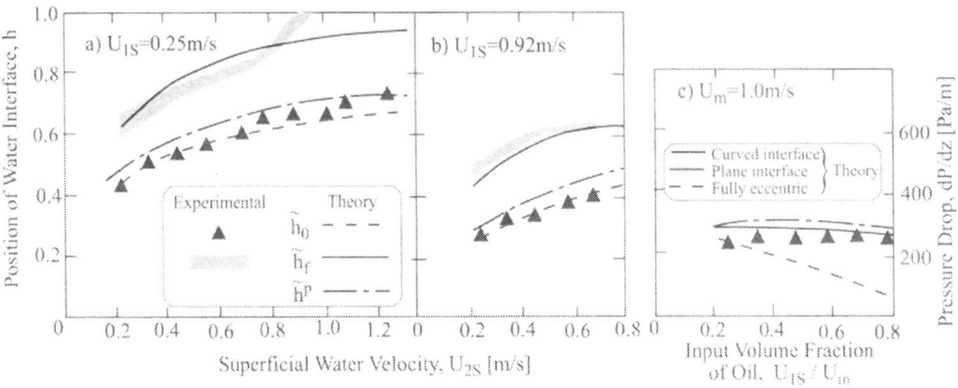

Figure 2.6 **Location of the oil-water interface and pressure drop in oil-water system-comparison of the TFM results with Valle and Kvandal (1995) experimental data ($\widetilde{\mu} = 2.26, \widetilde{\rho} = 0.792, Eo_D = 10$, glass tube, $\alpha = 0^0$).**

2.5 Two-Plate model (TPM)

Another simple model that can be useful to analyze the characteristics of laminar stratified concurrent and countercurrent inclined flows, however, with a plane interface, is the flow between two parallel plates. For this simple geometry the solution of the Stokes equations (eqs. 2.1) can be easily solved to obtain the velocity profiles. Using mass balances on the two fluids results in the following relations (Ullman et al., 2002a):

$$Y = \frac{1}{4}\frac{(1-\widetilde{h})^2[(1+2\widetilde{h})+(\widetilde{\mu}-1)\widetilde{h}(4-\widetilde{h})] - \widetilde{h}^2[(3-2\widetilde{h})+(\widetilde{\mu}-1)\widetilde{h}^2]\widetilde{\mu}\widetilde{Q}}{\widetilde{h}^3(1-\widetilde{h})^3[\widetilde{h}\widetilde{\mu}+(1-\widetilde{h})]\widetilde{\mu}\widetilde{Q}} \tag{2.25}$$

$$\widetilde{P} = \frac{dP/\partial z - \rho_1 g\sin\beta}{(-dP/dz)_{1s}} = \frac{1}{4}\frac{3(1-\widetilde{h})^2 - 4\widetilde{h}(1-\widetilde{h})\widetilde{Q} - \widetilde{h}^2\widetilde{\mu}\widetilde{Q}}{\widetilde{h}(1-\widetilde{h})^2[(1+2\widetilde{h})+(\widetilde{\mu}-1)\widetilde{h}(4-\widetilde{h}) - 3\widetilde{h}\widetilde{\mu}]\widetilde{Q}} \tag{2.26}$$

where $Y = \frac{(\rho_2-\rho_1)g\sin\beta}{(-dP/dz)_{1s}}$ and $(-dP/dz)_{1s} = 12\mu_1 Q_1/H^3$ is the superficial frictional pressure drop for single phase flow of the lighter phase. Given the parameters Y, $\widetilde{\mu}(=\mu_1/\mu_2)$ and $\widetilde{Q}(=Q_1/Q_2)$, eq. (2.25) can be solved for the *in situ* holdup of the heavier phase $\widetilde{h} = h/H$, which in turn can be substituted in eq. (2.26) to obtain $\widetilde{P}$. The total pressure drop is composed of the gravitational (hydrostatic) pressure drop and the frictional pressure drop $(dP/dz)_f$:

$$\left(\frac{dP}{dz}\right)_g = [\rho_2\widetilde{h} + \rho_1(1-\widetilde{h})]g\sin\beta = [\rho_1 + (\rho_2-\rho_1)\widetilde{h}]g\sin\beta \tag{2.27}$$

$$\left(\frac{dP_1}{dZ}\right)_f = -\frac{dP/dz - (dP/dz)_g}{(-dP/dz)_{1s}} = -\widetilde{P} + Y\widetilde{h} \tag{2.28}$$

The solution for the holdup vs. the flow rates is demonstrated in Figure 2.7a. This

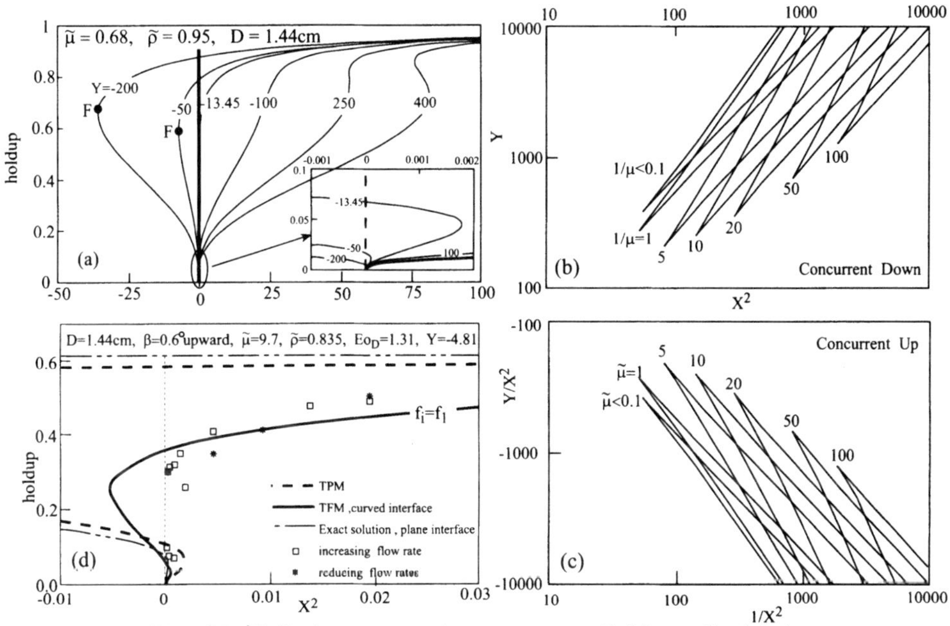

Figure 2.7 **Holdup in concurrent and counter current stratified flow predicted via the TPM: (a) Effect of Y, (b,c). Triple solution regions in downward/upward concurrent flows. (d) Experimental verification of multiple holdups.**

figure shows the TPM model predictions for the variation of the holdup curves as the inclination parameter changes form a negative to a positive value for a gravity dominated liquid-liquid system ($Eo_D >> 1$) of $\widetilde{\mu} = 0.68, \widetilde{\rho} = 0.95$. For counter-current flows ($\widetilde{Q} < 0$) double holdups are always predicted. Indeed, two-different holdups were observed in experiments on liquid-liquid counter-current stratified flows in inclined columns. These were found to be well predicted by the TPM and TFM models double solutions (Zamir, 2002). Point F represents the ultimate flooding conditions for a specified Y, as there is no solution for counter-current flow beyond this point. In concurrent flow, $\widetilde{Q} > 0$, Figure 2.7a shows that multiple (triple) holdups are predicted by the model in a limited range of operational conditions. Triple solutions for the upward flow ($Y < 0$) are typically obtained for small positive value of $X^2 = (\widetilde{\mu}\widetilde{Q})^{-1}$ corresponding to high flow rates of the light phase and/or low flow rates of the heavy phase (see the detailed picture inserted in the figure). This figure clearly shows that triple solutions for the holdup are obtained also in downward flows ($Y > 0$). The triple solutions in downward flows are in the range of relatively high X^2 and correspond to relatively high holdups of the heavy phase. These characteristics of the holdup curve in inclined flows are obtained also in the exact solution for laminar flow in inclined pipes (Goldstein,2002) and via the TFM. However, these are much easier to analyze using the TPM.

In view of Figure 2.7a, given a specified value of Y and $\widetilde{\mu}$, the range of $\widetilde{Q}$ corresponding to multiple solutions is bounded by the values of $\widetilde{Q}$ where $d\widetilde{Q}/d\widetilde{h} = 0$. Using eq. (2.25) for calculation of this derivative yields:

$$Y = \frac{1-(\widetilde{\mu}-1)^2\widetilde{h}^4+2(\widetilde{\mu}^2+2\widetilde{\mu}-1)\widetilde{h}^3+(2\widetilde{\mu}-1)\widetilde{h}}{2\widetilde{h}(\widetilde{h}-1)^3[1+(\widetilde{\mu}-1)^2\widetilde{h}^4-2(\widetilde{\mu}-1)(3\widetilde{\mu}-2)\widetilde{h}^3+2(2\widetilde{\mu}^2-6\widetilde{\mu}+3)\widetilde{h}^2+4(\widetilde{\mu}-1)\widetilde{h}]} \tag{2.29}$$

Equations (2.29) and (2.25) can be easily solved simultaneously to yield the X^2 vs. Y relationship that forms the boundaries of multiple solutions regions for a given liquid-liquid system (given $\widetilde{\mu}$). The effect of the viscosity ratio on the range of triple solution in upward concurrent flow is shown in Figure 2.7c in terms of Y/X^2 vs. $1/X^2$. Figure 2.7b is actually its mirror image and represents the ranges where triple solution exists in concurrent downward flows. Figures 2.7(b,c) demonstrate the complete similarity between upward and downward concurrent flows. They also show that there is a minimal X^2 (or minimal $1/X^2$) for concurrent downward (or concurrent upward) flows for which triple solutions can be obtained.

The identification of the parameter space associated with multiple solutions in the exact solution of the Stokes equations (2.1) for pipe geometry, or even in the TFM, requires a tedious search. However, the multiple solutions parameter space in the three models is similar (Goldstein,2002). Hence, the TPM is in fact a useful simple tool for analyzing inclined stratified flows. This model has been used to design an experimental set-up for concurrent upward oil-water flow in the range where multiple configurations are predicted (Gat, 2002). It was verified that multiple holdups are obtained in the experiments (Figure 2.7d), and are not just artifact of the models. Multi-holdups are obviously associated with multi-value pressure drops and other flow characteristics. Therefore, in the range of operational conditions where multiple solutions are suspected, the modelling and design of two-phase flow systems should be approached with extra care. Computational codes usually provide only one solution for specified operational conditions. Therefore, it is necessary to make sure that this solution indeed corresponds to a relevant, physical configuration. Moreover, the possibility of other relevant configurations should be examined. The procedure for using the TPM for identifying the multiple holdups region for turbulent pipe flows of either one of the layers or in both is detailed in Ullmann et al., (2002b).

2.6 Conclusion

The basic configuration in liquid-liquid pipe flow is two-layers separated by a curved interface, rather than a plane interface. Accounting for the interface curvature may have significant effects on the predicted holdup and pressure drop. Exact solutions exist only for laminar flows. However, the interface curvature can be handled also in the framework of the two-fluid model, which is a useful and simple tool for practical applications. In this model attention must be paid to the closure laws used for the wall and interfacial shear stresses. The two-plates model is a complementary simple tool for analyzing other aspects of liquid liquid inclined flows. In particular, the multi-holdup regions and the characteristic velocity profiles.

3 Core-Annular Flow

One of the flow patterns which appears most attractive from the view point of pressure loss reduction and power saving in the transport of viscous material is that of core annular flow (CAF). The viscous liquid (e.g. heavy crude oil or emulsion, waxy oils) forms the core phase, which is surrounded and lubricated by an immiscible low viscosity liquid (such as water) as the annular phase. A schematic description of CAF is shown in Figure 3.1. A stable CAF is a fully developed flow pattern, where the core and annular phases are distinct and continuous. The continued interest in core-flow resulted in many experimental and theoretical studies, which have been reviewed by Oliemans (1986), Oliemans and Ooms (1986) and Joseph and Renardy (1992). Core flow experiments are summarized in Table 1.2. These experiments proved that if stable core flow can be maintained, the pressure drop is almost independent of the oil viscosity and only slightly higher than for flow of water alone at the mixture flow rate. This flow pattern is promoted by minimizing the density difference between the oil and the lubricating aqueous phase, using additives and surface active agents for controlling and minimizing the emulsification of water into the oil, using hydrophilic pipe material to keep the oil from sticking to the wall and injecting the liquids into the pipe already in this desired configuration.

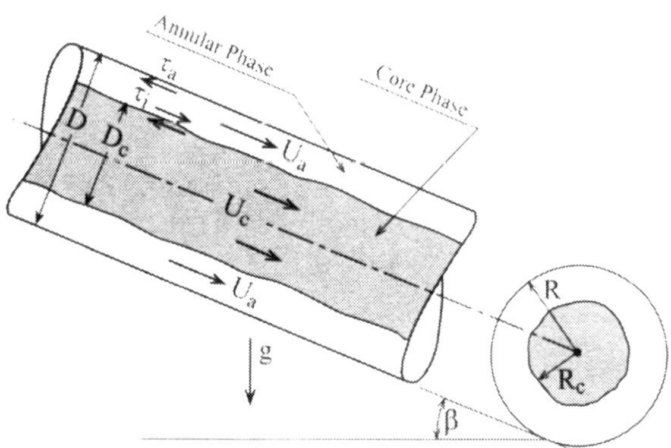

Figure 3.1 Schematic description of core-annular flow configuration.

Due to density difference between the core and annular liquids, the core may stabilize in an off-center position resulting in an eccentric core flow. A steady eccentric core flow is feasible when the overall vertical components of the viscous forces are in a dynamic equilibrium with the buoyancy force (due to the density differential). Stabilizing hydrodynamic forces may evolve due to core eccentricity and interfacial waviness (Ooms et al., 1984, 1985, Oliemans and Ooms, 1986). The development of a wavy core interface is believed to be a necessary condition for core flow stabilization. However, a critical (minimal) oil superficial velocity and water/oil ratio are required to maintain the core at a sufficiently 'safe' eccentricity, to avoid contamination of the upper tube wall by the waxy oil core. Below a critical oil velocity, a transition to stratified flow takes place with excursion of the pressure drop. The prediction of core-flow boundaries is discussed in Section 5.

3.1 Exact solution for laminar CAF

Models of laminar CAF are relevant for practical applications in particular for the case of CAF of highly viscous oils. The solution of the Stokes equations (2.1) for eccentric core annular flows in horizontal pipes was obtained in terms of Fourier Series in the bipolar coordinate system (Figure 3.2a). When dealing with eccentric core-annular flow, the tube wall is represented by $\xi = \gamma_{\mathrm{w}}$, while the two-fluid interface coincides with $\xi = \gamma_c$. Hence, the eccentric core-annular configuration in the x-y domain maps into a semi-infinite strip in the (ϕ, ξ) domain defined by:

$$\begin{aligned} &\text{Annular phase :} \qquad \gamma_{\mathrm{w}} < \xi \leq \gamma_c \,; \quad 0 \leq \phi \leq 2\pi \\ &\text{Core phase :} \qquad \gamma_c < \xi < \infty \,; \quad 0 \leq \phi \leq 2\pi \end{aligned} \tag{3.1}$$

$$\begin{aligned} \gamma_c &= \cosh^{-1}\left[\frac{(\xi_c+1) - E^2(\xi_c-1)}{2E}\right] \; ; \quad \xi_c = \frac{R}{R_c} \\ \gamma_{\mathrm{w}} &= \cosh^{-1}\left[\frac{(\xi_c+1) + E^2(\xi_c-1)}{2E\xi_c}\right] \; ; \quad E = \frac{e/R}{1-1/\xi_c} \end{aligned} \tag{3.2}$$

where R_c is the core radius and e is the core (dimensional) eccentricity. Given the core eccentricity, e, and diameter, R_c, the solution yields the non-dimensional velocity profiles for the core (u_c)and annular (u_a) phases (Bentwich et al., 1970), which can be used to compute the dimensionless wall shear stress and interfacial shear stress profiles:

$$\widetilde{u}_{a,c} = \frac{\widetilde{u}_{a,c}}{U_R} = \widetilde{u}\left(\frac{R_c}{R}, \frac{e}{R}, \widetilde{\mu}\right) ; \;\; \widetilde{\mu} = \frac{\mu_a}{\mu_c}; \;\; U_R = \frac{R^2}{4\mu_a}\left(\frac{-\partial P}{\partial z}\right) \; ; \; \widetilde{\tau}_a, \widetilde{\tau}_i, = \widetilde{\tau}(\widetilde{R}_c, \widetilde{e}, \widetilde{\mu}) \tag{3.3}$$

Integration of the velocity profiles over the phases flow cross section yields:

$$\widetilde{Q} = \frac{Q_a}{Q_c} = \widetilde{Q}(\widetilde{R}_c, \widetilde{e}, \widetilde{\mu}) \; ; \qquad \frac{d\widetilde{P}_c}{dZ} = \frac{(-\partial P/\partial z)}{(-\partial P/\partial z)_{cs}} = \frac{d\widetilde{P}_c}{dZ}(\widetilde{R}_c, \widetilde{e}, \widetilde{\mu}) \tag{3.4}$$

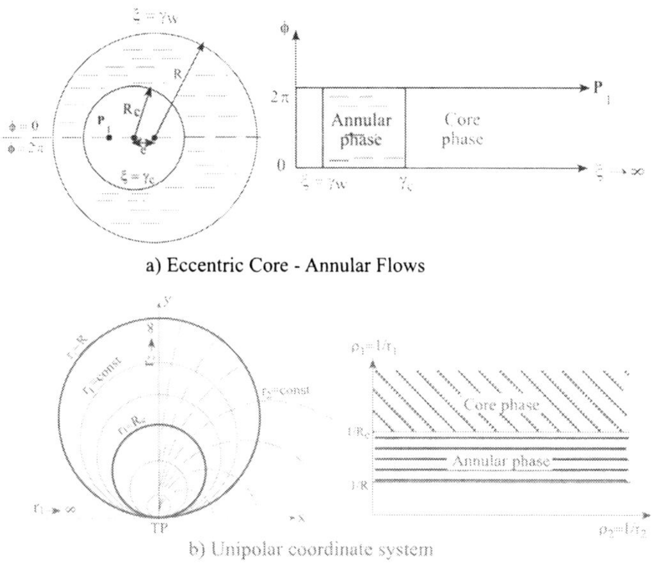

Figure 3.2 Coordinate systems used to solve the Stokes equations for core flow. (a) Eccentric core-bipolar coordinates. (b) Fully eccentric core-unipolar coordinates.

For concentric core, the solution for eccentric core converges to the simple explicit solution obtained by Russel and Charles, 1959:

$$\frac{\widetilde{A}_c}{1-\widetilde{A}_c} = \frac{1}{\widetilde{\mu}}\left[-1+\left(1+\frac{\widetilde{\mu}}{\widetilde{Q}}\right)^{\frac{1}{2}}\right]$$
$$\frac{d\widetilde{P}_c}{dZ} = \frac{1}{\widetilde{A}_c^2[1+2(\widetilde{A}_c^{-1}-1)/\widetilde{\mu}]} \tag{3.5}$$

which for highly viscous concentric core, $\widetilde{\mu} \ll 1$ yields:

$$\widetilde{D}_c^2 = \widetilde{A}_c = \frac{1}{2\widetilde{Q}+1}; \quad \frac{d\widetilde{P}_c}{dZ} = \frac{\widetilde{\mu}(2\widetilde{Q}+1)^2}{4\widetilde{Q}} \tag{3.6}$$

This solution indicates that in the limit of very viscous core flow, the pressure drop reduction factor achieved by the lubricating annular phase is proportional to $\widetilde{\mu}$ ($d\widetilde{P}_c/dZ \to 0$ as $\widetilde{\mu} \to 0$).

But the solution for eccentric core-annular flows fails in the other extreme of fully eccentric core. In this limit, both γ_c and γ_w are zero and the annular phase domain degenerates to a line, $\xi = 0$. The same problem arises when the bipolar coordinates are applied to curved stratified flows. In the limit of a fully eccentric core, the annular phase domain degenerates to an infinite line ($\phi = 2\pi$, for a core of the upper phase, $\phi = 0$ for a core of the lower phase). Thus, the bipolar coordinate system is not appropriate for solving the flow equations in the limit of a fully eccentric core. When the limit of the fully eccentric core-annular configuration is approached, calculations become tedious. The difficulties have been explained in Rovinsky et al (1997). Typically, the cut-off frequency of spectral functions (needed for carrying out the Fourier integrals in the bipolar coordinate system) is less than 50. However, when a configuration of a fully eccentric core is approached, the cut-off frequency increases by several orders of magnitude. This introduces convergence problems, thus increasing dramatically the computational effort and time. To handle the geometry of a fully eccentric core, a 'unipolar' coordinate system (Fig. 3.2b) has been introduced in Rovinsky et al (1997). Circles of constant r_1 are orthogonal to circles of constant r_2 and all circles are tangent to the single TP. In this coordinate system, the annular phase is described by an infinite strip and a solution for the velocity profiles can be worked out in the form of Fourier integrals. The velocity profiles have been integrated to yield $\widetilde{Q} = \frac{Q_a}{Q_c} = \widetilde{Q}(\widetilde{R}_c, \widetilde{\mu})$; and $\frac{d\widetilde{P}_c}{dZ} = \frac{(-\partial P/dz)_{TP}}{(-\partial P/dz)_{cs}} = \frac{d\widetilde{P}_c}{dZ}(\widetilde{R}_c, \widetilde{\mu})$. Given $\widetilde{Q}$ and $\widetilde{\mu}$, the equations are solved to yield $\widetilde{R}_c$ and then the pressure drop, the velocity profiles, as well as wall and interfacial shear stresses profiles for the limit case of fully eccentric core flow were obtained.

Comparison of the analytical solutions for concentric core flow and fully eccentric core flows can be used to evaluate the maximal effect of the core eccentricity on the annular flow characteristics. A detailed discussion has been presented in Rovinsky et al., (1997). Here only some of the results obtained for the effects of core eccentricity are briefly reviewed. Given a flow rate of the viscous phase Q_c(and $\tilde{\mu}$), the introduction of a small amount of the less viscous phase (low Q_a/Q_c) affects initially a decrease of the two-phase pressure drop, where $\frac{(\partial P/\partial z)}{(\partial P/\partial z)_{cs}} = \frac{d\widetilde{P}_c}{dZ} < 1$. However, eventually, increasing the flow rate of the lubricating phase yields an increase of the pressure drop, where the pressure drop factor exceeds the value of 1.0. The lubrication region is defined by the following range of the Martinelli parameter:

$$0 < \chi^2 = \frac{\mu_a Q_a}{\mu_c Q_c} < 1 \; ; \quad \text{Concentric core}$$
$$0 < \chi^2 < 0.65 \; ; \quad \text{Fully eccentric core} \tag{3.7}$$

Thus, the lubrication region is scaled with μ_c; given the flow rate of the viscous phase, Q_c, the range of the flow rates of the less viscous phase, which yields a lubricating effect, increases with increasing the oil viscosity. The potential for pressure drop reduction and power saving in core flows increases with increasing the core viscosity. But, increasing the core eccentricity reduces the potential of pressure drop reduction in lubricated core flow. Figure 3.3c shows that the pressure drop in concentric core-flow is always lower than that obtained with a fully eccentric core. Note that the pressure drop ratio is also the ratio of the pressure drop reduction factor that can be achieved in these two extremes. In concentric core flows, the pressure reduction factor is proportional to $\widetilde{\mu}$, while with a fully eccentric core, the pressure drop reduction factor is bounded (for concentric core: $\frac{d\widetilde{P}_c}{dZ} \longrightarrow 0$ as $\tilde{\mu} \rightarrow 0$, while for fully eccentric core: $\frac{d\widetilde{P}_c}{dZ} \longrightarrow 0.025$, see Figure 3.3a). Obviously, the results shown in Figure 3.3 for a fully eccentric core flow provide an upper bound for the effect of core eccentricity.

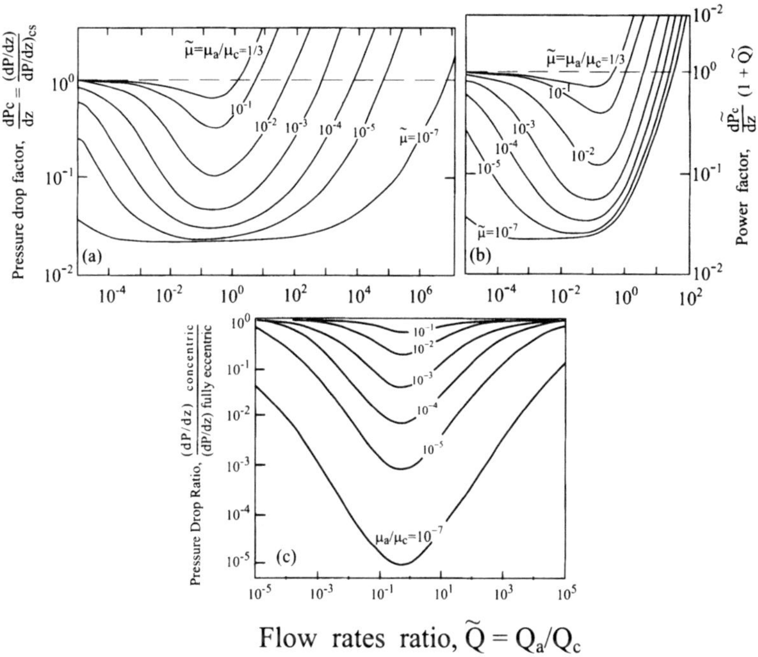

Figure 3.3 **The effect of core eccentricity on the potential of pressure drop reduction and power saving in laminar core flow (Rovinsky et al, 1997).**

The increase of the pressure drop in eccentric core flow evolves from the reduction of the annular-phase holdup and the increase of the wall shear stress. In fact, in concentric viscous core flow, the average velocity of the viscous core phase, U_c always exceeds the average velocity of the lubricating annular phase, U_a: The ratio U_c/U_a approaches a value of 2 for thin annular layer or $\tilde{\mu} \ll 1$. But, when the core approaches a fully eccentric position, it is slowed down (due to the proximity of the tube wall). Consequently, for eccentric core flow, the annular phase velocity may exceed the core phase velocity (for $\tilde{\mu} \ll 1$, the annular phase is the faster phase). As a result, given the flow rates, the viscous core holdup in concentric core flow represents a lower bound for that obtained with the core at eccentric position: $(Ac)_{con}/(Ac)_{ecc} < 1$. However, in the lubrication zone (where $(dP_c/dZ) < 1$), the effect of the core eccentricity on the holdup is moderate ($< 20\%$).

For the opposite case of viscous annulus, $\mu_a/\mu_c > 1$, the effect of core eccentricity on the flow characteristics is moderate. Generally, $(A_c)_{con}/(A_c)_{ecc} < 1$; $(U_c)_{con}/(U_c)_{ecc} > 1$; $(dP/dz)_{con}/(dP/dz)_{ecc} > 1$. The effect of the core eccentricity on the pressure drop is most pronounced around $\chi^2 = \tilde{\mu}\tilde{Q} = 10$, but is limited to about 35% for $\tilde{\mu} \gg 1$.

3.2 Two-Fluid model for CAF

A simple practical model for general annular concurrent liquid-liquid flow, which is not restricted to laminar flow regimes, can be obtained using the two-fluid approach (Brauner,

1991). The combined momentum equation for the core (c) and annular liquid (a) (obtained after eliminating the pressure drop) reads (see Figure 3.1):

$$\frac{4}{D(1-\widetilde{D}_c^2)}\left[-\tau_a+\frac{\tau_i}{\widetilde{D}_c}\right]+(\rho_a-\rho_c)g\sin\beta=0 \tag{3.8}$$

where $\beta > 0$ and $\beta < 0$ for downward and upward flow, respectively. In core flow $\rho_c = \rho_o$ and $\rho_a = \rho_w$, while for the inverted configuration $\rho_a = \rho_o$ and $\rho_c = \rho_w$. The wall shear stress τ_a and interfacial shear stress τ_i are expressed in terms of the phases average velocities u_a, u_c and the corresponding friction factors f_a, f_i. The appropriate structure for these closure relations has been identified as (Brauner, 1997):

$$\tau_a=\frac{1}{2}f_a\rho_a U_a^2 \quad ; \quad f_a=C_a\left[\frac{\rho_a D(1-\widetilde{D}_c^2)U_a}{\mu_a}\right]^{-n_a} \tag{3.9}$$

$$\tau_i=\frac{1}{2}f_i\rho_c(U_c-c_iU_a)U_c \quad ; \quad f_i=F_iC_c\left(\frac{\rho_c D_c U_c}{\mu_c}\right)^{-n_c} \tag{3.10}$$

where $c_i = u_i/U_a$ and u_i is the interfacial velocity. For laminar annular phase $c_i = 2$, while for turbulent annular phase $c_i \simeq 1.15 \div 1.2$. The constants $C_{a,c}$ and $n_{a,c}$ are set according to the flow regime in each phase ($C = 16$, $n = 1$ for laminar flow and $C = 0.046$, $n = 0.2$ for turbulent flow). The coefficient F_i denotes possible augmentation of the interfacial shear due to interfacial waviness. However, in core-flow, the liquids interface is characterized by long smooth waves and appears less roughened than in annular gas-liquid flows. Also, as the velocities of the two liquids in core flow are comparable, the modelling becomes even less sensitive to the estimation of the interfacial friction factor, and F_i can be set to 1. Using mass balances on the annular and core phases, $\widetilde{U}_c = \frac{U_c}{U_{cs}} = \frac{1}{\widetilde{D}_c^2}$; $\widetilde{U}_a = \frac{U_a}{U_{as}} = \frac{1}{(1-\widetilde{D}_c^2)}$, results in the following non-dimensional equation for the core diameter:

$$(1-\widetilde{D}_c^2)\widetilde{D}_c^{n_c-5}\left[1-\widetilde{D}_c^2(1+c_i\widetilde{Q})\right]-X^2+Y(1-\widetilde{D}_c^2)^3=0 \tag{3.11}$$

The dimensionless parameters are $\widetilde{Q}$, X^2 (Martinelli parameter) and Y:

$$X^2=\frac{C_a Re_{as}^{-n_a}}{C_c Re_{cs}^{-n_c}}\frac{\widetilde{Q}^2}{\widetilde{\rho}}=\frac{(dP/dz)_{as}}{(dP/dz)_{cs}} \quad ; \quad Y=\frac{1}{2}\frac{(\rho_a-\rho_c)}{\rho_c}\cdot\frac{Dg\sin\beta}{U_{cs}^2}\frac{1}{C_c Re_{cs}^{-n_c}} \tag{3.12}$$

where $\widetilde{\rho} = \rho_c/\rho_a$ and Re_{as}, Re_{cs} are the superficial Reynolds numbers of the annular and core liquids respectively. Obviously the physical solution for $\widetilde{D}_c$ is in the range $0 < \widetilde{D}_c \le 1$ and the corresponding core holdup is $\widetilde{A}_c = \widetilde{D}_c^2$. After solving eq. (3.8) for $\widetilde{D}_c$, the pressure gradient can be obtained by adding the momentum equations for the core and annular phases:

$$\frac{d\widetilde{P}_c}{dZ}=\frac{(-dP/dz)}{(-dP/dz)_{cs}}=\frac{X^2}{(1-\widetilde{D}_c^2)^2}-\frac{\rho_m}{\Delta\rho}Y \tag{3.13}$$

where ρ_m is the mixture density; $\rho_m = \rho_c\widetilde{D}_c^2 + \rho_a(1-\widetilde{D}_c^2)$.

For viscous oils, the flow in the core is laminar. Fortunately, for the case of horizontal laminar core (with either laminar or turbulent annular phase), simple explicit solutions for the in situ hold-up $\widetilde{D}_c^2$, and the resulting pressure drop are obtained. These are summarized in Table 3.1. Note that the solution obtained for laminar-laminar core flow via the above two-fluid model is identical to the exact solution obtained by Russel and Charles (1959).

For highly viscous oils, $\mu_c/\mu_a >> 1(X^2 \to 0)$, therefore the predicted insitu holdup is practically determined by the flow rates ratio and flow regime in the annular phase. The data and the model indicate that the water in situ holdup exceeds the input water cut by a few percent (e.g. Figure 12 in Brauner, 1998). Results of pressure drop in CAF are of the order of pressure loss for flow of water at the mixture flow rate (e.g. Figure 13 in Brauner, 1998). Both theory and data indicate that for each oil superficial velocity, there exists an optimum input water-cut (which yields minimum pressure drop) in the range of water-cut of $U_{ws}/U_m = 0.08 \div 0.12$ (compared to an optimal water-cut of 1/3 in L-L flows concentric core flows, e.g Russel and Charles, 1959).

Table 3.1 Core Diameter and Pressure Drop for Laminar Core

	Laminar core - Laminar annulus (L-L)	Laminar core - Turbulent annulus (L-T)
$X^2 =$	$\frac{\mu_a}{\mu_c} \cdot \widetilde{Q}$	$\frac{0.046}{16}\left(\frac{\mu_a}{\mu_c}\right)\widetilde{Q}Re_{as}^{0.8}$ or $\frac{0.046}{16}\left(\frac{\mu_a}{\mu_c}\right)^{0.2}\left(\frac{\rho_a}{\rho_c}\right)^{0.8}Re_{cs}^{0.8}\widetilde{Q}^{1.8}$
c_i	2	$1.15 \div 1.2$
$\widetilde{D}_c^2$	$\frac{1+\widetilde{Q}-(1+\widetilde{\mu}/\widetilde{Q})^{1/2}\widetilde{Q}}{1+2\widetilde{Q}-\widetilde{\mu}\widetilde{Q}}$	$\frac{1+\widetilde{Q}c_i/2[1-(1+4X^2/\widetilde{Q}^2c_i^2)^{1/2}]}{1+\widetilde{Q}c_i+X^2}$
$d\widetilde{P}_c/dZ$	$\frac{\widetilde{\mu}}{\widetilde{Q}}\left[\frac{1+2\widetilde{Q}-\widetilde{\mu}\widetilde{Q}}{1-\widetilde{\mu}+(1+\widetilde{\mu}/\widetilde{Q})^{1/2}}\right]^2$	$X^2\left[\frac{1+c_i\widetilde{Q}-X^2}{\widetilde{Q}c_i/2-X^2+(\widetilde{Q}c_i^2/4+X^2)^{1/2}}\right]^2$

3.3 Conclusion

The Two-fluid model for CAF is a simple practical tool for evaluating the potential pressure drop reduction and power saving in concentric CAF. However, the predicted pressure drop via this model may underestimate measured values in CAF operation. Possible reasons for deviations are the increase of the wall friction due to surface irregularities, fouling of pipe walls by a wavy core interface at high oil rates, and eccentric (rather than concentric) core flow, as discussed in Section 3.1. Accounting for these effects in the framework of the two-fluid model requires appropriate modifications of the closure laws used for the wall and/or interfacial shear stresses. The exact solutions for eccentric laminar CAF and numerical studies for the case of turbulent lubricating phase (e.g. Huang et al.,1994) can be used to test the validity of such closure laws.

4 Dispersed Flow

A dispersion of two immiscible liquids, where one of the liquids forms a continuous phase and the other is dispersed in it, is a flow pattern often observed in liquid-liquid systems. There are water-in-oil (w/o) and oil-in-water (o/w) dispersions. Emulsion is a stable dispersion, which usually involves the presence of surfactants that inhibit coalescence of the dispersed droplets. High viscous oil content emulsions are considered a lubricated regime of flow, since a dramatic decrease in the fluid viscosity and pressure drop can be achieved by emulsifying the oil into a continuous water phase, (e.g. McAuliffe, 1973, Pilehvari et al., 1988). Multiple emulsions (e.g. o/w/o, oil drops dispersed in aqueous droplets that are in turn dispersed in a continuous oil phase) can also be formed. Dispersions will always form in motions of two immiscible liquids which are sufficiently intense. However, relatively dilute dispersions can be also obtained at low velocities as a result of the entry device used to introduce the two liquids into the flow tube. In fact, dispersed flow is the basic flow pattern in upward vertical and off-vertical inclined flows.

For fully developed flow, the total pressure gradient, dP/dz is the sum of the frictional pressure gradient, dP_f/dz and the gravitational pressure gradient, dP_g/dz

$$(-dP/dz) = (-dP_f/dz) + (-dP_g/dz) = 2f_m\frac{\rho_m U_m^2}{D} - \rho_m g \sin\beta \tag{4.1}$$

where the z coordinate is attached to the direction of the continuous phase flow, $\beta > 0$ for downward inclination, and the mixture density $\rho_m = \rho_d\epsilon_d + \rho_c(1-\epsilon_d)$ is calculated based on the in situ holdup of the dispersed phase, ϵ_d. The friction factor, f_m is evaluated based on the mixture Reynolds number $DU_m\rho_m/\mu_m$. These require models for the in situ holdup and the mixture apparent viscosity, μ_m. The presence of droplets in a continuous fluid may affect the effective viscosity of the dispersion due to droplets interactions and modification of the continuous phase momentum transfer characteristics.

4.1 In-situ holdup

In vertical and off-vertical inclined systems the static head is a major contributor to the total pressure gradient. Therefore, good estimates for the in situ holdup and the corresponding mixture density are needed.

The simplest approach is the homogeneous model which neglects a possible difference between the in situ velocities of the two liquid phases (slippage). When the dispersed droplets move at the velocity of the surrounding continuous phase, $(U_c = U_d = U_m)$, the in situ holdup is determined by the input volumetric flow rates of the two liquids:

$$\epsilon_d = \frac{U_{ds}}{U_m} \quad ; \quad \epsilon_c = 1 - \epsilon_d = \frac{U_{cs}}{U_m} \; ; \; U_m = U_{ds} + U_{cs} \tag{4.2}$$

However, due to the density difference, drops of the dispersed phase tend to move at a different velocity than the continuous phase. The slippage between the phases was found to be negligible for Dw/o in viscous oils or for fine Do/w and Dw/o (e.g. Hassan and Kabir, 1990, Flores et al., 1997). For these flow regimes, the homogeneous model for estimating ϵ_d, eq. (4.2) is applicable even for inclined and vertical systems. However, when water forms the continuous phase and for low mixture velocities, relatively large oil droplets (bubbles) are formed, which may show a significant slippage.

The Zuber-Findlay (1965) drift flux model can be used to model the flow of oil-in-water dispersions. The average velocity of the dispersed drops, U_d is expressed in terms of the mixture velocity U_m and a drift velocity u_d:

$$U_d = C_o U_m + u_d = U_{ds}/\epsilon_d \quad ; \quad U_c = \frac{U_{cs}}{1-\epsilon_d} \tag{4.3}$$

where C_o is a distribution parameter, which accounts for the droplets velocity and concentration profiles. Typically, $C_o = 1$ for uniform droplet concentration, $C_o > 1$ when the droplets tend to flow at the center and $C_o < 1$ when the droplets concentration is higher near the wall. The drift velocity, u_d is evaluated based on the terminal rise (settling) velocity of a single droplet in the continuous phase, (u_∞) and corrected for the effect of the swarm of drops:

$$u_d = u_\infty (1-\epsilon_d)^{n_d} \mid \sin\beta \mid ; \quad 0 < n_d < 3 \tag{4.4}$$

The value of n_d depends mainly on the droplets size. For large drops (of the order of the tube diameter) $n_d \simeq 0$, whereas for liquid-liquid dispersions $n_d = 1.5 \div 2.5$ was recommended by Hassan and Kabir (1990) and Flores et al., (1997).

Equations (4.3) and (4.4) yield an implicit algebraic equation for ϵ_d:

$$\frac{U_{cs}}{U_{ds}} = \frac{1 - C_o\epsilon_d}{C_o\epsilon_d} - \frac{1}{C_o}\frac{u_\infty}{U_{ds}}(1-\epsilon_d)^{n_d} \mid \sin\beta \mid \tag{4.5}$$

This equation is applicable both for concurrent and counter current flows: $U_{cs}/U_{ds} > 0(< 0)$ for concurrent (counter current) flows, respectively. The sign of u_∞/U_{ds} depends on the direction of u_∞ with respect to U_{ds}. Note that in counter-current flows $u_\infty/U_{ds} > 0$ both for HPD and LPD modes (see Figure 1.3). For $u_d \neq 0$, eq. (4.5) predicts the existence of these two different modes in counter-current dispersed flows.

The drop shape characterization map given in Clift et al., (1978) (see Figure 16 in Brauner, 1998) can be used to extract the drop velocity u_∞. The graphical relation corresponds to $Re_d = Re_d(M,\ Eo_d)$, where $Re_d = \frac{u_\infty d}{\nu_c}$; $M = \frac{g\mu_c^4|\Delta\rho|}{\rho_c^2\sigma^3}$; $Eo_d = \frac{gd^2|\Delta\rho|}{\sigma}$ and d is the drop diameter. Recommended correlations u_∞ are also summarized in this reference. A widely used equation for u_∞ is the Harmathy's (1960) model for distorted drops:

$$u_\infty = 1.53\left[\frac{g\sigma|\Delta\rho|}{\rho_c^2}\right]^{1/4}, \tag{4.6}$$

which suggests u_∞ is independent of the drop diameter. It reflects an increase of the drag coefficient with an increase of the effective cross section of a distorted drop. For large drops, $d/D = O(1)$, u_∞ should be corrected for the reduction of the drop velocity due to pipe wall effects (Clift et al., 1978). For $d/D \simeq 1$ (large bubble or slug) and $Eo_d < 0.125$, $u_\infty \approx 0$, (Zukowski, 1966). Note that in concurrent flows, the sign of u_∞ is to be adjusted according to the sign of the buoyant force due to $\Delta\rho = (\rho_d - \rho_c)g\sin\beta$ with respect to the mixture flow direction. In counter-current flows, the sign of u_∞ is the same as U_{ds}.

4.2 Viscosity of emulsions

When the slippage between the dispersed and the continuous phase is significant, or in coarse dispersions, the mixture viscosity is normally taken as the viscosity of the continuous phase, $\mu_m = \mu_c$. On the other hand, a fine dispersion, or an emulsion, can be treated as a pseudo-homogeneous fluid of a viscosity μ_m, when $Re_c(d/D)^2\rho_d/\rho_c < 1, Re_c = \rho_c U_m D/\mu_c$ (e.g. Baron et al., 1953). Models for estimating the drops size are given in Section 4.4.

The viscosity of emulsion μ_m is defined as the ratio between the shear stress (τ) and the shear rate ($\dot{\gamma}$). The viscosity of the emulsion is proportional to the viscosity of the continuous phase μ_c. However, the emulsion viscosity depends upon several other factors, which include the volume fraction of the dispersed phase (ϵ_d), the droplets size (d) and viscosity (μ_d), the shear rate ($\dot{\gamma}$), temperature (T), the emulsifying agent used and its concentration (e.g Sherman, 1968, Schramm, 1992). At low to moderate holdup of the dispersed phase, emulsions generally exhibit Newtonian behavior.

The emulsion viscosity is affected mainly by the viscosity of the continuous phase and increases with increasing the holdup of the dispersed phase. Following Einstein's relation (1906) for the viscosity of suspensions in extreme dilution:

$$\mu^* = \frac{\mu_m}{\mu_c} = (1 + 2.5\epsilon_d); \quad \epsilon_d \ll 1, \tag{4.7}$$

other models/correlations which relate the emulsion viscosity to the volume fraction of the dispersed phase have been proposed in the literature, in the form of $\mu^* = f(\epsilon_d)$ or $\ln \mu^* = f(\epsilon_d)$. These include empirical fitting constants. A summary of various correlations for μ^* is given in Brauner, 1998, Tables 3a and 3b. In the presence of emulsifiers and/or impurities, the dispersed droplets behave like rigid particles and the emulsion viscosity is independent of the dispersed phase viscosity. In their absence, however, possible internal circulation within the droplets results in some decrease of the emulsion viscosity with reducing the dispersed phase viscosity. For high emulsion concentrations, the empirical correlations use a reduced dispersed phase concentration, $\epsilon_d/\hat{\epsilon}_d$, where $\hat{\epsilon}_d$ represents the maximum attainable dispersed phase concentration at phase inversion. This introduces in the correlations effects of additional parameters, such as emulsifier concentration, flow field and droplets sizes.

At high dispersed phase concentrations (approaching phase inversion conditions) emulsions behave as non-Newtonian shear-thinning (pseudoplastic) fluids (e.g. Pal, 1990). The relation between the shear stress and shear rate is modeled by the power law equation, (in terms of two constants k and n):

$$\tau = -k\dot{\gamma}\left[\dot{\gamma}^{n-1}\right]; \quad n < 1 \tag{4.8}$$

which indicates that the apparent emulsion viscosity $\mu_m = \tau/\dot{\gamma}$ decreases with increasing the shear rate. Concentrated emulsions can also exhibit a viscoelastic behavior.

The viscosity of emulsions decreases with increasing the temperature, $\mu_m = Ae^{-B/T}$, where T is the absolute temperature and A, B are constants dependent upon the specific emulsion and shear rate. Due to the sensitivity of the oil viscosity to temperature variations, the viscosity of Dw/o and the associated pressure drop are mostly affected by temperature.

Emulsion rheology may vary between different oils and emulsifiers. Even oils of similar properties may exhibit a different emulsion rheology. Therefore, it is recommended to experimentally study the rheology of the emulsion used in a particular application.

4.3 Friction factor

Given the mixture properties, one can apply the single phase flow equations. For laminar flow of Newtonian fluid - the friction factor is obtained from the Hagen-Poiseuille equation:

$$f_m = \frac{16}{Re_m} \quad ; \quad Re_m = \frac{\rho_m U_m D}{\mu_m} \leq 2100 \tag{4.9}$$

For turbulent flow of Newtonian fluids, the friction factor can be obtained from the Moody diagram, or calculated from one of the experimental correlations suggested in the literature. For instance, in smooth tubes, the Blasius correlation is applicable:

$$f_m = \frac{0.079}{Re_m^{0.25}} \; ; \quad 300 \leq Re_m \leq 100,000 \tag{4.10}$$

For rough walls, the Colebrook (1939) equation yields:

$$\frac{1}{\sqrt{f_m}} = -4\log_{10}\left(\frac{k^*}{3.71} + \frac{1.26}{Re_m\sqrt{f_m}}\right) \tag{4.11}$$

where $k^* = k/D$ is the nondimensional wall roughness scale. An explicit approximation to eq. (4.11) can be used (Zigrang and Sylvester, 1985):

$$\frac{1}{\sqrt{f_m}} = -4\log_{10}\left[\frac{k^*}{3.71} - \frac{4.518}{Re_m}\log\left(\frac{6.9}{Re_m} + \left(\frac{k^*}{3.7}\right)^{1.11}\right)\right] \tag{4.12}$$

For dense emulsions that behave as non-Newtonian pseudoplastic fluid, the frictional pressure drop can be estimated using Dodge and Metzner (1959) correlations:

$$\begin{aligned} f_m &= \frac{16}{R\acute{e}_m} \; ; \quad R\acute{e}_m = \frac{\rho_m U_m^{2-n'} D^{n'}}{k'(8)^{n'-1}} \; ; && \text{laminar flow, } R\acute{e}_m \leq R\acute{e}_{L-T} \\ \frac{1}{\sqrt{f_m}} &= \frac{4}{(n')^{0.75}}\log[R\acute{e}_m f_m^{(1-n'/2)}] - \frac{0.4}{n'^{1.2}} \; ; && \text{turbulent flow, } R\acute{e}_m > R\acute{e}_{L-T} \end{aligned} \tag{4.13}$$

where $n' = n \leq 1$, $k' = k\left(\frac{1+3n}{4n}\right)^n$ and n, k are the constants of the power-law model for the emulsion viscosity (eq. 4.8). The laminar-turbulent transitional Reynolds number is given by $R\acute{e}_{L-T} = 6464\,\frac{n(2+n)^{\frac{2+n}{1+n}}}{(1+3n)^2}$ (e.g. Hanks and Christeansen, 1962).

The variation of the pressure drop with the liquids flow rates is, however controlled by the phase inversion phenomenon (e.g. Figure 15 in Brauner, 1998). A sudden increase in the apparent dispersion viscosity (up to one order of magnitude higher than the single phase oil viscosity) occurs where the external phase invert from oil to water (or vice versa). When water forms the continuous phase, the mixture viscosity approaches the single phase water viscosity. The increase of the apparent mixture viscosity at phase inversion (compared to the pure oil viscosity) seems to be moderated with increasing the oil viscosity. The conditions under which phase inversion takes place are discussed in Section 4.5.

4.4 Drops sizes

The mechanisms of drop formation and their characteristic size are important for analyzing the hydrodynamic and transport phenomena in the flow of liquid-liquid dispersions. The main breakup mechanisms involve high shear stresses, turbulence in the continuous phase and rapid acceleration (Taylor, 1934, Kolmogorov, 1949, Hinze, 1955, 1959). The surface force which resists deformation and breakup is mainly due to surface tension and also due to internal viscous force (in the case of viscous drop). In dense dispersions, droplets coalescence and additional factors introduced when a swarm of droplets interact must be taken into account. These lead to an increase of the drop size.

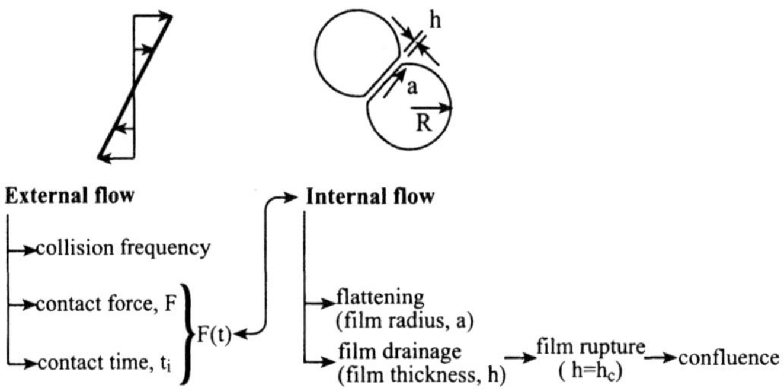

Figure 4.1 Conceptual framework for coalescence modeling (Chesters, 1991).

A substantial effort has been made to model the phenomenon of droplets coalescence in dense dispersions. Reviews of existing frameworks for analysis of droplets interactions with themselves and with the surrounding fluid can be found in Chesters (1991), Tsouris and Taularides (1994). Coalescence actually involves a number of coupled sub-processes (see Figure 4.1). Some are governed by the external flow field, due to the flow of the continuous phase (e.g. frequency of drops collisions, force and duration of collisions). These provide the boundary conditions for the internal flow (i.e. drop deformations, film drainage and rupture of the interfaces). However, the relationships that have been proposed for the various sub-processes involved include unknown parameters and therefore, at this time, they cannot be readily applied to general liquid-liquid flow systems.

In dilute dispersions, however, the characteristic drop size is governed by the drop breakup mechanism. In the following, models for evaluating the maximal drop size associated with the various breakup mechanisms and some extensions to dense dispersions are briefly reviewed. The maximum drop size, $d_{\max}$ provides an estimate to the drop volume-surface mean size (the Sauter mean diameter) $d_{32} = \sum n_i d_i^3 / \sum n_i d_i^2 \simeq d_{\max}/k_d$ with $k_d \simeq 1.5 \div 3$ (see also Azzopardi and Hewitt, 1997).

Shear flow – Drops deformation and splitting under the action of viscous shear (Couette flow and plane hyperbolic flow) was studied by Taylor (1934). The critical Weber number, defined based on the maximum velocity gradient in the flow field, $We_{crit} = \mu_c \dot{\gamma}_{\max} d_{\max}/\sigma$, was found to vary with μ_d/μ_c. It increases for $\mu_d/\mu_c \gg 1$ or $\mu_d/\mu_c \ll 1$. For $\mu_d/\mu_c \geq 20$ and Couette flow, breakup of drops was not observed. This evidence implies that it is difficult to disperse fluids of high viscosity ratio by the action of viscous shear.

For the case of viscous continuous phase, where $\mu_d/\mu_c \ll 1$, the model of Taylor (1934) and Arivos (1978) for breakup of long slender droplets in an axisymmetric straining motion can be used to estimate the drop size. When applied to laminar pipe flow, where the average value of $\dot{\gamma}$ is given by $\dot{\gamma} = 4U_m/D$, this model suggests that the drop size depends on the capillary number of the continuous phase, $\mu_c U_m/\sigma$:

$$\frac{d_{\max}}{D} = 0.296 \frac{\sigma}{\mu_c \dot{\gamma} D} \left(\frac{\mu_c}{\mu_d}\right)^{1/6} = 0.074 \frac{\sigma}{\mu_c U_m} \left(\frac{\mu_c}{\mu_d}\right)^{1/6} \quad ; \quad \frac{\mu_d}{\mu_c} \ll 1 \tag{4.14}$$

Turbulent flow – Most of the models for predicting the size of bubbles or drops in a turbulent flow field are based on the Kolmogorov (1949)-Hinze (1955) model for emulsification in a turbulent flow field. Using dimensional arguments, they showed that the splitting of a drop depends upon a critical Weber number, which yields the maximal drop size, $d_{\max}$ that can resist the stress due to dynamic pressure of turbulent eddies (τ). According to Hinze (1955):

$$We_{crit} = \frac{\tau d_{\max}}{\sigma} = C[1 + F(On)] \tag{4.15}$$

where C is a constant, On is the Ohnesorge (viscosity) number (ratio between the internal viscosity force and the interfacial force); $On = \frac{\mu_d}{\sqrt{\rho_d d_{\max} \sigma}}$ and F is a function that goes to zero as $On \to 0$. For pipe flow of a dilute dispersion, this model yields the maximal drop/bubble size, $d_{\max}$ in terms of the critical Weber number of the continuous phase, $We_c = \rho_c U_c^2 D/\sigma$ and the wall friction factor, f (e.g. Kubie and Gardner, 1977):

$$\left(\tilde{d}_{\max}\right)_o = \left(\frac{d_{\max}}{D}\right)_o = 0.55 We_c^{-0.6} f^{-0.4} \quad ; \quad \ell_k \ll d_{\max} < 0.1D \tag{4.16}$$

where ℓ_k is the Kolmogorov microscale and 0.1D represents the inertial subrange scale (length scale of energy containing eddies).

The Hinze model is applicable for dilute dispersions. It suggests that the maximal drop size, $(d_{max})_o$, can be evaluated based on a static force balance between the eddy dynamic pressure and the counteracted surface tension force (considering a single drop in a turbulent field). An extension of this model for dense dispersions was suggested by Brauner (2001). The idea is that in dense dispersions, where local coalescence is prominent, the maximal drop size, $(d_{max})_\epsilon$, is evaluated based on a local energy balance.

In the dynamic (local quasi-steady) breakage/coalescence processes, the turbulent kinetic energy flux in the continuous phase should exceed the rate of surface energy generation that is required for the renewal of droplets in the coalescing system. This energy balance yields:

$$\left(\tilde{d}_{max}\right)_\epsilon = \left(\frac{d_{\max}}{D}\right)_\epsilon = 2.22\tilde{C}_H\left(\frac{\rho_c U_c^2 D}{\sigma}\right)^{-0.6}\left[\frac{\rho_m}{\rho_c(1-\epsilon_d)}f\right]^{-0.4}\left(\frac{\epsilon_d}{1-\epsilon_d}\right)^{0.6} \quad (4.17)$$

where $\tilde{C}_H$ is a tunable constant, $\tilde{C}_H = O(1)$. In dilute systems, the energy balance is trivially satisfied for any finite drop size (as the rate of surface energy generation vanishes for $\epsilon_d \to 0$) thus, $(d_{max})_\epsilon < (d_{max})_o$. However, this is not the case in the dense system, where $(d_{max})_\epsilon > (d_{max})_o$. Thus, given a two-fluid system and operational conditions, the maximal drop size is taken as the largest of the two values:

$$\tilde{d}_{\max} = Max\left\{\left(\tilde{d}_{\max}\right)_o\left(\tilde{d}_{\max}\right)_\epsilon\right\} \quad (4.18)$$

Correlations for the friction factor in smooth or rough conduits can be used in eqs. (4.16) and (4.17). For instance, the Blasius equation ($f = 0.046/Re_c, Re_c = \rho_c DU_c/\mu_c$) yields:

$$\left(\frac{d_{\max}}{D}\right)_o = 1.88 We_c^{-0.6} Re_c^{0.08} \quad ; \quad (4.19)$$

$$\left(\tilde{d}_{max}\right)_\epsilon = 7.61\tilde{C}_H We_c^{-0.6} Re_c^{0.08}\left(\frac{\epsilon_d}{1-\epsilon_d}\right)^{0.6}\left[1+\frac{\rho_d}{\rho_c}\frac{\epsilon_d}{1-\epsilon_d}\right]^{-0.4} \quad ; \quad (4.20)$$

Eqs. (4.18) to (4.20) are the H-model in Brauner (2001), which is applicable provided $1.82Re_c^{-0.7} < \tilde{d}_{max} < 0.1$ and $Re_c > 2100$.

If the viscosity of the dispersed phase is much larger than that of the continuous phase, the viscous forces due to the flow inside the drop also become important and the effect of the On number in eq. (4.15) may turn to be non-negligible. Kolmogorov (1949) found that when $\mu_d/\mu_c \gg 1$, these viscous forces can be neglected only when $d_{max} \gg \ell_k(\nu_d/\nu_c)^{3/4}$. A correlation for $d_{\max}$, which accounts for viscous forces in the dispersed and continuous phase was suggested by Paul and Sleicher (1965):

$$\frac{\rho_c U_c^2 d_{\max}}{\sigma}\left(\frac{\mu_c U_c}{\sigma}\right)^{0.5} = C\left[1+0.7\left(\frac{\mu_d U_c}{\sigma}\right)^{0.7}\right] \quad (4.21)$$

with $C = 38 \div 43$. This correlation indicates no effect of the pipe diameter. Kubie and Gardner (1977) showed that a major part of Sleicher and Paul (1965) data correspond to drops that are larger than the scale of energy containing eddies ($\approx 0.1D$ for pipe flow). It was argued that for $d_{\max} > 0.1D$, the turbulent dynamic pressure force in Kolmogorov/Hinze analysis should be evaluated based on the fluctuating turbulent velocity ($\simeq 1.3u^*$ in pipe flow, Hughmark, 1971). In this case, the correlation that evolves for $d_{\max}$ (instead of eq. (4.16)) reads:

$$\frac{d_{\max}}{D} = 1.38\left(\frac{\rho_c u_c^2 D}{\sigma}\right)^{-1} f^{-1} \quad ; \quad d_{\max} > 0.1D \quad (4.22)$$

Accordingly, for $\widetilde{d}_{\max} > 0.1$, eqs. (4.19 - 4.20) are replaced by the K-Model in Brauner (2001):

$$\left(\widetilde{d}_{\max}\right)_0 = 30 We_c^{-1} Re_c^{0.2} \; ; \; \widetilde{d}_{\max} > 0.1 \tag{4.23}$$

$$\left(\widetilde{d}_{\max}\right)_\epsilon = 174 C_K We_c^{-1} Re_c^{0.2} \left(\frac{\epsilon_d}{1-\epsilon_d}\right) \; ; \tag{4.24}$$

where $C_K = O(1)$. However, in $D_{\mathrm{o/w}}$ of viscous oils, or in systems of low surface tension, additional stabilizing force due to the drop viscosity has to be considered, which affects an increase of $d_{\max}$ with μ_d. According to Hinze (1955), the effect of the dispersed phase viscosity is represented by the Ohnesorge number. For a non-vanishing On, the r.h.s. of eqs. (4.16) to (4.20) and (4.22)-4.24) are augmented by the term $[1+F(On)]^{0.6}$. Instead, the correction suggested by Davies (1987) can be applied by multiplying the R.H.S. of these equations by $(1+K_\mu \mu_d u_c'/\sigma)^{0.6}$, with $K_\mu = O(1)$, where u_c' is the characteristic turbulent fluctuation velocity in the continuous phase.

Accelerated Drops – Drops deformation and breakup due to rapid acceleration of drops bursting into a stream of a second fluid is the main mechanism for pneumatic atomization and has been studied extensively in the literature (e.g. Hinze, 1955, Clift et al., 1978, Brodkey, 1969, Cohen, 1991). This mechanism can be relevant to the formation of liquid dispersions in the entry region of the pipe, in particular, when nozzles are used for injection of the liquid, or for drop entrainment from the interface between a slow and a fast moving layers (as in wavy stratified flow). The following power-law empirical correlation for We_{crit} is often used to evaluate $d_{\max}$:

$$We_{crit} = \frac{\rho_c \Delta U_c^2 d_{\max}}{\sigma} = 12\left(1+1.077 On^{1.6}\right) \tag{4.25}$$

When ΔU_c is set to the initial velocity difference (between the drop and the continuous phase), eq. (4.25) may underestimate $d_{\max}$. Modified correlations which consider the breakup time and velocity history are given in the literature (see review by Azzopardi and Hewitt, 1997).

Rising (settling) drops – Even in a stagnant fluid ($U_c \to 0$), there is a limit to the size to which a bubble or a drop can reach while rising (or falling) freely through it. In the absence of external field disturbances, drop breakup has been attributed to Rayleigh-Taylor instability. Grace et al., (1978) showed that for $\mu_d/\mu_c > 0.5$, $d_{\max} = 4\sqrt{\sigma/g \mid \rho_c - \rho_d \mid}$ provides a reasonable estimate for the maximal drops size. For $\mu_d/\mu_c < 0.5$, it provides a lower bound to $d_{\max}$. Combining the Rayleigh-Taylor instability and Kelvin-Helmholtz instability (Kitsch and Kocamustafaogullari, 1989), the following equation was obtained for $d_{\max}$ of rising (falling) drops in stagnant fluids:

$$d_{\max}\sqrt{\frac{g \mid \rho_c - \rho_d \mid}{\sigma}} = \left\{ \frac{16}{1+0.5\left[4.5\left(\frac{\rho^*}{1+\rho^*}\right) - 0.35\left(\frac{2+3\rho^*}{1+\rho^*}\right)^{2.27}\left(1+M^{0.25}\right)^{0.36}\right]} \right\}^{1/2} \tag{4.26}$$

where $\rho^* = 0.993\rho_d/\rho_c$ and the Morton number, $M \leq 16$. Equation (4.30) predicts the experimentally observed increase of $d_{\max}$ with increasing M. Eq. (4.26) (with a lower numerical coefficient) represents the scale of highly deformable drops/bubbles in vertical and horizontal flows (Brodkey, 1969, Brauner and Moalem Maron, 1992c).

4.5 Phase inversion

The phase inversion refers to a phenomenon where with a small change in the operational conditions, the continuous and dispersed phase spontaneously invert. For instance, in oil-water systems, a dispersion (emulsion) of oil drops in water becomes a dispersion (emulsion) of water drops in oil, or vice versa.

The phase-inversion is a major factor to be considered in the design of oil-water pipelines, since the rheological characteristics of the dispersion and the associated pressure drop change abruptly and significantly at or near the phase inversion point (Pan et al (1995), Angeli and Hewitt (1996), Arirachakaran et al (1989)). Also, the corrosion of the conduit is determined to a large extent by the identity of the phase that wets it.

The inversion point is usually defined as the critical volume fraction of the dispersed phase above which this phase will become the continuous phase. Studies have been carried out in batch mixers, continuous mixers, column contractors and pipe flow, in attempt to characterize the dependence of the critical volume fraction on the various system parameters, which include operational conditions, system geometry and materials of construction. These have been reviewed by Yeo et al., 2000. In flow systems, phase inversion will not always occur as the holdup (say of water) is varied continuously from 0 to 1. It will occur only if U_m is high enough to have a good mixing of the liquids in both the pre- and post inversion dispersions.

Similarly to observations made in stirred tanks, also in pipe flows, data on dispersion inversion indicate a tendency of a more viscous oil to form the dispersed phase. It was found that the water-cut required to invert a dispersion decreases as the oil viscosity, μ_o increases. Based on the experimental results of various investigators on phase inversion, Arirachakaran et al (1989) proposed the following correlation for the critical water-cut, $\dot{\epsilon}_{\mathrm{w}}^I$:

$$\dot{\epsilon}_{\mathrm{w}}^I = \left(\frac{U_{\mathrm{ws}}}{U_m}\right)_I = 0.5 - 0.1108 \log_{10}(\mu_o/\mu_r)\,;\ \mu_r = 1 mPa \cdot s \qquad (4.27)$$

The trend is similar to that indicate by the Yeh et al (1964) model for the phase inversion point: $\epsilon_{\mathrm{w}}^I = \frac{1}{1+(\mu_o/\mu_{\mathrm{w}})^{0.5}}$. The later was developed with reference to a configuration of laminar flow in stratified layers, however, its validity was tested against the critical holdup data obtained in a flask (dispersion prepared by manual vigorous shaking of specified volumes of an organic and water phases).

Since phase inversion is a spontaneous phenomenon, it was proposed that its prediction can be based on the criterion of minimization of the total system free energy, (e.g. Luhning and Sawistowski 1971, Tidhar et al., 1986, Decarre and Fabre, 1997, Brauner and Ullmann, 2002). Under conditions where the composition of the oil and water phases and the system temperature are invariant with phase inversion, only the free energies of the interfaces have to be considered. The application of this criterion is, however, dependent on the availability of a model for characterizing the drop size in the initial and post-inversion dispersions, both are usually dense. This approach was recently followed by Brauner and Ullmann (2002).

According to this approach, when a dispersion structure (say $D_{o/w}$) is associated with higher surface energy than that obtained with an alternate structure (say $D_{w/o}$), it will tend to change its structure, and eventually to reach the one associated with the lowest surface energy. Hence, the phase inversion is expected under the critical conditions where both $D_{o/w}$ and $D_{w/o}$ are dynamically stable and the sum of surface energies obtained with either of these two configurations are equal.

Based on these considerations, the critical oil holdup can be obtained in terms of the liquid-solid surface wettability angle, α, and the Sauter mean drop diameter in pre-and post inversion dispersions (Brauner and Ullmann, 2002):

$$\epsilon_o^I = \frac{[\sigma/d_{32}]_{w/o} + \frac{s}{6}\sigma\cos\alpha}{[\sigma/d_{32}]_{w/o} + [\sigma/d_{32}]_{o/w}} \tag{4.28}$$

where s represents the surface wetted area per unit volume ($s = 4/D$ for pipe flow), $0 \le \alpha < 90^o$ corresponds to a surface which is preferentially wetted by water (hydrophilic surface), whereas for $90^o < \alpha \le 180^o$ the oil is the wetting fluid (hydrophobic surface). The Sauter mean drop size can be scaled with reference to the maximal drop size, $d_{32} = d_{\max}/k_d$. Using such a scaling, models for $d_{\max}$ in coalescing, dense $D_{o/w}$ or $D_{w/o}$ can be used in eq. (4.28) to evaluate the critical oil holdup at phase inversion. Applying the H-Model of Brauner (2001), eq. (4.20) yields:

$$\tilde{d}_o = 7.61\tilde{C}_H\left(\frac{\sigma}{\rho_w D U_m^2}\right)^{0.6}\left(\frac{\rho_w U_m D}{\mu_w}\right)^{0.08}\left(\frac{\rho_w}{\rho_m}\right)^{0.4}\frac{\epsilon_o^{0.6}}{(1-\epsilon_o)^{0.2}} \tag{4.29}$$

$$\tilde{d}_w = 7.61\tilde{C}_H\left(\frac{\sigma}{\rho_o D U_m^2}\right)^{0.6}\left(\frac{\rho_o U_m D}{\mu_o}\right)^{0.08}\left(\frac{\rho_o}{\rho_m}\right)^{0.4}\frac{(1-\epsilon_o)^{0.6}}{\epsilon_o^{0.2}} \tag{4.30}$$

where d_o and d_w represent the maximal drop size in $D_{o/w}$ and $D_{w/o}$ respectively. Under conditions where the oil-water surface tension in the pre-inversion and post-inversion dispersions is the same (no surfactants or surface contaminants are involved), $(k_d)_{o/w} \simeq (k_d)_{w/o}$ and solid-liquid wettability effects can be neglected ($\alpha = 90^o$ or $s \to 0$, as in large diameter pipes, where $d_o, d_w \ll D$), eqs. (4.28 - 4.30) yield:

$$\epsilon_o^I = \frac{\tilde{\rho}\tilde{\nu}^{0.4}}{1+\tilde{\rho}\tilde{\nu}^{0.4}} \tag{4.31}$$

where $\tilde{\nu}$ is the kinematic viscosity ratio, $\tilde{\nu} = \nu_o/\nu_w$.

Equation (4.31) provides an explanation for the observation made in many experimental studies, that the more viscous phase tends to form the dispersed phase. For a given holdup, and in the case of viscous oil, the characteristic drop size in $D_{o/w}$ is larger than in the reversed configuration of $D_{w/o}$. Hence, a larger number of oil drops must be present in order that the surface energy due to the oil-water interfaces would become the same as that obtained with the water dispersed in the oil. Therefore, with $\tilde{\rho}\tilde{\nu}^{0.4} > 1$, $\epsilon_o^I > 0.5$, and $\epsilon_o^I \to 1$ as $\tilde{\rho}\tilde{\nu}^{0.4} \gg 1$. The larger is the oil viscosity, the wider is the range of the oil holdup, $0 \leq \epsilon_o < \epsilon_o^I$, where a configuration of oil drops dispersed in water is associated with a lower surface energy. In this range of holdups, the flow pattern will be $D_{o/w}$ if the operational conditions are in range a the dynamic stability criterion is satisfied. Whereas, $D_{w/o}$ will be obtained in the range of $\epsilon_o^I \leq \epsilon_o \leq 1$, provided such a dispersion is dynamically stable (see boundaries 4 and 5, Section 5.1). Thus, when only the liquids' interfacial energy is involved, and the hydrodynamic flow field is similar in the initial and post inversion dispersions, the details of the flow field and the system geometry are not required for predicting the critical holdup at inversion.

Figure 4.2 shows a comparison of the critical oil holdup predicted via eq. (4.31), with experimental data of phase inversion in pipe flow which were used by Arirachkaran et al (1989) to obtain their experimental correlation, eq. (4.27) (line 2 in Figure 4.2). A lower variance is however obtained by correlating the data using the form of eq. (4.31). It is worth noting that for high critical oil holdup, corresponding to phase inversion of highly viscous oil dispersions, the water-in-oil dispersion is, in fact, dilute. It was shown by Brauner and Ullmann (2002) that in this range, if d_w is modelled by eq. (4.19) (rather than by eq. (4.20)), the critical oil holdup becomes practically independent on the viscosity ratio, in agreement with experimental findings. This phase inversion model was shown

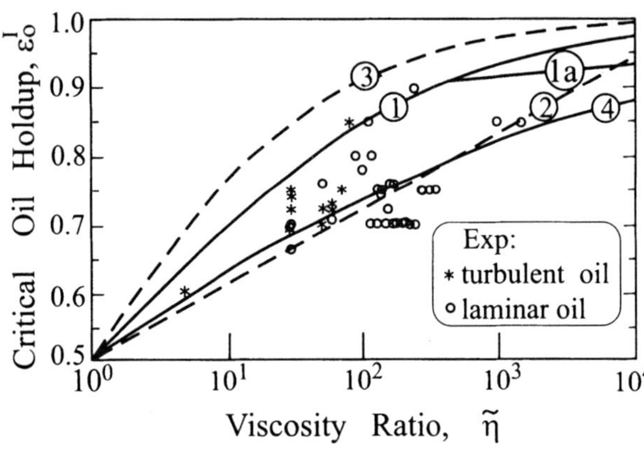

Figure 4.2 The critical oil-cut for phase inversion in pipe- flow - comparison of models/correlations predictions with experimental data: (1)Eq. (4.31); (1a) Eq. (4.28), dilute Dw/o $C_H = 1$); (2) Eq. (4.27); (3) $\epsilon_o^I = \sqrt{\tilde{\mu}}/(1+\sqrt{\tilde{\mu}})$, (4) Best fit $\epsilon_o^I = \tilde{\mu}^{0.22}/(1+\tilde{\mu}^{0.22})$

to be useful for explaining various experimentally observed features related to phase inversion in pipe flow and in static mixers. These include the effects of the liquids physical properties, liquid/surface wettability (contact angle), the existence of an ambivalent region and the associated hysteresis loop in pure systems and in contaminated systems (Brauner and Ullmann, 2002). Impurities or surfactant, and even entrained air bubbles, may have prominent effect on the critical holdup. Therefore, in many applications it is practically impossible to predict the conditions for phase inversion.

4.6 Conclusion

From the practical point of view, the main issue in predicting the pressure drop in homogeneous liquid-liquid dispersed flow is the modelling of the effective(apparent) mixture viscosity, μ_m. To this aim, the first decision to be made concerns the identity of the continuous phase. This decision is related to the phase inversion phenomenon. The second decision concerns the appropriate model to represent the variation of μ_m with the holdup in the particular system under consideration. The latter depends on the extent of mixing (emulsification) of the dispersed phase, which is a result of a combined effect of many factors (e.g. flow field, liquids physical properties, impurities and/or surfactant, liquid/wall wetting). This factors affect also the critical conditions for phase inversion. In any case, at the phase inversion point the liquids must be at intimate contact and models for emulsion viscosity are applicable to evaluate the pressure drop peak. However, so far, there are no general models or correlations for predicting the effective mixture viscosity for the variety of systems and operational conditions and much empiricism is still involved.

5 Flow Patterns Boundaries

Flow patterns characterization and transitions are usually related to the common parameters, which include the phases flow rates and physical properties. However, in dealing with liquid-liquid systems, the wide ranges of physical properties encountered generate a sort of ambiguity as to how to characterize liquid-liquid systems. It has been shown that it is beneficial to preliminary classify the system according to whether $Eo_D \gg 1$ *or* $Eo_D < 1$ (Brauner, 1998). Large Eotvös (gravity dominated) systems exhibit a similarity to gas-liquid systems, whereby density difference and inclination control flow pattern boundaries. On the other hand, in small Eo_D (surface tension dominated) systems, inclination does not play a role, whereas liquids wettability with the pipe material, entry conditions and start-up procedure are important. In this section some general guidelines for estimating the flow pattern that can be expected under specified operational conditions are outlined.

5.1 Horizontal Systems of $Eo_D >> 1$

Generally, these systems correspond to liquids with a finite density difference and sufficiently large tube diameter. In such systems the stratified flow configuration can be obtained in horizontal and slightly inclined tubes for some range of sufficiently low liquids flow rates. Models suggested for predicting flow patterns transition and guidelines for constructing flow patterns map for such systems are illustrated with reference to Figure (5.1).

1. Transition from (S) to (SM) or (SW) – This boundary defines transition from smooth stratified flow (S) to stratified flow with waves/mixing at the interface, (SW or SM, Figure 1.1b). The transitional criterion evolves from a linear stability analysis carried out on the transient formulation of the two-fluid model, and corresponds to the long-wave neutral stability boundary. It is given by (Brauner and Moalem Maron, 1993, Brauner, 1996).

$$J_1 + J_2 + J_h = 1 \tag{5.1}$$

$$J_1 = \frac{\rho_1}{\Delta\rho}\,\frac{U_{1s}^2}{Dg\cos\beta}\,\frac{\epsilon_2'}{(1-\epsilon_2)^3}\left[\left(\frac{C_{rn}}{U_1}-1\right)^2 + (\gamma_1 - 1)\left(1-2\frac{C_{rn}}{U_1}\right)\right] \tag{5.2.1}$$

$$J_2 = \frac{\rho_2}{\Delta\rho}\,\frac{U_{2s}^2}{Dg\cos\beta}\,\frac{\epsilon_2'}{\epsilon_2^3}\left[\left(\frac{C_{rn}}{U_2}-1\right)^2 + (\gamma_2 - 1)\left(1-2\frac{C_{rn}}{U_2}\right)\right] \tag{5.2.2}$$

$$J_h = C_h\,\frac{\rho}{\Delta\rho}\,\frac{(U_1-U_2)^2}{Dg\cos\beta}\,\frac{4S_i}{\pi\epsilon_2(1-\epsilon_2)D}\,; \tag{5.2.3}$$

where:

$$C_{rn} = \frac{\frac{U_2}{\epsilon_2}\frac{\partial\Delta F_{12}}{\partial U_2} - \frac{U_1}{(1-\epsilon_2)}\frac{\partial\Delta F_{12}}{\partial U_1} - \frac{\partial\Delta F_{12}}{\partial\epsilon_2}}{\left[\frac{1}{\epsilon_2}\frac{\partial\Delta F_{12}}{\partial U_2} - \frac{1}{(1-\epsilon_2)}\frac{\partial\Delta F_{12}}{\partial U_1}\right]}\,;\ \epsilon_2 = \frac{A_2}{A}\,;\ \epsilon_2' = \frac{d\epsilon_2}{d(h/D)} \tag{5.2.4}$$

$$\Delta F_{12} = \frac{\tau_2 S_2}{A_2} - \tau_i S_i\left(\frac{1}{A_2}+\frac{1}{A_1}\right) - \tau_1\frac{S_1}{A_1} + (\rho_2-\rho_1)\,g\sin\beta \tag{5.2.5}$$

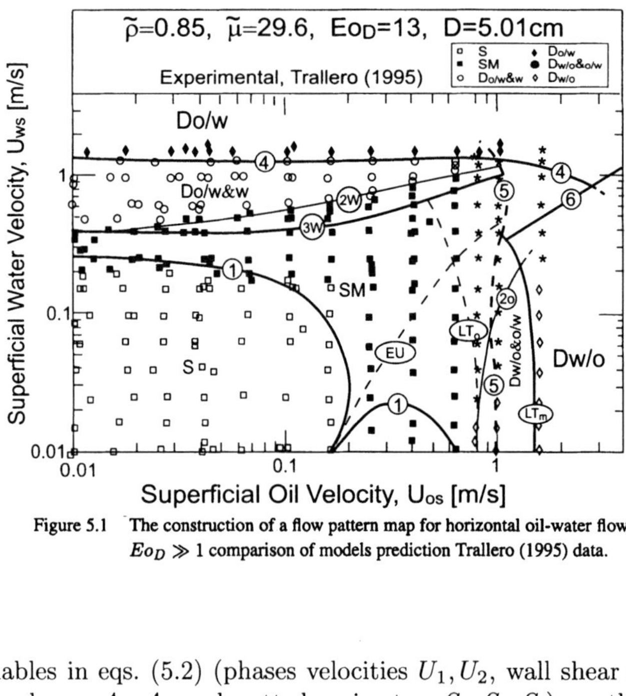

Figure 5.1 The construction of a flow pattern map for horizontal oil-water flow, $Eo_D \gg 1$ comparison of models prediction Trallero (1995) data.

All flow variables in eqs. (5.2) (phases velocities U_1, U_2, wall shear stresses τ_1, τ_2, τ_i flow cross-sectional area A_1, A_2 and wetted perimeters S_1, S_2, S_i) are those obtained for steady smooth stratified flow corresponding to superficial phases velocities U_{1s}, U_{2s} (see Section 2.4 and Figure 2.1). For $Eo_D \gg 1$, a plane interface can be assumed ($\phi^* = \pi$), whereby the flow geometry is determined by the lower layer depth, h. The shape factors γ_1, γ_2 (assumed constant) account for the velocity profiles in the two layers. For plug flow $\gamma_1 = \gamma_2 = 1$ and $\gamma > 1$ corresponds to a layer with a significant velocity gradient. Equation (5.1) represents a generalized stability criterion, which includes the Kelvin-Helmholtz mechanism and 'wave sheltering' mechanism. The destabilizing terms are due to the inertia of the two liquids (J_1, J_2 terms) and due to the dynamic interaction of the growing waves with turbulence in the faster layer, J_h. For laminar stratified layers $J_h = 0$. Otherwise a correlation for C_h is needed, but it is available only for gas-liquid systems. In liquid-liquid systems, the J_h term may be less significant (since the velocity difference is much smaller), and the stability criteria has been applied assuming $J_h = 0$ (Brauner and Moalem Maron, 1992a, 1992b).

Criterion (5.1) defines the combinations of U_{1s} and U_{2s} which corresponds to the evolution of interfacial disturbances (SW) and thus, possible entrainment of drops at the liquids layer interface (SM). This boundary is denoted by 1 in Figures 5.1 and 5.2, and is shown to predict the conditions for the evolution of interfacial disturbances in horizontal and inclined flows. Note that, in these Figures, the oil and water correspond to the lighter and heavier layer, respectively, $U_{1s} \equiv U_{os}$ and $U_{2s} \equiv U_{ws}$.

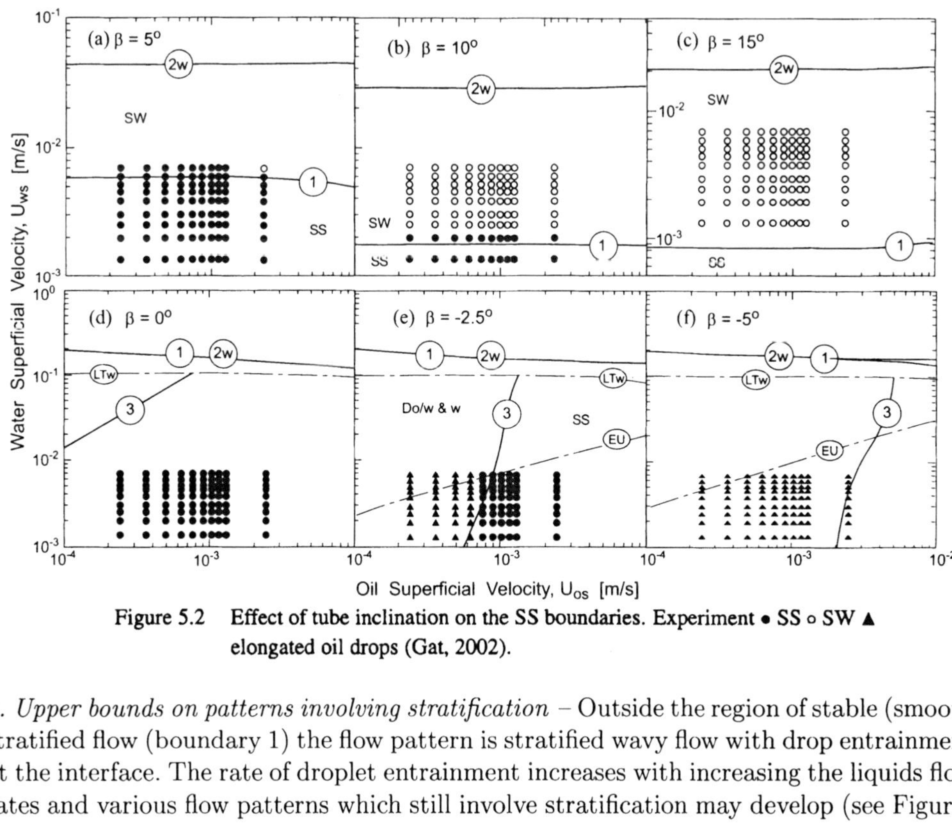

Figure 5.2 Effect of tube inclination on the SS boundaries. Experiment • SS ○ SW ▲ elongated oil drops (Gat, 2002).

2. Upper bounds on patterns involving stratification – Outside the region of stable (smooth) stratified flow (boundary 1) the flow pattern is stratified wavy flow with drop entrainment at the interface. The rate of droplet entrainment increases with increasing the liquids flow rates and various flow patterns which still involve stratification may develop (see Figures

1.1c to 1.1h). The stratified flow configurations are confined to a domain at whose boundaries the two-fluid formulation (for stratified configuration) becomes ill-posed (Brauner and Moalem Maron, 1991,1992d Brauner, 1996). The condition for ill-posedness is given by:

$$\widetilde{\rho}_2 U_2^2 \gamma_2(\gamma_2 - 1) + \widetilde{\rho}_1 U_1^2 \gamma_1(\gamma_1 - 1) - (\gamma_2 U_2 - \gamma_1 U_1)^2 + \\ + \frac{D}{\rho_{12}}[(\rho_2 - \rho_1) g \cos\beta - C_h \rho (U_1 - U_2)^2 S_i (A_1^{-1} + A_2^{-1})] \leq 0 \tag{5.3}$$

where $\widetilde{\rho}_2 = 1 + \frac{\rho_2}{\rho_1}\frac{A_1}{A_2}$, $\widetilde{\rho}_1 = 1 + \frac{\rho_1}{\rho_2}\frac{A_2}{A_1}$, $\rho_{12} = \frac{D(dA_2/dh)\rho_1\rho_2}{A_2[\rho_1 + \rho_2 A_1/A_2]}$. The ill-posedness boundary is indicated in Figure 5.1 by the two branches, 2w for a faster lower water layer and 2o for a faster upper oil layer. As shown in the figure, the ill-posedness boundary is always located in the region of amplified interfacial disturbances since the stable smooth stratified zone, which is confined by the stability boundary 1, is always a sub-zone of the well-posed region. Boundary 2w in Figure 5.1 was obtained with $\gamma_1 = 1.1$ for low U_{1s}, which was gradually reduced to $\gamma_1 = 1$ for higher oil rates where both layers are turbulent and $U_1 \simeq U_2$.

As shown in Figure 5.1 boundary 2w marks the location of SM to Do/w&w transition. The auxiliary lines, which provide useful information on the flow pattern that can be expected are the locus of $h/D = 0.5$; the locus of laminar/turbulent transition in the lower (water) layer LTw, laminar/turbulent transition in the oil layer, LTo (evolution of enhanced dispersive forces in either the water or oil layer) and the locus of $U_o = U_{\mathrm{w}}$, EU. Figure 5.1 points out an important difference between liquid-liquid systems and gas-liquid systems. In oil-water systems, the densities of the fluids are similar and therefore, the line of equal layers' velocity divide the zone of stable stratification into two regions, either faster oil layer or faster water layer. Entrainment of oil drops into the water layer takes place when $U_{\mathrm{w}} > U_o$ (left to the equal velocity curve, EU), whereas entrainment of water drops into the oil layer is associated with $U_o > U_{\mathrm{w}}$ (right to the EU curve). The dispersion of water drops into the oil layer is enhanced by transition to turbulent oil layer. However, as long as the water and oil flow rates are within the region where the transient stratified flow equations are well-posed (below curve 2w and below curve 2o), the flow patterns may involve a certain stratification, where in the upper layer, the oil forms the continuous phase and in the lower layer water is the continuous phase. It is worth noting that in contrast to gas-liquid systems, the entrained drops (water into oil, or oil into water) do not posses sufficient momentum to penetrate through the dense continuous phase and impinge on the tube walls. Therefore, the onset of drops entrainment in liquid-liquid systems is usually *not* associated with the formation of liquid film on the tube surface and the consequential transition to annular flow.

3w. Transition to Do/w&w – For $U_{\mathrm{w}} \gg U_o$ and outside boundary 1, the fragmentation of oil drops from the wavy oil-water interface is due to the inertia forces exerted by the faster water flow and is represented by the eq. (4.25) with $\Delta U_c = U_{\mathrm{w}} - U_o$. A dispersion of the entrained oil drops is stable provided $d_{\max} < d_{crit}$. The critical drop size, d_{crit} is taken as:

$$\frac{d_{crit}}{D} = Min\left(\frac{d_{c\sigma}}{D}, \frac{d_{cb}}{D}\right) \tag{5.4}$$

where $d_{c\sigma}$ represents the maximal size of drop diameter above which drops are deformed (Broodky, 1969):

$$\tilde{d}_{c\sigma} = \frac{d_{c\sigma}}{D} = \left[\frac{0.4\sigma}{\mid \rho_c - \rho_d \mid g \cos\beta' D^2}\right]^{1/2} = \frac{0.224}{(\cos\beta')^{1/2} Eo_D^{1/2}} \tag{5.5.1}$$

$$Eo_D = \frac{\Delta\rho g D^2}{8\sigma} \quad ; \quad \beta' = \begin{cases} \mid\beta\mid & ; \quad \mid\beta\mid < 45^o \\ 90 - \mid\beta\mid \;; & \mid\beta\mid > 45^o \end{cases} \tag{5.5.2}$$

and d_{cb} is the maximal size of drop diameter above which buoyant forces overcome turbulent dispersive forces in the continuous phase and therefore, migration of the drops towards the tube walls takes place (Barnea, 1987):

$$\tilde{d}_{cb} = \frac{d_{cb}}{D} = \frac{3}{8}\frac{\rho_c}{\mid\Delta\rho\mid}\frac{fU_c^2}{Dg\cos\beta} = \frac{3}{8}f\frac{\rho_c}{\Delta\rho g}Fr_c\,; \qquad Fr_c = \frac{U_c^2}{Dg\cos\beta} \tag{5.6}$$

with β denoting the inclination angle to the horizontal (positive for downward inclination). Equation (5.6) is relevant only in shallow inclinations and in case of turbulent flow in the faster (water) layer ($U_c \equiv U_{\text{w}}$). Moreover, in oil-water systems, where $\Delta\rho/\rho_c \ll 1, \widetilde{d}_{cb} > \widetilde{d}_{c\sigma}$, and in most practical cases $\widetilde{d}_{crit} = \widetilde{d}_{c\sigma}$ is used. In this case, the following transitional criterion evolves from eqs.(4.25) and (5.5.1) (Brauner, 2000):

$$\Delta U_c \equiv U_{\text{w}} - U_o \geq 4.36\left[\frac{\sigma\Delta\rho g\cos\beta'}{\rho_c^2}\right]^{1/4}\left\{1 + 1.443\,(N_{vd}\cos\beta')^{0.4}\right\}^{1/2} \tag{5.7}$$

where N_{vd} is the viscosity number of the dispersed oil phase, $N_{vd} = \frac{\mu_d^4 \Delta\rho g}{\rho_d^2 \sigma^3}$, $\mu_d \equiv \mu_o, \rho_d \equiv \rho_o, \rho_c \equiv \rho_{\text{w}}$. The constant coefficient (4.36) in eq. (5.7) may require some tuning when applied to a specific two-fluid system. According to this model, drops entrainment takes place when the velocity gap between the continuous (water) layer and the layer which is being dispersed (oil) exceeds a threshold value. This threshold value is given by the r.h.s. of eq.(5.7) and is independent of the tube diameter. For instance, a typical low viscosity oil-water system would be $\rho_c \simeq 1gr/cm^3$ $\Delta\rho = 0.1\rho_c$ and $\sigma = 30dyne/cm$, which yields a velocity gap of $0.3m/s$ that is required for significant entrainment. Boundary 3w in Figure 5.1 corresponds to eq. (5.7). It is worth noting that this figure is typical to low viscosity oil. For highly viscous oils ($\mu_0 > 1$ poise), the threshold value for the onset entrainment of oil drops into water increases (due to $N_{vd} >> 1$) in eq. (5.7). Also, the line of equal velocities of the oil and water layers is shifted to higher oil flow rates. Consequently, the zone outside the neutral stability boundary (up to boundary 2w may partially (or entirely) correspond to wavy stratified flow (SW), rather to stratified mixed flow (SM).

In systems which are not absolutely dominated by gravity ($Eo_D \simeq 1$), the Do/w&w pattern can be obtained instead of a continuous oil layer even for a small velocity gap. This can happen with a hydrophilic tube surface and when the largest oil drop that can occupy the upper part of the tube is smaller than the critical drop size. The criterion suggested for this transition (Brauner and Moalem Maron, 1992b,1992c):

$$A_1\cos\beta \leq \pi d_{crit}^2/4 \quad ; \quad d_{crit} = C\left[\frac{\sigma}{\Delta\rho g}\right]^{1/2} \tag{5.8}$$

provided the resulting $U_{1s}(\equiv U_{os})$ and $U_{2s}(\equiv U_{\text{ws}})$ are within the regions of stable stratification (below boundary 2w). While this criterion is irrelevant for predicting the flow patterns data in Figures 5.1, it is shown to predict the appearance of Do/w&w at low water and oil flow rates in upward inclined tubes (Figure 5.2e,f) and the gradual vanishing of the stratified flow pattern with increasing the upward inclination.

3o. Transition to Do/w&w – For $U_o \gg U_{\text{w}}$ and outside boundary 1, eq.(4.25), and thus eq.(5.7), is applied with $\Delta U_c = U_o - U_{\text{w}}, \mu_d \equiv \mu_{\text{w}}, \rho_d \equiv \rho_{\text{w}}$ and $\rho_c \equiv \rho_o$. This yields the critical velocity gap for dispersing the water layer into the oil layer. For the system studied in Figure 5.1, boundary 3o (not shown) is similar to 2o.

In systems of $Eo_D = O\{1\}$ and hydrophobic tube surface, eq. (5.8) with A_2 replacing A_1 signals transition to Dw/o&o due to capillary effects.

4. Transition to Do/w – A homogeneous oil-in-water dispersion (emulsion) can be maintained when the turbulence level in the continuous water phase is sufficiently high to disperse the oil phase into small and stable spherical droplets of $d_{\max} < d_{crit}$. Applying this criterion using the extended Hinze model, eqs. (4.18 to 4.21) with eqs. (5.4 to 5.6), yields a complete transitional criteria to dispersed flows (H-Model, Brauner, 2001). When the fluids flow rates are sufficiently high to maintain a turbulence level where $d_{\max} < d_{c\sigma}$ and $d_{\max} < d_{cb}$, spherical nondeformable drops are formed and the creaming of the dispersed droplets at the upper or lower tube wall is avoided. Thus, the fully dispersed flow pattern can be considered as stable. In these equations $U_c = U_m, U_{cs} = U_{\text{ws}}, U_{ds} = U_{os}, \rho_c = \rho_{\text{w}}$ and $\mu_c = \mu_{\text{w}}$. Hence, $We_c = \frac{\rho_{\text{w}} D U_m^2}{\sigma}$; $Re_c = \frac{D U_m}{\nu_{\text{w}}}$ and $\epsilon_d = U_{os}/U_m$. For instance, if $d_{crit} = d_{c\sigma}$, the transitional criterion reads:

$$C_{(\epsilon_d)} Eo_D^{1/2} We_c^{0.6} Re_c^{0.08} \geq 1 \tag{5.9}$$

The variation, $C_{(\epsilon_d)}$ with the dispersed phase holdup evolves from the H-model equations.

Curve 4 in Figure 5.1 predicts the transitional boundary from Do/w&w to Do/w. In systems of $Eo_D \simeq 1$, the K-model (eqs. 4.23 and 4.24) replaces the H-model in the evaluation of $d_{\max}$ (Brauner, 2001).

5. *Transition to* Dw/o – A homogeneous water-in-oil dispersion (emulsion) develops when turbulence level in the continuous oil phase is sufficiently high to disperse the water phase into stable small droplets. In this case eqs.(4.18 to 4.20) and (5.4 to 5.6) are applied with $U_c = U_m, U_{cs} = U_{os}, U_{ds} = U_{\text{ws}}, \rho_c = \rho_o$ and $\mu_c = \mu_o$. Hence, $\epsilon_d = U_{\text{ws}}/U_m$ and $We_c = \frac{\rho_o D U_m^2}{\sigma}$; $Re_c = \frac{D U_m}{\nu_o}$. In systems of $Eo_D \simeq 1$, eqs. (4.23 - 4.24) replace eqs. (4.19 - 4.20) for the evaluation of $\widetilde{d}_{\max}$. Boundary 5 in Figure 5.1 corresponds to the predicted transition to Dw/o. It is worth noting that for the critical flow rates along boundary 4, the mixture Reynolds number is already sufficiently high to assure turbulent flow in the water. However, when a viscous oil forms the continuous phase, the locus of the transition to $D_{\text{w}/o}$ may be constrained by the minimal flow rates required for transition to turbulent flow in the oil ($Re_c = 2100$ along boundary LT_m). The required turbulent dispersive forces exist only beyond the LT_m boundary, which therefore forms a part of the $D_{\text{w}/o}$ transitional boundary.

6. Transition from Do/w to Dw/o - This transition is associated with the phase inversion phenomena discussed in Section 4.5. Boundary 6 in Figure 5.1 was obtained by eq. (4.31). The phase inversion model is applicable for predicting this transition when the oil and water flow rates are sufficiently high to sustain both a homogeneous Do/w and Dw/o. As shown in Figure 5.1, boundaries 4 and 5 indeed define an ambivalent range where either of the oil or water phase can be homogeneously dispersed. It is the phase inversion phenomenon which eventually defines the boundaries of Do/w and Dw/o.

7. Core flow boundaries – In highly viscous oils, the laminar regime extends to high oil flow rates. In the absence of turbulent dispersive forces in the oil phase, it is possible to stabilize a viscous oil core which is lubricated by water annulus. The region were stable CAF is feasible is: (a) outside the boundaries of stable stratification (outside 2w and 2o), hence, sufficiently high oil rate (and water cut) to overcome the float-up tendency of the lighter oil core; (b) in the CAF configuration, the difference between the velocity of the oil in the core ($U_c \equiv U_o$) and the water velocity in the annulus ($U_a \equiv U_{\mathrm{w}}$), should not exceed the threshold value which would result in entrainment of the water film into the oil core. The threshold value is given by the r.h.s. of eq. (5.7), with $\rho_c = \rho_o$ ($N_{vd} = \mu_{\mathrm{w}}^4 \Delta\rho g/\rho_{\mathrm{w}}^2\sigma^3 \ll 1$ and can be ignored). The core annular model in section 3.2 can be used to evaluate U_c and U_a. However, since for turbulent water film, the slip between the phases is only few percents of the core velocity, this condition constrains the CAF only at high U_{os}; (c) water cut should not exceed a threshold value which results in disintegration of the oil core into oil globes by a thick wavy water annulus. Favorable conditions for wave bridging are $A_a/A_c > 1$. Using the annular flow model (Section 3.2) yields the flowing criterion for avoiding transition from core flow to oil slugs:

$$\frac{U_{os}}{U_{\mathrm{ws}}} \equiv \frac{U_{cs}}{U_{as}} \geq \frac{\mu_a}{\mu_c} + 2 \; ; \qquad \text{laminar core-laminar annulus} \qquad (5.10)$$

$$\frac{U_{cs}}{U_{as}} > 2.875 \times 10^{-3} \frac{\mu_a}{\mu_c} Re_{as}^{0.8} + 1.15 \; ; \qquad \text{laminar core turbulent annulus}$$

These criteria were shown to provide reasonable estimations of the oil and water flow rates where core flow is stable (see Figures 19 and 20 in Brauner,1998). The minimal water-cut needed to avoid stratification decreases with increasing the the oil core viscosity.

It is worth emphasizing the evolution of annular flow due to pure dynamical effects in systems of $Eo_D \gg 1$ (as in gas-liquid horizontal flows) is unlikely for oils of relatively low viscosity. Stabilization of the core requires sufficiently high velocity of the core phase: high mixture velocity and high input cut of the core phase. Under such conditions (and with low oil viscosity), dispersive forces are dominant and emulsification of the potential annular phase into the core phase results in a fully dispersed (emulsion) of the annular liquid within the core liquid and destruction of the CAF configuration.

5.2 Systems of $\mathrm{Eo_D} \ll 1$

Such systems exhibit flow patterns which are similar to microgravity systems (Brauner, 1990). The tube diameter is smaller than d_{crit} and in view of criterion (5.8) stratified flow will not be obtained even for low oil and water rates. The drift velocity of drops is negligible and the tube inclination has no effect on the flow patterns. Also, different flow patterns may result by changing the liquids/wall wettability properties (changing the tube material or the start-up procedure).

In hydrophilic tube, for low oil flow rate and high water cut, the flow pattern is oil droplets dispersed in water. With increasing the oil rate, enhanced droplets coalescence yields larger spherical oil drops (bubbles) with $d \simeq D$. This transition usually occurs for in situ oil holdup of about 0.15÷0.25 corresponding to $\widetilde{Q}^{-1} = \frac{U_{ws}}{U_{os}} = \frac{\epsilon_o}{1-\epsilon_o} = \simeq 0.17 \div 0.33$. For larger oil in situ holdup, the large spherical bubbles coalesce to form elongated oil bubbles (oil slugs). This transition takes place when the oil in situ holdup approaches the maximal volumetric packing, $\epsilon_o \simeq 0.4 \div 0.5$ corresponding to $\widetilde{Q} \simeq 1 \div 1.5$.

The slug/annular transition takes place for sufficiently low water cut, where stable thin water annulus can be maintained. This boundary can be calculated by eq. (5.10). For low viscosity oil and high oil rates, the oil core is turbulent. When the turbulent dispersive forces are sufficiently high to disperse the water annulus, transition to Dw/o takes place. It should be noted that in systems of $Eo_D \ll 1$, $d_{crit} = d_{c\sigma} > D$. Therefore, d_{crit} is scaled by D (e.g., $d_{crit} \simeq D/2$,and the K1-model in Brauner, 2001 is used to calculate the transition to Dw/o (boundary 5), or Do/w (boundary 4)). The entrainment of the water film due to the inertia of the core phase should also be considered. The corresponding critical velocity difference is given by eq. (5.7), with $\Delta U_c = U_o - U_w = U_c - U_a$ (using the CAF model in section 3.2 to calculate the velocity difference). The locus of Do/w to Dw/o transition at high oil and water rates is obtained by the phase inversion model (eq. 4.28). The application of these criteria for predicting flow pattern transition in systems of low Eo_D was demonstrated in Brauner,1998.

In a hydrophobic tube, there is evidence that for low oil water rates and high water cut inverted annular flow (Andreini, et al., 1997) with oil flowing in the annulus can be obtained (instead of Do/w). This flow pattern can be maintained as long as the level turbulence in the water core, as well as its inertia, are not sufficiently high to disperse the oil annulus. Also, the oil holdup in the annulus must be sufficiently low to avoid blockage of the water core ($A_c/A_a \geq 1$, A_c, A_a calculated via the annular flow model, Section 3.2).

5.3 Vertical upward systems

The construction of a flow pattern map for vertical upward oil-water flows is demonstrated in Figures 5.3 and 5.4. The basic flow configuration for low superficial oil velocity is Do/w. For low water rates, the oil is dispersed in the water in the form of relatively large bubbles. The criterion of $\epsilon_o \geq 0.25$ is usually suggested to mark transition from small spherical bubbles to large oil bubbles and slugs (e.g., Harsan and Kabir, 1990). The locus of $\epsilon_o = 0.25$ (as predicted via eq. (4.5) and (4.6)) is indicated by boundary 8. With increasing the water rate, transition to fine Do/w (o/w emulsion) takes place, which is predicted by transition 4. Similarly, the transition to fine Dw/o (w/o emulsion) takes place for sufficiently high oil superficial velocities (transition 5) which are higher than that required for establishing turbulent flow in the oil as a continuous phase (right to boundary LT_m). The phase inversion model yields the boundary between Do/w and Dw/o (transition 6).

The unstable region of churn flow is obtained for low water rates and intermediate oil rates. The oil flow rate is too high to sustain a stable configuration of Do/w. Large oil bubbles (of the order of $d \simeq d_{crit}$) coalesce and tend to form an oil core surrounded by a water annulus (CAF). Boundary 7 in Figure 5.3 is the locus of $\widetilde{D}_c = 0.5$ as predicted by the CAF model for vertical upward flow (a thinner core results in transition to oil bubbles dispersed in water). However, the oil velocity is too low to meet the dynamic requirements for stabilization of oil dominated flow patterns, namely, oil forms the continuous phase as in CAF flow or Dw/o. For stable CAF, the oil superficial velocity should be sufficiently high to suspend large water drops, $d \simeq D$ (say $d = D/2$), which are occasionally formed, whereby:

$$\frac{1}{8}\pi d^2 C_D \rho_o U_{os}^2 \geq \frac{1}{6}\pi d^3 \left(\rho_w - \rho_o\right) \; ; \text{ or } \; U_{os}\left[\frac{\rho_o}{\Delta\rho g D}\right]^{1/2} \geq \left(\frac{2}{3C_D}\right)^{1/2} \qquad (5.11)$$

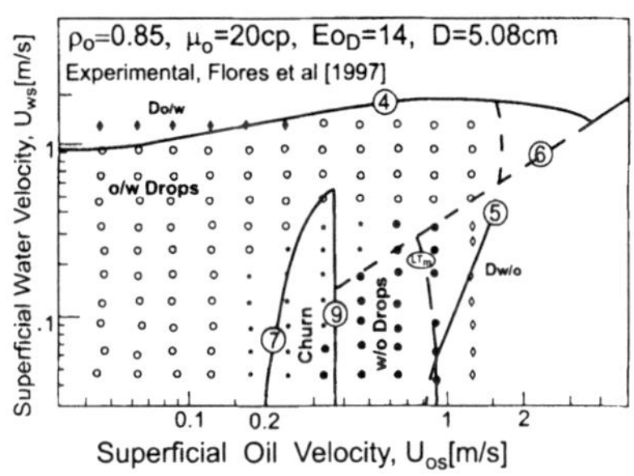

Figure 5.3 The construction of a flow pattern map for vertical oil-water system-comparison with Flores et al, (1997) data.

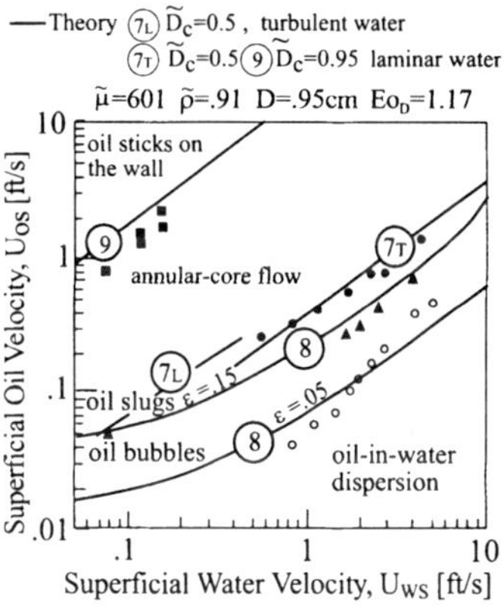

Figure 5.4 Flow pattern map for vertical upward flow of water and highly viscous oil (μ_0=601 cp)-comparison with Bai et al, (1992) data.

Boundary 9 in Figure 5.3 has been obtained by eq. (5.11) with $C_D = 0.44$. It is worth noting that criterion (5.11) is similar to that of flow reversal of the annular film, which is frequently used to estimate flooding conditions, as well as transition to annular flow in upward gas-liquid systems. Condition (5.11), however, introduces the effect of the oil viscosity (and drop size) through the variation of C_D. It is worth emphasizing that with relatively low viscosity oils (as is the case in Figure 5.3) the CAF configuration is eventually not obtained. The potential water annulus is dispersed into the oil phase to form the Dw/o pattern. The annular pattern in upward vertical flow has been observed only for highly viscous oils. The size of oil bubbles and slugs increases with the oil viscosity and for sufficiently large oil-cut, a continuous oil core surrounded by a water film may be formed (see Figure 5.4). Oil core flow lubricated by a water annulus was obtained for sufficiently high oil cut (instead of the churn regime observed with low viscosity oils). The core interface is wavy and water rate should be kept higher than a threshold value to prevent oil sticking on the tubes wall. Indeed, with highly viscous oils

($Re_d = \rho_o U_{os} d/\mu_o \ll 1$ and $C_D = 24/Re_d$) condition (5.11) is satisfied already for low oil velocities and the region of churn flow in Figure 5.3 is occupied by the CAF in Figure 5.4. As shown, the core flow region in Figure 5.4 can be estimated using the CAF model (Section 3.2) for calculation the core phase holdup. It extends from $\tilde{D}_c \simeq 0.5$ (transition to slug flow) to $\tilde{D}_c = 0.95$ (oil sticks to the wall).

5.4 Conclusion

The first step in the construction of a flow pattern map for a liquid liquid system is its classification according to its Eotvös number, to either being gravity dominated or surface tension dominated system. The guidelines and criteria for flow pattern transitions as outlined above, were found useful for estimating the flow pattern map for these two types of liquid liquid systems. However, these have still to be tested in view of more data in the variety of liquid-liquid systems, pipe diameters, materials and inclinations.

Table 1.1.1 Summery of Experimental Systems of Liquid -Liquid Horizontal Flow

Authors	D [cm] pipe material	μ_o/μ_w	ρ_o/ρ_w	σ dyne/cm	Additional Measurements	Observed Flow Patterns
Russell *et al.* (1959)	2.03 Cellulose Acetate-Butyrate	20.13	0.840		dP/dl ε_w	SM, Do/w, Bo
Charles *et al* (1961)	2.64 Cellulose Acetate-Butyrate	6.29 16.8 65	1 1 1	44 45 30	dP/dl ε_w	Do/w, ANw, SLo, Bo
Guzhov *et al.* (1973)	3.94 Steel	21.8	0.898	44.8	dP/dl	SM, Dw/o, Do/v Do/w & w, Dw/o & o/w
Malinowsky (1975)	3.84 Steel	3.33	0.850	22.3	dP/dl	SM, Do/w, Dw/ Dw/o & o/w
Laflin & Oglesby (1976)	3.84 Steel	4.12	0.830	22.3	dP/dl	SM, Do/w, Dw/ Dw/o & o/w
Oglesby (1979)	4.1 Steel	32 61 167	0.859 0.863 0.870	30.1 29.4 35.4	dP/dl	Do/w, Dw/o Dw/o & o/w
Cox (1985)	5.08 Acrylic	1.54	0.756		ε_w	S, Do/w Do/w & w
Scott (1985)	5.08 Acrylic	1.54	0.756		ε_w	S, o/w Do/w & w
Stapelberg & Mewes (1990)	2.38 5.9 Acrylic ,glass	30	0.852	50	dP/dl	SM, Do/w & w Dw/o & o/w
Fujii *et al* (1994)	2.5 Acrylic	61.5	0.98	29	dp/dl ε_w	Bo, Bw, SLo, SLw, ANo
Valle & Kvandal (1995)	3.75 Glass	2.55	0.792	37.3	dP/dl, ε_w (Conductivity & sampling probes)	S, SM Do/w & w Dw/o & o/w
Trallero (1995)	5.08 Acrylic	29.7	0.852	36	dP/dl ε_w	S, SM Do/w & w Do/w,Dw/o Dw/o & o/w Dw/o & w
Nadler & Mewes (1997)	5.9 Perspex	18-35	0.848		dP/dl Phase continuity (Conductivity probe)	S, SM Do/w & w Do/w, Dw/o Dw/o & o/w Dw/o & w
Vedapuri *et al.* (1997)	10.12 Plexi-glass	2			ε_w isokinetic probe	SM, Dw/o & o/v
Beretta *et al* (1997)	0.3 Borosilicate glass	9.9, 51.3, 71.2	0.87, 0.89, 0.87	37.4, 36, 31.5	dp/dl	Do/w, SLo, Bo ANw

Table 1.1.1 *Continued*

Kurban *et al* (1997)	2.43, 2.4 St. steel Acrylic	1.6	0.803	17	dP/dl (Conductivity & Impedance probe)	S, SM, Dw/o
Andreini *et al* (1997)	0.3, 0.6 Borosilicate glass, Steel, Copper, PVC	562, 920, 1307	0.886, 0.889, 0.893		dp/dl	Do/w, SLo, PLo, ANw
Valle & Utvik (1997)	7.62 Steel	1	0.741		dp/dl (Conductivity probe)	Do/w , Dw/o, S
Hapanowicz *et al* (1997)	1.2 1.6 2.2 Glass	40 40	1.2 0.915			Do/w, Dw/o, Dw/o&o, Do/w&w, Dw/o&w, S, DANo, PL, Foam
Angeli & Hewitt (1998)	2.43, 2.4 St. steel Acrylic	1.6	0.803	17	dP/dl (Conductivity and Impedance probe)	S, Do/w, Dw/o, Do/w &w, Dw/o &o, Dw/o &o/w
Soleimani (1999)	2.43 St. steel	1.6	0.803		dP/dl, ε_w Phase distribution (High Frequency Impedance probe & Gamma Densiometer)	SM, Dw/o &o/w
Angeli & Hewitt (2000)	2.43, 2.4 St. steel Acrylic	1.6	0.803	17	ε_w (Conductivity and Impedance probe)	S, Do/w, Dw/o, Do/w &w, Dw/o &o, Dw/o &o/w
Lovick *et al.* (2000)	3.8 Stainless Steel	5.25	0.828	44.7	dP/dl, ε_w (conductivity probe)	S, SM Do/w, Dw/o
Fairuzov *et al* (2000)	36.35 Steel	5.07	0.853		Volume fraction (Multi-Point Sampling Probe)	S Do/w & w
Simmons & Azzopardi (2001)	6.3 PVC	1.125	0.684	10	Drop size, velocity & distribution (Par-Tec 300C, Malvern 2600)	SM, Dw/o & w, Dw/o
Angeli *et al.* (2002)	3.8 Stainless Steel	5.25	0.828	44.7	dP/dl Phase distribution (Impedance probe, conductivity probe)	SW, Do/w, Dw/o &o/w, Dw/o

Table 1.1.2 Summery of Experimental Systems of Liquid -Liquid Core Annular Flow

Authors	D [cm] pipe material	Core Fluid	Annulus Fluid	μ_o/μ_w	ρ_o/ρ_w	σ dyne/cm	Additional Measurements
Charles *et al* (1961)	2.64 Cellulose Acetate-Butyrate	20.6% CTC in Marcol GX 18.7% CTC in Wyrol J 16.7% CTC in Teresso 85	Water Water Water	6.3 16.8 65	1 1 1	44 45 30	dP/dl
Kruyer *et al* (1967)	1.35, 3.19, 10.24 Copper, Acrylic, Steel, Aluminum	lubricating oil	Water	36, 28, 6, 17	0.86 0.86 0.83 0.85		dP/dl
Sinclair (1970)	1.905, 2.54, 6.35	Humble Fractol oil, water, emulsifier	Sea Water	1000	0.94		
Hasson & Nir (1970)	1.26 Glass	water	Kerosene-Perchloroethylene	1.2	1.02	17-17.5	film thickness
Hasson *et al* (1970)	1.26 Glass	water	Kerosene-Perchloroethylene	1.2	1.02	17-17.5	breakup mechanism
Hasson (1978)	0.9 Stainless Steel	water	Kerosene	1.6	0.803	17	Heat transfer & coefficients, scaling
Wu *et al* (1986)	5.08 Transparent	Zuata crude oil	water	2200, 10000	0.992 ,0.997		dp/dl specific energy consumption
Oliemans (1986)	5.08, 20.32 Perspex	oil	water	2584	0.97		dp/dl ε_w Wave characteristic
Guevara *et al* (1988)	20.3 Stainless Steel	viscous hydrocarbon	water	up to 110,000	0.995		dp/dl water fraction
Anon (1988)	20.27 Stainless Steel	Zuata crude oil	water	3,000 - 100,000			dp/dl water fraction
Bai *et al* (1992)	0.9525 Glass	oil	0.4% Sodium Silicate in Water	601	0.906	8.54	dp/dl ε_w
Miesen *et al* (1992)	5.08, 20.32	fuel crude	water water	13-25 30-42	0.96- 0.97 0.98- 0.99	20-50	dp/dl Wave characteristic
Arney *et al* (1993)	1.59 Glass	waxy crude oil No. 6 fuel oil	water water	600 2700	0.985 0.989	-- 26.3	dp/dl ε_w
Ho & Li (1994)	1.9, 7.3	10%v/v ECA polyamine surfactant in diesel	2%w/w KCl in water	109,091	0.836		dp/dl Annular size
Andreini *et al* (1997)	0.3, 0.6 Borosilicate glass, Steel, Copper, PVC	Miplar	water	562, 920, 1307	0.886, 0.889, 0.893		dp/dl

Table 1.2 Summery of Experimental Systems of Liquid -Liquid Inclined Flow

Authors	β(°)	D [cm] pipe material	μ_o/μ_w	ρ_o/ρ_w	σ dyne/cm	Additional Measurements	Observed Flow Patterns
Soøt & Knudsen (1972)	-90	1.89 Brass	0.98 8.6 180		40 -- 13	dp/dl ε_w	Do/w
Mukhopadhyay (1977)	±30 to ±90	3.81 Lexan	5-6	0.850		dP/dl ε_w	
Mukherjee *et al* (1981)	±30 to ±90	3.81 Lexan	5-6	0.852	22.3	dP/dl ε_w	Do/w & w/o
Hill & Oolman (1981)	+30 to +90	15.2, 21.6, 11.4 Steel, Acrylic	1.6	0.801	17	dP/dl ε_w	Bo, S
Cox (1985)	-15,-30	5.08 Plexiglass	1.54	0.756		ε_w	S, Do/w Do/w & w
Scott (1985)	+15,+30	5.08 Plexiglass	1.54	0.756		ε_w	S, Do/w Do/w & w
Vigneaux *et al* (1988)	+25 to +90	20		0.741		Phase distribution (Impedance probe)	Do/w
Zavareh *et al* (1988)	+85 and +75	18.41 Acrylic	2.46	0.783			Bo, dispersed bubbly
Tabeling *et al* (1991)	+15 to +90	20	5	0.782			
Ding *et al* (1994)	+30, +45, +60, +90	16.51 Transparent		0.811		ε_w	
Kurban (1997)	1+	7.79 Stainless Steel	45	0.865			S, SM, Do/w & w, Dw/o & o
Vedapuri *et al* (1997)	±2	10.16 Acrylic	2,96			ε_w	Semi-segragated Semi-dispersed
Flores (1997)	45,60,75,90	5.08 Acrylic	20	0.858	33.5	dP/dl ε_w (conductivity probe)	Do/w CT, Do/w PS, VFDo/w Dw/o CC, VFDw/o, Churn TF
Hassan & Kabir (1999)	+45, +75, +85	6.24, 12.7 Plexiglass	1.6	0.801	17	ε_w	Bo, Bw, SLo, SLw
Alkaya *et al* (2000)	0,±0.5,±1, ±2,±5	5.08 Acrylic	18	0.854	36	dP/dl ε_w	S, SM
Angeli *et al.* (2002)	0, +5	3.8 Stainless Steel	5.25	0.828	44.7	dP/dl Phase distribution (Impedance probe, conductivity probe)	S, Do/w, Dw/o, Do/w &w/o
Gat (2002)	0 - ±30	1.44 Glass	9.7	0.835	32	ε_w	S, SW, Bo, SLo

Table 1.3 Summery of Experimental Systems of Liquid -Liquid Vertical Upflow

Authors	D [cm] pipe material	μ_o/μ_w	ρ_o/ρ_w	σ dyne/cm	Additional Measurements
Govier *et al* (1961)	2.64 Cellulose Acetate-Butyrate	0.936 20.1 150	0.780 0.851 0.880	35.3 50.2 49.8	dp/dl ε_w
Brown & Govier (1961)	2.64 Cellulose Acetate-Butyrate	21.5	0.850	50.34	dp/dl, ε_w Bubble size & velocity
Vigneaux *et al* (1988)	20		0.741		Phase distribution (Impedance probe)
Zavareh *et al* (1988)	18.41 Acrylic	2.46	0.783	52.2	dp/dl ε_w
Hasan & Kabir (1990)	6.35, 12.7 Plexiglass	1.544	0.756		ε_w
Flores (1997)	5.08 Acrylic	20	0.858	33.5	dP/dl ε_w
Hamad *et al* (2000)	7.78 Perspex	1.6	0.803	17	Drop size, velocity & distribution (dual optical probe)
Simmons & Azzopardi (2001)	6.3 PVC	1.125	0.684	10	Drop size, velocity & distribution (Par-Tec 300C, Malvern 2600)

References

Acrivos, A. and Lo, T.S. (1978). Deformation and breakup of a single slender drop in an extensional flow. *Journal Fluid Mechanics* 86:641.

Alkaya, B., Jayawardena, S.S., and Brill J.P. (2000). Oil-water flow patterns in slightly inclined pipes. *In Proceedings 2000 ETCE/OMAE Joint Conference, Petroleum Production Symposium*, New Orleans, 14-17 February, 1-7.

Andreini, P.A., Greeff P., Galbiati, L., Kuklwetter, A. and Sutgia, G. (1997). Oil-water flow in small diameter tubes. *International Symposium on Liquid-Liquid Two-Phase Flow And Transport Phenomena.* Antalya, Turkey, 3-7.

Angeli, P., and Hewitt, G.F. (1996). Pressure-Gradient Phenomenon During Horizontal Oil-Water Flow, *ASME Proceedings OMAE* 5:287-295.

Angeli, P., and Hewitt, G.F. (1998). Pressure gradient in horizontal liquid-liquid flows. *International Journal Multiphase Flow* 24:1183-1203.

Angeli, P., and Hewitt, G.F. (2000). Flow structure in horizontal oil-water flow. *International Journal Multiphase Flow* 26:1117-1140.

Angeli, P., Lovick, S. and Lum, Y.L. (2002). Investigations on the Three-Layer Pattern During L-L Flows, *40th European Two-Phase Flow Group Meeting*, Stockholm, June 10-13.

Arirachakaran, S., Oglesby, K.D., Malinowsky, M.S., Shoham, O., and Brill, J.P. (1989). An Analysis of Oil/Water Flow Phenomena in Horizontal Pipes. *SPE Paper 18836, SPE Professional Product Operating Symposium* , Oklahoma.

Arney, M., Bai, R., Guevara, E., Joseph, D.D. and Liu, K. (1993). Friction factor and holdup studies for lubricated pipelining: I. Experiments and correlations. *International Journal of Multiphase Flow* 19:1061-1076.

Azzopardi, B.J. and Hewitt, G.F. (1997). Maximum drop sizes in gas-liquid flows. *Multiphase Science Technology* 9:109-204.

Bai, R., Chen, K. and Joseph, D.D. (1992). Lubricated pipelining: Stability of core-annular flow, Part V. experiments and comparison with theory, *Journal of Fluid Mechanics* 240:97-132.

Barnea, D. (1987). A Unified model for predicting flow-pattern transitions for the whole range of pipe inclinations. *International Journal Multiphase Flow* 11:1-12.

Baron, T., Sterling, C.S., and Schueler, A.P. (1953). Viscosity of Suspensions - Review and Applications of Two-Phase Flow. *Proceedings 3rd Midwestern Conference Fluid Mechanics.* University of Minnesota, Minneapolis 103-123.

Bentwich, M. (1964). Two-phase axial flow in pipe. *Trans. of the ASME.* Series D 84(4):669-672.

Bentwich, M., Kelly, D.A.I. and Epstein, N. (1970). Two-Phase Eccentric Interface Laminar Pipeline Flow. *J. Basic Engineering* 92:32-36.

Bentwich, M. (1976). Two-phase laminar flow in a pipe with naturally curved interface. *Chemical Engineering Sciences* 31:71-76.

Beretta, A., Ferrari, P., Galbiodi, L., Andreini, P.A. (1997). Oil-Water Flow in Small Diameter Tubes. Pressure Drop. *International Comm. Heat Mass Transfer* 24(2):231-239.

Beretta, A., Ferrari, P., Galbiodi, L., Andreini, P.A. (1997). Oil-Water Flow in Small Diameter Tubes. Flow Patterns. *International Comm. Heat Mass Transfer* 24(2):223-229.

Biberg, D., Halvorsen, G. (2000). Wall and interfacial shear stress in pressure driven two-phase laminar stratified pipe flow. *International Journal Multiphase Flow* 26:1645-1673.

Brauner, N., and Moalem Maron, D. (1989). Two-phase liquid liquid stratified flow, PCH Physico Chemica, *Hydrodynamics* 11(4):487-506.

Brauner, N. (1990). On the Relation Between Two-Phase Flow Under Reduced Gravity and Earth Experiment. *International Comm. Heat Mass Transfer* 17(3):271-282.

Brauner, N., and Moalem Maron, D. (1991). Analysis of stratified/nonstratified transitional boundaries in horizontal gas-liquid flows. *Chemical Engineering Science* 46(7):1849-1859.

Brauner, N. (1991). Two-Phase Liquid-Liquid Annular Flow. *International Journal Multiphase Flow*, 17(1):59-76.

Brauner, N., and Moalem Maron, D. (1992a). Stability analysis of stratified liquid-liquid horizontal flow. *International Journal of Multiphase Flow* 18:103-121.

Brauner, N., and Moalem Maron, D. (1992b). Flow pattern transitions in two phase liquid-liquid horizontal tubes *International Journal of Multiphase Flow* 18:123-140.

Brauner, N., and Moalem Maron, D. (1992c). Identification of the range of small diameter conduits regarding two-phase flow patterns transitions. *International Comm. Heat Mass Transfer* 19:29-39.

Brauner, N., and Moalem Maron, D. (1992d). Analysis of stratified/nonstratified transitional boundaries in inclined gas-liquid flows. *International Journal of Multiphase Flow* 18(4):541-557.

Brauner, N., and Moalem Maron, D. (1993). The role of interfacial shear modelling in predicting the stability of stratified two-phase flow. *Chemical Engineering Science* 8(10):2867-2879.

Brauner, N., and Moalem Maron, D. (1994). Stability of two-phase stratified flow as controlled by laminar turbulent transition. *International Comm. Heat Mass Transfer*, 21:65-74.

Brauner, N., Rovinsky, J. and Moalem Maron, D. (1995a). Analytical Solution of Laminar-Laminar Stratified Two-Phase Flows with Curved Interfaces, *Proceedings of the 7th International Meeting of Nuclear Rector Thermal-Hydraulics* NURETh-7(1):192-211.

Brauner, N. (1996). Role of Interfacial Shear Modelling in Predicting Stability of Stratified Two-Phase Flow, in Encyclopedia of Fluid Mechanics, edited by N.P. Cheremisinoff. Advances in Engineering Fluid Mechanics: Boundary Conditions Required for CFD Simulation, 5:317-378.

Brauner, N., Rovinsky, J., and Moalem Maron, D. (1996a). Analytical solution for laminar-laminar two-phase stratified flow in circular conduits. *Chemical Engineering Comm.* 141-142, 103-143.

Brauner, N., Rovinsky, J. and Moalem Maron, D. (1996b). Determination of the Interface Curvature in Stratified Two-Phase Systems by Energy Considerations. *International Journal Multiphase Flow* 22:1167-1185.

Brauner, N. (1997). Consistent Closure Laws for Modelling Two-Phase Annular Flow Via Two-Fluid Approach. *Internal Report*, Tel-Aviv University, Faculty of Engineering, October.

Brauner, N., Moalem Maron, D. and Rovinsky, J. (1997). Characteristics of Annular and Stratified Two-Phase Flows in the Limit of a Fully Eccentric Core Annular Configuration, *Proc. of the ExHFT-4*, Brussels, 2:1189-1196.

Brauner, N. (1998). Liquid-Liquid Two-Phase Flow, Chap. 2.3.5 in HEDU - Heat Exchanger Design Update, edited by G.F. Hewitt 1:40.

Brauner, N., Moalem Maron, D. and Rovinsky, J. (1998). A Two-Fluid Model for Stratified Flows with Curved Interfaces. *International Journal Multiphase Flow.* 24:975-1004.

Brauner, N. (2000). The onset of drops atomization and the prediction of annular flow boundaries in two-phase pipe flow. Internal Report-5101, Faculty of Engineering, Tel-Aviv, Israel.

Brauner, N. (2001). The Prediction of Dispersed Flows Boundaries in Liquid-Liquid and Gas-Liquid Systems, *International Journal Multiphase Flow* 27(5):911-928

Brauner, N. and Ullmann, A. (2002). Modelling of Phase Inversion Phenomenon in Two-Phase Pipe Flow, *International Journal Multiphase Flow.* In print.

Brodkey, R.S. (1969). *The Phenomena of Fluid Motions.* Addison-Wesley, Reading, MA.

Brown, R.A.S., and Govier, G.W. (1961). High-Speed Photography in the Study of Two-Phase Flow. *Canadian Journal Chemical Engineering* 159-164.

Chesters, A.K. (1991). The Modelling of Coalescence Process in Fluid-Liquid Dispersions. *Chemical Engineering Res. Des., Part A* 69(A4):259-270.

Cohen, R.D. (1991). Shattering of Liquid Drop due to Impact. *Proceedings Royal Society, London* A435:483-503.

Charles, M.E., Govier, G.W., and Hodgson, G.W. (1961). The horizontal flow of equal density oil-water mixtures, *Canadian Journal of Chemical Engineering*, 39:287-36.

Charles, M.E., and Lilleleht, L.U. (1966). Correlation of Pressure Gradients for the Stratified Laminar-Turbulent Pipeline Flow of Two Immiscible Liquids. *Canadian Journal Chemical Engineering* 44:47-49.

Clift, R., Grace, J.R., and Weber, M.E. (1978). Bubbles, Drops and Particles. Academic Press.

Colebrook, C. (1938-39). Turbulent Flow in Pipes with Particular Reference to the Transition Region Between the Smooth and Rough Pipe Laws. *Journal Inst. Cir. Engineering* 11:133-156.

Cox, A.L. (1986). A Study of Horizontal and Downhill Two-Phase Oil-Water Flow, M.S. Thesis, The University of Texas.

Davies, J.T. (1987). A physical interpretation of drop sizes in homogenizers agitated viscous oils. *Chemical Engineering Science* 42(7):1671-1676.

Decarre, S., and Fabre, J. (1997). Phase inversion behavior for liquid-liquid dispersions, *Revue Institution Francais du Petiale.* 52:415-424.

Ding, Z.X., Ullah, K., and Huang, Y. (1994). A comparison of predictive oil/water holdup models for production log interpretation in vertical and deviated wellbores, in In Proceedings SPWLA 35th Annual Logging Symposium, Tulsa, OK, USA, June 19-22, 1-12.

Epstein, N., Bianchi, R.J., Lee, V.T.Y., and Bentwich, M. (1974). Eccentric Laminar Couette Flow of Long Cylindrical Capsules, *Canadian Journal of Chemical Engineering.* 52:210-214.

Fairuzov, Y.V., Medina, P.A., Fierro, J.V. Islas, R.G. (2000). Flow pattern transitions in horizontal pipelines carrying oil-water mixtures: full-scale experiments. *Journal Energy Resources Technology-Trans. ASME* 122:169-176.

Flores, J.G. (1997). *Oil-Water Flow in Vertical and Deviated Wells.* Ph.D. Dissertation, The University of Tulsa, Tulsa, Oklahoma.

Flores, J.G., Chen, X.T., Sarica, C., and Brill, J.P. (1997). Characterization of Oil-Water Flow Patterns in Vertical and Deviated Wells. *1997 SPE Annual Technical Conference and Exhibition.* San Antonio, Texas, SPE paper 38810 1-10.

Fujii, T., Otha, J., Nakazawa, T., and Morimoto, O. (1994). The Behavior of an Immiscible Equal-Density Liquid-Liquid Two-Phase Flow in a Horizontal Tube. *JSME Journal Series B, Fluids and Thermal Engineering,* 30(1):22-29.

Garner, R.G., and Raithby, G.D. (1978). Laminar Flow Between a Circular Tube and a Cylindrical Eccentric Capsule, *Canadian Journal of Chemical Engineering.* 56:176-180.

Gat S., (2002). *Two-Phase Liquid-Liquid Concurrent flow in Inclined Tubes.* M.Sc. Thesis, Faculty of Engineering, Tel-Aviv University.

Goldstein, A. (2000). *Analytical Solution of Two-Phase Laminar Stratified Flow in Inclined Tubes,* M.Sc. Thesis.

Gorelic, D. and Brauner, N. (1999). The Interface Configuration in Two-Phase Stratified Flow, *International Journal Multiphase Flow.* 25:877-1007.

Govier, G.W., Sullivan, G.A., and Wood, R.K. (1961). The Upward Vertical Flow of Oil-Water Mixtures. *Canadian Journal of Chemical Engineering* 9:67-75.

Govier, G.W., and Aziz, K. (1972). The Flow of Complex Mixtures in Pipes, Robert E. Krieger Publishing Company, 1st ed., 326-327, New York.

Grace, J.R., Wairegi, T., and Brophy, J. (1978). Break-up of Drops and Bubbles in Stagnant Media. *Canadian Journal of Chemical Engineering* 56:3-8.

Guevara, E., Zagustin, K., Zubillaga, V., and Trallero, J.L. (1988). Core-Annular Flow (CAF): The Most Economical Method for the Transportation of Viscous Hydrocarbons, 4th UNITAR/U.N. Dev. Program AOSTRA-Petro-Can-Pet. Venez., S.A.-DOE Heavy Crude Tar Sands. *International Conference Edmonton.* 5:194.

Guzhov, A., Grishin, A.D., Medredev, V.F. and Medredeva, O.P. (1973). Emulsion formation during the flow of two immiscible liquids. *Neft. Choz.* (in Russian). 8:58-61.

Hall, A.R., and Hewitt, G.F. (1993). Application of two-fluid analysis to laminar stratified oil-water flows. *International Journal Multiphase Flow.* 19:4 711-717.

Hamad, F.A., Pierscionek, B.K., Brunn, H.H. (2000). A Dual Optical Probe for Volume Fraction, Drop Velocity and Drop Size Measurements in Liquid-Liquid Two-Phase Flow. *Meas. Science Technology* 11:1307-1318.

Hanks, R. and Christianson, E.B. (1962). The Laminar-Turbulent Transition in Normothermia Flow of Pseudoplastic Fluids in Tubes. *AIChE Journal* 8(4):467-471.

Hapanowicz, J., Troniewski, L., and Witczak S. (1997). Flow Patterns of Water-Oil Mixture Flowing in Horizontal Pipes, *International Symposium on Liquid-Liquid Two-Phase Flow and Transport Phenomena,* Antalya, Turkey, 3-7 Nov.

Harmathy, T.Z. (1960). Velocity of Large Drops and Bubbles in Media of Infinite or Restricted Extent. *AIChE Journal* 6(2):281-288.

Hasan, A.R., and Kabir, C.S. (1990). A New Model for Two-Phase Oil/Water Flow; Production Log Interpretation and Tubular Calculations. *SPE Production Engineering*, 193-199.

Hasan, A.R. and Kabir, C.S. (1999). A simplified model for oil/water flow in vertical and deviated wellbores, *SPE In Proceedings and Facilities*, 141:56-62.

Hasson, D., Mann, U., and Nir A. (1970). Annular Flow of Two Immiscible Liquids: I, Mechanisms, *Canadian Journal of Chemical Engineering*. 48:514-520.

Hasson, D. and Nir, A (1970). Annular Flow of Two Immiscible Liquids: II, *Canadian Journal of Chemical Engineering*. 48:521-526.

Hasson, D. (1978). Scale Prevention by Annular Flow of an Immiscible Liquid Along the Walls of a Heated Tube, *Proc. 6th International Heat Toronto*. 4:391-397.

Hill, A.D., and Oolman, T. (1982). Production Logging Tool Behavior in Two-Phase Inclined Flow. *JPT* 2432-2440.

Hinze, J. (1955). Fundamentals of the Hydrodynamic Mechanism of Splitting in Dispersion Process. *AIChE Journal* 1(3):289-295.

Hinze, J.O. (1959). *Turbulence*. McGraw-Hill, New York.

Ho, W.S., and Li,N.N. (1994). Core Annular Flow of Liquid Membrane Emulsion, *AIChE Journal*, 40:1961-1968.

Huang, A., Christodoulou, C., and Joseph, D.D. (1994). Friction Factor and Holdup Studies for Lubricated Pipelining. *International Journal Multiphase Flow* 20:481-494.

Hughmark, G.A. (1971). Drop Breakup in Turbulent Pipe Flow. *AIChE Journal* 4:1000.

Joseph, D.D., and Renardy, Y.Y. (1992). Fundamentals of Two Fluids Dynamics Part I and II (edited by F. John, et al), Springer-Verlag.

Kitscha, J. and Kocamustafaogullari, G. (1989). Breakup Criteria for Fluid Particles. *International Journal Multiphase Flow* 15:573:588.

Kolmogorov, A.N. (1949). On the Breaking of Drops in Turbulent Flow. *Doklady Akad. Nauk.* 66:825-828.

Kruyer, J., Redberger, P.J., and Ellis, H.S. (1967). The Pipeline Flow of Capsules - Part 9, *Journal Fluid Mechanics*. 30:513-531.

Kubie, J. and Gardner, G.C. (1977). Drop Sizes and Drop Dispersion in Straight Horizontal Tubes and in Helical Coils. *Chemical Engineering Science* 32:195-202.

Kurban, A.P.A. (1997). *Stratified Liquid-Liquid Flow*. Ph.D. Dissertation, Imperial College, London, U.K.

Laflin, G.C., and Oglesby, K.D. (1976). An Experimental Study on the Effect of Flow Rate, Water Fraction, and Gas-Liquid Ratio on Air-Oil-Water Flow in Horizontal Pipes. B.S. Thesis, University of Tulsa.

Luhning, R.W. and Sawistowki, H. (1971). Phase inversion in stirred liquid-liquid systems. *Proceedings International Solvent Extr. Conference*, The Hague, Society of Chemical Industry, London. 883-887.

Malinowsky, M.S. (1975). An Experimental Study of Oil-Water and Air-Oil-Water Flowing Mixtures in Horizontal Pipes, M.S. Thesis, University of Tulsa.

Masliyah, H. & Shook C.A. (1978). Two-phase laminar zero net flow in circular inclined pipe. *The Canadian Journal Chemical Engineering*, 56:165-175.

McAulifee, C.D. (1973). Oil-in-Water Emulsions and Their Flow Properties in Porous Meida. *Journal Petroleum Technology* 727-733.

Miesen, R., Beijnon, G., Duijvestijn, P.E.M., Oliemans, R.V.A., and Verheggen, T. (1992). Interfacial Waves in Core-Annular Flow, *Journal Fluid Mechanics*, 238(97).

Moalem-Maron, D., Brauner, N., and Rovinsky, J. (1995). Analytical Prediction of the Interface Curvature and its Effects on the Stratified Two-Phase Characteristics. *Proceedings of the International Symposium Two-Phase Flow Modelling and Experimentation* 1:163-170.

Mukherjee, H.K., Brill, J.P. and Beggs H.D. (1981). Experimental Study of Oil-Water Flow in Inclined Pipes, *Transactions of the ASME*, 103:56-66.

Mukhopadhyay, H. (1977). An Experimental Study of Two-Phase Oil-Water Flow in Inclined Pipes, M.S. Thesis, U. of Tulsa.

Nadler, M. Mewes, D. (1997). Flow induced emulsification in the flow of two immiscible liquids in horizontal pipes. *International Journal Multiphase Flow* 23(1):55-68.

Ng, T.S., Lawrence, C.J., Hewitt, G.F. (2001). Interface shapes for two-phase laminar stratified flow in a circular pipe. *International Journal Multiphase Flow* 27:1301-1311.

Ng, T.S., Lawrence, C.J., Hewitt, G.F. (2002). Laminar stratified pipe flow. *International Journal Multiphase Flow* 28(6):963-996.

Oglesby, K.D. (1979). An Experimental Study on the Effects of Oil Viscosity Mixture Velocity, and Water Fraction on Horizontal Oil-Water Flow, M.S. Thesis, University of Tulsa.

Oliemans, R.V.A. (1986). *The Lubricating Film Model for Core-Annular Flow.* Ph.D. Dissertation, Delft University Press.

Oliemans, R.V.A., Ooms, G. (1986). Core-Annular Flow of Oil and Water Through a Pipeline. *Multiphase Science and Technology.* vol. 2, eds. G.F. Hewitt, J.M. Delhaye, and N. Zuber, Hemisphere Publishing Corporation, Washington.

Ong, J., Enden, G. & Popel A.S. (1994). Converging three dimensional Stokes flow of two fluids in a T-type bifurcation, *Journal Fluid Mechanics* 270:51-71.

Ooms, G., Segal, A., Van der Wees, A.J., Meerhoff, R., and Oliemans, R.V.A. (1984). Theoretical Model for Core-Annular Flow of a Very Viscous Oil Core and a Water Annulus Through a Horizontal Pipe. *International Journal Multiphase Flow*, 10:41-60.

Ooms, G., Segal, A., Cheung, S.Y., and Oliemans, R.V.A. (1985). Propagation of Long Waves of Finite Amplitude at the Interface of Two Viscous Fluids. *International Journal Multiphase Flow* 10:481-502.

Pal, R. (1990). On the Flow Characteristics of Highly Concentrated Oil-in-Water Emulsions. *The Chemical Engineering Journal* 43:53-57.

Pan, L., Jayanti, S., and Hewitt, G.F. (1995). Flow Patterns, phase inversion and pressure gradients in air oil water flow in horizontal pipe, *Proceedings of the ICMF'95*, Kyotò, Japan, paper FT2.

Paul, H.I. and Sleicher Jr., C.A. (1965). The Maximum Stable Drop Size in Turbulent Flow: Effect of Pipe Diameter. *Chemical Engineering Science* 20:57-59.

Pilehvari, A., Saadevandi, B., Halvaci, M. and Clark, P.E. (1988). Oil/Water Emulsions for Pipeline Transport of Viscous Crude Oil. *Paper SPE 18218, SPE. Annual Technology Conference & Exhibition* Houston.

Ranger, K.B. & Davis A.M.J. (1979). Steady pressure driven two-phase stratified laminar flow through a pipe. *Canadian Journal of Chemical Engineering* 57:688-691.

Rovinsky, J., Brauner, N. and Moalem Maron, D. (1997). Analytical Solution for Laminar Two-Phase Flow in a Fully Eccentric Core-Annular Configuration, *International Journal Multiphase Flow* 23:523-542.

Russell, T.W.F. and Charles, M.E. (1959). The Effect of the Less Viscous Liquid in the Laminar Flow of Two Immiscible Liquids. *Canadian Journal Chemical Engineering* 37:18-34.

Russell, T.W.F., Hodgson, G.W., and Govier, G.W. (1959). Horizontal pipeline flow of oil and water, Can. Journal of Chemical Engineering, 37: 9-17.

Semenov, N.L. & Tochigin, A.A. (1962). An analytical study of the separate laminar flow of a two-phase mixture in inclined pipes. *Journal Engineering Physics* 4:29.

Shacham, M., and Brauner, N. (2002). Numerical solution of non-linear algebraic equations with discontinuities. Comp. and Chemical Engineering. in print.

Schramm, L.L. (1992). Emulsions Fundamentals and Applications in the Petroleum Industry. *Advances in Applications in the Petroleum Industry, Advances in Chemistry Series 231, American Chemical Society.*

Scot, P.M. and Knudsen, J.G. (1972). Two-Phase Liquid-Liquid Flow in Pipes. *AIChE Symposium Series* 68(118):38-44.

Scott, G.M. (1985). A Study of Two-Phase Liquid-Liquid Flow at Variable Inclinations, M.S. Thesis, The University of Texas.

Sherman, P. (1968). *Emulsion Science*, (editor), Academic Press, New York.

Simmons, M.J.H. (2001). Drop size Distribution in Dispersed Liquid-Liquid Pipe Flow. *International Journal Multiphase Flow* 23:843-859.

Sinclair, A.R. (1970). Rheology of Viscous Fracturing Fluids. *Journal Petroleum Technology* 711-719.

Soleimani, A. (1999) . *Phase distribution and associated phenomena in oil-water flows in horizontal tubes.* Ph.D. Dissertation. Imperial College, University of London.

Stalpelberg, H.H., and Mewes, D. (1990). The flow of two immiscible liquids and air in horizontal gas-liquid pipe, Winter Annual Meeting of the ASME, 89-96.

Tabeling, P., Pouliquen, O., Theron, B., and Catala G. (1991). Oil water flows in deviated pipes: experimental study and modelling. *In Proceedings of the 5th International Conference on Multiphase Flow Production*, Cannes, France, June 19-21, 294-306.

Tang, Y.P., Himmelblau, D.M. (1963). Velocity Distribution of Isothermal Two-Phase Cocurrent Laminar Flow in Horizontal Rectangular Duct. *Chemical Engineering Science.* 18:143-144.

Taylor, G.I. (1934). The Formation of Emulsions in Definable Fields of Flow, *Proceedings Royal Society London A* 146:501-523.

Theissing, P.A. (1980). A Generally Valid Method for Calculating Frictional Pressure Drop in Multiphase Flow. *Chemical Ing. Technik.* 52:344-355. (In German).

Tidhar, M., Merchuk, J.C., Sembira, A.N., Wolf, D. (1986). Characteristics of a motionless mixer for dispersion of immiscible fluids – II. Phase inversion of liquid-liquid systems. *Chemical Engineering Science*, 41(3):457-462.

Trallero, J.L. (1995). *Oil-Water Flow Patterns in Horizontal Pipes.* Ph.D. Dissertation, The University of Tulsa.

Tsouris, C. and Tavlarides, L.L. (1994). Breakage and Coalescence Models for Drops in Turbulent Dispersions. *AIChE Journal* 40(3):395-406.

Ullmann, A., Zamir, M. Ludmer, Z., Brauner, N. (2000). Characteristics of Liquid-Liquid Counetr-Current Flow in Inclined Tubes - Application to PTE Process, Proc. of the International Symp on Multiphase Flow and Transport Phenomenon, Antalya, Turkey, Nov. 5-10. ICHMT 112-116.

Ullmann, A., Zamir, M., Ludmer L. and Brauner, N. (2001a). Flow Patterns and Flooding Mechanisms in Liquid-Liquid Counter-current Flow in Inclined Tubes, ICMF-2001, New Orleans, Louisiana, May 27-June 1.

Ullmann, A., Zamir, M., Gat S., and Brauner, N. (2001b). Multi-Value Holdups in Stratified Co-current and Counter-current Inclined Two-Phase Flows, 39th European Two-Phase Flow Group Meeting, Aveiro, Portugal, 17-20.

Ullman, A., Zamir, M., Ludmer, Z., and Brauner, N. (2002a). Counter-current Flow of Two Liquid Phases in an Inclined Tube: Part I. Submitted.

Ullman, A., Zamir, M., Gat, S., and Brauner, N. (2002b). Counter-current Flow of Two Liquid Phases in an Inclined Tube: Part II. Submitted.

Valle, A., and Kvandal, H.K. (1995). Pressure drop and dispersion characteristics of separated oil-water flow. In Celata, G.P., and Shah, R.K. Edizioni ETS, eds., In Proceedings of the International Symposium on Two-Phase Flow Modelling and Experimentation, Oct. 9-11, Rome, Italy, 583-591.

Valle, A., and Utvik, O.H.(1997). Pressure drop, flow pattern and slip for two phase crude oil/water flow: experiments and model predictions, International Symposium on Liquid-Liquid Two-Phase Flow and Transport Phenomena, Antalya, Turkey, 3-7 Nov.

Vedapuri, D., Bessette, D. and Jepson, W.P. (1997). A segregated flow model to predict water layer thickness in oil-water flows in horizontal and slightly inclined pipelines, in In Proceedings Multiphase'97, Cannes, France June 18-20, 75-105.

Vigneaux, P., Chenois, P. and Hulin, J.P. (1988). Liquid-Liquid Flows in an Inclined Pipe. *AIChE Journal* 34:781-789.

Wu, H.L., Duijrestijn, P.E.M. (1986). Core-Annular Flow: A Solution to Pipeline Transportation of Heavy Crude Oils. *Review Tec. INTERVER*, 6(1):17-22.

Yeh, G. Haynie Jr., F.H., Moses, R.E. (1964). Phase-Volume Relationship at the Point of Phase Inversion in Liquid Dispersion. *AIChE Journal.* 10(2):260-265.

Yeo, L.Y., Mater, O.K., Perez de Ortiz, E.S., Hewitt, G.F. (2000). Phase inversion and associated phenomena. *Multiphase Science & Technology*, 12:51-116.

Zamir, M. (1998). *Multistage Extraction in an Inclined Column Based on a Phase Transition of Critical-Solution Mixtures.* Ph.D. Dissertation.

Zavareh, F., Hill, A.D. and Podio, A.L. (1988). Flow Regimes in Vertical and Inclined Oil/Water Flow in Pipes, Paper SPE 18215, Presented at the 63rd Annual Technical Conference and Exhibition, Houston, Texas, Oct. 2-5.

Zigrang D.I. and Sylvester, N.D. (1985). A Review of Explicit Friction Factor Equations. *Journal Energy Res. Technology* 107:280-283.

Zuber, N. and Findlay, I. (1965). *Trans. J. Heat Transfer.* ASME 87:453-468.

Zukoski, E.E. (1966). Influence of Viscosity, Surface Tension and Inclination Angle on Motion of Long Bubbles in Closed Tubes. *Journal Fluid Mechanics* 25:821-837.

Two-Phase Flow Measurement Techniques

Volfango Bertola

Laboratoire de Physique Statistique, Ecole Normale Supérieure, Paris, France

Abstract. In these notes the most common measurement techniques for two-phase flows are reviewed. The working principles and the configurations of instruments for void fraction measurements, flow visualization and velocity measurements are presented; in detail: radiation attenuation, optical and electrical impedance techniques for void fraction measurement; tomographic and time-average visualization techniques; velocity measurements from signal cross-correlation, hot film anemometry, particle image velocimetry.

1 Introduction

Two-phase flow is one of the most common flows in nature as well as in industrial applications; it covers gas-solid, liquid-liquid, solid-liquid and gas-liquid flows. Among these, gas-liquid flow, which also includes the whole subject of boiling and condensation, is probably the most relevant topic, and can be encountered in a wide range of industrial applications including evaporators, boilers, distillation towers, chemical reactors, condensers, oil pipelines, nuclear reactors, etc. Thus, the early research on two-phase flow was carried out on gas-liquid flow; anyway, the theoretical framework developed for gas-liquid flow can be often extended to other two-phase systems, such as liquid-liquid flow.

Advancements in the knowledge of two-phase flow need a close working relationship among experimentalists, theoreticians and numerical analysts. In particular, experiments are necessary to verify the reliability and the limitations of both mathematical models and numerical codes; furthermore, they allow capturing the physics of phenomena, and therefore provide a basis for their theoretical description.

Measurement techniques in two-phase flows are quite different from those of single-phase flows: in fact, there exist peculiar quantities of this kind of flows, such as the void fraction and the interfacial area concentration, which require a specifically conceived instrumentation. On the other hand, measurement techniques which have been developed for single-phase flow measurements, such as hot wire anemometry or particle image velocimetry, generally cannot be used in two-phase flows as they are, but require some modifications.

2 Void Fraction Measurements

Void fraction is a dimensionless quantity indicating the fraction of a geometric or temporal domain occupied by the gas phase, and it is probably the most significant quantity one can measure in two-phase flow. Although the first void fraction measurements are dated back to the 1940s, they still play the most important role in today's experiments: a recent statistical study of the experimental research on two-phase flow shows that almost any journal article on this subject presents void fraction data (Chanson, 2002). The importance is even greater in applications, where the void fraction is often a key parameter in design as well as in financial issues (just think of the petroleum industry, where one of the main problems is the oil-water-gas flow metering). On the other hand, void fraction measurements are essential to validate the predictions of mathematical models.

According to the different definitions of void fraction (related to the choice of the spatial or temporal domain), different instruments have been developed; in particular, one can distinguish:

- local instruments, which directly measure the phase density function or the local, time-average void fraction;
- integral instruments, which measure space-averaged void fractions.

Void fraction measurement techniques are based on various principles: usually, instruments are sensitive to some physical property which is different for the two phases, such as the fluid density or the electrical conductivity.

2.1 Void Fraction Definition

The different definitions of void fraction can be placed in a common mathematical framework introducing the so-called phase density function, which is the characteristic function of the geometrical subsets of the controlled volume occupied by one of the two phases. Consider a three dimensional geometrical domain occupied by a two-phase mixture (e.g. gas-liquid) at time t; the set D of the internal points of such domain, identified by their position **x**, can be divided into two subsets D_G and D_L, occupied by the gas phase and by the liquid phase, respectively. The phase density function P_k ($k = G, L$) is the characteristic function of these subsets:

$$P_k(\mathbf{x},t) = \begin{cases} 1 & if\ \mathbf{x} \in D_k \\ 0 & if\ \mathbf{x} \notin D_k \end{cases} \tag{1}$$

By averaging the gas density function over different spatial or temporal domains one can obtain four definitions for the void fraction.

Local void fraction. The local void fraction in a given point **x** of space represents the ratio between the time during which the point is occupied by the gas and the total duration of the observation T; from Eq. (1), it is the time average over T of the gas density function $P_G(\mathbf{x}, t)$:

$$\varepsilon_G(\mathbf{x},T) = \frac{1}{T}\int_T P_G(\mathbf{x},t)\,dt \tag{2}$$

Since two-phase flow is intrinsically inhomogeneous, the value of ε_G a priori is a function of the integration time T; yet, in case of steady flow conditions, ε_G can be considered constant with respect to T if T is large enough, so that:

$$\varepsilon_G(\mathbf{x}) = \frac{1}{T}\int_T P_G(\mathbf{x},t)\,dt \tag{3}$$

Figure 1 shows an example of the local void fraction fluctuations with respect to the integration time for a steady air-water flow in a horizontal pipe.

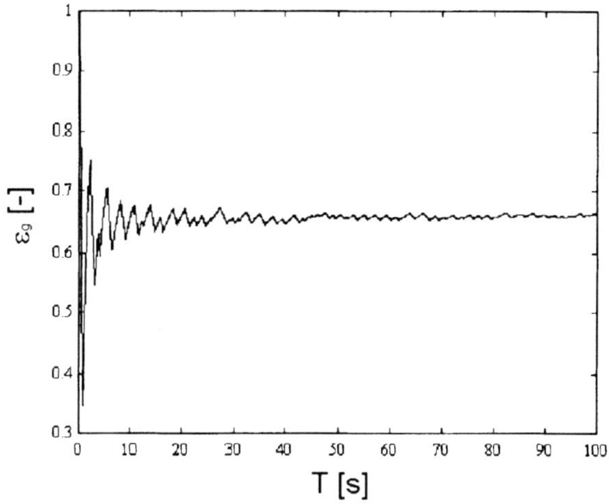

Figure 1. Local void fraction fluctuations vs. integration time.

One-dimensional void fraction. The one-dimensional void fraction comes from averaging the phase density function along a straight line at a given time:

$$\varepsilon_{G,l}(t) = \frac{1}{L}\int_L P_G(\mathbf{x},t)\,dl = \frac{L_G(t)}{L} \tag{4}$$

Thus, it represents the ratio between the length occupied by the gas and the total length L. Very often, the value of $\varepsilon_{G,1}$ depends both on the length, and on the orientation of the line: therefore, the local flow structure should be always taken into account in order to avoid misunderstanding.

Two-dimensional void fraction. The two-dimensional void fraction, or cross-sectional average void fraction, comes from averaging the phase density function over a given area A at a given time:

$$\varepsilon_{G,2}(t) = \frac{1}{A}\int_A P_G(\mathbf{x},t)\,da = \frac{A_G(t)}{A} \tag{5}$$

It is the ratio between the area occupied by the gas and the total area A at time t. For the flow in pipes (or rectangular channels), a natural choice is the cross-sectional area perpendicular to the pipe axis.

Three-dimensional void fraction. The three-dimensional void fraction, or average void fraction over a control volume, is the ratio between the volume of the gas and the whole control volume, and is defined as:

$$\varepsilon_{G,3}(t) = \frac{1}{V}\int_V P_G(\mathbf{x},t)\,dv = \frac{V_G(t)}{V} \tag{6}$$

Double-averaged void fractions. Space and time averages can be combined to give double-averaged void fractions. For a well-known property of double integrals of measurable functions, space and time average are commutative: in other words, changing the sequence of the operations does not change the result (Ishii, 1975; Delhaye and Achard, 1976). Therefore, one can define three double-averaged void fractions:

$$\varepsilon_1 = \frac{1}{T}\int_T \left\{\frac{1}{L}\int_L P_G(\mathbf{x},t)\,dl\right\} dt = \frac{1}{L}\int_L \left\{\frac{1}{T}\int_T P_G(\mathbf{x},t)\,dt\right\} dl = \frac{L_G}{L} \tag{7}$$

$$\varepsilon_2 = \frac{1}{T}\int_T \left\{\frac{1}{A}\int_A P_G(\mathbf{x},t)\,da\right\} dt = \frac{1}{A}\int_A \left\{\frac{1}{T}\int_T P_G(\mathbf{x},t)\,dt\right\} da = \frac{A_G}{A} \tag{8}$$

$$\varepsilon_3 = \frac{1}{T}\int_T \left\{\frac{1}{V}\int_V P_G(\mathbf{x},t)\,dv\right\} dt = \frac{1}{V}\int_V \left\{\frac{1}{T}\int_T P_G(\mathbf{x},t)\,dt\right\} dv = \frac{V_G}{V} \tag{9}$$

Unless otherwise declared, the quantity which is usually referred to as "void fraction" in the literature is the two dimensional (cross-sectional) and time average void fraction, and is generally denoted by the symbols ε or, less frequently, α.

2.2 Radiation Attenuation Techniques

These methods are based on the interaction between condensed matter (in our case, fluids) and either radiation or particles. Since particles (or, alternatively, the radiation wavelength) are very small as compared with atoms, interaction with matter occurs only if they hit or at least pass very close to an atom; as a result, the particle may deviate, bounce, or hit and eventually destroy the atom. For each fundamental interaction (*strong*, *electro-weak* and *gravitational*) one can define an area surrounding the atomic nucleus where interactions occur (Krane, 1988), which is known as *cross-section*, and is measured in barn (1 barn = 10^{-28} m^2).

For simplicity, let's consider a uniform, steady-state flux Φ of particles entering a volume Adx perpendicular to the surface A, so that $N = \Phi A$ is the number of entering particles per unit time; the number of interactions after a length dx (that is, the beam attenuation) is:

$$dN = -\Phi A n \sigma\, dx \tag{10}$$

where n is the number of nuclei per unit volume and σ the cross-section of the interaction. This leads to the well known exponential decay law of a monochromatic radiation:

$$N = N_0 e^{-x/\lambda} \tag{11}$$

where λ is the so-called interaction length, which is characteristic of each material, and can be easily expressed as a function of the density ρ, of the atomic mass M_A and of Avogadro's number N_{AV}:

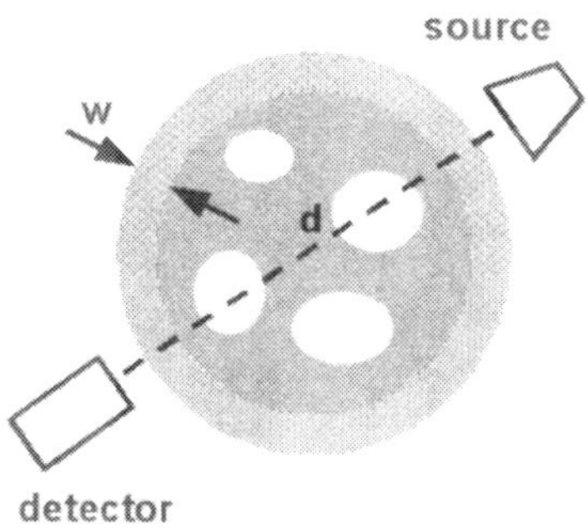

Figure 2. Measurement of one-dimensional void fraction by radiation attenuation.

$$\lambda = \frac{1}{\sigma n} = \frac{M_A}{\sigma \rho \mathrm{N}_{\mathrm{AV}}} \tag{12}$$

Since the interaction length is different for different fluids, a radiation beam can be used to measure the one dimensional void fraction in two-phase flow (either instantaneous or time-average), as shown in figure 2. In this case, there are three contributions to the beam attenuation, those of the two fluids and that of the pipe walls:

$$N = N_0 \exp\left(-\frac{2w}{\lambda_w}\right) \exp\left(-\frac{(1-\varepsilon)d}{\lambda_L}\right) \exp\left(-\frac{\varepsilon d}{\lambda_G}\right) \tag{13}$$

where w is the wall thickness and d the pipe diameter. This kind of instrument generally uses a γ-ray source, and therefore is known as γ-densitometer.

For gas-liquid flow at relatively low pressure, Eq. (13) simplifies because the last exponential can be neglected. Let N_G and N_L indicate respectively the number of emerging particles one can measure in case of gas flow ($\varepsilon = 1$) and in case of liquid flow ($\varepsilon = 0$):

$$\begin{aligned} N_G &= N_0 \exp\left(-\frac{2w}{\lambda_w}\right) \exp\left(-\frac{d}{\lambda_G}\right) \\ N_L &= N_0 \exp\left(-\frac{2w}{\lambda_w}\right) \exp\left(-\frac{d}{\lambda_L}\right) \end{aligned} \tag{14}$$

Then, the void fraction can be calculated after one measurement in two-phase flow as follows:

$$\varepsilon = \frac{\mathrm{Ln}\left(\dfrac{N}{N_L}\right)}{\mathrm{Ln}\left(\dfrac{N_G}{N_L}\right)} \tag{15}$$

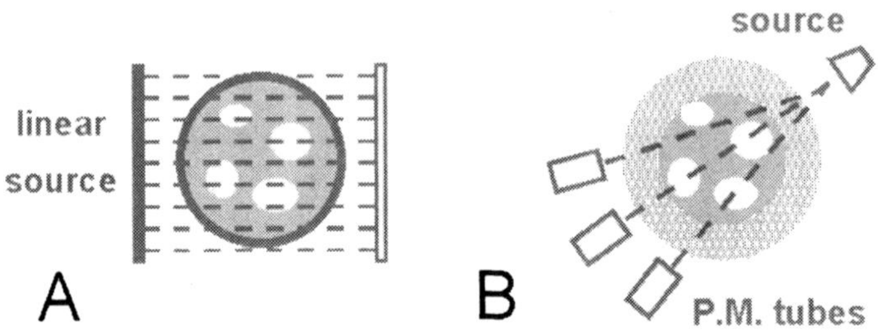

Figure 3. Configurations of radiation sources for two-dimensional void fraction measurements.

By replacing the beam with a linear source (a radiation sheet, as shown in figure 3a), one can measure the cross sectional average (two-dimensional) void fraction. Unfortunately, in this case there is no specific absorption law like Eq. (11), because absorption depends both on the pipe geometry and on the flow pattern: thus, a preliminary calibration with plastic mockups that simulate the different flow regimes and void fractions is necessary. The cross-sectional void fraction can also be obtained by integration of one-dimensional void fraction values measured simultaneously by a multi-beam device (figure 3b). Although in this case Eq. (13) holds for the single beams, the precision of this method strongly depends on the number of beams used: since commercial devices usually have 3 or 5 beams, a precise calculation of the cross-sectional void fraction is impossible if the flow pattern is unknown.

Both the source and the detectors must be properly shielded in order to protect the operators from exposure to radiations. On the other hand, the thickness of shielding is often considerable, so that γ-densitometers cannot be too small: for example, the most common radiation source is ^{137}Cs, which generates γ-rays at 662 keV, requiring a lead shielding 10 cm thick; a less common (and more expensive) source is ^{241}Am, which releases a weaker radiation (60 keV), and needs only a 2 mm thick lead shielding. The choice of the appropriate radiation source for a given two-phase mixture generally represents a compromise between sensitivity (which depends on the number of interactions in the flow) and accuracy (which depends on the number of particles interacting inside the detector): if sensitivity is high, most of the interactions take place in the flow, so that fewer particles are left for the detector, and the instrument has a low accuracy. This problem could be solved by using X-rays instead of γ-rays, because they can be produced with much higher intensity ($10^3 \div 10^4$ times); on the other hand, X-ray generators are less stable and, more important, less monochromatic, so that they are preferred for flow visualization purposes.

Radiation is generally detected by means of a scintillator coupled to a photomultiplier. The scintillators absorb radiation and emit visible light by fluorescence; they are usually classified on the basis of the sensitive element, which can be either solid or liquid, organic or inorganic. Organic scintillators are for instance anthracene and some polymeric materials, while inorganic scintillators are usually ionic crystals added with impurities to increase the probability of photon emission and reduce self-absorption, such as NaI(Tl).

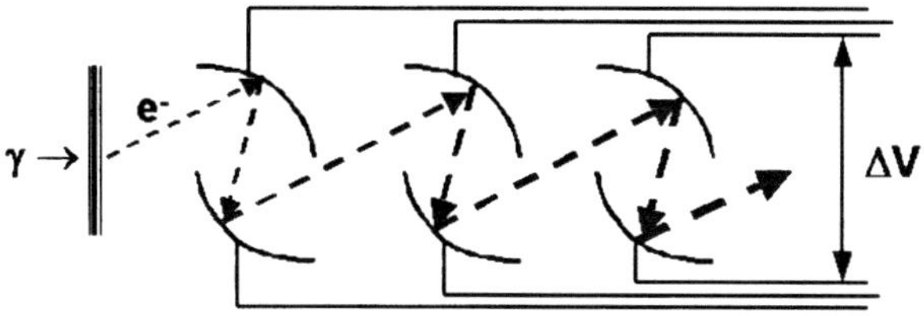

Figure 4. Schematic of a photomultiplier tube.

The photomultiplier is a device that is able to convert very weak luminous pulses into electric signals. Figure 4 shows the typical layout of a photomultiplier tube, which consists of a photocathode and of a cascade of pairs of electrodes at different potentials. A small number of electrons (smaller than the number of incident photons) is released at the photocathode, then multiplied and focused by a series of electrodes called dynodes. The dynodes are connected to a voltage chain produced by a high-voltage supply and a series of voltage dividers. The typical potential difference between adjacent dynodes is about 100 V, and each successive dynode is at a higher potential than the previous one. A typical tube might have 10 dynodes: at each stage, the number of electrons increases by a factor of the order of 5, so that the overall gain is 5^{10} (about 10^7).

Since radiation detectors measure the radiation intensity by counting the number of particles received in a given time interval, one cannot have "instantaneous" measurements; on the other hand, Eq. (15) states that the relationship between the number of detected particles and the void fraction is logarithmic: thus, the void fraction given by Eq. (15) is not a time-average value, because obviously $\langle N \rangle \neq \langle Ln(N) \rangle$. In order to overcome this problem, one might reduce the duration of the time interval in which particles are counted, but this reduces the instrument accuracy because less particles are detected; a better solution is represented by logarithmic amplifiers, which convert the instantaneous signal into a quantity proportional to the void fraction, on which the true time average can be calculated.

2.3 Optical Probes

Optical instrumentation is extremely common in two-phase flow diagnostics, and is used to measure several different quantities, including void fraction. In spite of their versatility, optical instruments are essentially based on the same principle, that is, the well-known behavior of electromagnetic waves on the interface between two different media, where both reflection and refraction occur simultaneously. The classical theory of these phenomena is described in detail in optics textbooks, such as in Guenther (1990), and is briefly summarized here.

As it is well known, electromagnetic waves describe the propagation either in vacuum or in a dielectric medium of the electric field and the associated magnetic field (figure 5a); for simplicity, let's consider the electric field in a plane sinusoidal wave of frequency $f = \omega/2\pi$:

$$\mathbf{E} = \mathrm{Re}\left[\mathbf{E}_0 e^{i(\omega t - \mathbf{k}\cdot\mathbf{x})}\right] \tag{16}$$

where $\mathbf{k}$ is the propagation vector.

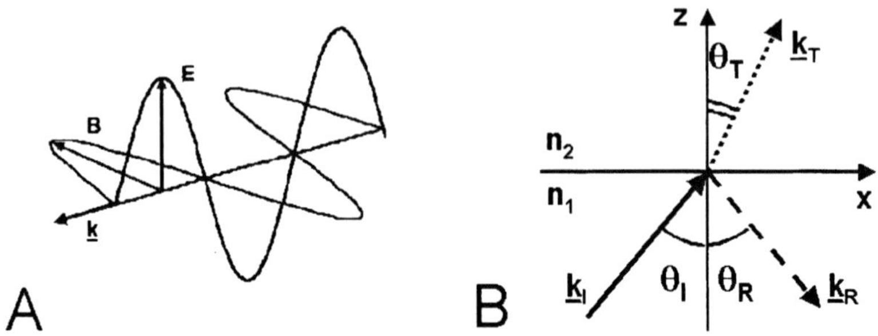

Figure 5. Electromagnetic waves; (a) propagation in space of the electric (E) and magnetic (B) fields; (b) at the interface between two media of different refractive index the incident wave $\mathbf{k}_I$ is decomposed into a reflected wave $\mathbf{k}_R$ and a transmitted (refracted) wave $\mathbf{k}_T$.

Maxwell's equations yield $|\mathbf{k}| = \omega\, n / c$, also called *dispersion relationship*, where n is the refractive index of the medium where the wave propagates (in vacuum, n = 1), and c is the speed of light. When the e.m. wave reaches the interface between two media characterized by different refractive indexes n_1 and n_2 (the plane z = 0 in figure 5b), it is decomposed into a reflected wave, that goes back into the first medium, and a refracted (or transmitted) wave, that proceeds into the second medium. One finds that the three waves have the same frequency, and their propagation vectors $\mathbf{k}_I$, $\mathbf{k}_R$ and $\mathbf{k}_T$ lay in the same plane, called *plane of incidence*; furthermore, if θ_I, θ_R and θ_T are the angles defined by the direction normal to the interface and $\mathbf{k}_I$, $\mathbf{k}_R$, $\mathbf{k}_T$, respectively (see figure 5b), one can obtain the well-known Snell's laws:

$$\begin{aligned} \theta_I &= \theta_R \\ n_1 \sin\theta_I &= n_2 \sin\theta_T \end{aligned} \qquad (17)$$

The reflection coefficient r is defined as the ratio between the reflected electric field and the incident electric field; it can be calculated separately for the parallel and normal components of the electric field with respect to the plane of incidence:

$$\begin{aligned} r_P &= \frac{n_2 \cos\theta_I - n_1 \cos\theta_T}{n_2 \cos\theta_I + n_1 \cos\theta_T} \\ r_N &= \frac{n_1 \cos\theta_I - n_2 \cos\theta_T}{n_1 \cos\theta_I + n_2 \cos\theta_T} \end{aligned} \qquad (18)$$

where θ_T can be expressed as a function of θ_I thanks to the second of Eqs. (17). The quantities r_P and r_N represent the fraction of the incident field reflected by the interface between the two media. This is not the same as the fraction of reflected energy, which is proportional to the square of these reflection coefficients (this statement is rigorous only for r_P, while for r_N it is an approximation).

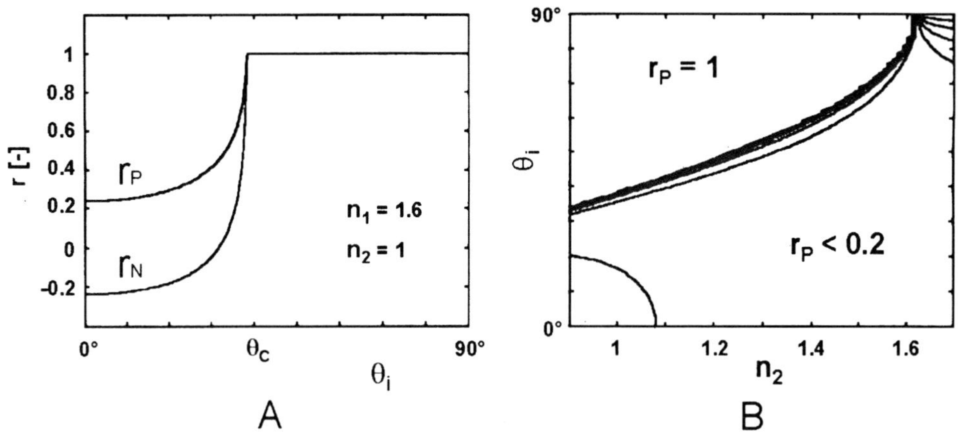

Figure 6. Reflection coefficient; (a) parallel and normal reflection coefficient as a function of the incidence angle; (b) contour plot of the parallel reflection coefficient as a function of the incidence angle and of the refractive index of the second medium n_2 ($n_1 = 1.6$).

The reflection coefficients r_P and r_N are plotted in figure 6a with respect to the angle of incidence, for an e.m. wave passing from glass ($n_1 = 1.6$) to air ($n_2 = 1$). When θ_I is larger than a critical value θ_C, both r_P and r_N become complex numbers with a modulus equal to 1: thus, the amplitude of the reflected field is the same as that of the incident field, and the only difference is in the phase. For $\theta_I < \theta_C$ both the coefficients drop to about 20%: in terms of reflected energy, this corresponds to 4% of the incident energy. Figure 6b shows the contour plot of the parallel reflection coefficient r_P as a function of θ_I and of the refractive index of the second medium n_2, assuming that the first medium is glass with n = 1.6; the parallel reflection coefficient r_N is analogous.

The optical probe is an instrument that allows to distinguish two fluids of different refractive index, thanks to the abrupt change of the reflection coefficient that can be achieved when n_2 changes with constant θ_I, as shown in figure 6b for θ_I greater than about 30°. The probe consists of a glass optical fiber (refractive index $n_1 \approx 1.6$) with a diameter of about 100 μm, connected at one end to an infra-red source and to a photo-diode, whereas the free end is placed inside the fluid; this technology has been known for more than 30 years (Hinata, 1972; Jones and Delhaye, 1976; Cartellier, 1990; Cartellier and Achard, 1991). Figure 7a shows the working principle of a typical single-fiber optical probe: when the light emitted inside the optical fiber reaches the tip, it is completely reflected back if the free end is immersed in air, while the reflection coefficient drops to almost zero when the probe is immersed in water; the reflected light is converted by the photo-diode into an electric signal, which ideally consists of two voltage levels, and corresponds to the phase density function. The difference in the signal level for the liquid and the gas phases has been found to vary with the void fraction and phase velocities, resulting in threshold or discrimination level uncertainties, as shown for example by Abuaf et al. (1978).

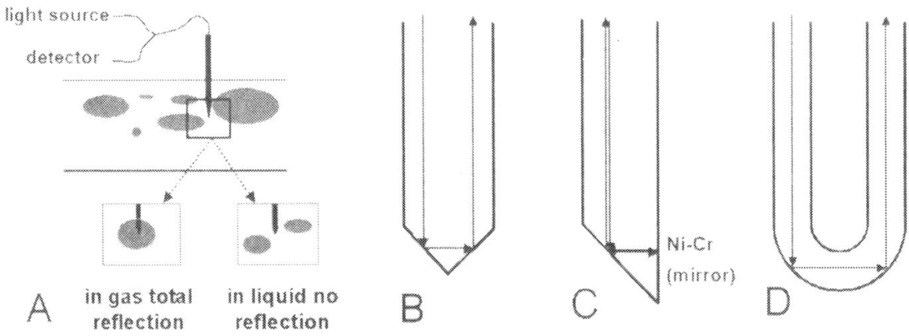

Figure 7. Optical probes; (a) working principle; (b) conical probe; (c) probe; (d) U-shaped probe, no longer in use.

Inside optical fibers the inclinations of light beams (which correspond to the propagation vectors) are statistically distributed around their most probable direction, coincident with the fiber axis; for simplicity, we can assume that all the beams have this direction: thus, we can also define a single angle of incidence θ_I, which is determined only by the geometry of the free end. For instance, for a conical tip θ_I is equal to the cone half angle. The choice $\theta_I = 45°$ represents an optimal solution to distinguish gases from liquids with a glass optical fiber ($n_1 \approx 1.6$), because in this case we have that total internal reflection occurs for $n_2 < 1.15$ (this value can be determined graphically with the aid of figure 6b, or calculated imposing $\theta_T = 90°$ in the second of Eqs. 17) while for $n_2 > 1.15$ reflection is not significant. Thus, this system can properly distinguish gases at low pressure ($n_2 \approx 1$) from most liquids, which generally have $n_2 > 1.15$ (for water, $n_2 \approx 1.33$; for kerosene, $n_2 \approx 1.44$; freons have $n_2 \approx 1.25$).

In order to be detected by the photo-diode, the beam must be deflected of 180°; since $\theta_R = \theta_I$, this can be accomplished by means of two total reflections with $\theta_I = 45°$: figure 7 shows three possible geometries of the probe free end. The conical tip (figure 7b) is the simplest geometry, which can be obtained for instance by etching a cleaved fiber with a corrosive liquid; due to the very small dimensions (the cone tip may be as small as 20 μm), manufacturing conical tips is extremely difficult, and finding two probes which have exactly the same shape is almost impossible: Cartellier and Barrau (1998a) showed that probes supplied by different manufacturers have different responses in the same operating conditions, so that a careful calibration of each probe is necessary. The cleaved probe (figure 7c) has a flat surface inclined at 45° with respect to the fiber axis; the end of the fiber is covered with a reflecting coating (e.g. a Ni-Cr paint), so that after a first total internal reflection the light beams hit the coating and are reflected back to the active surface, where they undergo a second total internal reflection back to the photo-diode. This probe, described in detail by Fordham et al. (1999a,b), can be manufactured very easily by cutting at 45° a bundle containing several hundreds of fibers. Another possible geometry is represented by the U-shaped probe (figure 7d), in which the double total internal reflection is obtained by bending the fiber of 180°; although this configuration has been very popular in the past years, its performances cannot be compared with those of the conical probe, and nowadays it is no longer in use.

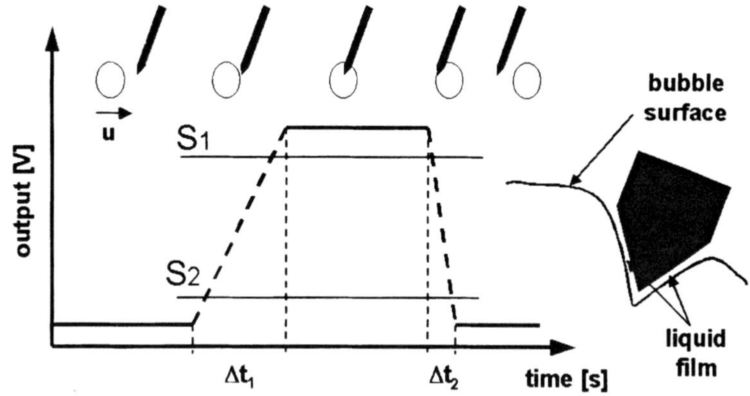

Figure 8. Dynamic response of an optical probe to the transition from one phase to another; the detail shows the liquid film still wetting the probe surface after the tip has entered a gas bubble.

The dynamic response of the optical probe to a change of phase is schematically depicted in figure 8, which describes the passage of a gas bubble through a probe initially surrounded by liquid. In the liquid phase the voltage output has a constant value corresponding to no reflection; when the probe enters the gas phase, the reflected beam activates the photo-diode, increasing the voltage output to the higher value. The growth of the output value is not instantaneous, but requires a time Δt_1, because the smaller contact angle of liquids on glass causes the probe tip to be covered by a thin liquid film, which is dried out only when the tip has already advanced into the gas phase; as the film becomes thinner, the reflection coefficient grows, until the output reaches the maximum value. Since wetting phenomena are extremely complex (see e.g. the review article by Bonn and Ross, 2001), the time delay is sensitive to a great number of parameters (probe geometry, orientation with respect to the bubble direction, impurities in the fluid, manufacturing procedure, etc.), and can be determined only empirically. A similar phenomenon occurs when the probe passes from the gas to the liquid: the output value drops only after the tip has entered the liquid phase. In this case, the time delay Δt_2 is shorter than Δt_1, because the liquid tends to spread over the probe surface, accelerating its wetting. These signal transients make the proper identification of the bubble 'front' and 'back' interfaces difficult. If the bubble velocity is not too small, it is inversely proportional to the time delay Δt_1 (Cartellier and Barrau, 1998a): thus, the quantity $L = u\, \Delta t_1$ (the so-called *latency length*) is a constant, and can be used to characterize the probes quantitatively.

In order to discriminate the two levels of the output signal S, a threshold value S_0 could be chosen such that one phase corresponds to $S < S_0$ and the other to $S > S_0$. Due to the time delays Δt_1 and Δt_2 this method is not very precise: as one can verify qualitatively on the plot in figure 8, a value of S_0 close to the lower value of the signal (liquid phase) allows to detect with precision the transitions from liquid to gas, whereas the detection of passages from gas to liquid is affected by an error; similarly, if S_0 is close to the higher value of the signal (gas phase) only gas/liquid interfaces are detected precisely; if S_0 has an intermediate value between the two levels, both liquid/gas and gas/liquid interfaces are missed.

Figure 9. Cleaved optical probe: total internal reflection is never possible.

A better discrimination technique uses two threshold levels S_1 and S_2 (figure 8), the former for gas/liquid and the latter for liquid/gas interfaces: the passage from liquid to gas corresponds to $S = S_2$, while $S = S_1$ denotes the passage from gas to liquid. Past experience shows that the probe orientation with the flow direction has little effect on the measurement accuracy, provided that the probe support does not affect the flow past the tip (e.g. Sene, 1984).

Unlike the probes described above, the cleaved probe (figure 9) is not based on the phenomenon of total internal reflection. In fact, this phenomenon occurs when the reflection coefficient equals 1, which is possible only if θ_I is not too small, as one can verify for r_P on the plot of figure 6b (for a glass probe θ_I should be at least 30-35°). In the cleaved probe, beams are almost perpendicular to the probe end ($\theta_I \approx 0°$), so that the reflection coefficient is always smaller than 0.2, corresponding to a fraction of reflected energy smaller than 4%, and total internal reflection never takes place; this kind of reflection, which is known as Fresnel reflection, is the cause of the faint images reflected by window panes, for instance. Eqs. (17) and (18) show that the reflection coefficients change in a continuous fashion with the refractive index n_2 of the second medium: thus, the cleaved probe can be used to distinguish fluids with different refractive indexes, even if they cannot de discriminated by conical probes. This happens, for instance, in liquid-liquid flows, in which all the components have $n_2 > 1.15$, so that conical probes give a constant output signal for both phases.

Hamad et al. (1997) built a cleaved probe from a glass fiber with $n_1 = 1.48$, which could be used to distinguish the different phases of a water/kerosene flow. The optical probe cable had a 100 μm core diameter and a 140 μm cladding diameter; the cladding provided a constant step in the refractive index at the core-cladding interface, which guided the light inside the fibre. The current through the photo-diode was directly proportional to the strength of the light received. The energy fraction reflected by this probe was 0.28% in water ($n_2 = 1.33$) and 0.024% in kerosene ($n_2 = 1.44$), and grew to 3.5% in air. As one can see from these values, only a very small proportion of the transmitted light is reflected and delivered to the photo-diode receiver, resulting in a very small operational signal for phase detection: consequently, considerable attention must be paid to the signal/noise aspect of the system.

If the diameter of the probe is much smaller than the size of the dispersed phase, the use of a flat-tipped probe has distinct interface detection advantages compared with the use of a conical probe: in fact, surface tension effects will cause different responses for the two probes. When first contact occurs between an interface and a flat tip, with the contact angle being very close to zero (Farrar and Bruun, 1989), the interface will virtually instantaneously deform so that the whole of the tip comes into contact with the interface, causing a steep change in the probe signal. On the contrary, when using a conical probe a meniscus will be formed, which will gradually creep up the conical tip as the bubble or drop interface moves past the tip, which results in a much longer transition time (Bruun 1995).

Another technique to distinguish the phases in liquid-liquid mixtures consists in adding a fluorescent dye to one of the two fluids, and excite fluorescence with a laser light source at the appropriate wavelength; some liquids, such as light crude oil, have a little natural fluorescence, and need no additives. Ramos et al. (2001) applied this technique to distinguish the components of a three phase mixture (oil/water/gas) by a combination of reflectance and fluorescence methods (Wu et al., 2000). The signal reflected due to the contrast in refractive index provided by gas is directly detected by a photo-diode, while the fluorescence signal is separated from the excitation signal by a long-wavelength-pass filter and detected by a second photo-diode to provide the detection of oil.

2.4 Electrical Impedance Methods

Electrical impedance is the measure of the degree to which an electric circuit resists a flow of electric current when a voltage is impressed across its terminals: in particular, impedance is the ratio of the voltage impressed across a pair of terminals to the current flow between those terminals; the SI unit for this quantity is the Ohm (Ω). In direct-current (DC) circuits, impedance corresponds to resistance; in transient or alternating current (AC) circuits, impedance is a function of resistance, inductance, and capacitance. Inductors and capacitors build up voltages that oppose the flow of current: this opposition, called reactance, is combined with resistance to find the impedance. The reactance produced by inductance is proportional to the frequency of the alternating current, whereas the reactance produced by capacitance is inversely proportional to the frequency. In AC circuits, both the voltage V and the current I can be expressed as rotating vectors at a frequency $\omega/2\pi$ and phase φ; with this representation, electrical impedance is a complex quantity, where the real part is the resistance R, and the imaginary part the reactance X:

$$Z = \frac{\mathbf{V}}{\mathbf{I}} = \frac{Ve^{i(\omega t+\varphi_1)}}{Ie^{i(\omega t+\varphi_2)}} = \frac{V}{I}e^{i\Delta\varphi} = \frac{V}{I}\cos\Delta\varphi + i\frac{V}{I}\sin\Delta\varphi = R + iX \tag{19}$$

Electrical impedance is also a measurement of how electricity travels through a given material: every material has different electrical impedance determined by its molecular composition. In case of a heterogeneous material, impedance depends both on the composition and on the spatial distribution of the different components. Thus, the electrical impedance exhibited by an assigned volume of a two-phase mixture (in particular, gas-liquid) depends both on the void fraction and on the flow pattern: this provides a way to measure the void fraction once the response for different flow patterns is known, for instance by means of a preliminary calibration.

As shown by Ceccio and George (1996) in their extensive review analysis on the impedance method, there are many different possibilities to arrange a system of electrodes for void fraction measurement purposes. Regardless of the electrode arrangement, the main problem to tackle when employing the impedance technique is to obtain the proper calibration curve since the probe response to void fraction variations depends upon phase distribution within the volume where the measurement is taken. The problem is described by Laplace's equation of the electrical potential with the proper boundary conditions:

$$\nabla^2 V = 0 \tag{20}$$

Depending upon the geometry of the probe and the form of the interface between the phases, Laplace's equation can be solved either analytically or numerically. The probe response is generally nonlinear, and it depends on the electrode geometry (dimensions, shape, distance). As a consequence, one should know a priori the flow pattern established inside the system under observation in order to apply the proper calibration curve to the measurements.

According to the physical properties of the mixture components and to the frequency of the current applied to the electrodes, the electrical impedance may exhibit a resistive or a capacitive behavior: in particular, for air-water mixtures the resistive component dominates in the frequency range between 10 kHz and 100 kHz, while for higher frequencies (typically 1 MHz) the capacitive component is prevailing.

Electrical resistance. These instruments are based on the measurement of the electrical resistance R of a two-phase mixture, where at least one of the fluid is conductive. Since the resistance of conductive fluids is very small, the electrical conductance G = 1/R is often used instead. In general, electrical conductance measurement devices require simple instrumentation and provide reliable results; the main limitation of this kind of probes is that the electrodes must be in contact with a conductive fluid, so that the probe geometry must be designed taking into account the flow pattern; furthermore, the measured conductance is strongly influenced by undesired factors due to the electrode fouling or to temperature changes. The measurement of the unknown resistance is usually done by means of a Wheatstone bridge configuration (figure 10), which provides an easy way to compensate bias errors due for instance to a thermal drift of the water conductance. As shown in figure 10, if the bridge is initially balanced ($R_1R_3 = R_2R_4$) the changes of the resistances placed on the same side of the bridge with respect to the power supply have different signs in the expression of the output voltage V_{out}, so that if they both have the same bias error this will not affect the measurement. Practically, these two resistances are represented by the measuring probe and by a slave probe, which is immersed in water kept at the same temperature as the water in the flowing mixture.

In principle, conductance measurements can be performed indifferently by means of DC or AC current; since DC current applied to a conducting fluid induces electrolytic phenomena, AC current in the range 10 ÷100 kHz (where the reactive component of impedance is not significant) is used to avoid the corrosion of the electrodes. A remarkable exception is represented by local probes such as those described by Kim et al. (2000): they used gold-plated electrodes to increase the life of the sensor.

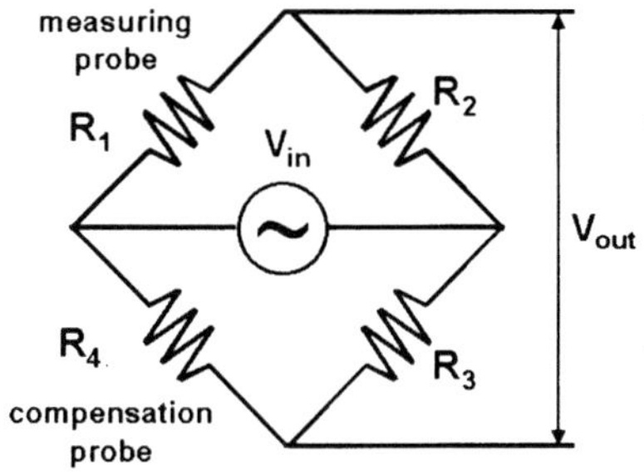

$$\frac{V_{out}}{V_{in}} = \frac{R_2R_4 - R_1R_3}{(R_1 + R_2)(R_3 + R_4)}$$

$$\frac{dV_{out}}{V_{in}} = \frac{1}{4}\left[\frac{dR_1}{R_1} - \frac{dR_2}{R_2} + \frac{dR_3}{R_3} - \frac{dR_4}{R_4}\right]$$

Figure 10. Wheatstone bridge configuration for resistance measurements with error compensation.

There are many different possibilities to arrange a system of electrodes for either local or integral void fraction measurement, as described in a comprehensive review of the technique by Hewitt (1978). A system of electrodes assembled to constitute an impedance probe has to meet several functional and practical requirements. First of all, the probe must ensure an effective confinement of the electrical field in order to minimize the edge effects and enhance the spatial resolution. Furthermore, the shape of the sensor should produce a linear response to the variations of the observed quantity. Finally the electrodes have to be easily manufactured and installed in pipes.

Local electrical conductance sensors were first proposed by Neal and Bankoff (1963). These sensors (figure 11a) are one of the most widely used measurement techniques in obtaining local two-phase flow parameters, such as the phase density function and the interfacial area concentration. When immersed in a mixture of one conducting and one non-conducting fluid, the characteristic rise/fall of conductance signals can be acquired as the two phases pass through the sensor. Depending on the application, single-sensor probes, double-sensor probes (Kataoka et al., 1986) or four-sensor probes (Kim et al., 2000) can be used. Single-sensor probes consist of two coaxial metallic electrodes, divided by a layer of insulating material; the diameter is generally smaller than 1 mm.

The studies by Brown et al. (1978), Karapantios et al. (1989), Koskie et al. (1989), Ruder and Hanratty (1990) concern the use and calibration of parallel wire probes. As shown in figure 11b, these sensors consist of two parallel wires perpendicular to the flow direction; wires are generally in copper or platinum, with a diameter of about 0.2 mm. The electrical conductance between the two wires is proportional to the wetted length: therefore, these sensors are useful to determine the liquid layer thickness in separated flows (stratified, annular), from which the cross-sectional void fraction can be inferred. Wire probes are also used to detect the passage of liquid slugs in intermittent flows, because when a slug passes through the sensor the output signal switches to a value different from that corresponding to the stratified region: in particular, slug frequency can be estimated by counting the number of slugs per unit time or by Fourier analysis of the output signal, while slug velocity can be inferred by cross-correlating the signals of two probes placed at a known distance along the pipe.

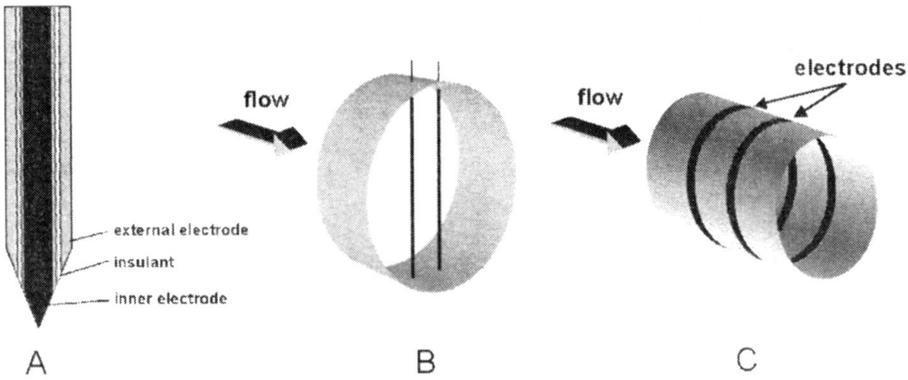

Figure 11. Electrical conductance probes; (a) local measurements (phase density function) can be performed with two coaxial electrodes divided by insulating material; (b) in wire probes, electrodes are parallel metallic wires; (c) in ring probes electrodes are flush mounted on the pipe wall.

Probes with flush-mounted electrodes provide a reliable non-intrusive method for integral (cross-sectional or volume averaged) void fraction measurements. Flush-mounted ring electrodes were employed by Asali et al. (1985), Andreussi et al. (1988), Tsochatzidis et al. (1992), Fossa (1998); a typical arrangement is represented by two metallic rings annealed in the pipe inner wall, as shown in figure 11c. The bases for integral measurement devices were given by Coney (1973), who described the theoretical behavior of flat electrodes wetted by a liquid layer. In particular, he obtained that the conductance between plane electrodes of length L covered by a liquid layer of thickness h is given by:

$$G = G_L L \frac{K(m)}{K(1-m)} \tag{21}$$

where:

$$K(m) = \int_0^{\pi/2} \left(1\text{-}m \sin^2 \alpha\right)^{-\frac{1}{2}} d\alpha$$
$$m = \frac{\sinh^2\left(\frac{\pi s}{2h}\right)}{\sinh^2\left[\frac{\pi\left(s + D_e\right)}{2h}\right]} \tag{22}$$

In Eqs. (21) and (22) G_L is the conductance of the liquid, s is the electrode width and D_e the distance between the electrodes. Therefore, from an impedance measurement one can calculate the liquid layer thickness.

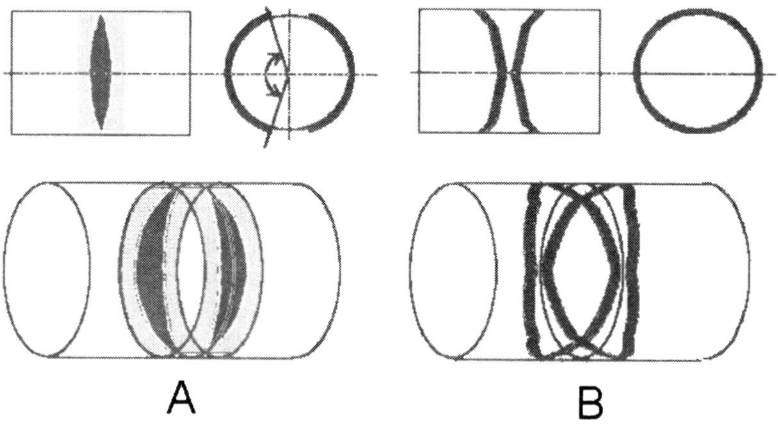

Figure 12. Optimized geometries of flush-mounted electrical conductance probes; (a) half-ring probes with variable width (black) between guard electrodes (gray); (b) ring probes with variable axial distance (courtesy of Marco Fossa).

Andreussi et al. (1988) showed that Coney's solution holds even for a cylindrical geometry, provided a proper definition of the geometrical parameters is used: in particular, the liquid layer thickness must be replaced by the ratio between the cross-sectional area of the liquid and the wetted perimeter. Ring electrodes have also been considered by Tsochatzidis et al. (1992), who theoretically solved Laplace's equation in cylindrical coordinates for the ring-probe response to a conducting annulus of thickness h.

When ring electrodes are employed to determine the void fraction in uniformly dispersed two-phase systems (e.g. bubble flow or fluidized beds), the apparent conductance of the two-phase mixture is expected to be proportional to the mean volumetric liquid fraction (1-ε). Bruggeman (1935) suggested the following expression:

$$G = G_L(1-\varepsilon)^{\frac{3}{2}} \tag{23}$$

Eq. (23) well agrees (the maximum difference is of a few percent for $\varepsilon < 0.3$) with the solution obtained by Maxwell (1882):

$$G = G_L \frac{2(1-\varepsilon)}{2+\varepsilon} \tag{24}$$

For other electrode configurations (e.g. plates), no theoretical solutions have been developed: a few data are available in the form of experimental probe responses obtained with reference to the uniformly dispersed flow pattern. This is the case of the probe employed by Ma et al. (1991) and later by Costigan and Whalley (1997) which consists of a pair of measuring half-rings located along the pipe circumference; the electrodes are placed between two pairs of guard electrodes (half-rings, also) at the same potential as the corresponding measuring electrodes.

Among the various configurations proposed in the literature, the flush-mounted ring and half-ring probes seem to have the best performances. The main difference between these families is that the electrical field is oriented mainly along the pipe axis in ring probes, while in half-ring probes it is oriented perpendicular to the axis. Starting from these reference configurations, the electrode geometry can be modified in order to achieve better performances in terms of response linearity and spatial resolution. Fossa and Devia (1999) evaluated the response of the probes under different flow conditions by solving Laplace's equation with a finite-element commercial code (Ansys 5.5). On the basis of numerical simulations, two new geometries have been developed, which are characterized by variable electrode width (figure 12a) or variable distance between the electrodes (figure 1b). Preliminary tests (Devia and Fossa, 2003) confirm that these geometries greatly improve the probe response, both in terms of linearity and in terms of dynamic response to step changes in the void fraction.

Electrical capacitance. Electrical capacitance sensors provide a fully non-intrusive way to measure the average void fraction over a given volume of a two-phase mixture: in fact, unlike resistance sensors, the electrodes need not to be in contact with the fluids. Another advantage is that, unlike conductivity, capacitance is not much sensitive to temperature drifts and does not depend on the liquid ionization. In general, the measured capacitance represents the amount and the configuration of the phases within the sensor; for a gas-liquid mixture the capacitance C can be expressed as:

$$C = K\left[\xi_G \varepsilon + \xi_L (1-\varepsilon)\right] \tag{25}$$

where ξ_G and ξ_L are the dielectric constants of the two fluids, ε the volume average void fraction and K is a factor which takes into account the sensor volume and geometry, as well as the flow pattern. Capacitance measurement is possible by integrating over the sensor volume the well-known relationship $C(\mathbf{x}) = dq/dE$, where q is the electric charge and the potential E is a solution of Laplace's equation (Eq. 20). For symmetrical electrodes surrounding a homogeneous dielectric the analytical solution of this problem exists, and can be easily extended to a multi-layer cylindrical geometry (which is the case of annular flow in vertical pipes). If the distribution of the fluids inside the sensor volume is not axisymmetric, no analytical solution exists, so that only approximate or empirical solutions are possible.

Capacitance sensors have been widely investigated, for instance by Merilo et al. (1977), Masuda et al. (1980), and Heerens (1986) among others. Unfortunately, especially for gas-liquid flow measurements capacitances are very small, generally in the range of 0.1 to 10 pF: thus, proper shielding against stray capacitances and a good signal to noise ratio are mandatory. Stray capacitance can occur between circuit wires, wires and the chassis, electronic components, and even between the sensor terminals if they are not properly designed.

A variety of electrode configurations has been designed by a number of researchers in the past few decades, ranging from flat plate, concave, helical and multiple helical wound (figure 13), where either the electrodes were in contact or isolated from the fluid (Huang et al., 1988). Among all of these configurations, best performances were obtained with the helical wound sensor (figure 13c), described for instance by Gregory and Mattar (1973), Albouelwafa and Kendall (1979) and Geraets and Borst (1988), and with the concave plate sensor (figure 13d), described among others by Elkow and Rezkallah (1996, 1997) and Lowe and Rezkallah (1999).

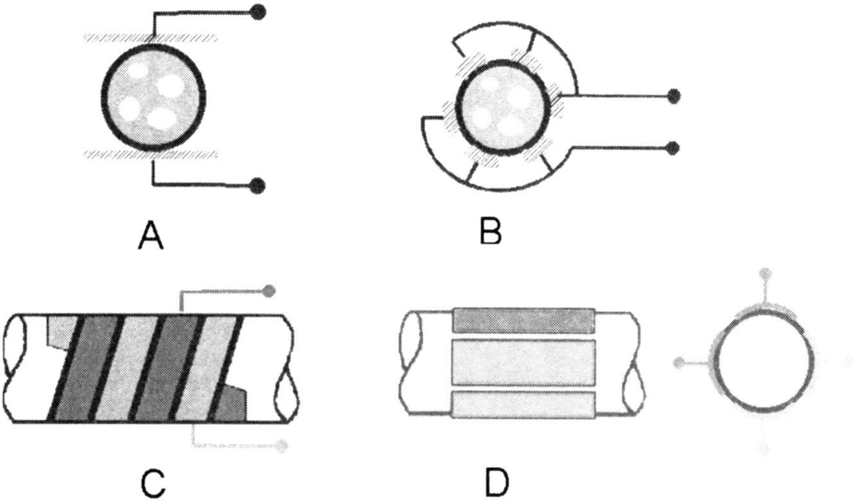

Figure 13. Configuration of electrodes for electrical capacitance probes; (a) flat plate electrodes; (b) multi-electrode probe; (c) helical wound probe; (d) concave plates probe.

For the helical wound sensor, empirical design criteria are:

$$\frac{w}{p} = 0.1 \div 0.136 \tag{26}$$

and

$$\frac{p}{D} = 1.83 \div 3.66 \tag{27}$$

where w is the width of the electrode, p is the pitch and D is the outside diameter of the pipe. Besides an undoubted complexity, which requires an extremely careful design and construction, the helical wound sensor has two intrinsic limitations: the presence of stray capacitance between the sensor electrodes, which cannot be completely eliminated by guard electrodes, and nonlinear response at the ends of the sensor, which is due to the different length of the electrodes and makes the instrument more sensitive to the flow pattern.

The concave parallel plate sensor represents a great improvement over the helically wounded sensor, provided guard electrodes are placed at the sensor ends, and appropriate shielding against stray capacitances with other electronic equipments is provided. The sensor length is a key design parameter: in fact, the measured capacitance is proportional to the sensor volume, so that stray capacitances are less important for long sensors; on the other hand, if the sensor is too long, the void fraction fluctuations cannot be detected. Typical values for the sensor length range between 2 and 10 pipe inner diameters.

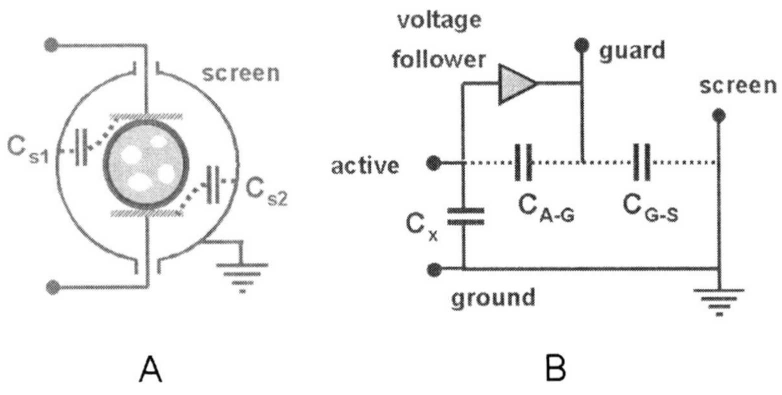

Figure 14. Stray capacitance; (a) schematic description of stray capacitances between the electrodes and the shielding; (b) active-guard circuit for stray capacitance compensation.

Possible electronic schemes for this kind of instrument are discussed in detail by Huang et al (1988). While an acceptable resolution (0.01 pF) can be easily achieved with standard electronic components, the major problem is represented by the stray capacitance between the electrodes and the screen (figure 14a), which may be of the same order of the capacitance to be measured. Moreover, if one of the electrodes is grounded (which is a common practice when accurate, high-frequency measurements are required), the measurement error is not distributed homogeneously inside the sensor's volume. This problem can be solved thanks to the so-called "active guard method", described schematically in figure 14b: in this case, the stray capacitance between the live electrode and the screen is replaced by the electrode-guard and the guard-screen capacitances; thus, if the guard is driven to the same potential as the live electrode (which can be easily done by a commercial voltage-follower), the electrode-guard capacitance behaves like an open circuit, so that the measurement is no longer affected by stray capacitances.

3 Flow Visualization

Two-phase flow visualization techniques are extremely important in two-phase flow modeling, because they allow one to quantify the distribution of the phases inside a controlled volume or over a given cross-section: thus, they provide essential information about the flow pattern. Such techniques are quite different from those used to visualize single-phase flows, which are conceived to be used in a homogeneous medium. In particular, the complex interface between the two phases causes multiple scattering of visible and near-visible radiation, on which several visualization techniques, such as laser interferometry, are based. Therefore, like for void fraction measurements, the instruments must be able to discriminate between some physical property which is different for the two phases (e.g. attenuation coefficient, electrical impedance, etc.). A common feature to these devices is that, in addition to the instantaneous or time-average distribution of the two phases in a given geometrical domain, they also give the value of the space-averaged void fraction inside the same domain, which comes from simple integration of the phase density function values.

3.1 Radiography

Radiography provides a very efficient tool for investigations in the field of non - destructive testing as well as for many applications in fundamental research. A beam penetrating a specimen is attenuated by the sample material and detected by a two dimensional imaging device. The image contains information about material and structure inside the sample because the beam is attenuated according to the basic exponential law of radiation attenuation.

X-ray radiography has a long tradition as an important tool for medical applications as well as in the field of non destructive testing. While penetrating matter, X-rays interact with the electron clouds of the atoms in the object. Therefore the attenuation of X-rays depends on the charge density of the electron cloud and increases with the atomic number of the sample material. In general, as the atomic number increases, X-ray attenuation increases for lower X-ray energies and/or higher density materials.

Neutron radiography is similar in principle to X-ray radiography, and is complementary in the nature of information supplied. The interactions of X-rays and neutrons with matter are fundamentally different, forming the basis of many unique applications using neutrons. While X-rays interact with the electron cloud surrounding the nucleus of an atom, neutrons interact with the nucleus itself. Thus, the total neutron cross-section subject to the neutron energy depends on the properties of the nuclei and varies from element to element, even from isotope to isotope. Contrary to X-rays, neutrons are attenuated by some light materials, as i.e. hydrogen, boron and lithium but penetrate many heavy materials (e.g. lead, iron, chromium, etc.). Neutrons are able to distinguish between different isotopes, and neutron radiography is an important tool for studies of radioactive materials.

Neutron radiography is a powerful technique for the investigation of two-phase flows with a growing number of applications (Hibiki et al., 1995; Mishima and Hibiki, 1996). In particular, it is useful when the operating conditions do not allow to place the instrumentation in contact with the fluids and the insertion of a transparent section is impossible; this happens, for instance, in thick walled pipes containing fluids at high pressures and temperatures. An example of application of neutron radiography to boiling flows can be found in Kureta et al. (1999).

3.2 Tomography

The word tomography denotes a number of experimental techniques that allow one to visualize a two-dimensional image of a given cross-section of the measured object (figure 15). Tomographic techniques are based on the simultaneous measurement of a physical property in several positions on the measurement plane. Such techniques, which are commonly used in the medical and material science fields, have been recently applied to the process industry.

Process tomography has been developing rapidly in recent years, and now has come to the stage of industrial applications, as it is reported in several exhaustive review papers (Beck and Williams, 1996; Chaucki et al., 1997; George et al., 1998; Mewes and Schmitz, 1999). Tomographic techniques provide novel means of visualizing the internal behavior of two phase flow in industrial processes, such as gas/liquid three-component flows in oil pipelines (gas/oil/water), gas/solid flows in pneumatic conveyors and processes of mixing or separation in plant vessels. These systems can provide more valuable information for measurement, on-line monitoring and control of industrial processes than can traditional equipment, as shown among others by Scott and Williams (1995), Xie et al. (1995) and by Thorn et al. (1999).

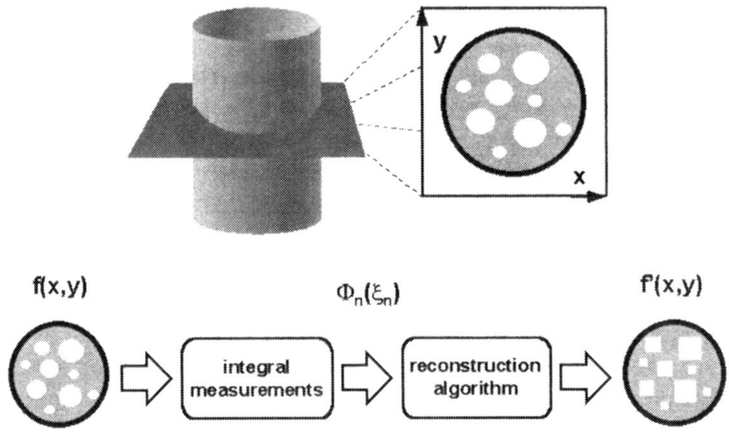

Figure 15. Principle of process tomography and schematic of the operating sequence.

The tomographic measurement is performed in two steps, which are illustrated schematically in figure 15. In the first step, a number of integral, independent measurements $\phi_n(\xi_n)$ of the property field f(x,y) are taken in the measurement plane by the tomographic sensor. After measurements have been completed, the local properties of the flow are calculated by means of the so-called reconstruction algorithm, which returns a distribution function f'(x,y) of the measured quantity; of course f'(x,y) might not be identical to f(x,y). Figure 16a depicts a typical arrangement for integral measurements by X-ray or γ-ray tomography: the measurement object, placed on the plane (x,y), is projected in each direction thanks to several beams, that are attenuated according to its composition. Linearly independent projections $\phi_n(\xi_n)$ can be obtained by rotating the measurement device or by using more beam sources. The integral measurements implicitly contain the information about the measured property f(x,y): the larger the number of measurements, the more detailed will be the reconstruction. In practice, the number of independent projections cannot be smaller than three in order to avoid ambiguities: in the example of figure 16b, the two orthogonal projections may correspond to any of the two couples of cylinders (1) and (2), placed on the diagonals of the measurement plane.

Measurement devices are based on the interaction of the multiphase flow with electromagnetic waves, particle radiations or acoustic fields, as shown in detail by Mewes and Schmitz (1999); according to the type of interaction between the sensor and the flow, they can be classified into three categories:

- transmission techniques: the flows changes some physical property of the emitted field, which can be measured by the sensor receiver; such properties include e.g. the polarization of e.m. waves or the intensity of electric and acoustic fields (see e.g. Yang, 1996);

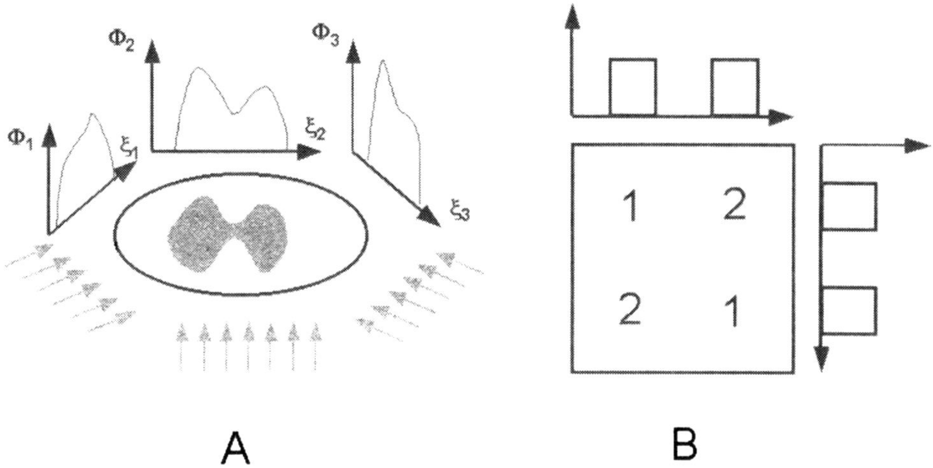

Figure 16. Integral projections in process tomography; (a) example of linearly independent integral measurements in three different directions; (b) schematic description of ambiguities arising when $n < 3$.

- emission techniques: the sensor measures the radiation emitted by the material inside the measurement cross-section, which can be either spontaneous or induced by the sensor itself; these techniques include infrared termography, magnetic resonance imaging (Gladden, 1995; Le Gall et al., 2001) and positron emission tomography (Parker et al., 1994);
- deviation techniques: the flow changes the direction of the radiation emitted by the sensor; these techniques include reflection, refraction (deviations are due to an interface), diffraction (deviations occur in a continuous medium), scattering.

Reconstruction algorithms are used to extract the information about the measurement object from the integral measurements. Their theoretical bases have been developed in the last decades, (Herman, 1980; Natterer, 1986; Kak and Slaney, 1988). The algorithms are usually divided into inversion techniques, in which measurements are mathematically inverted after some preprocessing of data, and series expansion techniques (Censor, 1983), for which the discretization of the measured function f(x,y) is required. In series expansion techniques, the sensor is described by a linear system of equations expressing the (discretized) measurements as a function of the (unknown) discrete values of f(x,y), which can be reconstructed by solving the system. If the number of measurements equals the number of unknown elements of the measured property, the direct analytical inversion of the measurement field is possible; examples are direct inversion techniques such as convolution, 2-D and 3-D Fourier transforms, etc. (Mersereau, 1976). On the other hand, in process tomography the number of projections is generally small: thus, iterative techniques are used in order to improve the quality of the final reconstruction.

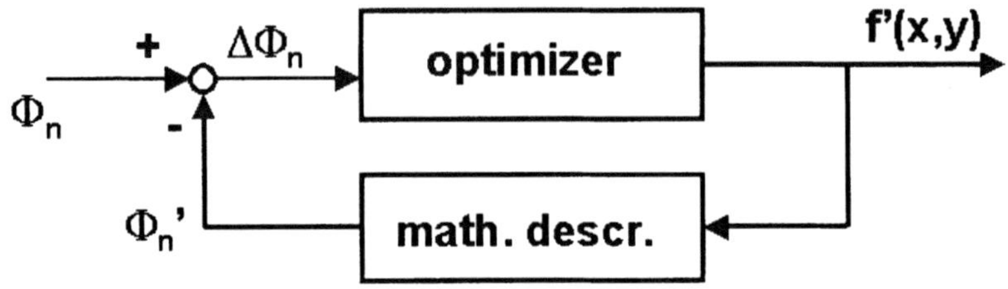

Figure 17. Feedback loop of iterative reconstruction methods.

The output property distribution f'(x,y) obtained from the integral measurements $\phi_n(\xi_n)$ is introduced into a mathematical model of the sensor, that yields a new set of measured values $\phi_n'(\xi_n)$, which are supplied back to the reconstruction algorithm until convergence is reached, as shown schematically in figure 17. Examples of these methods are given for instance by Rowland (1979) and Herman (1980).

Tomographic measurement systems are required to have a high temporal resolution, in order to detect the evolution of transient flows; moreover, a sufficiently high spatial resolution is necessary in order to recognize the flow structure. Of course, these characteristics are strongly influenced by the working principle of the sensor: in particular, classical techniques such as X-ray or nuclear magnetic resonance tomography have a high spatial resolution, suitable for biomedical applications, but a rather low time resolution (1 ÷ 500 s), so that they cannot be used in transient multiphase flows of process industry. For these applications, better solutions are electrical capacitance or acoustic tomography, which allow a high temporal resolution together with a low (but still acceptable) spatial resolution.

Electrical conductance tomography. This technique has been recently developed (Xie et al., 1995; Reinecke et al., 1998), and probably represents the most efficient, low-cost method for two-phase flow diagnostic, where one of the fluids is electrically conductive. As it is schematically depicted in figure 18a, the sensor consists of three sets of parallel wires placed in three adjacent parallel planes perpendicular to the flow direction, so that they create a mesh through which the two-phase mixture flows. The distance between two wires in the same plane is of the order of 5 mm, while the axial distance between two planes is about 1.5 times larger; stainless steel wires with a diameter of 0.1 mm were used, so that the free cross-sectional area through the sensor is greater than 95%, and consequently the additional pressure drop is not significant. This sensor, though, is intrusive, and might somewhat alter the flow structure; furthermore, in the three-array configuration measurements cannot be performed over the whole circular cross-section, but only in the exagonal area depicted in figure 18a.

The working principle of the sensor is based on the conductance measured between two parallel wires, which is a function of the electrical properties and of the distribution of the two phases. Depending on the flow pattern, the liquid bridges the gap between each couple of wires in several positions, as shown in figure 18b: the liquid bridges, which have different lengths L_i, can be described by a set of parallel ohmic resistors.

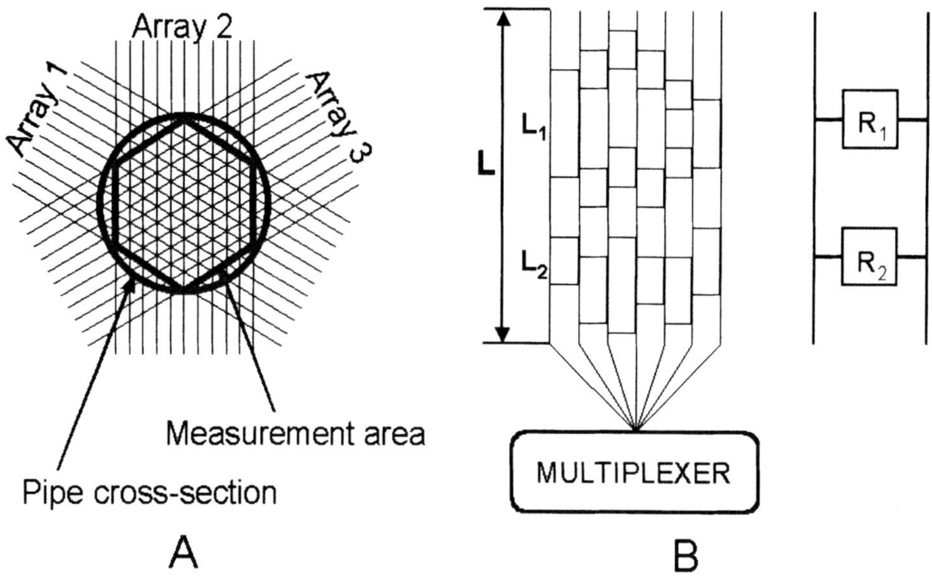

Figure 18. Electrical conductance tomography; (a) arrangement of the electrical conductors in the measurement cross-section: note that the axial distance between two arrays is of the same order of the distance between two wires; (b) schematic representation of the conductive fluid bridges of length L_i between each couple of wires, which are equivalent to a succession of parallel electrical resistances (conductances).

Neglecting the conductance of the gas, the measured conductance is:

$$\frac{1}{R}=\sum_{i=1}^{N}\frac{1}{R_i}=\frac{1}{LR_{LO}}\sum_{i=1}^{N}L_i \tag{28}$$

where R_{LO} is the electrical resistance of the liquid, and L is the total length of the wire in contact with the mixture. The ratio of the wetted length to the total length equals the one-dimensional liquid holdup ($1-\varepsilon_1$). The electrical conductance represents the integral measurement to be supplied to the reconstruction algorithm; each set of wires provides a linearly independent projection of the measurement plane. If conductance measurements are performed simultaneously, the conductance between two wires will be affected by the electric field created by the other active wires: thus, a multiplexing device that scans in rapid succession all the couples is necessary. Of course, such sequential measurements require a time shift correction strategy in order to be considered as if they were simultaneous.

The parallel wire electrical conductance sensor allows a time resolution of about 0.01 s, which means that cross-sectional images of the phase distribution can be acquired at a rate of about 100 Hz. Three dimensional representations of the flow structure are also possible: in fact, by stacking the reconstructed images, an Eulerian representation of the phase distribution can be obtained.

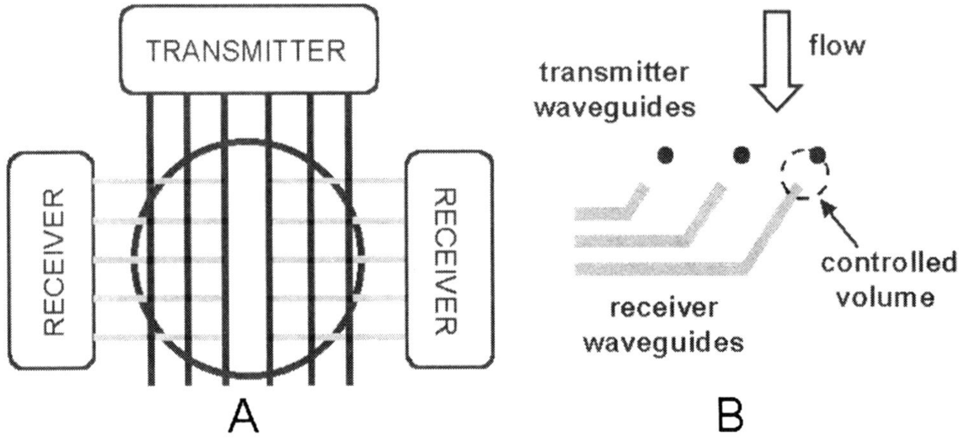

Figure 19. Ultrasound acoustic tomography; (a) configuration of the acoustic waveguides in the two-phase flow; (b) detail of the arrangement of the transmitter and the receiver waveguides in the controlled volume of fluid.

Acoustic tomography. These systems, based on the measurement of the acoustic conductivity of gas-liquid mixtures, have been recently proposed as a reliable diagnostic technique for two-phase flows at high pressures and temperatures; examples are reported by Melnikov and Nigmatulin (1994), Schlaberg et al. (1996), Melnikov and Kontelev (1999). The ultrasound acoustic waves generated by piezoelectric transducers are irradiated in the mixture by means of an acoustic waveguide (for instance, metallic rods or thin-walled tubes), and reach another waveguide connected to a receiver, which converts the acoustic fluctuations into an electrical signal. If the fluid between the transmitter and the receiver waveguides is a liquid, the acoustic waves attenuation is not significant, while if the fluid is a gas the ultrasound fades and does not reach the receiver; in fact, the acoustic energy transmitted by the fluid is proportional to the acoustic impedance $Z = \rho c$, where ρ is the fluid density and c the sound velocity: for water, $Z \approx 1.4 \times 10^6$ kg m^{-2}s^{-1} whereas for air at atmospheric pressure we have that $Z \approx 4 \times 10^2$ kg m^{-2}s^{-1}.

Figure 19 shows the scheme of the acoustic sensor proposed by Melnikov and Kontelev (1999): the irradiation of ultrasound waves is carried out by means of separate, parallel waveguides, along which the free ends of the receiver waveguides are located at a distance of 1 mm. The transmitter and the receiver waveguides create an orthogonal grid that covers the whole cross-section of a pipe, and allow a spatial resolution of about 1 cm.

Acoustic tomography has two important advantages with respect to other tomographic techniques: it can be used in extreme operating conditions (temperature up to 350°C and pressure up to 20 MPa), and there are no specific requirements about the fluid properties, whereas for electric conductance tomography at least one of the fluids must be conductive. The high resistance of this kind of sensors is due to the fact that only the waveguides are placed in contact with the fluids, while all the electronic components and conductors are kept outside, in a safer environment. The main drawbacks are the relatively low temporal and spatial resolution: the former is related to the path of acoustic pulses and to signal processing, while the latter depends on the size of the waveguides, which is of the order of 1 mm, and on the distance between waveguides, which must be optimized in order to avoid interferences.

3.3 Time-Average Flow Visualization

While the instantaneous visualization of the whole flow field gives the most detailed information about the flow structure, deducing from it the overall behavior of the flow is often difficult: therefore, sometimes a time average description of the flow structure can be much more useful. One such representation is provided for by the local void fraction distribution, which can be obtained for instance by means of optical probe measurements in several points uniformly distributed over a cross-section of the flow (Bertola, 2002a). Although this technique is time-consuming, because measurements must cover the whole cross-section with an acceptable spatial resolution, very often there exist symmetries that allow a considerable reduction of the number of measurements: for instance, in horizontal pipe flow gravity causes the flow to be symmetrical with respect to the vertical, so that the void fraction can be measured only over half cross-section. The two-dimensional polynomial interpolation of the experimental data allows to obtain a surface, which represents the time average flow structure of the gas-liquid flow.

The cross-sectional average void fraction can be calculated by numerical quadrature of the local measurements over the discretized pipe cross-section:

$$\varepsilon \approx \frac{1}{A}\sum_{i=1}^{N} \varepsilon_{Gi} A_i \tag{29}$$

where A is the cross-sectional area, A_i the discrete cell area and ε_{Gi} the local void fraction value corresponding to the i-th cell.

This technique can be useful to study the flow structure evolution near pipe fittings (such as contractions, enlargements, bends, valves, orifices, etc.). Figure 6 shows the air-water intermittent flow structure in several cross-sections along a horizontal pipe with a sharp-edged sudden area contraction: the void distribution was determined from local void fraction measurements, as described above, in seven cross-sections placed 27.5, 15, 1.1 diameters upstream and 1.2, 6, 20, 48 diameters downstream of the contraction; the linear interpolation of these distributions allowed calculating the void distribution over cross-sections placed at different axial positions.

The sudden contraction considerably affects the gas distribution in both the upstream and the downstream pipe, and its effect grows more and more as the flow approaches the singularity. The obstacle encountered by the flow causes the level of the incompressible phase in the stratified region between two slugs to raise, so that the cross-sectional area available for the gas flow gets smaller; this can be observed for a gas fraction of volume flow up to about 70%, when the gas flow rate becomes too large to allow an area reduction. In the downstream pipe, the situation is symmetrical to the upstream pipe, from a qualitative standpoint: the area available on average for the gas flow is minimum just after the fitting, and grows as the flow goes on. Indeed, a migration of the air towards the lower part of the cross-section can be noticed.

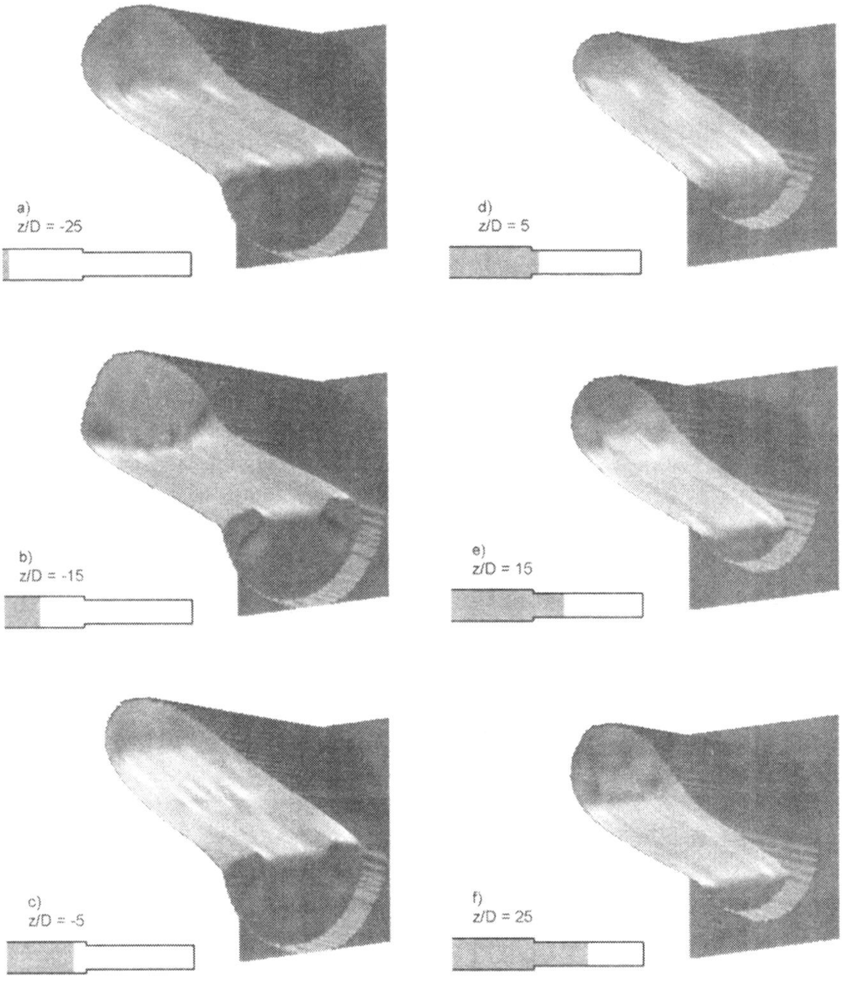

Figure 20. Axial evolution of the cross-sectional void fraction distribution for an intermittent air-water flow in a horizontal pipe with abrupt area contraction; images are reconstructed by linear interpolation of local void fraction measurements performed in seven reference cross-sections.

Far away from the contraction, the void fraction is almost constant in the upper part of the pipe, and abruptly slopes down to zero in the lower part: this is due to the fact that in this cross-section plug or slug flow patterns are well established, and no more feel the contraction influence. The same things can be observed in the cross-section 27.5 diameters upstream. The change in flow structure along the downstream duct is accompanied by a deceleration of the lower density phase; hence a void fraction increase and a velocity ratio decrease can be observed.

4 Velocity Measurements

Velocity measurements in two-phase flows are generally much more difficult than in single-phase flows: in fact, except for the trivial case of homogeneous flow, the phases have different velocities, so that an ideal instrument should be able to measure both of them. Furthermore, very often one is not interested in the velocity of one of the phases, but rather in the velocity of some peculiar structure of the flow pattern (for example, a gas bubble or a liquid slug).

The simplest approach suggests to use the same techniques developed for single-phase flow; these techniques include laser Doppler velocimetry (LDV), hot wire (or hot film) anemometry, and particle image velocimetry (PIV). Alternatively, methods specifically conceived for two-phase flows can be used, such as the well known cross-correlation technique or the various and rapidly developing digital image processing techniques. Among the techniques for single-phase flows, LDV, which has been popular since the 1970s as a non-intrusive velocimetry technique, is recognized to give the most reliable and accurate results; this method measures the frequency shift of a laser beam scattered by a moving object (Doppler effect), and is described in detail, for instance, by Durst et al. (1981), Heitor et al. (1993), Adrian (1996). Unfortunately, in two phase flows the laser beam is also scattered by the interfaces between the two fluids, so that the efficiency of LDV is greatly reduced; thus, its applications are limited to dispersed flows with a low concentration of spherical bubbles or droplets, as summarized by Chigier (1983, 1991).

4.1 Velocity in Two-Phase Flows

Let us denote by $u_k(\mathbf{x})$ the local, time-average velocity component of the phase k in the mean flow direction. One can define an average velocity over the cross-sectional area occupied by each phase:

$$U_k = \frac{1}{A_k}\int_{A_k} u_k(\mathbf{x})\,dA \tag{30}$$

If the gas (liquid) velocity is multiplied by the local void (liquid) fraction, one obtains the so-called *superficial velocities*:

$$\begin{aligned} j_G &= \varepsilon_G u_G \\ j_L &= \varepsilon_L u_L = (1-\varepsilon_G)u_L \end{aligned} \tag{31}$$

The physical meaning of the superficial velocities can be inferred from their average values on the pipe cross-section:

$$J_k = \frac{1}{A}\int_A j_k(\mathbf{x})\,dA = \frac{1}{A}\int_A \varepsilon_k(\mathbf{x})u_k(\mathbf{x})\,dA \tag{32}$$

Thus, superficial velocities indicate the volume flux of each phase; the cross-sectional average superficial velocity can also be interpreted as the mean velocity each phase would have if the other phase disappears:

$$J_k = \frac{\dot{V}_k}{A} \tag{33}$$

If one considers the local or the cross-sectional total volume flow rate, the so-called mixture velocity can be defined:

$$\begin{aligned} j &= j_G + j_L \\ J &= J_G + J_L \end{aligned} \tag{34}$$

Sometimes, a coordinate system moving at the mixture velocity can be useful (Zuber and Findlay, 1965); in this reference frame, the two phases will have a drift velocity:

$$\begin{aligned} u_{Dk} &= u_k - j \\ U_{Dk} &= U_k - J \end{aligned} \tag{35}$$

Finally, the behavior of two-phase mixtures is often characterized by the velocity ratio, or slip ratio, which is defined as:

$$S = \frac{U_G}{U_L} = \frac{\dfrac{\dot{V}_G}{A_G}}{\dfrac{\dot{V}_L}{A_L}} = \frac{\dot{V}_G}{\dot{V}_L}\frac{1-\varepsilon}{\varepsilon} \tag{36}$$

Introducing the gas fractions of mass flow, x, and of volume flow, ε^*, the slip ratio can be written as:

$$S = \frac{\rho_L \dot{M}_G}{\rho_G \dot{M}_L}\frac{1-\varepsilon}{\varepsilon} = \frac{\rho_L}{\rho_G}\frac{x}{1-x}\frac{1-\varepsilon}{\varepsilon} = \frac{\varepsilon^*}{1-\varepsilon^*}\frac{1-\varepsilon}{\varepsilon} \tag{37}$$

4.2 Cross-Correlation

The cross-correlation R_{12} is a measure of the similitude between two signals S_1 and S_2. Assuming that S_1 and S_2 can be observed simultaneously as a function of time, we can multiply the values the two functions have at each instant, and then add together all the results over a given time interval: thus, if S_1 and S_2 are identical the products are always positive and their sum is large, while if they are different some of the products are positive and some negative, so that the absolute value of the result is smaller. If S_1 and S_2 are identical, but S_2 is shifted of an unknown quantity τ^*, that is, $S_1 = S_2(t + \tau^*)$, we can calculate the sum of the products $S_2(t)S_2(t + \tau)$ for any τ, and of course the sum will be maximum for $\tau = \tau^*$. From the mathematical standpoint, the definition of cross-correlation (normalized with respect to time) is:

$$R_{12}(\tau) = \lim_{T \to +\infty} \frac{1}{T} \int_{-T/2}^{+T/2} S_1(t) S_2(t+\tau)\, dt \tag{38}$$

If the same signal is contained both in S_1 and in S_2, it is strengthened by the cross-correlation, while any other uncorrelated component is damped; for instance, figure 21 shows the cross-correlation of two signals containing the same square pulse at different times: R_{12} shows a maximum when τ equals the time lag between the pulses, and rapidly drops to zero for other values: if the pulses correspond to the passage of an object through two different positions at a known distance, then the object velocity can be calculated.

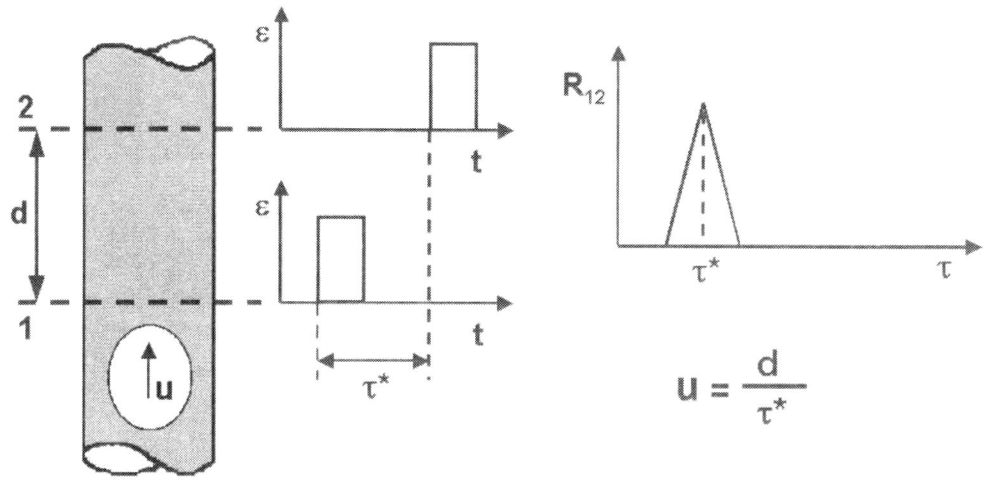

Figure 21. Principle of the cross-correlation technique for velocity measurements.

The application of the cross-correlation technique to two-phase flows is quite straightforward (see e.g. Galaup and Delhaye, 1976), because the passage of the flow structures (e.g. bubbles or drops in dispersed flow) can be detected by two sensors placed along the direction of the desired velocity component. The sensors detect some signal in empathy with the flow structures, and can be based on various techniques, such as ultrasound, optical beams, electrical conductivity or electrical capacitance.

In an ideal case, if a signal detected by the upstream sensor reappears after a certain period τ^* at the downstream sensor and the distance d between the two sensors is known, then the velocity u can be calculated from $u = d/\tau^*$. In practice, the signals will gradually change while moving downstream; however, if the position of the downstream sensor is reasonably close to that of the upstream sensor, the patterns or signals would be sufficiently similar to be recognized by the correlator, allowing the transit time τ to be measured.

Usually, void fraction signals are to be preferred because they are directly related to the flow structure: in particular, either space-averaged void fractions or the local phase density function can be used. However, it must be kept in mind that the cross-correlation technique provides only the most probable propagation velocity of the signal itself, which may have different physical meanings (or no physical meaning at all) depending on a very large number of parameters.

In general, the signal reflects the passage of gas/liquid interfaces: although the interfaces represent the boundary of some flow structure (e.g. gas bubbles), their velocity equals that of the considered flow structure only if the relative (drift) velocity is zero, that is, if the flow structure appears as "frozen" to a fixed observer. In several circumstances, measurement errors are due to deformations of the gas/liquid interfaces and/or deviations of the trajectories; while deformations affect the measurements performed both by integral and by local sensors, trajectory deviations have worse consequences on local instruments.

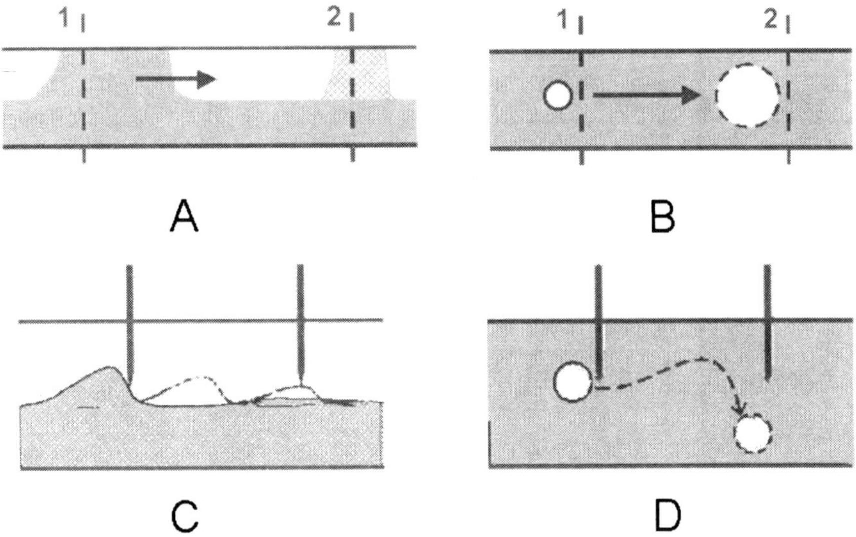

Figure 22. Examples of possible errors in cross-correlation velocity measurements; (a) deformation of developing slugs; (b) expanding or collapsing gas bubbles; (c) flattening of the crest of waves on the liquid surface; (d) trajectory deviation due to turbulent fluctuations of the bubble velocity.

If the flow structure deforms while passing from the upstream to the downstream sensor, the measured velocity takes into account both the true velocity and the velocity of deformation; on the other hand, if two local sensors (e.g. optical probes) are used, trajectory deviations might cause the structures detected by the upstream sensor to pass far away from the downstream one, and vice-versa. Figure 22 shows four examples among the possible error sources: deformations can be observed in developing slugs (figure 22a), where the interfaces at the two ends have different velocities, or in expanding/collapsing gas bubbles (figure 22b); deviations can be observed for instance when the crest of waves in stratified flow flatten (figure 22c), or when a bubble or a drop exhibits turbulent velocity fluctuations.

The cross-correlation technique efficiency is also affected by the number of gas-liquid interfaces detected, which must be large enough to be statistically significant: in particular, the number of interfaces is small when the void fraction is close to zero or to one, that is, when one phase is prevailing on the other. Therefore, one can use the local void fraction as a parameter to preliminary check the quality of velocity measurements (Bertola, 2002b).

In order to reduce errors, one must carefully select the rate at which the signals are sampled by the acquisition system and the distance between the sensors. The sampling rate s should be high enough to allow the detection of the structures of interest: if L is a characteristic dimension of these structures (e.g. the mean diameter of bubbles), Nyquist's sampling theorem yields $s > 2u/L$. The sensors, as stated above, should be placed at a small distance D from each other to avoid flow structure changes in between; on the other hand, the distance cannot be too small, because it is inversely proportional to the velocity measurement uncertainty:

$$\left|\frac{\Delta u}{u}\right| = \left|\frac{u\Delta t}{D}\right| \tag{39}$$

where Δt = 1/s. A thumb rule suggests that the optimal distance between the sensors is 3÷5 times smaller than the characteristic dimension of the flow structures L.

4.3 Hot wire anemometry

The hot wire anemometer, allows one to estimate the velocity U of a fluid from the heat transfer between a hot wire and the fluid itself. Wires are usually made of platinum or tungsten with diameters of the order of 5 μm, and active lengths of the order of 1 mm. The surrounding fluid can be considered as a continuum even for very thin wires, since the ratio between the free path of the fluid molecules and the wire diameter, also known as the Knudsen number, is always very small (Kn < 0.01 for d > 5 μm). The wire temperature is generally not greater than 200°C, so that heat transfer is essentially due to forced convection and heat conduction through the wire supports, while free convection and radiation are not significant.

According to Newton's law, the heat flux between the wire and the surrounding fluid is given by:

$$\dot{Q} = h\pi\, d\, L(T_w - T_f) \tag{40}$$

where d is the wire diameter, L the wire length, T_w and T_f the wire and fluid temperatures, respectively. The quantity h [$Wm^{-2}K^{-1}$] is the so-called heat transfer coefficient, which is a function of the fluid properties (the thermal conductivity k, the specific heat at constant pressure c_p, the density ρ and the viscosity η) and of the relative velocity between fluid and wire; this functional relationship is usually expressed in dimensionless form as:

$$\mathrm{Nu} = C\,\mathrm{Re}^{\alpha}\,\mathrm{Pr}^{\beta} \tag{41}$$

where Nu = hd/k, Re = $\rho Ud/\eta$, Pr = $c_p\eta/k$, and C, α, β are determined empirically or semi-empirically. In steady-state conditions, the heat flux equals the electric power dissipated in the wire:

$$R_w I^2 = h\pi\, d\, L\,(T_w - T_f) \tag{42}$$

where I is the electric current and R_w is the wire electrical resistance, which is a function of the wire temperature, and can be expressed with reference to the resistance at the fluid temperature:

$$R_w = R_f\left[1 + a(T_w - T_f) + b(T_w - T_f)^2 + \ldots\right] \tag{43}$$

The combination of Eqs. (41,42,43) allows in principle to find a relationship between the wire temperature and the fluid velocity; unfortunately, the exact estimation of all the coefficients is extremely difficult, and leads to inaccurate velocity measurements. Thus, it is preferred to assume that over a given range of velocities the following expression holds:

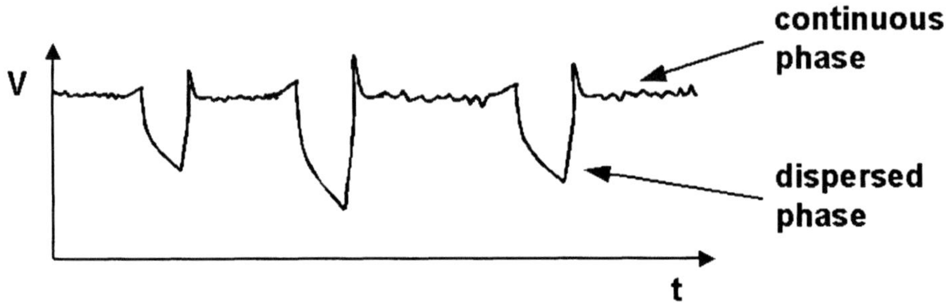

Figure 23. Typical voltage output signal of a hot-wire probe in a two-phase dispersed flow; large, U-shaped fluctuations are related to the passage of the dispersed phase (either bubbles or drops), whereas smaller fluctuations are related to turbulence in the continuous phase.

$$\frac{R_w I^2}{R_w - R_f} = A + BU^n \tag{44}$$

where the coefficients A, B and n are determined from calibration with a flow of known velocity. Since these coefficients are extremely sensitive to small changes of the wire conditions (e.g. due to wire fouling), the calibration must be extremely careful, and should be frequently verified. Eq. (44) suggests that the hot wire anemometer can be used in two ways: the constant-current anemometer, and the constant-temperature anemometer.

In the constant-current mode one measures the change of the wire electrical resistance thanks to a Wheatstone bridge circuit, whereas in the constant-temperature mode a feedback loop circuit allows to regulate the current in order to keep the wire temperature (and, consequently, the wire resistance) steady. The main difference between the two operating modes lies in the dynamic response, which is much faster for the constant-temperature anemometer. This can be intuitively understood because the response of the constant current anemometer essentially depends on its thermal inertia (the wire must change its temperature before a change in the resistance can be detected), while for the constant temperature anemometer the response depends only on the inertia of the electric circuitry, which is much smaller.

Attempts to use the hot wire anemometer in two-phase flows are dated back to 1960s, as summarized in the review article by Jones and Delhaye (1976); these works show that the hot wire anemometer can be used in two-component two-phase flow, whereas in one-component two-phase flow with phase change the results are not consistent with the calibrated liquid velocity measurements. Hot wire anemometry is now an established method for the measurements of volume fraction and statistical velocity parameters of the continuous phase in two-phase dispersed flows and it can, in principle, be applied to the flow of either a liquid/gas mixture or two immiscible liquids (see, for example, Serizawa et al., 1975, Farrar and Bruun, 1989, Liu and Bankoff, 1993a, Al-Deen et al., 1996).

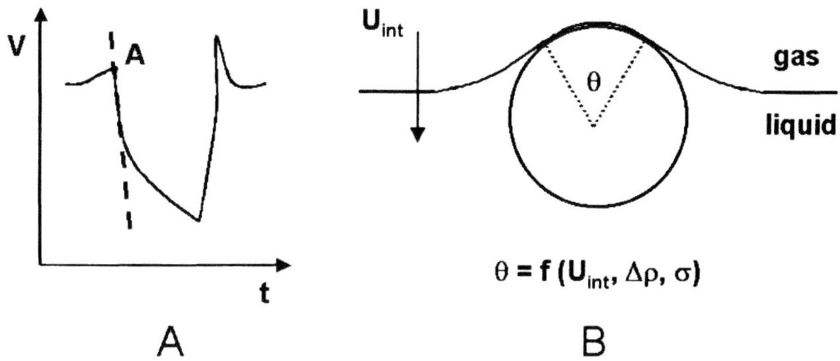

Figure 24. Single-probe measurement of the velocity of the dispersed phase; (a) output analog signal, where the dashed line represents the slope when the gas-liquid interface reaches the probe; (b) deformation of the gas-liquid interface as a bubble touches the hot wire.

When used for two-phase flow velocity measurements, of course, the response of the instrument is different for the two phases. The anemometer output V(t) from a cylindrical hot-film probe placed in a gas-liquid dispersed flow will be of the form reported in figure 23, as shown by Farrar and Bruun (1989) and Farrar et al (1995). The signal displays two distinct features: a fluctuating signal caused by the mean velocity and turbulence in the continuous phase interspersed with a U-shaped transient signal due to the passage of bubbles over the probe. Provided that the individual bubble influences are correctly eliminated from the whole signal, then the remaining signal only contains information about the velocity fluctuations in the continuous phase and it is possible to analyze this part of the signal using statistical techniques to obtain information related to the corresponding mean velocity and turbulent quantities. In addition to this, to a limited extent the bubble signals can currently be analyzed to provide statistical information about the bubbles.

Based on existing techniques it is possible to evaluate the volume fraction and the bubble frequency. For instance, Delhaye (1969) proposed to distinguish the liquid and the gas signals in air-water flow from the probability density function of the output signal amplitude, while Resch and Leutheusser (1972) identified phase changes in the analog output of the instrument when the amplitude between two successive extremes was found to exceed a threshold value.

Hamad and Bruun (2000) proposed a signal processing technique that extends the applicability of a single hot-film probe to measure the bubble/drop velocity in addition to the continuous phase velocity. In particular, they studied the impact of a gas bubble (or liquid drop) against the anemometer wire, and observed that due to surface tension, the gas-liquid interface progressively wets the wire before the instrument can be immersed in the dispersed phase. As the interface wraps the wire, the output signal drops (figure 24a); the wrapping angle θ depends both on the physical properties of the two fluids and on the interface velocity U_{int}, as shown in figure 24b. For small θ, the interface velocity is shown to be proportional to dθ/dt; on the other hand, the slope of the output signal is also proportional to dθ/dt, so that:

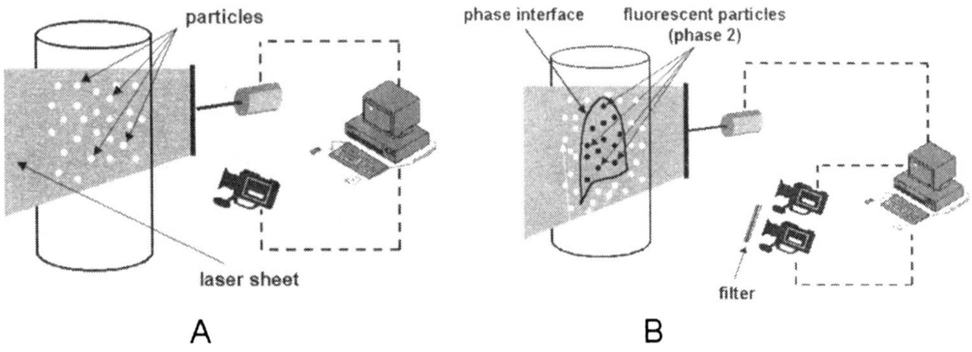

Figure 25. Particle image velocimetry (PIV) setup: a laser sheet illuminates the measurement plane, on which small tracer particles embedded in the flow are visualized; (a) for single-phase flow, particle images are recorded on film or CCD camera at right angles to the laser sheet; (b) in two-phase flows, different seeding is different for the two phases: one kind of seeds simply scatters the light from the laser source, whereas the other one emits light at a different wavelength, which can be detected by a second camera equipped with a suitable filter.

$$U_{\text{int}} = K\left(\frac{dV}{dt}\right)_A \tag{45}$$

where K can be determined from a calibration. However, due to difficulties in measuring the slope accurately, the uncertainty in this method was estimated to be about ± 10%, which is larger than for other methods, such as cross-correlation techniques.

4.4 Particle Image Velocimetry (PIV)

Particle image velocimetry (PIV) is a well-known technique that is used for performing whole field velocity measurements. In the PIV technique, the velocity is measured by recording the displacement of microscopically small neutral particles, from which the particles velocity can be inferred.

Figure 25a schematically shows the PIV setup used for velocity field measurements in single-phase flows: the measurement plane is visualized by a laser sheet that illuminates the flow in the test section. The flow is seeded with tracer particles, which exactly follow the fluid velocity: in particular, the tracer particles are considered as ideal when they (1) exactly follow the motion of the fluid, (2) do not alter the flow or the fluid properties and (3) do not interact with each other. This kind of seeding is quite different from that used for flow visualization purposes. For flow visualization, the aim is to make certain flow structures or flow regions visible. This can be accomplished by introducing the seeding at a particular location (i.e. inhomogeneous seeding). On the other hand, a single-exposure PIV recording of an incompressible flow with (ideal) homogeneous seeding appears featureless: any flow structure only becomes visible when the velocity field is evaluated.

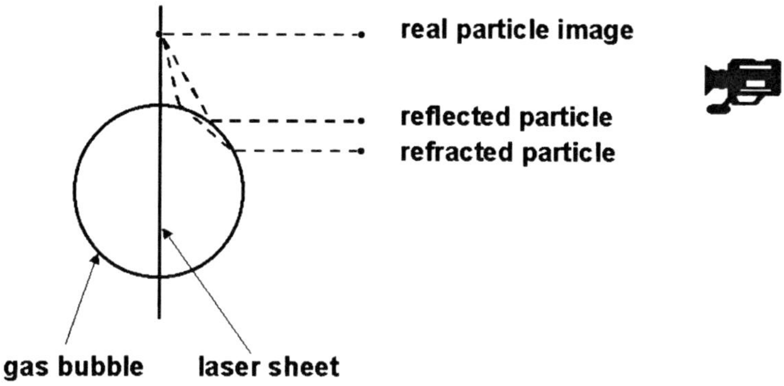

Figure 26. Particle image velocimetry (PIV) in two-phase flows: generation of virtual particle images in the vicinity of gas-liquid interfaces.

Two or more short laser light pulses with a known time separation illuminate the tracer particles: thus, particles appear with spacing proportional to the local velocity vector. The image is captured on film at right angles to the laser sheet, and the velocity field is extracted after image post-processing of highly resolved PIV photographs (Keane and Adrian 1990; Adrian, 1991). However, nowadays PIV has developed towards the use of electronic cameras for direct recording of the particle images (see e.g. Willert and Gharib, 1991).

Different tracking methods may be used to process the data. These include techniques such as auto-correlation, cross-correlation and particle tracking velocimetry (Hassan et al., 1992). In the autocorrelation technique, post-processing is performed on a single, double exposed image of the flow; in other words, the same image contains both the particles illuminated by the first laser pulse and those illuminated by the second one.

The image is divided into small subregions, each containing a number of the seeded particles (at least 9-10); then, a spatial autocorrelation function (Bendat and Piersol, 1971) of the region with itself is calculated:

$$R_{11}(\delta x, \delta y) = \lim_{\substack{x \to +\infty \\ y \to +\infty}} \int_{-\infty}^{+\infty} dx \int_{-\infty}^{+\infty} F(x,y) \mathrm{F}(x+\delta x, y+\delta y) dy \tag{46}$$

where F(x,y) is a scalar property of the image (e.g. luminance). R_{11} has a peak in (0,0), because there is a perfect overlapping of the image with itself, but also a (smaller) peak for $(\delta x^*, \delta y^*)$, where overlapping between the particles illuminated by the two different pulses occurs: thus, δx^* and δy^* represent the components of the most probable displacement of the particles in the interrogation window, from which the local velocity vector can be calculated if the time separation between the laser pulses is known. By iterating this procedure over all the subregions, the whole velocity field is retrieved. In a more recent and popular development, the two images are captured on separate frames, and a cross-correlation between corresponding subregions is performed.

Autocorrelation and cross-correlation based PIV algorithms have a limited spatial resolution, which is mainly determined by the size of the interrogation window. This drawback represents an important issue especially after the use of digital images has become widespread: as the resolution and image format of electronic cameras is several orders of magnitude lower than that of a photographic medium, digitization cannot be ignored (Westerweel 1997). In any situation, the best spatial resolution can be achieved using particle tracking velocimetry (PTV), which consists of the determination of the displacements of individual tracers (Stitou and Riethmuller, 2001). Recently, new algorithms based on pattern recognition are becoming popular; among them neural networks (Hassan and Philip, 1997; Cenedese et al., 1992), genetic algorithms and fuzzy logic techniques seem to have the greatest potential for particle tracking.

The PIV technique can easily be used for multiphase investigations if the different phases can be distinguished. Several methods are reported to separate the images resulting from different phases. The use of fluorescent tracers together with a double-camera image acquisition setup is one way of solving this problem (Philip et al. 1994; Hilgers et al. 1995). With the aid of optical filters, one may separate the radiation scattered by fluorescent and neutral particles, as shown schematically in figure 25b: one camera captures both phases, and the other camera with the filter only captures the seed particle images. Then, by subtracting the images from both cameras, the two phases can be distinguished.

The use of fluorescent particles allows to suppress the intense reflections near the wall and interface regions and to determine velocity vectors very close to the interface and to the wall. On the other hand, when cross illuminated by a laser sheet, gas/liquid interfaces originate important spurious optical effects that must be properly taken into account in order to determine accurately the surface coordinates from the PIV images (Dias and Riethmuller, 1998; Nogueira et al., 2001). In particular, these optical effects generate on the PIV image a number of virtual particles, which have no real counterpart in a corresponding position of the real flow.

As shown in figure 26, the light scattered by particles illuminated by the laser sheet can follow three different paths before reaching the camera:

- the first path is the direct path that forms the real particle images;
- the second path is due to the a total reflection of light at the gas/liquid interface: it also reaches the camera and produces spurious particle images seen inside the other phase;
- the third path is due to a double refraction of light beams: if the gas/liquid interface is curved, the refracted light may reach the interface in a different position, where it if refracted again back into the first medium; these virtual particles are seen inside the second phase, too.

The result of these optical effects is that it appears essential to use a second method for determining the gas/liquid interface, since the PIV image itself cannot be trusted.

References

Abuaf, N., Jones, O.C., Zimmer, G.A. (1978), Optical probe for local void fraction and interface velocity measurements, *Rev. Sci. Instrum. 49*: 1090-1094.

Adrian, R.J. (1991), Particle-Imaging Techniques for Experimental Fluid Mechanics, *Annu. Rev. Fluid Mech. 22*: 261-304.

Adrian, R.J. (1996), Laser Velocimetry. In: R.J. Goldstein (Ed.), *Fluid Mechanic Measurements*, New York: Taylor&Francis.

Albouelwafa, M.S.A., Kendall, E.J.M. (1979), Analysis and design of helical capacitance sensors for volume fraction determination, *Rev. Sci. Instrum. 50*: 872-878.

Al-Deen, M.F.N., Samways A.L., Bruun, H.H. (1996), Water flow studies using split-film anemometry *Meas. Sci. Technol. 7*: 1529-1535.

Andreussi, P., Di Donfrancesco, A., and Messia, M. (1988), An Impedance Method for the Measurement of Liquid Hold-Up in Two Phase Flow, *Int. J. Multiphase Flow, 14*: 777-785.

Asali, J.C., Hanratty, T.J. and Andreussi, P. (1985), Interfacial Drag and Film Height for Vertical Annular Flow, *AIChE J. 31*: 895-902.

Beck M.S., Williams R.A. (1996), Process tomography; a European innovation and its applications, *Meas. Sci. Technol. 7*: 215-224.

Bendat J.S., Piersol A.G. (1971), *Random Data: Analysis and Measurement Procedures*, New York: Wiley.

Bertola, V. (2002a), Optical Probe Visualization of Air-Water Flow Structure through a Sudden Area Contraction, *Exp. Fluids 32(4)*: 481-486.

Bertola, V. (2002b), Slug Velocity Profiles in Horizontal Gas-Liquid Flow, *Exp. Fluids 32(6)*: 722-727.

Bonn, D., Ross, D. (2001), Wetting transitions, *Rep. Prog. Phys. 64*: 1085-1163.

Brown, R.C., Andreussi, P. and Zanelli, S. (1978), The Use of Wire Probes for the Measurement of Liquid Thickness in Annular Gas-Liquid Flows, *Can. J. Chem. Eng. 56*: 754-757.

Bruggeman, D.A.G. (1935), Calculation of Different Physical Constants of Heterogeneous Substances, *Ann. Phys. 24*: 636-679.

Bruun, H.H. (1995), *Hot-wire Anemometry*, Oxford: Oxford University Press.

Bruun H H, Samways A L and Ali J (1995), A hot-film study of oil/water flow in vertical pipes. Proc. 2nd Int. Conf. on Multiphase Flow (Kyoto, Japan, 3–7 April) paper P1, pp 61–67.

Cartellier, A. (1990), Optical probes for local void fraction measurements: characterization of performance, *Rev. Sci. Instrum. 61*: 874–886

Cartellier, A. (1992) Simultaneous void fraction measurement, bubble velocity and size estimate using a single optical probe in gas–liquid two-phase flows, *Rev. Sci. Instrum. 63*: 5442-5453

Cartellier, A., Achard, J.L. (1991), Local phase detection probes in fluid/fluid two-phase flows, *Rev. Sci. Instrum. 62*: 279-303.

Cartellier, A., Barrau, E. (1998a), Monofiber optical probes for gas detection and gas velocity measurements: Conical probes, *Int. J. Multiphase Flow 24(8)*: 1265-1294.

Cartellier, A., Barrau, E. (1998b), Monofiber optical probes for gas detection and gas velocity measurements: Optimised sensing tips, *Int. J. Multiphase Flow 24(8)*: 1295-1315.

Ceccio, S.L., George, D.L. (1996), A Review of Electrical Impedance Techniques for the Measurement of Multiphase Flow, *ASME J. Fluid Eng. 118*: 391-399.

Censor, Y. (1983), Finite series-expansion reconstruction methods, *Proc. IEEE 71(3)*: 409-419.

Chanson, H. (2002), Air-Water Flow Measurements with Intrusive, Phase-Detection Probes: Can We Improve Their Interpretation? *J. Hydr. Eng. 128*: 252-255.

Chaucki, J. Laracki, F., Dudukovic, M.P. (1997), Noninvasive tomographic and velocimetric monitoring of multiphase flows, *Ind. Eng. Chem. Res. 36*: 4476-4503.

Chigier, N. (1983), Drop size and velocity instrumentation, *Prog. Energy Combustion Sci. 9*: 155-177.

Chigier, N. (1991), Optical imaging of sprays, *Prog. Energy Combustion Sci. 17*: 211-262.

Coney, M.W.E. (1973), The Theory and Application of Conductance Probes for the Measurement of Liquid Film Thickness in Two Phase Flow, *J. Phys. E: Scient. Instrum. 6*: 903-910.

Costigan, G., Whalley, P.B. (1997), Slug Flow Regime Identification from Dynamic Void Fraction Measurement in Vertical Air-Water Flows, *Int. J. Multiphase Flow 23*: 263-282.

Cartellier, A., Barrau, E. (1998a), Monofiber optical probes for gas detection and gas velocity measurements: Conical probes, *Int. J. Multiphase Flow 24(8)*: 1265-1294.

Cartellier, A., Barrau, E. (1998b), Monofiber optical probes for gas detection and gas velocity measurements: Optimised sensing tips, *Int. J. Multiphase Flow 24(8)*: 1295-1315.

Cenedese, A., Romano, G.P., Paglialunga, A., Terlizzi, M. (1992), Neural net for trajectories recognition in a flow, *Sixth Int. Symp. on Applications of Laser Techniques to Fluid Mechanics* (Lisbon) pp 27.1.1-27.1.6.

Delhaye, J.M. (1969), Hot-film anemometry, in: Le Tourneau, B.W., Bergles, A.E. (Eds.), Two-Phase Flow Instrumentation, New York: ASME International.

Devia, F., Fossa, M. (2003), Design and optimisation of impedance probes for void fraction measurements, *Flow Measurement and Instrumentation 14(4-5)*: 139-149.

Dias, I., Riethmuller, M.L. (1998), PIV in two-phase flows: Simultaneous bubble sizing and liquid velocity measurements, *Proc. 9th Symposium on Laser techniques to fluid mechanics*, New York: Springer-Verlag.

Durst, F., Melling, A., Whitelaw, J.H. (1981), *Principles and practice of Laser Doppler Anemometry* (2nd ed.), New York: Academic Press.

Elkow, K.J., Rezkallah, K.S. (1996), Void fraction measurements in gas-liquid flows using capacitance sensors, *Meas. Sci. Technol. 7*: 1153-1163.

Elkow, K.J., Rezkallah, K.S. (1997), Void fraction measurements in gas-liquid flows under 1-g and μ-g conditions using capacitance sensors, *Int. J. Multiphase Flow 23*: 815-829.

Farrar, B., Bruun, H.H. (1989), Interaction effects between a cylindrical hot-film anemometer probe and bubbles in air/water and oil/water flow, *J. Phys. E: Sci. Instrum. 22*: 114-123.

Farrar B., Samways A.L., Ali J., Bruun, H.H. (1995), A computer based hot-film technique for two-phase flow measurements, *Meas. Sci. Technol. 6*: 1528-1537.

Fordham, E.J., Holmes, A., Ramos, R.T., Simonian, S., Huang, S.-M., Lenn, C.P. (1999a), Multi-phase-fluid discrimination with local fibre-optical probes: I. Liquid/liquid flows, *Meas. Sci. Technol. 10*: 1329-1337.

Fordham, E.J., Simonian, S., Ramos, R.T., Holmes, A., Huang, S.-M., Lenn, C.P. (1999b), Multi-phase-fluid discrimination with local fibre-optical probes: II. Gas/liquid flows, *Meas. Sci. Technol. 10*: 1338-1346.

Fossa, M. (1998), Design and Performance of a Conductance Probe for Measuring the Liquid Fraction in Two-Phase Gas-Liquid Flows, *J. Flow Meas. Instrum. 9*: 103-109.

Fossa, M., Devia, F. (1999), Theoretical Performance of Impedance Probes for Void Fraction Measurements, *Proc. XX International Congress of Refrigeration*, Sydney, Australia.

Galaup, J.P., Delhaye, J.M. (1976), Utilisation des sondes optiques miniatures en écoulement diphasique gaz-liquide, application à la mesure du taux de présence local et de vitesse local de la phase gazeuse, *La Houille Blanche 1*: 17-30.

George, D.L., Ceccio, S.L., O'Hern, T.J., Skollenberg, K.A., Torczynski, J.R. (1998), Advanced material distribution measurement in multiphase flows: a case study, *Proc. ASME-FED 247*: 31-42.

Geraets, J.J.M., Borst, J.C. (1988), A capacitance sensor for two-phase void fraction measurement and flow pattern identification, *Int. J. Multiphase Flow 14*: 305-320.

Gladden, L.F. (1995), Industrial applications of NMR imaging, *Eng. J. 56*: 149-158.

Gregory, G.A., Mattar, L. (1973), An in-situ volume fraction sensor for two-phase flow of non-electrolytes, *J. Canad. Petrol. 12-13*: 48-52.

Guenther, R. (1990), *Modern Optics*, New York: Wiley.

Hamad, F.A., Imberton, F., Bruun, H.H. (1997), An optical probe for measurements in liquid–liquid two-phase flow, *Meas. Sci. Technol. 8*: 1122–1132.

Hamad, F.A., Bruun, H.H. (2000), Evaluation of bubble/drop velocity and slip velocity by a single normal hot-film probe placed in a two-phase flow, *Meas. Sci. Technol. 11*: 11–19.

Hassan, Y.A., Blanchat, T.K., Seeley, C.H. Jr. (1992), PIV flow visualization using particle tracking techniques, *Meas. Sci. Technol. 3*: 633–642.

Hassan, Y.A., Philip, O.G. (1997), A new artificial neural network tracking technique for particle image velocimetry, *Exp. Fluids 23*: 145-154.

Heerens W.C. (1986), Application of capacitance techniques in sensor design, *J. Phys. E: Sci. Instrum 19*: 897-906.

Heitor M.V., Starner, S.H., Taylor, A.M.K.P., Whitelaw, J.H. (1993), Velocity, Size, and Turbulent Flux measurements by Laser Doppler Velocimetry. In: A.M.K.P. Taylor (Ed.), *Instrumentation for flows with combustion*, New York: Academic Press.

Herman, G.T. (1980), *Image reconstruction from projections - the fundamentals of computerized tomography*, New York: Academic Press.

Hewitt, G.F. (1978), *Measurements of Two-Phase Flow Parameters*, New York: Academic Press.

Hibiki, T., Mishima, K., Matsubayashi, M. (1995), Application of High-Frame-Rate Neutron Radiography with a Steady Thermal Neutron Beam to Two-Phase Flow Measurements in a Metallic Rectangular Duct, *Nuclear Technology 110*: 422-435.

Hilgers, S., Merzkirch, W., Wagner, T. (1995), PIV measurements in multiphase flow using CCD- and photo-camera flow visualization and image processing of multiphase flow systems, *Proc. 1995 ASME/JSME Fluids Engineering and Laser Anenometry Conf. Exh. FED 209*: 151-154.

Hinata, S. (1972), A study on the measurement of the local void fraction by the optical fibre glass probe, *Bull. JSME 15*(88): 1228-1235.

Huang, S.M., Stott, A.L., Green, R.G., Beck, M.S. (1988), Electronic transducers for industrial measurement of low value capacitances, *J. Phys. E. Sci. Instrum. 21*: 242-250.

Jones, O.C., Delhaye, J.M. (1976), Transient and Statistical Measurement Techniques for Two-Phase Flows: a Critical Review. *Int. J. Multiphase Flow 3*: 89-116.

Kak, A.C., Slaney, M. (1988), *Principles of computerized tomography imaging*, New York: IEEE Press.

Kang, H.C. Kim, M.H. (1992), The Development of a Flush Wire Probes and Calibration Method for Measuring Liquid Film Thickness, *Int. J. Multiphase Flow 18*: 423-437.

Karapantios, T.D., Paras, S.V., Karabelas, A.J. (1989), Statistical Characteristics of Free Falling Films at High Reynolds Numbers, *Int. J. Multiphase Flow 15*: 1-21.

Kataoka, I., Ishii, M., Serizawa, A. (1986), Local formulation and measurements of interfacial area concentration in two-phase flow, *Int. J. Multiphase Flow 12*(4): 505-529.

Keane R.D., Adrian R.J. (1990) *Meas. Sci. Technol.* 1 1202.

Kim, S., Fu, X.Y., Wang, X., Ishii, M. (2000), Development of the miniaturized four-sensor conductivity probe and the signal processing scheme, *Int. J. Heat Mass Transfer 43*: 4101-4118

Koskie, J.E., Mudawar, I., Tiederman, W.G. (1989), Parallel Wire Probes for Measurements of Thick Liquid Films, *Int. J. Multiphase Flow 15*: 521-530.

Krane, K.S. (1988), *Introductory Nuclear Physics*. New York: Wiley.

Kureta, M., Hibiki, T., Mishima, K., Akimoto, H. (1999), Visualization and void fraction measurement of subcooled boiling water flow in a narrow rectangular channel using high-rate neutron radiography, in: *Two-Phase Flow Modelling and Experimentation 1999* (ed. Celata, G.P., Di Marco, P., Shah, K.): 1509-1514, Pisa: Edizioni ETS.

Le Gall, F., Pascal-Ribot, S., Leblond, J. (2001), Nuclear magnetic resonance measurements of fluctuations in air–water two-phase flow: Pipe flow with and without "disturbing" section, *Phys. Fluids 13(5)*: 1118-1129.

Liu T.J., Bankoff S.G. (1993a) Structure of air–water bubbly flow in a vertical pipe-I. Liquid mean velocity and turbulence measurements, *Int. J. Heat Mass Transf. 36*: 1049-1060.

Liu T J and Bankoff S G 1993b Structure of air–water bubbly flow in a vertical pipe-II. Void fraction, bubble velocity and bubble size distribution Int. J. Heat Mass Transf. 36 1061-1072

Lowe, D., Rezkallah, K.S. (1999), A capacitance sensor for the characterization of microgravity two-phase liquid-gas flow, *Meas. Sci. Technol. 10*: 965-975.

Ma, Y, Chung, N., Pei, B., Lin, W. (1991), Two Simplified Methods to Determine Void Fractions for Two-Phase Flow, *Nucl. Technology 94*: 124-133.

Masuda, Y., Nishikawa, M., Ichijo, B. (1980), New methods of measuring capacitance and resistance of very high loss materials at high frequancies, *IEEE Trans. Instrum. Meas. 29*: 28-36.

Maxwell J.C. (1882), *A Treatise on Electricity and Magnetism*, Oxford: Clarendon Press.

Melnikov, V.I., Kontelev, V.V. (1999), Two-phase flow diagnostic acoustic system based on ultrasound waveguides, in: *Two-Phase Flow Modelling and Experimentation 1999* (ed. Celata, G.P., Di Marco, P., Shah, K.): 1515-1519, Pisa: Edizioni ETS.

Melnikov, V.I., Nigmatulin, B.I. (1994), The newest two-phase control devices in LWR equipment based on ultrasonic and WAT technology, *Nucl. Eng. Des. 149*: 349-355.

Merilo M., Dechene, R.L., Cicowlas, W.M. (1977), Void fraction measurement with a rotating electric field conductance gauge, *J. Heat Transfer Trans. ASME 99*: 330-331.

Mersereau, R.M. (1976), Direct Fourier transform techniques in 3-D image reconstruction, *Comput. Biol. Med. 6*: 247-258.

Mewes, D., Schmitz, D. (1999), Tomographic methods for the analysis of flow patterns in steady and transient flows, in: *Two-Phase Flow Modelling and Experimentation 1999* (ed. Celata, G.P., Di Marco, P., Shah, K.): 29-42, Pisa: Edizioni ETS.

Mishima, K., Hibiki, T. (1996), Quantitative Method to Measure Void Fraction of Two-Phase Flow Using Electronic Imaging with Neutrons, *Nucl. Sci. Eng. 124*: 327-338.

Natterer, F. (1986), *The mathematics of computerized tomography*, Stuttgart: Teubner Verlag.

Neal, L.G., Bankoff, S.G. (1963), A high resolution resistivity probe for determination of local void properties in gas-liquid flow, *AIChE J. 9*: 490-494.

Nogueira S., Dias, I., Pinto, A.M.F.R., Riethmuller, M.L. (2001), Liquid PIV measurements around a single gas slug rising through stagnant liquid in vertical pipes, *Proc. 4th Int. Conf. On Multiphase Flow*, New Orleans.

Okamoto, K., Schmidl, W., Hassan, Y. (1995), New tracking algorithm for particle image velocimetry, *Exp. Fluids 19*: 342–347.

Parker, D.S., Hawkesworth, M.R., Broadbent, C.J., Fowles, P., Fryer, T.D., McNeal, P.A. (1994), Industrial positron-based imaging: principles and applications, *Nucl Instr. Meth. A349*: 583-592.

Philip, O.G., Schmidl, W.D., Hassan, Y.A. (1994), Developments of a high speed particle image velocimetry technique using fluorescent tracers to study steam bubbles collapse. *Nucl. Eng. Des., 149*: 375-385.

Ramos, R.T., Holmes, A., Wu, X., Dussan, E. (2001), A local optical probe using fluorescence and reflectance for measurement of volume fractions in multi-phase flows, *Meas. Sci. Technol. 12*: 871-876.

Reinecke, N., Boddem, M., Petritsch, P., Mewes, D. (1998), Tomographic imaging of the phase distribution in two-phase slug flow, *Int. J. Multiphase Flow 24(4)*: 617-634.

Resch, F.J., Leuthesser, Leutheusser, J.H. (1972), Le ressaut hydraulique: mesures de turbulence dans la region diphasique, *Houille Blanche 4*: 279-293.

Rowland, S.W. (1979), Computer implementation of image reconstruction formulas, in: *Image Reconstruction from Projections Implementation and Applications* (ed. Herman, G.T.): 9-80, Berlin: Springer-Verlag.

Ruder, Z., Hanratty, T.J. (1990), A Definition of Gas-Liquid Plug Flow in Horizontal Pipes, *Int. J. Multiphase Flow 16*: 233-242.

Schlaberg, H.I., Yang, M., Hoyle, B.S. (1996), Real time ultrasonic process tomography for two-component flows, *Electronic Letters 32(17)*: 1571-1572.

Scott D.M., Williams R.A. (eds) (1995), *Frontiers in Industrial Process Tomography*, New York: Engineering Foundation.

Sene, K.J. (1984), Aspects of bubbly two-phase flow, PhD thesis, Trinity College, Cambridge, U.K.

Serizawa, A., Kataoka, I., Michiyoshi, I. (1975), Turbulence structure of air-bubbly flow II. Local properties *Int. J. Multiphase. Flow 2*: 235–246.

Stitou, A., Riethmuller, M.L. (2001), Extension of PIV to super resolution using PTV, *Meas. Sci. Technol. 12*: 1398-1403.

Thorn, R., Johansen, G.A., Hammer, E.A. (1999), Three-phase flow measurement in the offshore oil industry - is there a place for process tomography? *Proc. 1st World Congress on Industrial Process Tomography* Buxton (UK) pp 228–235.

Tsochatzidis, N.A., Karapantios, T.D., Kostoglou, M.V., Karabelas, A.J. (1992), A Conductance Method for Measuring Liquid Fraction in Pipes and Packed Beds, *Int. J. Multiphase Flow 5*: 653-667.

Westerweel, J. (1997), Fundamentals of digital particle image velocimetry, *Meas. Sci. Technol. 8*: 1379-1392.

Willert, C.E., Gharib, M. (1991), Digital Particle Image Velocimetry, *Exp. Fluids 10(4)*: 181-193.

Wu, X., Fordham, E.J., Mullins, O.C., Ramos, R.T. (2000), Single point optical probe for measuring three-phase characteristics of fluid flow in a hydrocarbon well, *USA Patent* 6 023 340.

Xie, C.G., Reinecke, N., Beck, M.S., Mewes, D., Williams, R.A. (1995), Electrical tomographic techniques for process engineering applications, *Chem. Eng. J. 56*: 127-133.

Yang W.Q. (1996), Hardware design of electrical capacitance tomography systems *Meas. Sci. Technol. 7*: 225-232.

Zuber, N. Findlay, J.A. (1965), Average volume concentration in two-phase flow systems, *Journal of Heat Transfer (Transactions ASME)C*: 453-468.

Critical Heat Flux, Post-CHF Heat Transfer and Their Augmentation*

Gian Piero Celata and Andrea Mariani

ENEA, Institute of Thermal Fluid-Dynamics, Rome, Italy

Abstract. The present work reports on the state-of-the-art review on the critical heat flux and the post-dryout heat transfer. The first two sections are somewhat tutorial, and are featured in a similar way. They provide, after a brief introduction, with information on parametric trends, i.e. on the influence of the thermal-hydraulic and geometric parameters on the thermal crisis. After that, the most widely used correlations are described in detail, either in terms of reliability and simplicity of use. Eventually, the various approaches for a modelling of the critical heat flux are reported. The third section describes correlations and models available for the prediction of the post-dryout heat transfer, trying also to highlight the main drawbacks. Finally, the fourth section describes the passive techniques for the enhancement of the critical heat flux and the post-dryout heat transfer, together with available correlations. The present work is a merge of original researches carried out at the Institute of Thermal Fluid Dynamic of ENEA and a thorough review of the recent literature.

0 Introduction

The term *critical heat flux* (CHF) indicates an abrupt worsening of the heat transfer between a heating wall and a coolant fluid, generally with undesired consequences. This is typically due to the presence on the heated wall of a vapour layer which strongly reduces the heat transfer rate from the heater to the coolant. In systems where the heat transfer is temperature-controlled (i.e., when a variation in the coolant thermal-hydraulic conditions implies only a variation in the heat flux, and not in the wall temperature) the sudden decrease of the heat transfer coefficient leads to a reduction in the performance of the *heat exchanger* and may cause chemical consequences for the wall (fouling, etc.) or safety consequences for the plant. This is obvious once we consider Eq. (1):

$$\dot{q} = \alpha \, (T_W - T_l) \qquad (1)$$

As the wall-to-fluid temperature difference is imposed, a reduction in the heat transfer coefficient a will cause a decrease in the heat flux $\dot{q}$. A typical temperature controlled system is that where the wall is heated by a condensing fluid on one side and cooled on the other side.

In systems with imposed heat flux (i.e., when a variation in the coolant thermal-hydraulic conditions implies only a variation in the wall temperature and not in the heat flux) the sudden decrease in the heat transfer coefficient leads to a sharp increase in the wall temperature, as given by Eq. (1). This latter may lead to the wall melting or its deterioration. A nuclear reactor core, an electrically heated rod or channel are typical heat flux controlled systems.

The term CHF, which is the limiting phenomenon in the design and operating conditions of water-cooled nuclear reactors as well as of much other thermal industrial equipment, will be used to represent the heat transfer deterioration above described, although different mechanisms of the thermal crisis might also suggest different names. Under subcooled or low-quality saturated flow boiling conditions, being nucleate boiling the main boiling mechanism, the onset of thermal crisis is following the *departure from nucleate boiling* (DNB), and this is often the name used in this case. Under high-quality saturated flow boiling conditions, typically characterized by the annular flow regime, the *dryout* of the liquid film adjacent to the heated wall is the leading mechanism to the thermal crisis, which is therefore named dryout.

In a heat flux-controlled situation (which will be the only one treated here), the rapid wall temperature rise may cause rupture or melting of the heating surface, which is termed as *physical burnout*. The burnout heat flux is generally different from the DNB or the dryout heat flux. Only in the case of extremely high heat fluxes under subcooled flow boiling conditions (expected to be faced in some components of the thermonuclear fusion reactor), the CHF is characterized by extremely high temperature differences. Failure of the heating wall is very often experienced and therefore the heat flux causing the DNB is practically identical with the physical burnout heat flux (Celata (1996)). This is absolutely not the case of situations where higher heat transfer coefficients and lower critical heat fluxes give rise to only reduced temperature excursions at the DNB or dryout (Hewitt (1978), Bergles et al. (1981), Hsu & Graham (1986), Weisman (1992), Collier & Thome (1994), Katto (1994)). If the temperature rise does not cause failure of the heating surface, a post-CHF heat transfer is thus possible, although the heat transfer rate will be much lower than that before the CHF occurring.

In the following sections, the CHF in subcooled and saturated flow boiling will be discussed, together with the post-CHF heat transfer. Finally, a section will deal with existing methods for CHF and post-CHF heat transfer augmentation.

1 CHF in Subcooled Flow Boiling.

Simply speaking, forced convective subcooled boiling involves a locally boiling liquid, whose bulk temperature is below the saturation value, flowing over a surface exposed to a heat flux. Under such conditions the critical heat flux is always of the DNB type, resulting in a significant increase in the wall temperature, the larger the higher the heat flux.

Although relevant to the thermal-hydraulic design of Pressurized Water Reactor cores and therefore studied since the far past (Hewitt (1978), Bergles et al. (1981), Hsu & Graham (1986), Weisman (1992), Collier & Thome (1994), Katto (1994), Gambill (1968), Bergles (1977)), the CHF in subcooled flow boiling received a renewed attention in the recent past due to the possible use of water in subcooled flow boiling for the cooling of some components of the thermonuclear fusion reactor believed to be subjected to operating conditions characterized by extremely high thermal loads (Celata (1996), Boyd (1985a)). Hereafter the parametric trends experimentally observed, together with available correlations and theoretical models will be discussed.

1.1 Parametric Trends.

The magnitude and the occurrence of the CHF are affected by many parameters such as thermal-hydraulic, geometric and external parameters. Among thermal- hydraulic parameters we have subcooling, mass flux, pressure, binary component fluids, while important geometry parameters are channel diameter, heated length, channel orientation, tube wall thickness and material. External parameters of interest are heat flux distribution and content of dissolved gas.

Influence of subcooling. As reported by Boyd (1985a), most of the early experimental studies reveal that the relationship between subcooling and CHF is almost linear, even though Bergles (1963) indicated that for very large subcooling at moderate to large liquid velocity (1 to 10 m/s) the relationship between CHF and subcooling is nearly linear, but becomes highly nonlinear as the subcooling decreases, showing a minimum at small positive subcooling. Recent experiments under conditions of high liquid subcooling confirmed the almost linear relationship between CHF and subcooling (Celata et al. (1993a), Nariai et al. (1987), Vandervort et al. (1992)). Figure 1.1 shows the CHF versus inlet subcooling for data carried out by Celata et al. (1994b) in 2.5 mm I.D. stainless steel tubes, 0.25 mm wall thickness, 10 cm long, uniformly heated by Joule effect, with vertical upflow of water. The functional dependence of the CHF on the subcooling is practically linear, up to very high subcooling and very high liquid velocity. The CHF versus $\Delta T_{sub,in}$ curves, plotted at different liquid velocities, result parallel among each other, and no inter-relation between u and $\Delta T_{sub,in}$ would seem to exist.

Influence of mass flux. The CHF is an increasing function of the mass flux (or fluid velocity) with a less than a linear fashion. This was observed up to very high values of mass flux (90 Mg/m^{2}s). Figure 1.2 shows the results of experiments carried out by Boyd (1988, 1989, 1990) using water as a fluid in horizontal test sections of amzirc (copper-zirconium alloy) with an inner diameter of 3.0 mm, wall thickness around 0.5 mm, and a heated length of 0.29 m (Boyd (1988, 1989)), or 10.2 mm I.D., 0.125 mm wall thickness, 0.5 m long, and copper as a material (Boyd (1990)). Tests were performed at a constant inlet temperature of 20 °C. Similar results were obtained by Celata et al. (1993a).

Influence of pressure. Recent experiments (Celata et al. (1993a, 1994b), Nariai et al. (1992), Vandervort et al. (1992)) showed that in the range 0.1-5.0 MPa, direct influence of the pressure on the CHF is weak, other conditions being equal (i.e., for same subcooling and liquid velocity). This is demonstrated in Figure 1.3a, where the CHF is plotted versus exit pressure p, for Vandervort et al. data (1992), obtained with stainless steel tubes of 1.07 mm I.D., 26.75 mm long. Virtually, no pressure effect was noted; in fact, there seemed to be a very slight decrease of the CHF with increasing pressure. Figure 1.3b shows the results of Celata et al. (1994b) obtained with stainless steel tubes of 8.0 mm I.D., 10 cm long, with uniform heating. The CHF versus subcooling data lie on a unique curve independent of the pressure, evidencing the negligible effect of this parameter. Boyd (1985a) reported how other researchers found a maximum in the CHF versus pressure trend in the vicinity of a reduced pressure of 0.75, being this value somewhat variable with the mass velocity.

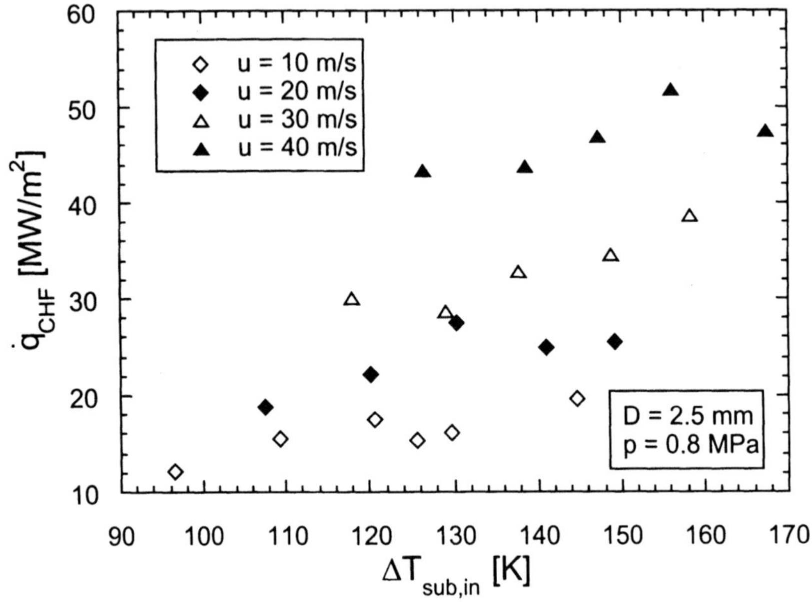

Figure 1.1. CHF versus inlet subcooling, Celata et al. (1993a).

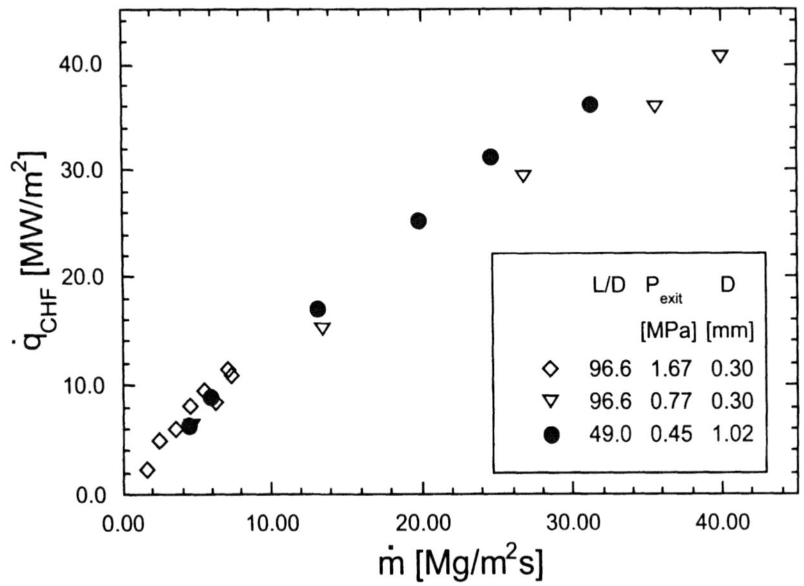

Figure 1.2. CHF versus mass flux, Boyd (1988, 1989, 1990).

Binary component fluids. Tolubinsky & Matorin (1973) used ethanol-water, aceton-water, ethanol-benzene, ethylene-glycol-water with a 4 mm I.D., 60 mm long tube; Andrews et al. (1968) tested acetone-tuolene and benzene-tuolene with an annulus 6.35 mm I.D. 20.9 mm O.D. 76 mm long; Sterman et al. (1968) used mono-iso-propyldiphenyl-benzene with an annulus 10 mm I.D. 16 mm O.D. 110 mm long; Naboichenko et al. (1965) tested the same fluids as Sterman et al. (1968) using an annulus 6 mm I.D. 16 mm O.D. 80 mm long; Carne (1963) used acetone-tuolene and benzene-tuolene with an annulus 6.35 mm I.D. 19.05 mm O.D. 76.2 mm long; and finally Bergles & Scarola (1966) tested water-1-pentanol using a 6.26 mm I.D. 170 mm long tube. Typical trends of CHF are shown in Figure 1.4; the CHF tends to reach a maximum value increasing the mole fraction and increases with subcooling and velocity. The maximum corresponds to the maximum difference between the vapour and liquid composition of the more volatile component *(y-x)*. As the difference between the more volatile component concentration in the vapour and the liquid phase increases (in absolute value), a reduction occurs in the vapour bubble departure diameter, in the bubble rate of growth, and in the number of active nucleation sites. This results in a reduction of the vapour content of the wall layer of the boiling fluid and, therefore gives rise to an increase in the CHF (Tolubinsky & Matorin (1973)).

Influence of channel diameter. Works to identify the dependence of the CHF on the channel diameter have been conducted up to the recent past (Vandervort et al. (1992), Bergles (1963), Kramer (1976), Celata e al. (1993c), Nariai & Inasaka (1992)). It is well established that the CHF is inversely related to the channel diameter. Figure 1.5 shows the CHF versus the channel diameter *D*, for Vandervort et al. data (1992). As observed by previous researchers, for given values of exit thermal hydraulic conditions, heated length and liquid velocity, the CHF increases with the decrease of the tube inside diameter, but the effect was less significant for decreased mass flux. A threshold is observed beyond which the effect of the tube inside diameter may be considered negligible, that is a function of the channel geometry and thermal hydraulic conditions. To explain the observed dependence of the CHF on the tube inside diameter it is worth reporting here three different reasons proposed by Bergles (1963). For a tube with a smaller inside diameter we have: (1) a small bubble diameter, (2) an increased velocity of the bubbles with respect to the liquid, and (3) the fluid subcooled bulk closer to the growing bubbles (collapsing in the bulk). From the analysis of experimental data of void fraction in narrow tubes, Nariai & Inasaka (1992) concluded that, as tube inside diameter decreases and mass velocity increases, the diameter of generated bubbles or, better, the thickness of the two-phase boundary layer becomes smaller due to the intense condensation effect by subcooled water at core region, and the void fraction becomes smaller, making the CHF higher. The decrease in the diameter gives rise to an increase in the slope of the velocity profile in the two-phase boundary layer, making the detachment of growing bubbles and the consequent condensation in the core region easier. The higher the mass flux the most consistent the effect.

Influence of channel heated length. The heated length of the channel seems to be inversely related to the CHF. Generally, investigators use the ratio of the heated length to the inside (or equivalent) diameter of the channel *L/D*, as the characteristic non-dimensional length, but this still needs to be established. Recent experiments were carried by Nariai et al. (1987) and by Vandervort et al. (1992). Figure 1.6 reports the results of Nariai et al. (1987) showing the CHF versus

L/D. The CHF increases as *L/D* decreases, and the effect is more significant for smaller channel diameter. As the effect seems to be greatest for *L/D* < 20 (depending on the diameter), this would indicate that the CHF is related to the state of development of the bubble-boundary layer. Vandervort et al. (1992) verified that the functional dependence between CHF and *L/D* is independent of mass flux. As for the case of the channel diameter, experiments showed the presence of a threshold beyond which the CHF is practically independent of *L/D* and this limit (between 20 and 40) is related to flow parameter since *L/D* is related directly to the flow development.

Influence of channel orientation. The effect of flow orientation (e.g., horizontal versus vertical upflow) may be significant if the buoyant force is a nonnegligible percentage of the axial inertial force in flow boiling. Quantitatively, this can be evaluated by considering the modified Froude number *Fr*, defined as:

$$\mathrm{Fr} = \frac{\dot{m}\cos\phi}{\rho_l\left[gD\left(\frac{\rho_l-\rho_g}{\rho_l}\right)\right]^{1/2}} \tag{1-1}$$

where $\phi = 0$ represents the horizontal case. For modified Froude number greater than 5-7, effects of stratification and orientation may disappear. Wherever flow orientation plays a relevant role, the CHF for horizontal flow is always less than the value for vertical flow (Merilo (1977), Cumo et al. (1978)). Recent experiments carried out by Celata et al. (1994b) using water under conditions relevant to the NET/ITER divertor (*p* around 3.5 MPa) showed that for a liquid velocity greater than 5.0 m/s, horizontal and vertical data do not show any remarkable difference (at 5.0 m/s the modified Froude number is greater than 20).

Influence of tube wall thickness and material. Celata et al. (1997) tested a number of SS 304 tubes having almost the same inner diameter but different wall thicknesses (from 0.25 to 1.75 mm) and found a slight effect of the tube wall thickness on the CHF: a slight decrease in the CHF as the wall thickness increased was observed, but within 20% passing from the smallest to the largest thickness. Vandervort et al. (1992) used five different materials in their experiments, such as SS 304, SS 316, nickel 200, brass 70/30 and inconel 600, and, under very similar geometric and thermal-hydraulic conditions, did not observe any significant effect of the tube material on the CHF.

Influence of dissolved gas. On the basis of previous literature it is reasonable to conclude that dissolved gas has no effect on CHF. However, for experiments carried out with small diameter tubes, the bubble boundary layer may be smaller, and it is conceivable that even small amounts of dissolved air coming out of solution could affect the CHF. Specific tests were performed by Vandervort et al. (1992) using 1.07 mm I.D. channels, at a mass flux of 25 Mg/m^{2}s, an exit pressure of 0.6 MPa and an exit subcooling of 100 K. No significant change was observed in the CHF results over the range of dissolved gas concentration in water from near zero (2 ppm) up to the saturation level (~ 9.5 ppm).

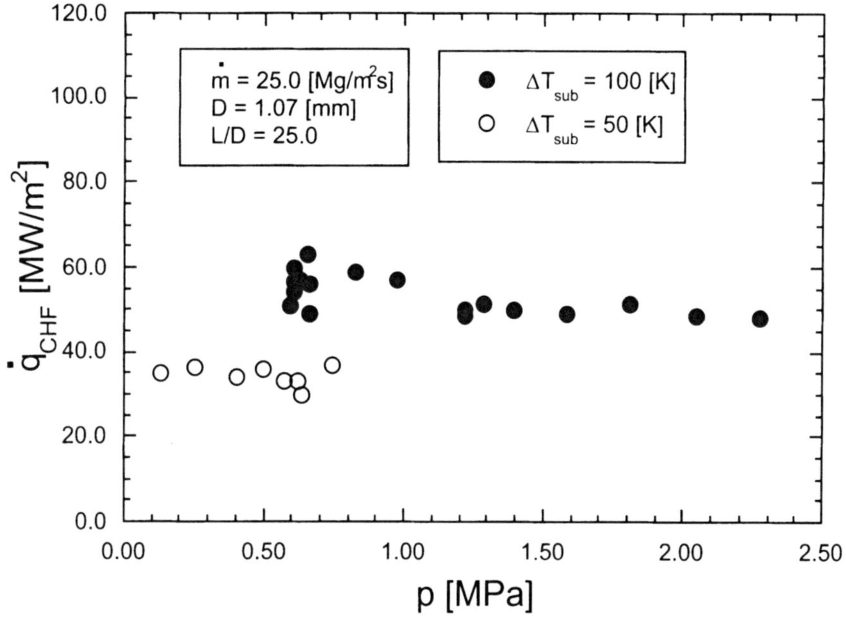

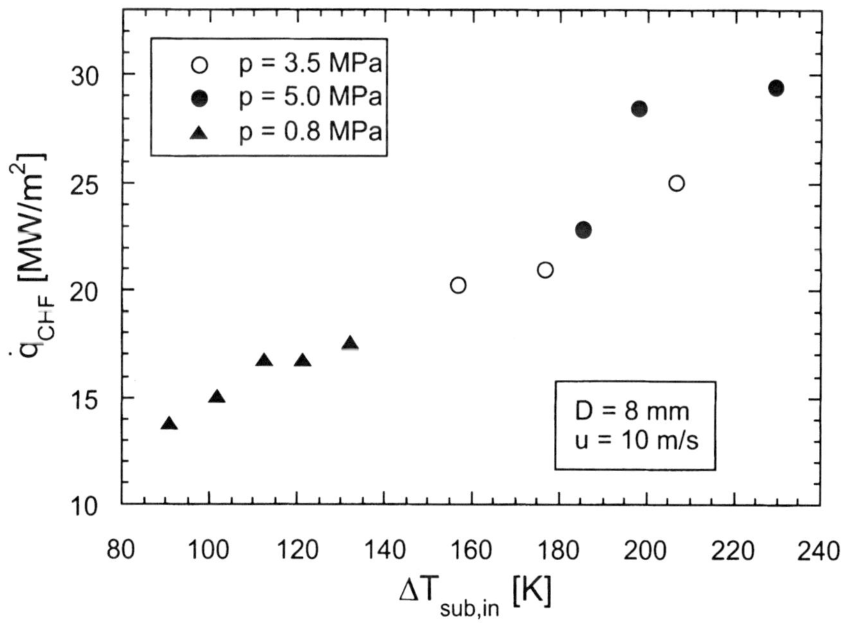

Figure 1.3. CHF versus pressure, Vandervort et al. (1992) (above graph) and versus inlet subcooling, Celata et al. (1994b) (below graph).

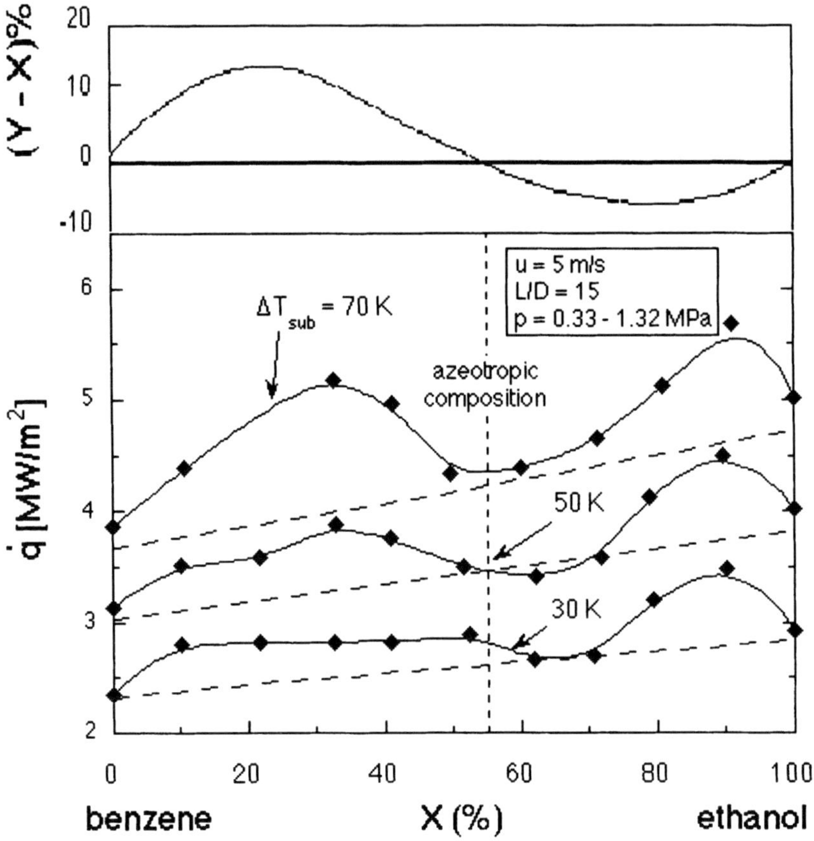

Figure 1.4. CHF versus mixture composition for forced convection boiling of benzene/ ethanol mixtures, Tolubinsky & Matorin (1973). In the top figure, Y-X represents the difference between the composition of the vapour phase, Y, and the liquid phase, X, for the more volatile component.

Influence of heat flux distribution. The optimum axial heat flux distribution for subcooled flow boiling is one where the peak heat flux occurs near the inlet (Boyd (1985a)). Groeneveld (1981) notes that a short pulse spike has a significant effect on subcooled flow boiling CHF, finding a CHF increase but a critical power decrease. Doroschuk et al. (1978) found that the CHF was lower for cosine distribution than for uniform ones. Ad hoc experiments were recently performed by Nariai et al. (1992) and by Gaspari (1993) to investigate the effect of the circumferential heat flux distribution on the CHFIn particular, Gaspari made a comparison between peripherally full and half-heated tubes, straight flow, analysing the CHF at both inlet and exit thermal hydraulic conditions. Using a 10 mm I.D. channel, 0.15 m long, Gaspari observed that, under constant inlet liquid subcooling, higher CHF values were observed for half-heated tubes. Plotting the CHF versus exit liquid subcooling such a difference tends to disappear, as reported in Figure 1.7, where the CHF is plotted versus inlet/exit subcooling.

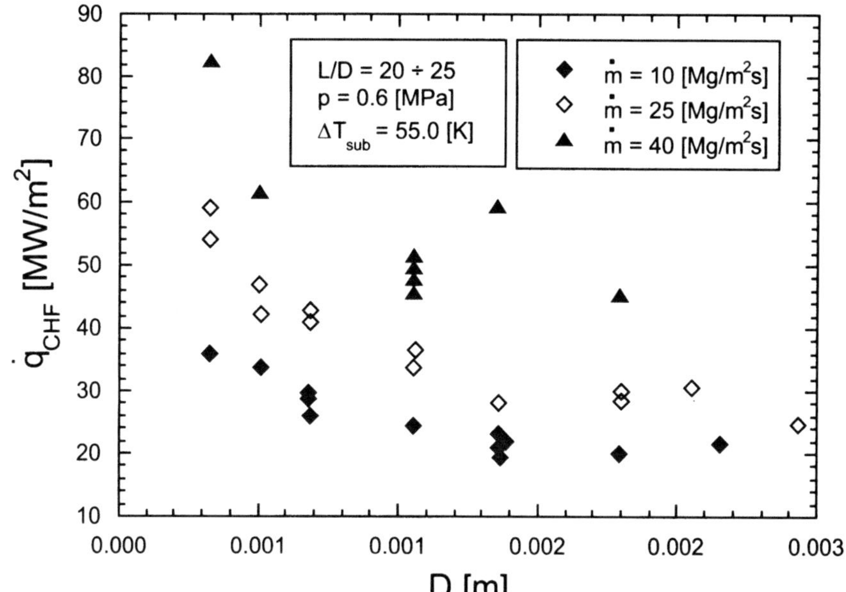

Figure 1.5. CHF versus channel diameter, Vandervort et al. (1992).

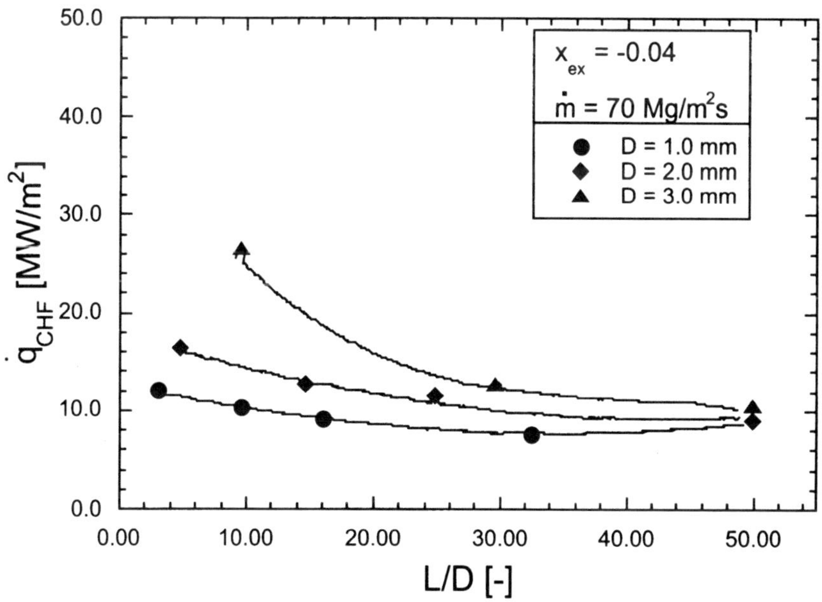

Figure 1.6. CHF versus channel heated length, Nariai et al. (1987).

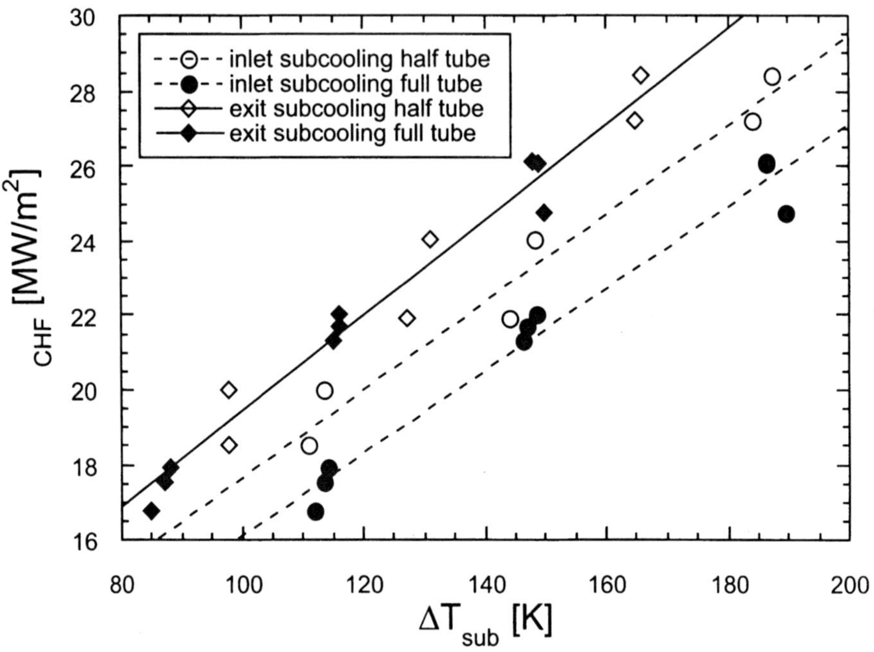

Figure 1.7. Influence of circumferential heat flux distribution on CHF, Gaspari (1993)

1.2 Available Correlations for the Prediction of Subcooled Flow Boiling CHF

Many different types of correlational approaches have been proposed. These include empirical, dimensional analysis or similitude-based, analytical, tabular, and graphical, being the first two categories the most widely used. A thorough review of them has been given by Boyd (1985b), listing as many as 38 correlations. We just report here the most widely used, also on the basis of their possible extrapolation to conditions different from the originating ones, (Celata et al. (1994a), Inasaka & Nariai (1996)) although this must be done with great care.

- *Gunther (1951)*

$$\dot{q}_{CHF} = 71987 u^{0.5}\, \Delta T_{sub,\,ex} \tag{1-2}$$

(recommended ranges: p = 0.1 - 1.1 MPa; u = 1.5 - 12.1 m/s; CHF = 0.4 - 11.4 MW/m^2; ΔT_{sub} = 11-139 K)

- *W -2, Tong et al. (1968)*

$$\dot{q}_{CHF} = (0.23\ 10^6 + 0.094\ \dot{m})\ (3 + 0.018\ \Delta T_{sub})\ [0.435 + 1.23\ \exp(-\ 0.0093\ L/D)]\ *$$

$$\left\{1.7 - 1.4\ \exp\left[-\ 0.532\left(\frac{h_l - h_{in}}{h_{lg}}\right)^{3/4}\left(\frac{\rho_g}{\rho_l}\right)^{-1/3}\right]\right\} \tag{1-3}$$

(recommended ranges: p = 5.5 - 19.0 MPa; u = 0.3 - 12.1 m/s; CHF = 0.4 - 4 MW/m^2; L/D = 21 - 365; ΔT_{sub} = 0 - 126.7 K)

- *Tong-75, Tong (1975)*

$$\dot{q}_{CHF} = 0.23\, f\, \dot{m}\, h_{lg} \left[1 + 0.0216\, (p_{ex}/p_c)^{1.8}\, Re^{0.5}\, Ja \right] \tag{1-4}$$

where

$$f = 8\,(D/D_o)^{0.32}\, Re^{-0.6}, \quad Ja = \frac{c_p\,(T_b - T_{sub})}{h_{lg}} \frac{\rho_l}{\rho_g}, \quad Re = \frac{\dot{m}\, D}{\eta_l\,(1 - \varepsilon)}$$

with D_o = 1.27 10^{-2} m being ε (void fraction) evaluated using the Thom's correlation (Collier & Thome (1994), Thom et al. (1965)) (recommended ranges: p = 6.8 - 13.6 MPa; u = 0.68-5.9 m/s; void fraction at CHF < 0.35; D = 3 - 10 mm; L/D = 5 - 100).

- *Tong-68, Tong (1968)*

$$\frac{\dot{q}_{CHF}}{h_{lg}} = C \frac{\dot{m}^{0.4}\, \eta_l^{0.6}}{D^{0.6}} \tag{1-5}$$

with $C = 1.76 - 7.433\, x_{ex} + 12.222\, x_{ex}^2$

(recommended ranges: p > 7 MPa). The Tong-68 correlation can also be written as:

$$Bo = \frac{C}{Re^{0.6}}$$

where Bo and Re are Boiling number and Reynolds number, respectively.

A modification of the Tong-68 correlation for pressure lower than 7.0 MPa has been proposed by Celata et al. (1994a):

$$Bo = \frac{C'}{Re^{0.5}} \tag{1-6}$$

where:

$C' = (0.216 + 4.74\ 10^{-2}\, p)\, \psi$ (p in MPa)

$\psi = 1$ if $x_{ex} < -0.1$

$\psi = 0.825 + 0.986\, x_{out}$ if $0 > x_{ex} \geq -0.1$

(recommended ranges: p ≤ 5.5 MPa; u = 2.2 - 40 m/s; $\Delta T_{sub,ex}$ = 15 - 190 K; D = 0.3 - 15 mm).

In using the above reported correlations, two methods are generally followed: the so-called *heat balance method*, HBM, which requires an iterative procedure, and the so-called *direct substitution method*, DSM (Inasaka & Nariai (1986), Groeneveld et al. (1986)). The two methods lead to different results and their use has been deeply debated in the recent past, with the possible conclusion that the HBM would give better results and should be therefore preferred, Theofanous (1996).

For binary mixtures (Collier & Thome (1994), Celata & Cumo (1996)) the CHF in subcooled flow boiling may be expressed as the sum of two terms: the first term $\dot{q}_{CHF,i}$ is the ideal value evaluated from the CHF value of the two components at the same pressure, velocity and subcooling (linear combination), and the second term $\dot{q}_{CHF,E}$ is an additional CHF connected to the increasing in the CHF due to mass transfer effects. Thus, the final expression is given by:

$$\dot{q}_{CHF} = \dot{q}_{CHF,i} + \dot{q}_{CHF,E} = \dot{q}_{CHF,i}\,(1 + C_{11}) \tag{1-7}$$

where $\dot{q}_{CHF,i} = \left[x\,\dot{q}_{CHF,1} + (1-x)\,\dot{q}_{CHF,2} \right]$
being *1* and *2* referred to the more and the less volatile component, respectively, and C_{11} the mole fraction of the more volatile component in the liquid phase.

Sterman et al. (1968) verified in their experiments that C_{11} varies between 0 and 0.8 and proposed an expression for C_{11}, as:

$$C_{11} = A\frac{(C_{21}-x)^3}{Re_2} + B\frac{(C_{21}-x)^{1.5}}{Re_2^{0.4}}\left[\frac{T_{sat,1}}{T_{sat,m}-T_{sat,1}}\right] \tag{1-8}$$

with $A = 3.2\ 10^5$; $B = 6.9$
being C_{21} the mole fraction of the more volatile component in the vapour phase. This correlation was proved valid also for refrigerant mixture, Celata et al. (1994d).

Tolubinsky & Matorin (1973) gave the following expression of C_{11}, as:

$$C_{11} = 1.5\left|C_{21}-x\right|^{1.8} + 6.8\left|C_{21}-x\right|\left[\frac{T_{sat,m}-T_{sat,1}}{T_{sat,1}}\right] \tag{1-9}$$

Equation (1-9) is applicable to ethanol-water, acetone-water, ethanol-benzene and ethylene-glycol-water mixtures with a ± 20% error.

1.3 Available Models for the Prediction of Subcooled Flow Boiling CHF.

As is known, correlations have the drawback to be not reliable outside the recommended ranges of application. In this respect, models may have the advantage to characterize not only the existing and developing data base, but also to predict CHF beyond the established data base. Recent reviews about CHF modelling were given by Katto (1994, 1995), Weisman (1992) and Celata (1997).

Major theoretical approaches to CHF can be categorized into five groups, according to the basic mechanism assumed by relative authors to be the main cause of the CHF occurrence.

(1) Liquid layer superheat limit model. The difficulty of heat transport through the bubbly layer causes a critical superheat in the liquid layer adjacent to the wall, giving rise to the occurrence of the CHF, Tong et al. (1965).

(2) Boundary layer separation model. This model is based on the assumption that an *injection* of vapour from the heated wall into the liquid stream causes a reduction in the velocity gradient close to the wall. Once the vapour effusion increases beyond a critical value, the consequent flow stagnation is assumed to originate the CHF (Kutateladze & Leontiev

(1966), Tong (1966, 1975), Purcupile & Gouse (1972), Hancox & Nicoll (1973), Thorgerson et al. (1974)). The weak physical basis of the model has been demonstrated by the studies above reported (Fiori & Bergles (1970), van der Molen & Galjee (1978), Hino & Ueda (1985), Mattson et al. (1973)).

(3) Liquid flow blockage model. It is assumed that the CHF occurs when the liquid flow normal to the wall is blocked by the vapour flow. Bergel'son (1980) considers a critical velocity raised by the instability of the vapour-liquid interface, while Smogalev (1981) considers the effect of the kinetic energy of vapour flow overcoming that of the counter motion of liquid.

(4) Vapour removal limit and near-wall bubble crowding model. It is assumed that the turbulent interchange between the bubbly layer and the bulk of the liquid may be the limiting mechanism leading to the CHF occurrence. The CHF occurs when bubble crowding near the heated wall prevents the bulk cold liquid from reaching the wall (Hebel et al. (1981). Weisman & Pei (1983), and Weisman & Ying (1983) postulate that the CHF occurs when the void fraction in the bubbly layer, calculated under the assumption of homogeneous two-phase flow in the bubbly layer in Weisman & Pei (1983), and using the slip model in Weisman & Ying (1983), just exceeds the critical value of 0.82. The void fraction in the bubbly layer is determined through the balance between the outward flow of vapour bubbles and the inward liquid flow at the bubbly layer-bulk liquid flow interface. Weisman & Ileslamlou model (1988) is an improvement of Weisman & Pei model, for subcooled exit conditions. A research work carried out by Styrikovich et al. (1970), showed that measured void fraction at the CHF ranges from as low as 0.3 to as high as 0.95, making the validity of the near-wall bubble crowding models questionable. In addition, the models are quite empirical in the determination of the turbulent exchange in the bubbly layer.

(5) Liquid sublayer dryout model. The model is based on the dryout of a thin liquid sublayer underneath a vapour blanket or elongated bubble, due to coalescent bubbles, flowing over the wall. (Lee & Mudawar (1988), Katto (1990), Celata et al. (1994c)).

At present, the liquid sublayer dryout theory is being received significant attention, is well developed, and is able to provide good predictions over a wide range of conditions. Lee & Mudawar (1988) are the first in developing and proposing a mechanistic model based on the liquid sublayer dryout theory, which was assessed for data at a pressure above 5.0 MPa.

Following the same principles as Lee & Mudawar, Katto (1990a, 1990b) developed a generalised CHF model applicable to not only water but also non aqueous fluids (water, nitrogen, helium, R 11, R 12, and R 113). Then Katto extended his model so as to cover the CHF of water boiling at low pressure also, (Katto (1992)).

Lee and Mudawar, and Katto models make use of empirical constants determined through the experimental data. This limits somehow the use of these models within the data base on which they are assessed. Further, the Katto model is applicable only to those cases where the local void fraction at the CHF in the near-wall bubbly layer is lower than 0.7. The most recent model developed in the frame of the liquid sublayer dryout theory was proposed by Celata et al. (1994c), without making use of any empirical constant, yet being capable of predicting the CHF of water boiling in a wide range of conditions for the subcooled flow boiling, Celata et al. (1995b).

Briefly, to describe the Celata et al. model, let us consider the situation at the tube exit (locus of the CHF for axial uniform heating) approaching the CHF, that may be presumably that sketched in Figure 1-8 (Celata et al. (1995a)): a thin vapour clot or blanket forms in the vicinity of the heated wall due to small bubbles coalescence, holding a liquid sublayer between the vapour clot and the wall surface.

The occurrence of the CHF is determined by the evaporation of the liquid sublayer during the passage time of the blanket which insulates the liquid sublayer between the heating surface and the bulk of the liquid:

$$\dot{q}_{CHF} = \frac{\delta \rho_l h_{lg}}{L_B} u_B \tag{1-10}$$

where δ is the initial liquid layer thickness, ρ_l is the liquid layer density, L_B and u_B are the blanket length and velocity, respectively. The vapour blanket length, L_B, is assumed to be given by the Helmholtz instability wavelength at the interface facing to the liquid sublayer. The vapour blanket velocity, u_B, is evaluated considering the velocity distribution of the main stream in the tube under the assumption of homogeneous flow. The Celata et al. model considers the temperature distribution of the main stream in the tube under the assumption of homogeneous flow, determining the thickness s^* of the superheated layer (distance from the heated wall at which the liquid temperature is equal to the saturation value), beyond which vapour blanket cannot develop or exist due to subcooled conditions. Vapour blanket can develop and exist only in the near-wall region where the local liquid temperature is above the saturation value.

As the temperature distribution is linked to the inside tube wall temperature, this latter is obtained by equating the local cross-section average fluid temperature given by the coolant heat balance with that provided by the temperature profile. Then δ can be determined as the difference between the superheated layer s^* (where the vapour clot can exist only, and as close as possible to the saturation line) and the vapour blanket thickness, D_B.. This latter is calculated from the Staub model (1968), under the assumption (common with the Lee & Mudawar model) that the circumferential growth of a vapour blanket is strongly limited by adjacent blankets and by the steep velocity gradient in case of high liquid velocity. It is therefore assumed that the equivalent diameter of each blanket (i.e., its thickness) may be approximated by the diameter of a bubble at the departure from the wall. In other words, it is assumed that departing bubbles may coalesce into a distorted blanket that stretches along the fluid flow direction (due to vapour generation by sublayer evaporation) and keeps almost a constant equivalent diameter (thickness). Equations used in the mathematical description of the Celata et al. model are reported in the Appendix. A comparison between Katto and Celata et al. models is reported in Figure 1.9, for the data set published in Celata & Mariani (1993) (about 1900 data). The figure reports the percentage of data point calculated with a given error band (%). The Celata et al. model, unlike the other liquid sublayer dryout models, can be also used for peripheral non uniform heating simply by considering the total thermal power delivered to the fluid in the coolant heat balance for the calculation of the local average coolant temperature (Celata et al. (1995b)).

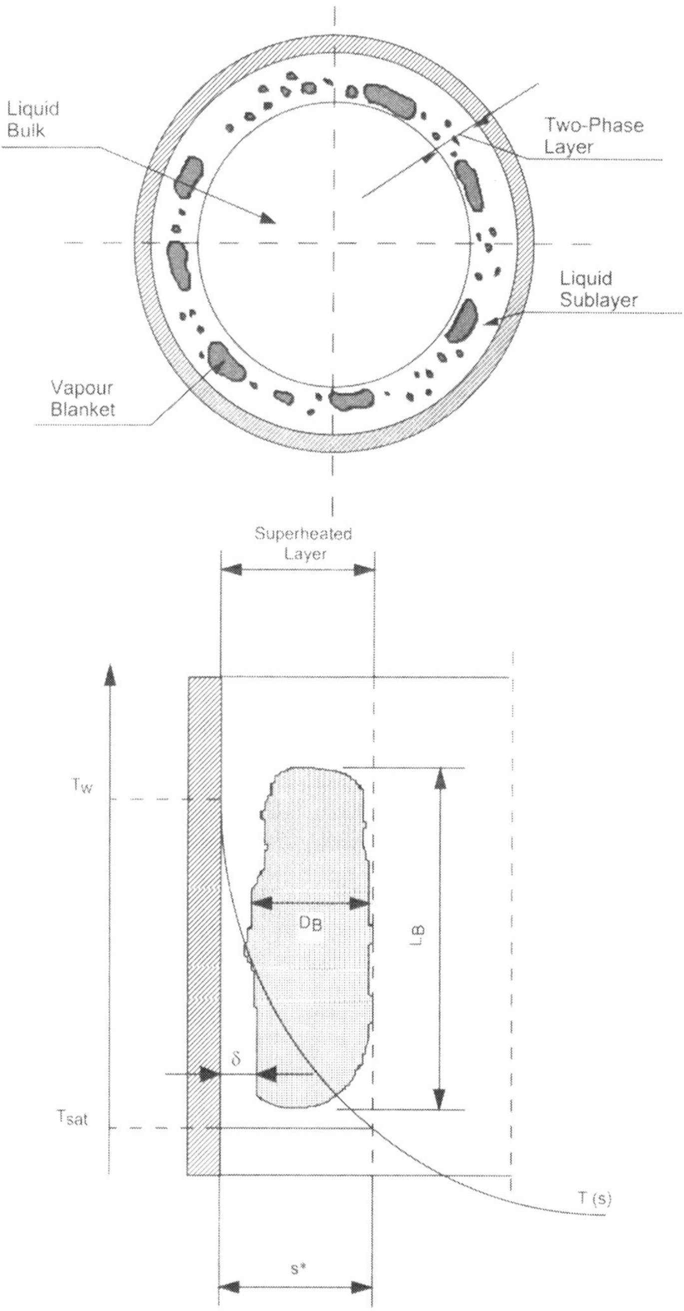

Figure 1.8. Schematic of the liquid sublayer dryout theory, Celata et al. (1994c).

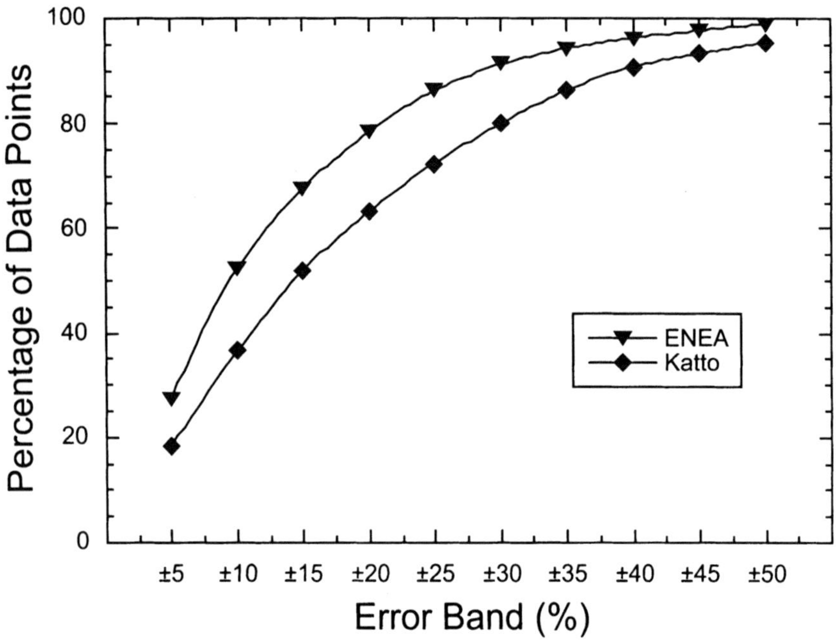

Figure 1.9. Comparison between Katto (1990) and Celata et al. (1994c) models for the prediction of water subcooled flow boiling CHF.

1.4 CHF Calculation Procedure in the Celata et al. Model (1994c)

Input parameters $\dot{m}$, p_{ex}, D, L, T_{in}. Assume a value of $\dot{q}_{\,1}$. Necessary physical properties are: c_{pl}, λ_l, η_l, h_{lg}, ρ_l, ρ_g σ. Where not specified, physical properties are calculated at the saturated state at p_{ex}.

$$T_{in} + \frac{\dot{q}A}{\dot{M}\, c_{pl}} = \frac{5}{s^+(R)}\ T_{m1} + \frac{25}{s^+(R)}\ T_{m2} + \frac{s^+(R) - 30}{s^+(R)}\ T_{m3}$$

where c_{pl} is calculated at $(T_m + T_{in})/2$ and T_{m1}, T_{m2} and T_{m3} are calculated from the temperature distributions:

$$T_w - T = QPr\, s^+ \qquad 0 \leq s^+ < 5$$

$$T_w - T = 5Q\left\{Pr + \ln\left[1 + Pr\left(\frac{s^+}{5} - 1\right)\right]\right\} \qquad 5 \leq s^+ < 30$$

$$T_w - T = 5Q\left[Pr + \ln(1+5Pr) + 0.5\ln\left(\frac{s^+}{30}\right)\right] \qquad s^+ \geq 30$$

$$Q = \frac{\dot{q}}{\rho_l c_{pl} u_\tau}$$

In the above temperature distribution equations, c_{pl} is calculated at saturated conditions at p_{ex}, s^+ is the non-dimensional distance from the wall, and u_τ is the friction velocity. From the above calculation the wall temperature T_W is obtained. Using the above temperature distribution equations it is possible to calculate s*, that is the value of the distance from the heated wall, s, at which the fluid temperature is equal to the saturation value at p_{ex}. Calculation of D_B:

$$D_B = \frac{32}{f} \frac{\sigma\, f(\beta)\, \rho_l}{\dot{m}^2} \qquad \frac{1}{\sqrt{f}} = 1.14 - 2.0 \log\left(\frac{0.72\, \sigma \rho_l}{f\, D\, \dot{m}^2} + \frac{9.35}{Re\sqrt{f}}\right)$$

where $f(\beta) = 0.03$. Calculation of δ and C_D

$$\delta = s^* - D_B \qquad C_D = \frac{2}{3} \frac{D_B}{\left(\frac{\sigma}{g(\rho_l - \rho_g)}\right)^{0.5}}$$

Calculation of u_B and L_B (linked each other) through an iterative procedure:

$$u_B = \left(\frac{2 L_B\, g\, (\rho_l - \rho_g)}{\rho_l C_D}\right)^{0.5} + 0.125\left(\delta + \frac{D_B}{2}\right)\frac{f \dot{m}^2}{\rho_l \eta_l}$$

$$u_B = \left(\frac{2 L_B\, g\, (\rho_l - \rho_g)}{\rho_l C_D}\right)^{0.5} + 1.768\sqrt{f}\frac{\dot{m}}{\rho_l}\left\{\ln\left[0.354\frac{\dot{m}}{\eta_l}\sqrt{f}\left(\delta + \frac{D_B}{2}\right)\right] - 0.61\right\}$$

$$u_B = \left(\frac{2 L_B\, g\, (\rho_l - \rho_g)}{\rho_l C_D}\right)^{0.5} + 0.884\sqrt{f}\frac{\dot{m}}{\rho_l}\left\{\ln\left[0.354\frac{\dot{m}}{\eta_l}\sqrt{f}\left(\delta + \frac{D_B}{2}\right)\right] + 2.2\right\}$$

where L_B is given by

$$L_B = \frac{2\pi\sigma(\rho_l + \rho_g)}{\rho_l \rho_g u_B^2}$$

Calculation of $\dot{q}_2$:

$$\dot{q}_{CHF} = \frac{\rho_l\, \delta\, h_{lg}}{L_B} u_B$$

The condition of critical heat flux, $\dot{q}_{CHF}$, is reached when $\dot{q}_1 = \dot{q}_2$.

2 CHF in Saturated Flow Boiling

Forced convection saturated flow boiling involves a boiling liquid, whose average bulk temperature is at the saturation temperature, flowing over a surface exposed to a heat flux. The critical heat flux always occurs with a positive quality at the CHF. Generally speaking, under saturated conditions, we may have two different types of CHF: i) the DNB type, typically occurring at low quality conditions, and ii) the dryout type, which is encountered in high quality flow. Although the two different types of CHF are much different each other from the phenomenological point of view, this kind of classification is somewhat schematic, the threshold being very difficult to be established. As the quality at the CHF increases we gradually pass from DNB to dryout. An interesting simple method to identify *a priori* the CHF type has been recently given by Lombardi & Mazzola (1998).

None the less, although DNB and dryout types of the CHF are associated with different mechanisms leading to the onset of thermal crisis, parametric trends of the CHF in saturated flow boiling may be more or less independent of the CHF mechanisms, and the general trends can be given for the CHF in saturated flow boiling.

2.1 Parametric Trends

The magnitude and the occurrence of the CHF are affected by many parameters such as thermal-hydraulic geometric and external parameters. Among thermal-hydraulic parameters we have subcooling, mass flux, pressure, while important geometry parameters are channel diameter, heated length, channel orientation, tube wall thickness. External parameters of interest are heat flux distribution and binary component fluids.

Influence of subcooling. For fixed mass flux $\dot{m}$, tube length L, and tube diameter D, the CHF increases almost linearly with inlet subcooling, but the effect decreases with decreasing mass flux, as reported in Figure 2.1, where data of Weatherhead (1963) are plotted. At a mass flux of 500 kg/m^{2}s Moon et al. (1996) observed that the inlet subcooling effect on the CHF is very small, suggesting that it can be negligible at much lower mass fluxes (Mishima (1984), Chang et al. (1991)). If we plot the same data of Figure 2.1 in terms of exit conditions, see Figure 2.2, we find an interesting feature, which accounts for the inter-relation between exit quality and mass flux effects on the CHF. In the subcooled region ($x < 0$) the CHF increases as mass flux increases for a given exit quality x. In the saturated region ($x > 0$) we may find a cross-over, and the CHF decreases with increased mass flux, for a given x. It is therefore important to establish which variables are kept constant when considering the influence of a specific variable on the CHF, also specifying if we refer to inlet or exit condition.

Influence of mass flux. For fixed inlet conditions and geometry, the CHF increases with increasing mass flux. At low values of $\dot{m}$, the CHF rises approximately linearly with $\dot{m}$, but then rises much less rapidly for higher $\dot{m}$ values. The effect of mass flux on the CHF depends on the pressure, being stronger at lower pressures. The influence of mass flux on the CHF for fixed exit conditions has been already outlined in the previous sections: the CHF increases with $\dot{m}$ for $x < 0$, while decreases with $\dot{m}$ for $x > 0$, being x the exit quality.

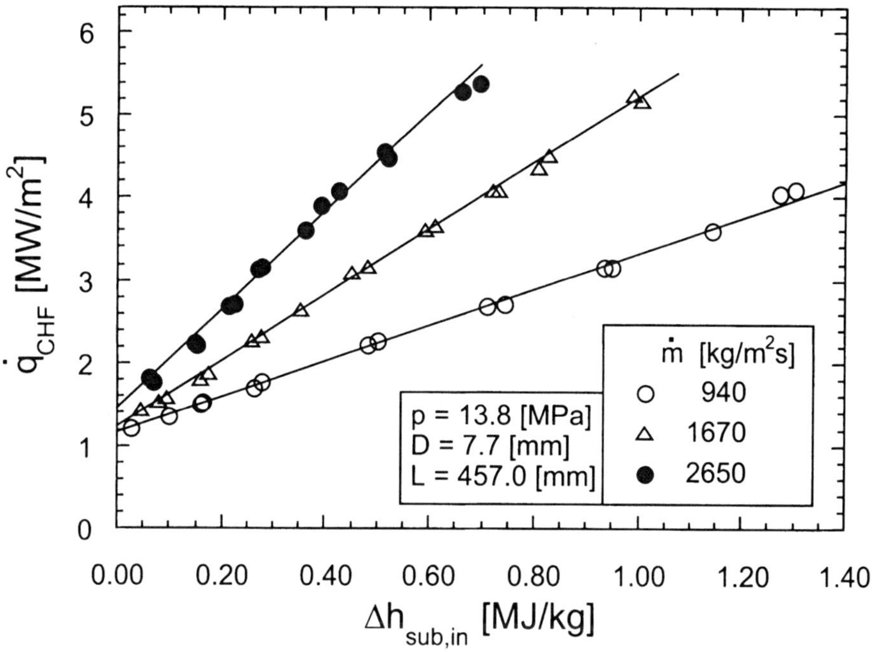

Figure 2.1. Critical heat flux versus inlet subcooling, for different mass fluxes.

Influence of pressure. The influence of the pressure on the CHF is very complex as indicated by Collier & Thome (1994), and reported in Figure 2.3, where data of Alekseev et al. (1965) are plotted. In overall, for fixed inlet conditions the CHF increases with increasing pressure at low pressure, passes through a maximum, at around 3.0 MPa, and then decreases at higher pressures. Yin et al. (1988) experienced a secondary maximum at 19.0 MPa for $\dot{m}$ = 2040 kg/m^{2}s and inlet subcooling of 33 and 55 K. For fixed exit conditions, Moon et al. (1996) report a clearer trend than that for fixed inlet conditions. As the pressure increases the CHF sharply increases, passes a maximum, then gradually decreases. The pressure corresponding to the maximum CHF decreases as quality increases.

Influence of diameter. The effect of tube diameter on the CHF for fixed inlet and exit conditions is shown in Figs. 2.4 and 2.5, respectively. For fixed inlet conditions, the CHF increases with increasing tube diameter, the effect increasing with the inlet subcooling. For fixed exit conditions, the CHF is a decreasing function of tube diameter. It appears that the diameter effect strongly depends on the flow regime due to the difference in CHF mechanisms.

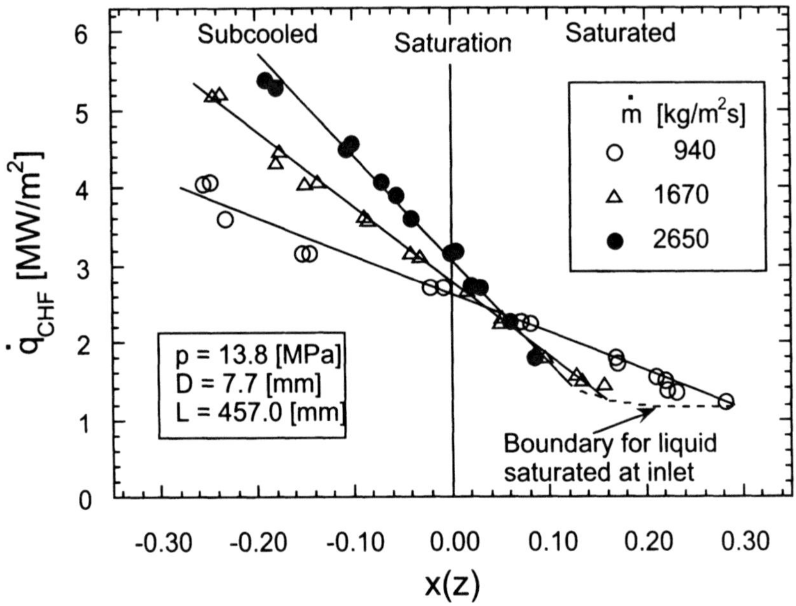

Figure 2.2. Critical heat flux versus exit quality, for different mass fluxes.

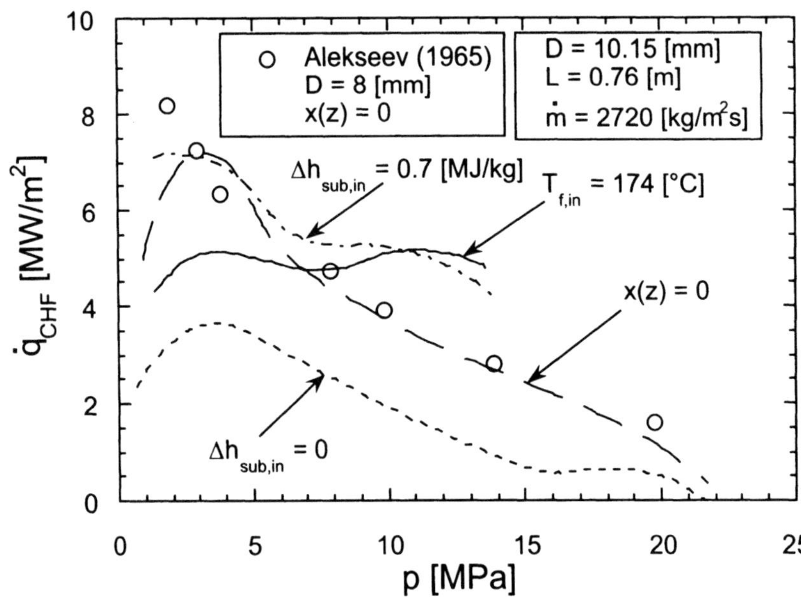

Figure 2.3. Influence of the pressure on the critical heat flux.

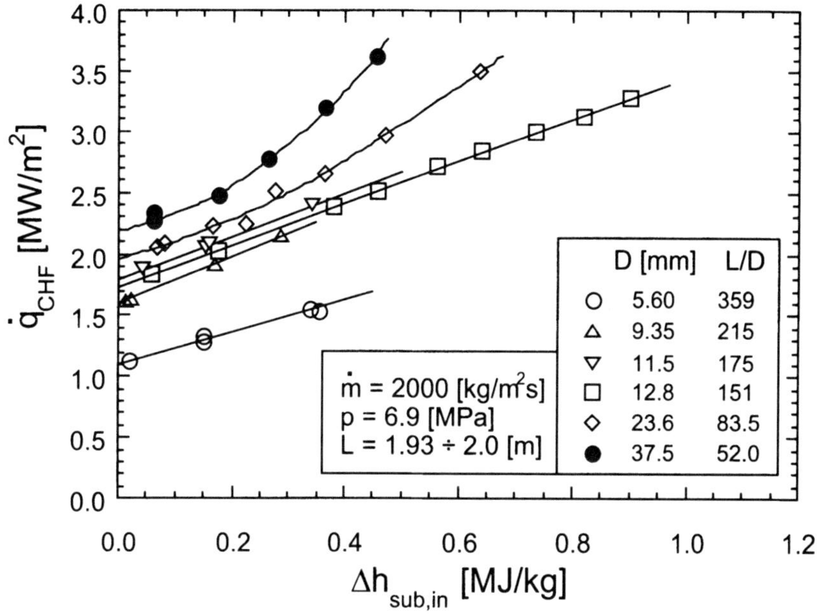

Figure 2.4. Effect on tube diameter on the CHF for fixed inlet conditions.

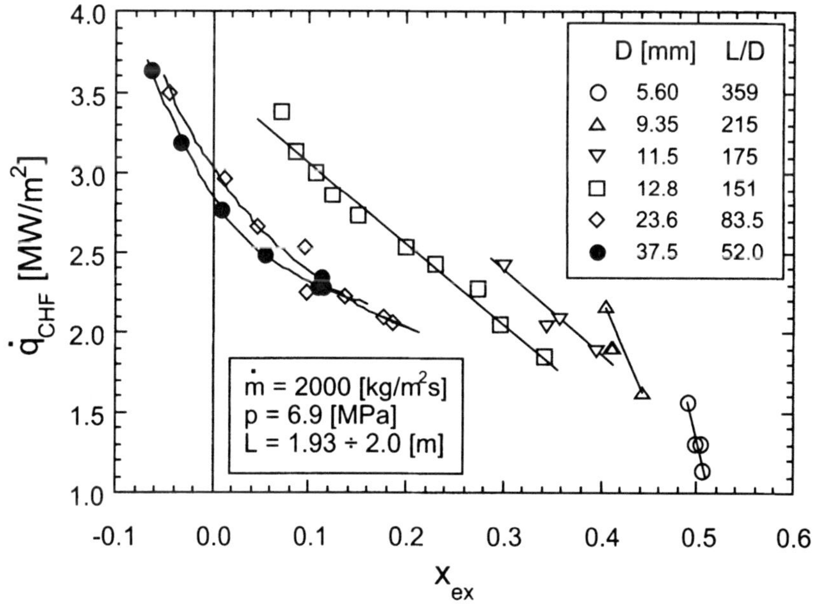

Figure 2.5. Effect on tube diameter on the CHF for fixed exit conditions.

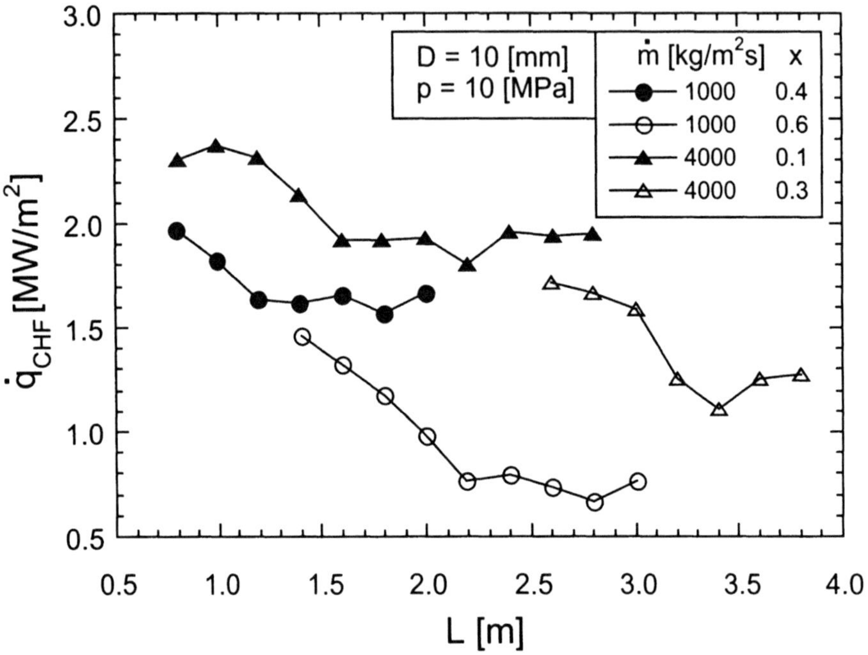

Figure 2.6. Effect on tube length on the CHF for fixed exit conditions.

Influence of heated length. For fixed inlet conditions there is a common evidence (Collier & Thome (1994), Hewitt (1982) and Chang et al. (1991)) that the CHF decreases with increasing heated length. For fixed exit conditions, from the interesting study of Moon et al. (1996), reported in Figure 2.6, we may say that for short tubes the CHF decreases with the heated, while for heated lengths above a threshold the heated length effect would seem to disappear. The threshold length is a function of other system parameters.

Effect of channel orientation. Vertical downflow against upflow CHF studies have been performed among others by Papell et al. (1966) using liquid nitrogen, Kirby et al. (1967) using water, and Bertoni et al. (1976) using R-12. Generally speaking, downflow CHF was found to be 10-30% lower than upflow, buoyancy effects playing the main role in the reduction. The buoyancy effect was found to be an inverse function of pressure and subcooling, and was proved to be small if the liquid downflow velocity is significantly above the bubble rise velocity.

Among other, Becker (1971) found that the CHF for horizontal tubes results lower than that experienced for vertical upflow if the mass flux is lower than a critical value. This is because bubbles formed in the nucleate boiling regime move upwards due to gravity and concentrate in the upper region of the tube, thus causing a premature burnout, with respect to vertical upflow, as the void increases. Larger diameter tubes require larger critical mass fluxes to avoid the separation of the phases. Cumo et al. (1978) carried out experiments using R-114 at different pipe inclinations between horizontal and vertical upflow conditions included. The tube inclination has a significant influence on the CHF, which varies up to a factor of two passing from horizontal to upward vertical flow. Authors found that the buoyancy effect on the CHF may be neglected when the modified Froude number, as given by Eq. (1-1), is greater than 5-7.

Influence of wall thickness. Relatively little information is available on the effect of wall thickness. As reported by Collier & Thome (1994), some experiments on the wall thickness effect were performed by Aladyev et al. (1961), Barnett (1963), Lee (1965), and Tippets (1962). Results are quite contradictory, as Aladyev et al. (1961) did not find any effect in the range 0.4 to 2.0 mm, Lee (1965) observed a 5% reduction as the tube wall thickness is decreased from 2.1 to 0.86 mm, and Tippets (1962) found up to 20% decrease as a 0.254 mm ribbon heater was replaced by a 0.152 mm thick ribbon.

Influence of heat flux distribution. The effect of the axial heat flux distribution has been investigated, for example, by Keeys et al. (1972) and by Cumo et al. (1980). Authors found a considerable difference in heat flux for burnout at a given quality for the uniform and non-uniform heating mode, noting that, with the non-uniform heating burnout can occur first upstream of the end of the tube.

Influence of mixture composition. The effect of composition on the CHF in the case of binary mixtures has been studied by Auracher & Marroquin (1995), Celata et al. (1994d), and Mori et al. (1990). The composition of the binary mixture has little or no effect on the CHF for long tubes, i.e., $L/D > 30$. For shorter tubes, as also reported by Collier & Thome (1994), the CHF increases with the mixture composition, passes through a maximum, and then decreases, all with respect to the ideal linear behaviour between the values of the pure fluids, for same thermal hydraulic conditions.

2.2 Available Correlations for the Prediction of Saturated Flow Boiling CHF

For given fluid, thermal-hydraulic and geometric conditions, and for a given heat flux, axially uniform, experimental data are usually found to lie approximately on a single curve in a CHF versus burnout quality representation, being the CHF located at the end of the channel. This implies that the local quality conditions govern the magnitude of the CHF, and is termed as *local conditions hypothesis*.

We can plot the same data in terms of burnout quality and boiling length at burnout, this latter being the length between the location where the saturation condition is reached and the CHF location. The boiling length is easily obtained form a heat balance knowing heat flux, quality, mass flux and tube geometry. This type of plot can be regarded a indicating the possibility of some *integral* rather than *local* phenomenon.

Existing correlations are given in one of the two above reported forms and, for uniform heat flux, can be converted easily to the other, providing with equivalent results. When the heat flux is non-uniform, the two forms give quite different results, and this will be discussed later.

Referring the reader also to other sources collecting CHF correlations, such as Lee (1977), Katto (1986), Whalley (1987) and Collier & Thome (1994), some widely used correlations for uniform heat flux are reported hereunder, for which great care is recommended in their application. As usually such correlations are not based on a physical background, they should be regarded as mathematical interpolation for the data range they cover. Their use outside this range can give high inaccuracy in the prediction.

- CISE, Bertoletti et al. (1965)

$$\frac{\dot{q}CHF}{\pi\, D\, L\, \dot{M}\, h_{lg}} = \frac{a - x_{in}}{1 + \frac{b}{L}} \tag{2-1}$$

where $\dot{q}\ CHF$ is the critical heat flux in kW/cm^2, D and L are the tube internal diameter and length, respectively, in cm, $\dot{M}$ the mass flow rate in g/s, and

$$a = \left(1 - \frac{p}{p_c}\right)\left(\frac{\dot{m}}{\dot{m}_o}\right)^{-0.33} \qquad \dot{m}_o = 100 \text{ g/cm}^2\text{s}$$

$$b = 0.315\left(\frac{p_c}{p} - 1\right)^{0.4} D_h^{1.4}\, \dot{m}$$

being p_c the water critical pressure, and D_h the equivalent hydraulic diameter in cm (recommended ranges: p = 45 - 150 kg/cm^2; $100\,(1 - p/p_c)^3 \leq \dot{m} \leq 400$ g/cm^{2}s; $x_{in} \leq 0.2$; $D > 0.7$ cm; L = 20.3 - 267 cm).

- W-3, Tong (1969)

$$\frac{\dot{q}CHF}{10^6} = \{(2.022 - 0.0004302\, p) + (0.1722 - 0.0000984\, p) \exp[(18.177 - 0.004129\, p)x]\}$$

$$[(0.1484 - 1.596x + 0.1729\, x\,|x|\,)\, \dot{m}/10^6 + 1.037]\,(1.157 - 0.869\, x)\,[0.2664 + 0.8357 \exp(-3.151\, D_h)]\,[0.8258 + 0.000784\,(h_l - h_{in})] \tag{2-2}$$

The heat flux $\dot{q}\ CHF$ is in Btu/(hr)(ft^2) (recommended range and units of the parameters are: p = 1000-2300 psia; $\dot{m}$ = 1.0 10^6 - 5.0 10^6 lb/(hr)(ft^2); D_h = 0.2 - 0.7 in; x = - 0.15 to + 0.15; $h_{in} \geq 400$ Btu/lb; L = 110 - 144 in; heated perimeter/wetted perimeter = 0.88 - 1.0).

- Bowring (1972)

$$\dot{q}\ CHF = \frac{A + 0.25\, D\, \dot{m}\, (\Delta h_{sub})_{in}}{F + L} \tag{2-3}$$

$$A = \frac{2.317\left[\frac{D\,\dot{m}\,h_{lg}}{4}\right]F_1}{1.0 + 0.0143\,F_2\,\dot{m}\,D^{1/2}}\ ;\ F = \frac{0.077\,F_3\,D\,\dot{m}}{1.0 + 0.347\,F_4\,(\dot{m}/1356)^n}\ ;\ n = 2.0 - 0.00725\,p$$

where $\dot{q}_{CHF}$ is the critical heat flux in W/m^2, $(\Delta h_{sub})_{in}$ is the inlet subcooling expressed in J/kg, L is the tube length expressed in m, D is the internal tube diameter in m, $\dot{m}$ the mass flux in kg/m^{2}s, h_{lg} is the latent heat of vaporization in J/kg, and p is the system pressure in bar. Parameters F_1, F_2, F_3, and F_4 are given by:

$p' = p/69$

$p' < 1$

$$F_1 = \frac{\{p'^{\,18.942}\exp[20.8\,(1 - p')]\} + 0.917}{1.917}\ ;\ \frac{F_1}{F_2} = \frac{\{p'^{\,1.316}\exp[2.444\,(1 - p')]\} + 0.309}{1.309}$$

$$F_3 = \frac{\{p'^{\,17.023}\exp[16.658\,(1 - p')]\} + 0.667}{1.667}\ ;\ \frac{F_4}{F_3} = p'^{\,1.649}$$

$p' > 1$

$$F_1 = p'^{\,-0.368}\exp[0.648\,(1 - p')]\ ;\ F_1 = p'^{\,-0.448}\exp[0.245\,(1 - p')]$$

$$F_3 = p'^{\,0.219}\ ;\ \frac{F_4}{F_3} = p'^{\,1.649}$$

(recommended ranges: p = 2 - 190 bar; D = 0.002 - 0.045 m; L = 0.15 - 3.7 m; $\dot{m}$ = 136 - 18600 kg/m^{2}s).

- *Katto & Ohno (1984)*

a) In the case of $\rho_g/\rho_l < 0.15$

$$\frac{\dot{q}_{CHF}}{\dot{m}h_{lg}} = \frac{C}{(l_b/D)}\left(\frac{\sigma\rho_l}{\dot{m}^2 l_b}\right)^{0.043} \tag{2-4}$$

$$\frac{\dot{q}_{CHF}}{\dot{m}h_{lg}} = 0.10\,(\rho_g/\rho_l)^{0.133}\left(\frac{\sigma\rho_l}{\dot{m}^2 l_b}\right)^{1/3}\frac{1}{(1 + 0.0031\,l_b/D)} \tag{2-5}$$

$$\frac{\dot{q}_{CHF}}{\dot{m}h_{lg}} = 0.098\,(\rho_g/\rho_l)^{0.133}\left(\frac{\sigma\rho_l}{\dot{m}^2 l_b}\right)^{0.433}(l_b/D)^{\,0.27}\frac{1}{(1 + 0.0031\,l_b/D)} \tag{2-6}$$

where C is given as $C = 0.25$ for $l_b/D < 50$, $C = 0.25 + 0.0009\,[(l_b/D) - 50]$ for $l_b/D = 50$ - 150, and $C = 0.34$ for $l_b/D > 150$, being l_b the boiling length. Roughly speaking, Eqs. (2-4) and (2-5) correspond to the CHF in annular flow, and Eq. (2-6) to the CHF in froth or bubbly flow. With increasing $\dot{m}$ (i.e., with decreasing $\sigma\rho_l/\dot{m}^2 l_b$), the above equations are employed in the order of the first, second, and third equation so as to connect the value of the CHF continuously.

b)In the case of $\rho_g/\rho_l > 0.15$

$$\frac{\dot{q}_{CHF}}{\dot{m}h_{lg}} = C\left(\frac{\sigma\rho_l}{\dot{m}^2 l_b}\right)^{0.043}(l_b/D) \tag{2-7}$$

$$\frac{\dot{q}_{CHF}}{\dot{m}h_{lg}} = 0.234\,(\rho_g/\rho_l)^{0.513}\left(\frac{\sigma\rho_l}{\dot{m}^2 l_b}\right)^{0.433}(l_b/D)^{0.27}\frac{1}{(1+0.0031\,l_b/D)} \tag{2-8}$$

$$\frac{\dot{q}_{CHF}}{\dot{m}h_{lg}} = 0.0384\,(\rho_g/\rho_l)^{0.6}\left(\frac{\sigma\rho_l}{\dot{m}^2 l_b}\right)^{0.173}\frac{1}{(1+0.28\,(\sigma\rho_l/\dot{m}^2 l_b)^{0.233}\,l_b/D)} \tag{2-9}$$

where C takes the same value as in Eq. (2-4) (recommended ranges: $L = 0.01$ - 8.8 m; $D =$ 0.001 - 0.038 m; $L/D = 5$ - 880; $\rho_g/\rho_l = 0.00003$ - 0.41; $(\sigma\rho_l/\dot{m}^2 L) = 3\ 10^{-9}$ - $2\ 10^{-2}$).

The Katto & Ohno (1984) correlation has been tested for water, ammonia, benzene, ethanol, helium, hydrogen, nitrogen, R12, R21, R22, R113, and potassium.

Correlations for the CHF in binary mixtures. The above reported correlations have been developed for pure fluids such as water (CISE, W-3, and Bowring, 1972) or more fluids (Katto & Ohno, 1984). Much different is the case where we have to face with binary mixtures. An exhaustive description of the CHF in binary mixtures can be found in Collier & Thome (1994), while Celata et al. (1994d), Auracher & Marroquin (1995) and Celata & Cumo (1996) dealt specifically with refrigerant binary mixtures. Upon results obtained with mixtures of refrigerants and on the basis of the parametric trends described in 2.1.9, it is possible to say here that for short tubes, i.e., $L/D \leq 30$, the CHF can be calculated using the Tolubinsky & Matorin (1973) correlation, given by Eqs. (1-7) and (1-9). For long tubes, i.e., $L/D > 30$, Celata et al. (1994d) found that the CISE correlation, proposed by Bertoletti et al. (1965), provides quite good results. Also the Katto & Ohno (1984) correlation may be directly applied to binary mixtures in long tubes, although the accuracy is less than the CISE correlation.

Correction for axial non-uniform heat flux. For non-uniform heat flux single channels, Tong et al. (1966) recommends to use a shape factor F_c so that:

$$\dot{q}_{CHF,nu} = \frac{\dot{q}_{CHF,u}}{F_C} \qquad (2\text{-}10)$$

where subscript *nu* indicates the non-uniform heating and subscript *u* indicates uniform heating supply, and where F_C is expressed as:

$$F_C = \frac{C}{\dot{q}_{loc}\left[1 - \exp\left(-C\, l_{CHF,u}\right)\right]} \int_{l_{OB}}^{l_{CHF,nu}} \dot{q}(z) \exp\left[-C\left(l_{CHF,nu} - z\right)\right] dz \qquad (2\text{-}11)$$

with

$$C = 0.44 \frac{\left(1 - x_{CHF,nu}\right)^{7.9}}{\left(\dot{m}/10^6\right)^{1.72}} \quad (\text{in.}^{-1})$$

In Eq. (2-11) $\dot{q}_{loc}$ is calculated using one of the available correlations for uniform heat flux, and $l_{CHF,nu}$ is the axial location at which the CHF occurs for non-uniform heat flux, in., $l_{CHF,u}$ is the axial location at which the CHF occurs for uniform heat flux, in., l_{OB} is the axial location at which nucleate boiling begins, in., $x_{CHF,nu}$ is the quality at the CHF location under non-uniform heat flux, and $\dot{m}$ is in lb/(hr)(ft^2). The term F_C is a *memory effect parameter* which accounts for the thermal history of the fluid along the tube. F_C is small in the subcooled region and local heat flux determines the boiling crisis. At high qualities, C is small, the memory effect is high, and the average heat flux, or enthalpy rise, primarily determines the boiling crisis.

2.3 The Artificial Neural Network as a CHF Predictor

An advanced information processing technique such as artificial neural networks (ANNs) (Wasserman, 1989) might provide a valuable alternative to the current techniques for estimating the CHF, since there exists a large number of experimental data for the CHF. Yapo et al. (1992), Moon & Chang (1994), Moon et al. (1996), Mazzola (1997) applied the ANNs to the CHF prediction, showing promising results. An artificial neural network is composed of elements that are analogous to the elementary functions of biological neurons. ANNs have the characteristic of tolerance against experimental noise owing to the massive internal structure of the network. Also, it is easy to update the performance of the ANN for new experimental data.

Although the ANNs do not require accurate information about physical phenomena, however, their main drawbacks are the loss of model transparency (*black-box* character) and the lack of any indicator for evaluating the accuracy and reliability of the ANN answer when *never-seen* patterns are presented. From applications to CHF of Moon et al. (1996) and Mazzola (1997), it appears, none the less, that the ANNs are able to predict CHF data within ± 20-25% for most of data points, providing a consistent alternative method to empirical correlations.

2.4 The Tabular Method for the Prediction of Saturated Flow Boiling CHF

Another interesting method for the prediction of the CHF in saturated flow boiling is that proposed first by Doroshchuk et al. (1975) which consists in a series of standard tables of CHF values as a function of the local bulk mean water condition and for various pressures and mass fluxes for a fixed tube diameter of 8 mm. Correction factors for tube length and for tube diameters other than 8 mm must be used. The latest updating of these *look-up table*, (Groeneveld et al. 1996) consists of 22946 data points covering the range 0.1 to 20.0 MPa, up to 8.0 Mg/m^2s and -0.5 to 1.0 for discrete values of pressure, mass flux and CHF quality, respectively. For tube diameters other than 8 mm, the CHF is given by the approximate equation:

$$\dot{q}\,\mathrm{CHF} = \dot{q}\,\mathrm{CHF,\ 8\ mm}\left(\frac{D}{0.008}\right)^{k} \qquad (2\text{-}12)$$

being $k = -1/2$ the best parameter found by Groeneveld et al. (1996) in the range of tube diameter from 3 to 25 mm. Other researchers propose $k = -1/3$, such as Smith (1986) and Groeneveld et al. (1986).

The CHF look-up table method has become a widely accepted prediction technique. It has the following advantages over correlations or semi-analytical CHF models: i) accurate prediction; ii) the widest range of applications; iii) ease of use (no fluid properties are needed); iv) ease of updating; and v) correct parametric and asymptotic trends. Main drawbacks are the complexity of their use in a computer code with respect to a correlation, providing more or less the same accuracy.

2.5 Available Models for the Prediction of Saturated Flow Boiling CHF

The main advantage of mechanistic methods, is that, as they are based on the physical mechanisms leading to the CHF, in principle their validity should not be confined to the range of the available experimental data on which they are assessed. The models should be only linked to the range of validity of the mechanisms identified, which should result of much more general application. As a matter of fact, sometimes some models for the mathematical description of bubble dynamics, rely on empirical constants or correlations which restrict their general validity. As a model is strictly linked to the mechanisms which can be responsible of the CHF occurrence, it is necessary a grouping of existing models in DNB and dryout models.

DNB type critical heat flux. Different CHF mechanisms have been postulated for the DNB type thermal crisis, in order to develop reliable correlations or predicting methods for the CHF calculation, or to identify possible methods to avoid the CHF occurrence. Typically, for low quality flow, the flow regime consists in an agglomeration of vapour in the near-wall region, and a prevailing presence of liquid in the centre of the channel. The governing heat transfer mechanisms is the bubble growing and detachment at the wall, and their migration in the liquid bulk. Among the many mechanisms proposed, see, for instance, detailed reviews by Tong & Hewitt (1972), Hewitt (1980), Weisman (1992), and Katto (1994), those which appear to be somehow established experimentally are the following:

a) Hot spot formation under a growing bubble. As observed by Kirby et al. (1967), a dry patch forms between the growing bubble and the nucleation cavity as the micro-layer of liquid under the bubble evaporates. The dry patch may be rewetted at the bubble departure and the process can go on. Before the rewetting of the dry patch, the wall temperature rises due to the heat transfer deterioration. However, if the dry temperature exceeds a critical temperature (often called Leidenfrost temperature), then rewetting does not happen readily, thus causing local overheating and hence burnout. A schematic of this mechanism is drawn in Figure 2.7.

b) Near-wall bubble crowding model. Tong et al. (1966) first started from the idea that a *bubble boundary layer* takes place on the surface and vapour generated by boiling at the heated wall must leave the near-wall region through this two-phase boundary layer. Burnout occurs when vapour escape through the layer is prevented because of a critical crowding of the boundary layer with bubbles. More recently, Hebel et al. (1981) and Weisman & Pei (1983) and Weisman & Ying (1983) assumed that the turbulent interchange between the bubbly layer and the bulk of the liquid may be the limiting mechanism leading to the CHF occurring. CHF occurs when bubble crowding near the heated wall prevents the bulk cold liquid from reaching the wall. This mechanism is discussed in more detail below.
c) Dryout under a slug or vapour clot. Fiori & Bergles (1968, 1970) observed that in plug flow, the thin liquid film around the large bubble may dry out causing burnout. Alternatively, a stationary vapour clot can form on the heated wall, being a thin liquid film present between the clot and the wall. In this case the local drying out of the film causes wall overheating and then burnout. A schematization of this mechanism is shown in Figure 2.8.
d) Liquid sublayer dryout theory. This mechanism has been already discussed for the understanding of the CHF in subcooled flow boiling, chapter 1.3. The Lee & Mudawwar (1988) liquid sublayer dryout model was developed for subcooled flow boiling, on the basis of the Helmholtz instability at the microlayer/vapour interface as trigger condition for microlayer dryout. Such a model has been extended to low-quality flow by Lin et al. (1989) under pressurized water reactor conditions. Basically, the main improvements of Lee & Mudawwar's model include the following: 1) The homogeneous two-phase flow model is assumed to be suitable for high-pressure, high-mass flux conditions. Fluid properties are calculated using the effective homogeneous flow rather than single-phase fluid properties. 2) The liquid enthalpy flowing into the microlayer is assumed to be independent of bulk subcooling and is approximated by the saturated liquid enthalpy for maintaining the local boiling.

Among the models listed above, it is interesting to give few details on the Weisman & Pei (1983) model, above described in b), which is currently the only theoretically based CHF prediction procedure that has been shown to give good accuracy with fluids other than water, especially with refrigerants.

The Weisman & Pei (1983) model, the schematic of which is drawn in Figure 2.9, assumes that: a) During low-quality boiling, the bubbly layer builds up along the channel until it fills the region near the wall where the turbulent eddies are too small to transport bubbles radially. At the CHF site, the bubbly layer is assumed to be at this maximum thickness. b) CHF occurs when the volume fraction of steam in the bubbly layer just exceeds the volume fraction (critical void fraction) at which an array of slightly flattened ellipsoidal bubbles can be maintained without significant contact between the bubbles. c) The volume fraction of steam in the bubbly layer is determined by a balance between the outward flow of vapor and the inward flow of liquid at the bubbly layer-core interface.

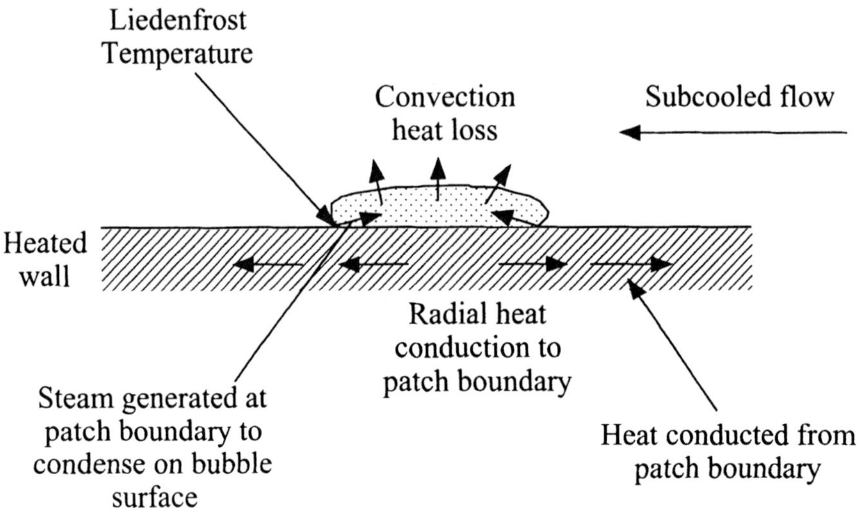

Figure 2.7. Schematic of the hot spot formation under a growing bubble model, Kirby et al. (1967).

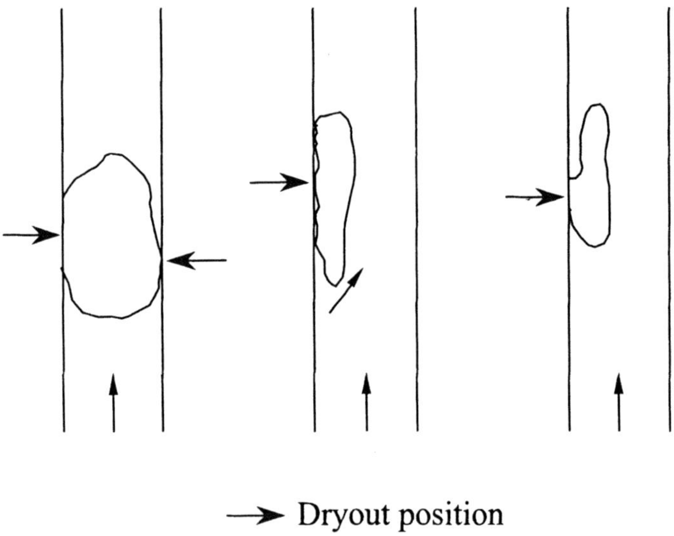

Figure 2.8. Schematic of the dryout under a slug or vapour clot model, Fiori & Bergles (1968, 1970).

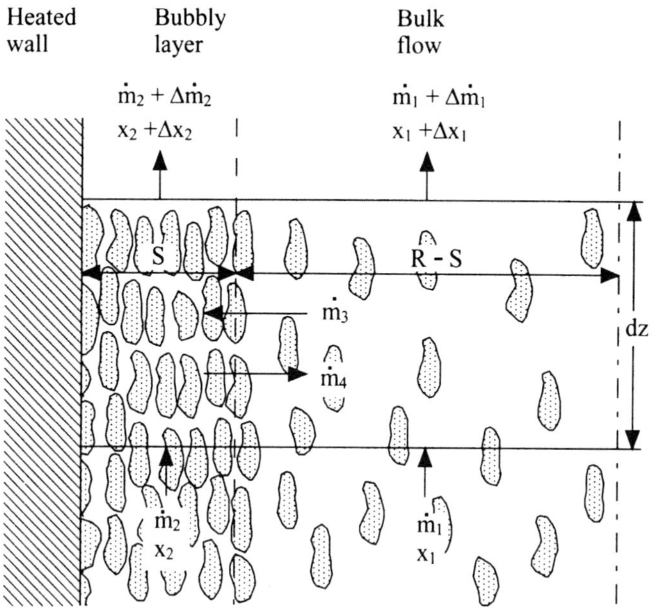

Figure 2.9. Schematic of the near-wall bubble crowding model, Weisman & Pei (1983).

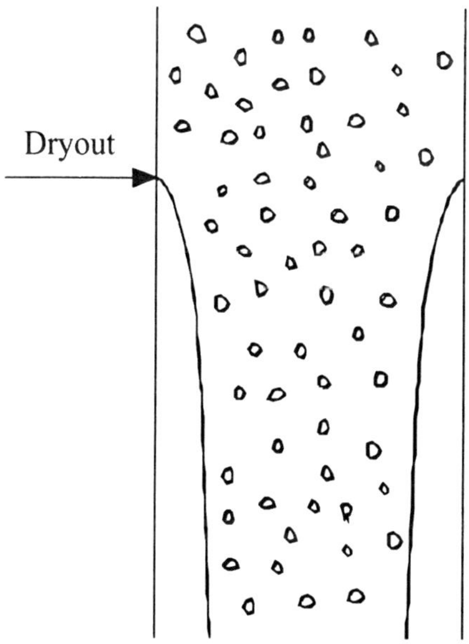

Figure 2.10. Schematic of the dryout type critical heat flux.

Considering a bubbly layer control volume, they can write the total mass balance on the bubbly layer taking into account the total flow rate from core to bubbly layer, which must be equal to the total flow rate from bubbly layer to core plus the axial flow in and out of the bubbly layer control volume. From a simple mass balance over the bubbly layer they obtain:

$$\frac{\dot{q}_{CHF}}{h_{lg}\dot{m}'} = \frac{x_2 - x_1}{F} \tag{2-13}$$

where $\dot{m}'$ represents the mass flow rate into the bubbly layer.

This mass flow rate is determined by the turbulent velocity fluctuations at the bubbly layer edge. The distance from the edge of the bubbly layer to the wall is taken as the distance at which the size of the turbulent eddies is k times the average bubble diameter. Only a fraction of the turbulent velocity fluctuations produced are assumed to be effective in reaching the wall. The effective velocity fluctuations are those in which the velocity exceeds the average vapour velocity away from the wall produced by the vapour being generated at the wall. The quantities x_1 e x_2 represent the vapour qualities in the core region and bubbly layer, respectively, at the CHF (these are actual values and not thermodynamic equilibrium qualities).

The factor F represents the fraction of the heat flux producing vapour that enters the core region, given by the ratio between the difference of the enthalpy of saturated liquid and that at bubble detachment point, and the difference between the enthalpy of liquid at given axial location and that at bubble detachment point. The occurrence of the CHF is for that quality in the bubbly layer that corresponds to the maximum void fraction that is possible in a bubbly layer of independent bubbles just prior agglomeration. For slightly flattened elliptically shaped bubbles with a length-to-diameter ratio of 3/1, this void fraction is estimated as 0.82.

Dryout type critical heat flux. This type of CHF mechanism consists in the gradual depletion of the liquid film wetting the heating wall, until the liquid film flow rate is zero and consequent drying of the wall. It is evident that the dryout type is linked to the annular flow regime in convective flow boiling, as reported in the sketch of Figure 2.10. Observations of transparent test sections and flow pattern maps show that, for most CHF cases where we have an exit quality greater than 10%, the flow pattern is annular. And this is probably the most frequent situation in steam generation apparatuses.

Many studies have suggested that the CHF may occur when the liquid film flow rate goes to zero due to the combined effects of: i) liquid droplet entrainment from the liquid film, produced by the gas flow in the core (droplets are mainly entrained from liquid waves on liquid film surface); ii) liquid droplet deposition on the liquid film (some droplets initially entrained by the gas flow hit the liquid film and are *captured*); and iii) evaporation of the liquid film because of the heat flux delivered from the wall.

The first evidence showing that dryout occurs at the point where the film flow rate becomes zero was due to the measurement of the film flow rate at the end of a heated channel as a function of power input to the channel, performed by Hewitt et al. (1963, 1965) and detailed in Hewitt & Hall-Taylor (1970). The results are drawn in Figure 2.11, where it is possible to observe that the critical heat flux point occurs at the power delivered to the fluid for which the film flow rate at the tube outlet is zero.

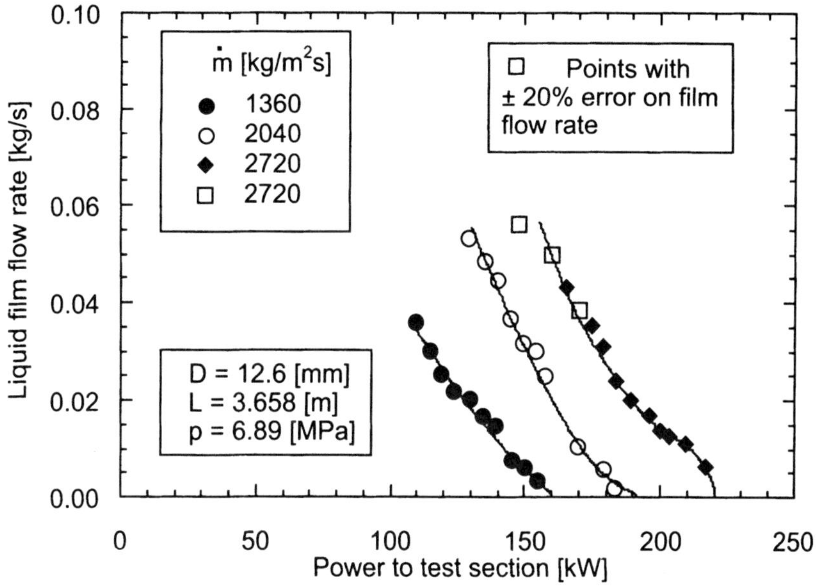

Figure 2.11. Measurement of the film flow rate at the end of a heated channel as a function of power input to the channel, Hewitt et al. (1963, 1965).

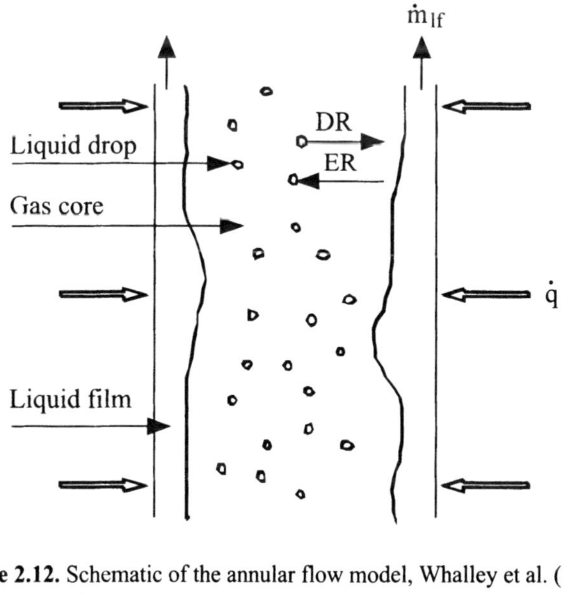

Figure 2.12. Schematic of the annular flow model, Whalley et al. (1974).

More exactly, the occurrence of dryout should happen when the liquid film flow rate becomes smaller than the minimum value which is necessary to wet the whole heating wall, and the liquid film breaks. Also the so-called *cold patch* experiments by Bennet et al. (1967) represent a further evidence of this CHF mechanism.

The first attempt to use an annular flow model for the prediction of dryout is due to Whalley et al. (1974), while the model has been recently updated by Govan et al. (1988) and by Hewitt and Govan (1989). For the complexity of the model description, the reader is referred to the original sources, while a brief review will be given here. Figure 2.12 shows the postulated mechanisms, in which dryout occurs when the liquid film flow rate falls smoothly to zero as a result of entrainment and evaporation. A mass balance, which also accounts for deposition, gives:

$$\frac{d\dot{m}_{lf}}{dz} = \frac{4}{D}\left(DR - ER - \frac{\dot{q}}{h_{lg}}\right) \tag{2-14}$$

Where $\dot{m}_{lf}$ is the liquid film mass flux, *DR* the deposition rate, and *ER* the entrainment rate. In order to integrate this equation, it is required:

i) a value for $\dot{m}_{lf}$ at the start of annular flow. Typically, it is assumed that at the start of annular flow $x_1 = 0.01$ and $\dot{m}_{lf1} = 0.99\ \dot{m}_1$. Govan (1984) found that the predicted CHF was sensitive to $\dot{m}_{lf1}$ but not to x_1. However, very little information exists on the transition to annular flow in a boiling channel.
ii) a means to calculate the entrainment rate *ER*. Whalley et al. (1974) expressed this as a function of surface tension, interfacial shear and liquid film thickness. Govan (1984) tried using various entrainment correlations but found that the CHF predictions were not greatly affected, mainly because the entrainment becomes small as dryout is approached.
iii) a means to calculate the deposition rate *DR*. Whalley et al. (1974) assumed a simple proportionality between *DR* and the droplet concentration in the gas core, the constant of proportionality depending on surface tension. Govan (1984) found that the predicted CHF is sensitive to *DR*.

This mechanism of dryout is widely accepted though there is some debate about the details. Anyway, recent updatings by Govan et al. (1988) and Hewitt & Govan (1989) demonstrated that comparison with 5300 CHF data points shows a mean error of –9.7% with a standard deviation of 16%, provided the CHF mechanism is dryout, for a wide range of fluids

3 Post-CHF Heat Transfer

Post-CHF heat transfer is of interest in all cases where the CHF condition can be reached or exceeded and the heating wall temperature is still low in comparison with the melting temperature or that value for which the wall material failure may happen. Heat transfer knowledge in these areas is required in many engineering applications such as in the design of once-through steam generators (where complete evaporation of the feedwater occurs), or of very high pressure recirculation boilers (where the CHF levels are low). The thermal-hydraulic design of pressurized water reactors has also called for an intensive investigation of heat transfer rates beyond the CHF point for transient and accident analyses.

Main heat transfer regimes in post-CHF heat transfer are film boiling and liquid deficient region. Film boiling typically occurs after the CHF in subcooled flow boiling, with low-quality CHF or in pool boiling. A schematic representation of such a heat transfer regime is given in Figure 3.1. The liquid deficient region or dispersed flow boiling, which occurs after the high-quality CHF is schematically drawn in Figure 3.2.

3.1 Film Boiling

In pool boiling or after the subcooled flow boiling CHF we may have the occurrence of the film boiling heat transfer regime once the CHF has been exceeded. The heat is transferred by conduction through the vapour film, and evaporation takes place at the liquid-vapour interface. Nucleation is absent and, in general, the problem may be simply treated as an analogy to film-wise condensation. Many theoretical solutions can be obtained for horizontal and vertical flat surface, and also inside and outside tubes under both laminar and turbulent conditions with and without interfacial stress. The simplest solution may be obtained for laminar flow and linear temperature distribution. For a flat vertical surface the local heat transfer coefficient is given by:

$$\alpha(z) = C\left[\frac{\lambda_g^3 \rho_g (\rho_l - \rho_g) g h_{lg}}{z \Delta T \eta_g}\right]^{1/4} \tag{3-1}$$

where C is dependent on boundary conditions; for zero interfacial stress we have $C = 0.707$, while for zero interfacial velocity we have $C = 0.5$. For film boiling outside a cylinder of diameter D we have $C = 0.62$ and Eq. (3-1) calculated for $z = D$.

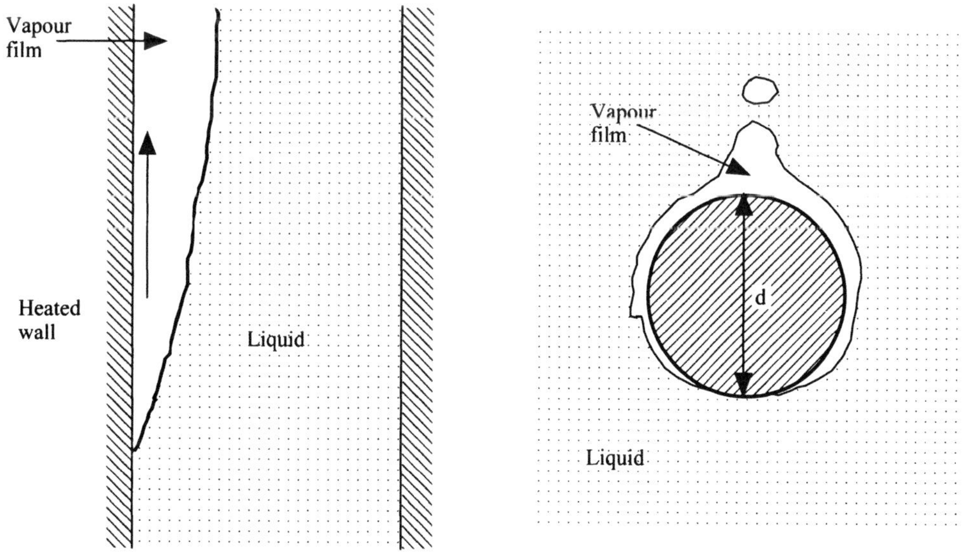

Figure 3.1. Film Boiling (a) on a vertical flat plate and (b) on a horizontal cylinder.

Wallis & Collier (from Collier & Thome (1994)) for turbulent flow in the vapour film found (vertical flat surface):

$$\frac{\alpha(z)}{\lambda_g} = 0.056\, Re_g^{0.2} \left[Pr Gr^*\right]^{1/3} \tag{3-2}$$

where:

$$Gr^* = \frac{z^3\, g\, \rho_g (\rho_l - \rho_g)}{\eta_g^2}$$

Fung et al. (1979) developed a model which covers both the laminar and the turbulent flow.

Although Eq. (3-1) gives good predictions in some cases (see Figure 3.3, where the Costigan et al. (1984) data for water in an 8 mm diameter vertical tube are compared with theoretical predictions), the vapour film is not smooth in reality (Dougall & Rohsenow (1963)), and more-refined equations are therefore necessary for a better physical description of the phenomenon (Bailey (1971), Denham (1984)). Further experimental evidences (Bromley et al. (1953), Motte & Bromley (1957), Liu et al. (1992), Papell (1970, 1971), Newbold et al. (1976)) can be summarized as follows: classical laminar film boiling may be a valid approximation up to 5 cm downstream of the CHF front; the heat transfer coefficient is an increasing function of the velocity and a decreasing function of the channel diameter (for film boiling inside and on tubes); the heat transfer coefficient in downflow is generally lower (up to 3-4 times) than in upflow. Information on hydrocarbons can be found in Glickstein & Whitesides (1967).

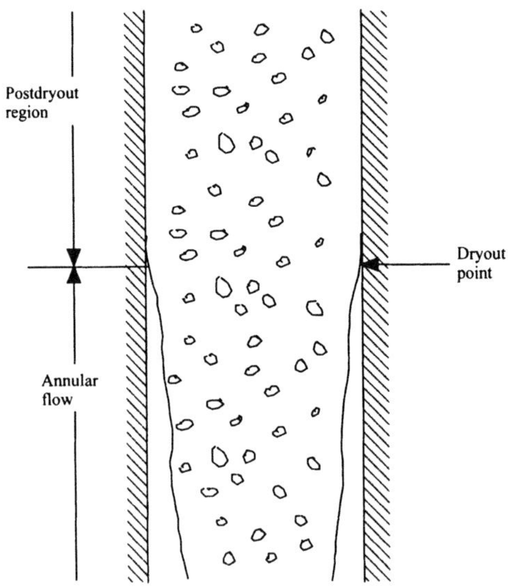

Figure 3.2. High quality-post CHF flow.

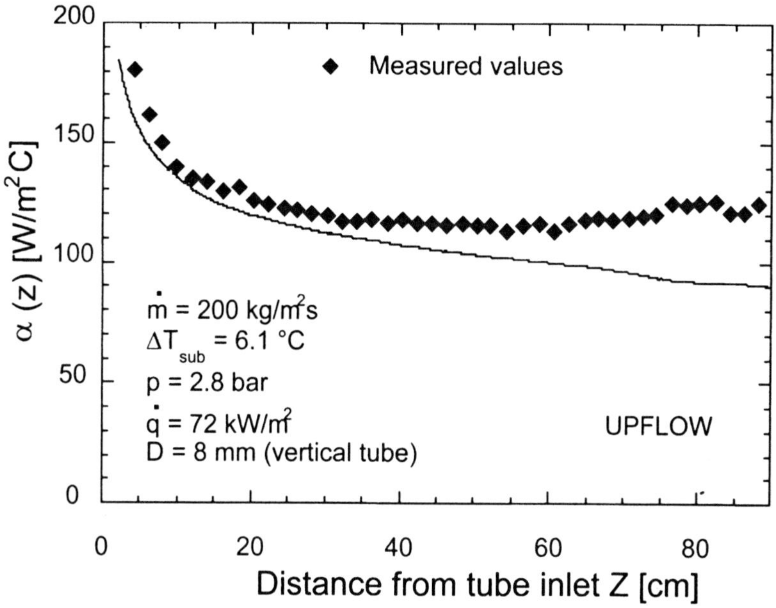

Figure 3.3. Film boiling heat transfer for water, Costigan et al. (1984).

3.2 Heat Transfer in the Liquid Deficient Region

This heat transfer regime is sketched in Figure 3.2, and its knowledge is important in the design of high-pressure once-through steam generators and recirculation boilers. Experimental data for steam-water mixtures, up to 25 MPa, have been produced in the past (Schmidt (1959), Swenson et al. (1961), Herkenrath et al. (1967), Bahr et al. (1969)). The liquid deficient region heat transfer in circular bends has been recently experimented (Lautenschlager & Mayinger (1986), Wang & Mayinger (1995)), together with the use of refrigerants (Lautenschlager & Mayinger (1986), Wang & Mayinger (1995), Nishikawa et al. (1986), Obot & Ishii (1988), Yoo & France (1996)). Kefer et al. (1989) studied the post-CHF heat transfer in inclined evaporator tubes, while Burdunin et al. (1987) and Unal et al. (1988) investigated complex geometries.

Three types of predictive tools have been adopted for the calculation of the heat transfer coefficient (generally through wall temperature calculation), as reviewed by Groeneveld (1972), and Wang & Weisman (1983):

a) empirical correlations (no theoretical background behind, but only functional equations between the heat transfer coefficient and independent variables);
b) correlations which take into account the thermodynamic non-equilibrium and calculate the true vapour quality and temperature; and
c) theoretical or semi-theoretical models.

Empirical correlations. Many empirical correlations have been proposed for the calculation of the heat transfer coefficient, mostly based on modifications of the well-known Dittus-Boelter type equation for liquid single-phase flow. None of them takes into account non-equilibrium effects. One of the most accurate among available correlations is that proposed by Groeneveld (1973):

$$Nu_g = a\left\{Re_g\left[x + \frac{\rho_g}{\rho_l}(1 - x)\right]\right\}^b \; Pr_{g,w}^c \; Y^d \tag{3-3}$$

where:

$$Y = 1 - 0.1\left(\frac{\rho_l}{\rho_g} - 1\right)^{0.4} (1 - x)^{0.4}$$

For tubes $a = 1.0910^{-3}$; $b = 0.989$, $c = 1.41$, and $d = -1.15$, while for annuli $a = 5.2\ 10^{-2}$, $b = 0.688$, $c = 1.\ 26$, and $d = -1.06$. For tubes and annuli $a = 3.27\ 10^{-3}$, $b = 0.901$, $c = 1.32$, and $d = -1.5$. The range of data on which correlations are based is reported in Table 3.1. Improvements of Eq. (3-3) have been given by Slaughterback et al. (1973a, 1973b).

Correlations accounting for thermodynamic non-equilibrium. These correlations account for thermodynamic non-equilibrium. Theoretically, two extreme conditions would be possible, i.e.:

a) all the heat is transferred to liquid drops until their complete evaporation (complete equilibrium, hypothesis valid for very high pressure, nearly critical, and mass flux > 3000 kg/m^2s);
b) all the heat is transferred to the vapour phase, causing its superheating (complete non-equilibrium, hypothesis acceptable for low pressure and low flow rate).

As generally real situations will be in between, we may think to split the heat flux in two components:

$$\dot{q}_{tot} = \dot{q}_g + \dot{q}_l \tag{3-4}$$

where $\dot{q}_g$ is the component of the heat flux delivered to the vapour (which raises its temperature) and $\dot{q}_l$ is the heat flux absorbed by liquid drops (which causes their evaporation). Usually, correlations provide an evaluation of:

$$\varepsilon = \frac{\dot{q}_l}{\dot{q}_{tot}} \tag{3-5}$$

through which is possible to obtain the vapour and wall temperature with thermodynamic calculations. Such correlations have been proposed by a variety of investigators (Plummer et al. (1977), Groeneveld & Delorme (1976), Jones & Zuber (1977), Chen et al. (1977)) and that proposed by Plummer et al. (1977) is reported here:

$$\varepsilon = C_1 \ln\left[G\left(\frac{D_h}{\rho_g \sigma}\right)^{0.5} (1 - x_{CHF})^5\right] + C_2 \tag{3-6}$$

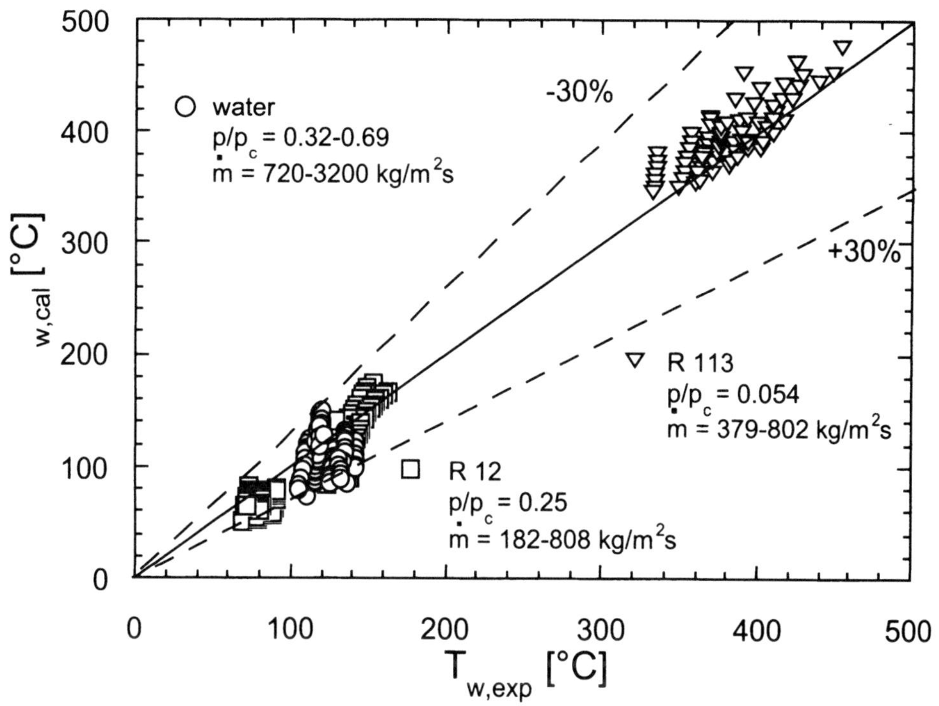

Figure 3.4. Wall temperature prediction in post-CHF heat transfer using Eq. (3-4), Bennett et al. (1968).

where D_h is the hydraulic diameter, and the constants C_1 and C_2 have been given by authors for nitrogen, water, and R 12. More recently Yoo & France (1996) have proposed C_1 and C_2 for R 113, showing that the parameter C_2 could be correlated using the molecular weight. C_1 and C_2 values are given in Table 3.2, while Figure 3.4 shows the prediction of experimental data obtained using Eq. (3-6).

Nishikawa et al. (1986) proposed also a method based on a non-dimensional parameter representing the ratio of the heat capacitance of the vapour flow to the thermal conductance from the vapour to the liquid droplets. Such a parameter was described as a function of non-dimensional thermodynamic parameters. Prediction of experimental data with the Nishikawa et al. correlation is shown in Figure 3.5.

Theoretical models. Many theoretical models have been proposed with different levels of complexity (Groeneveld (1972), Chen et al. (1977), Bennett et al. (1968), Iloeje et al. (1974), Ganic & Rohsenow (1976), Moose & Ganic (1982), Whalley et al. (1982), Hein & Köhler (1984), Kirillov et al. (1987), Yagov et al. (1987) Rohsenow (1988)), accounting in a more or less detailed way , for the various paths by which heat is transferred from the heating surface to the bulk vapour phase. Namely, models should account for: a) the heat transferred to liquid droplets impacting on he wall; b) the heat transferred to liquid droplets entering the thermal boundary layer without *wetting* the surface; c) the heat transferred from the heating surface to the vapour bulk by convection; d) the heat transferred from the vapour bulk to suspended droplets in the vapour core by convection; e) the heat transferred from the heating surface to liquid droplets by radiation; and f) the heat transferred from the surface to the vapour bulk by radiation. Nonetheless, following the starting assumption, not all of the above mechanisms are generally considered in the proposed models. Because of the complexity of the general mathematical description of existing models, the reader is referred to original papers reported in the bibliography.

Table 3.1. Range of data for the Groeneveld correlation (1973).

	Geometry	
	Tube	Annulus
Flow direction	Vertical and horizontal	Vertical
D_h, cm	0.25 to 2.5	0.15 to 0.63
p, Mpa	6.8 to 21.5	3.4 to 10.0
m, kg/m^{2}s	700 to 5300	800 to 4100
x, fraction by weight	0.1 to 0.9	0.1 to 0.9
q, kW/m^2	120 to 2100	450 to 2250
Nu_g	95 to 1770	160 to 640
$Re_g\ (x + (1-x)\rho_g/\rho_l)$	6.6 10^4 to 1.3 10^6	1.0 10^5 to 3.9 10^5
$Pr_{g,w}$	0.88 to 2.21	0.91 to 1.22
Y	0.706 to 0.976	0.61 to 0.963

Table 3.2. Constant for Plummer et al. correlation (1977).

Fluid	C1	C2
Nitrogen	0.082	0.290
Water	0.07	0.400
R 12	0.078	0.255
R 113[(a)]	0.078[(a)]	0.13[(a)]

[(a)] values given by Yoo & France (1996)

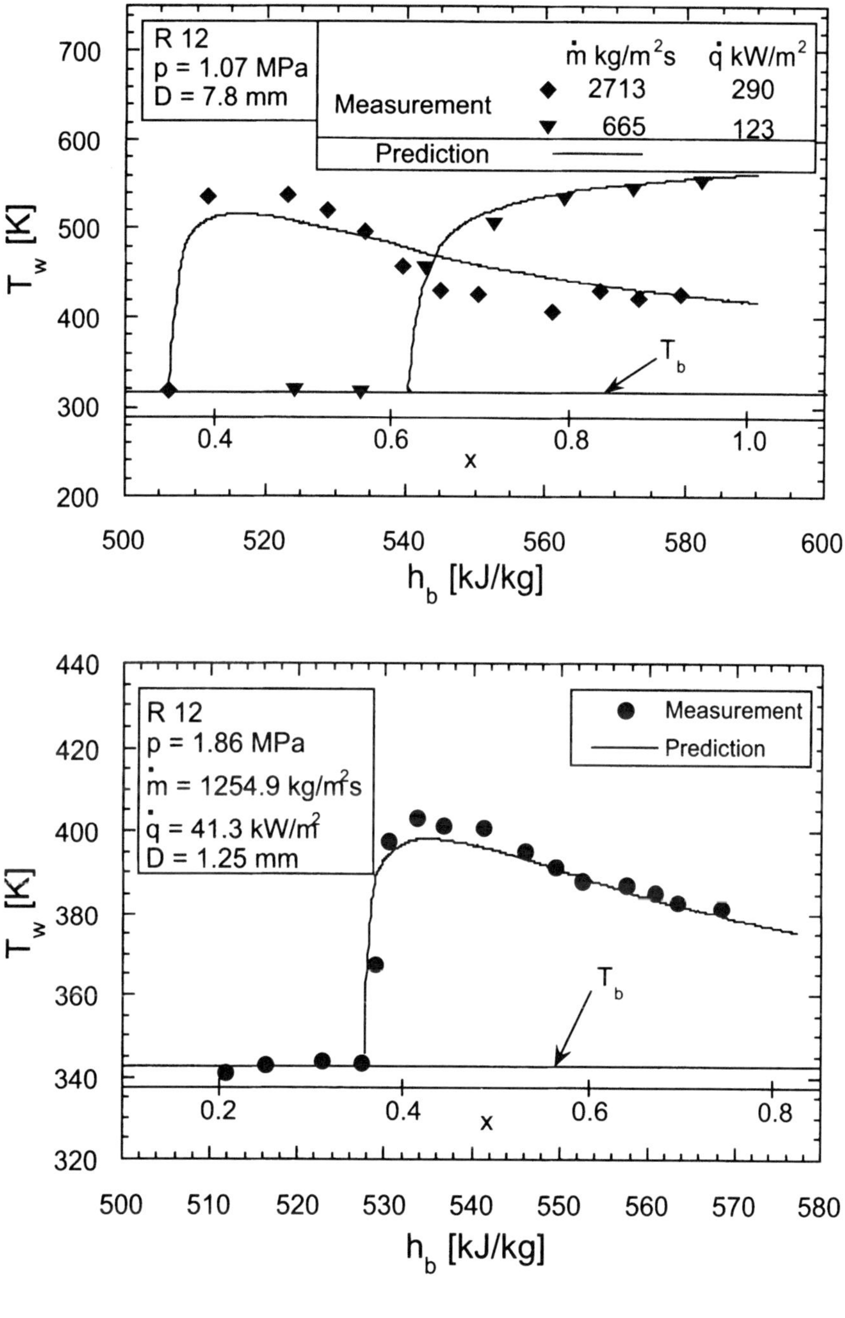

Figure 3.5. Prediction of wall temperature in post-CHF heat transfer using the correlation by Nishikawa et al. (1986).

4 Augmentation of CHF and Post-CHF Heat Transfer

In the thermal-hydraulic design of a heat exchanger, a steam generator, or a thermal equipment where the critical heat flux (CHF) is the limiting parameter or where the designer has to face with post-CHF heat transfer, it can be necessary to obtain a higher CHF value or a better post-CHF heat transfer coefficient than that allowed by the process thermodynamic and geometry conditions. It is therefore necessary to make use of enhancement techniques in order to have a higher CHF or a higher post-CHF heat transfer rate, similarly to what pursued in single and two-phase flow heat transfer (before the thermal crisis) (Thome (1990) and Bergles (1992)).

4.1 CHF Enhancement Techniques

Recent reviews of CHF enhancement techniques have been given by Boyd (1985a) and by Celata (1996). Among the possible techniques, we may have passive devices, such as swirl flow (twisted tapes and helically coiled wires), extended surfaces (hypervapotron), and helical coiled tubes, and active techniques, such as electrical fields, pressure wave generation and tangential injection. Only passive techniques will be discussed here; for active techniques see Boyd (1985a).

Swirl flow. Swirl flow is obtained using twisted tapes or helically coiled wires inside the flow channel to induce secondary radial and circumferential velocity components in the fluid, in order to obtain a better heat transfer rate and therefore a higher CHF value. A considerable increase in the pressure drop with respect to smooth tubes is observed, in general, with the use of swirl flow promoters. The use of twisted tapes as swirl flow promoters in the augmentation of the CHF in subcooled flow boiling has been studied by Gambill & Greene (1958), Gambill et al. (1961), Nariai et al. (1991), Cardella et al. (1992), and Achilli et al. (1993). An increase in the CHF typically by a factor of 2 over that for tubes without twisted tapes was generally obtained. Results by Gambill et al. (1961) and by Nariai et al. (1991) are plotted in Figure 4.1, where the ratio between the CHF obtained with the twisted tape and the value obtained with the smooth tube is reported versus pressure for different values of the non-dimensional centrifugal acceleration, ϑ:

$$\vartheta = \frac{a_t}{g} = \frac{\pi^2 u^2}{2gD\,TTR} \tag{4-1}$$

where *TTR* is the twisted tape ratio; ϑ is defined as the ratio between the tangential centrifugal acceleration (due to the twisted tape) and the standard gravitational acceleration. The thermal efficiency of the twisted tape decreases as pressure increases and becomes insignificant when pressure is above 2.0 MPa. This effect is probably due to the presence of a gap between the wall and the twisted tape. In fact, the clearance allows steam trapping in the tube-tape gap (which is an increasing function of pressure) which may results in premature CHF. Cardella et al. (1992) and Achilli et al. (1993) did not find any effect of the system pressure on the thermal efficiency of the twisted tape.

A correlation for the prediction of the CHF with twisted tapes has been given by Nariai et al. (1992):

$$\frac{\dot{q}_{CHF,tt}}{\dot{q}_{CHF,st}} = \{1 + 10^{-2}\, \vartheta \exp[(-10^{-6}\, p)^2]\}^{1/6} \qquad (4\text{-}2)$$

with ϑ given by Eq. (4-1). Also Eq. (1-4) can be used to predict the CHF with twisted tapes, using the resultant water velocity at the inner tube wall u_r, in place of u, as given by Schlosser et al. (19):

$$\frac{u_r}{u} = \left(\frac{1 + \pi^2}{4\,(TTR)^2}\right)^{1/2} \qquad (4\text{-}3)$$

The Celata et al. model (1994, 1995) presented in section 1.3 can be also used to predict the CHF in subcooled flow boiling with twisted tapes.

The use of twisted tapes to enhance the CHF under saturated flow boiling conditions has been recently investigated by Lee et al. (1995). Authors found that in the low-quality region the effect of the twisted tape on the CHF is negligible. In the middle-quality region, the CHF of the twisted tape inserted tube increased with mass velocity, which was contrary to the trend observed for the empty one. Besides, the CHF was found to increase by insertion of the twisted tape except for cases of very small flow rate and large twist ratio. The clearance effect was weak as compared to the subcooled region. Finally, in the high-quality region, the CHF decreased with exit quality, the decreasing rate being slower with the twisted tape than without. Also, the CHF enhancement was most remarkable in this region. A correlation for the prediction of the CHF with twisted tape in saturated flow boiling has been proposed by Jensen (1985):

$$\frac{\dot{Q}_{tt}}{\dot{Q}_{CHF}} = (4.597 + 0.09254\,(TTR) + 0.004154\,(TTR)^2 \left(\frac{\rho_l}{\rho_g}\right)^{-0.7012} + 0.09012 \ln\vartheta \qquad (4\text{-}4)$$

with ϑ given by Eq. (4-1). The critical power $\dot{Q}_{CHF}$ can be obtained using a suitable correlation or model.

Helically coiled wires as swirl flow promoters have been used by Celata et al. (1994b) for CHF in subcooled flow boiling. Authors used wires of spring steel having a diameter of 0.5, 0.7 and 1.0 mm and a pitch from 1.5 to 20.0 mm in 8.0 mm I.D. tubes. Results are presented in Figure 4.2, where an increase in the CHF up to 50% using a 1.0 mm wire at 3.5 MPa can be observed. Contrarily to the twisted tape performance, where the increase in the thermal efficiency and the associate pressure drop increase are strictly inter-related, the thermal efficiency of wires is practically independent of the wire pitch, while pressure drop is inversely related to it. This latter can therefore be properly reduced decreasing the pitch, without affecting the thermal performance. The effect of the pressure on the wires efficiency is observed to be negative, in the sense that at a pressure of 5.0 MPa it drops to only 30%.

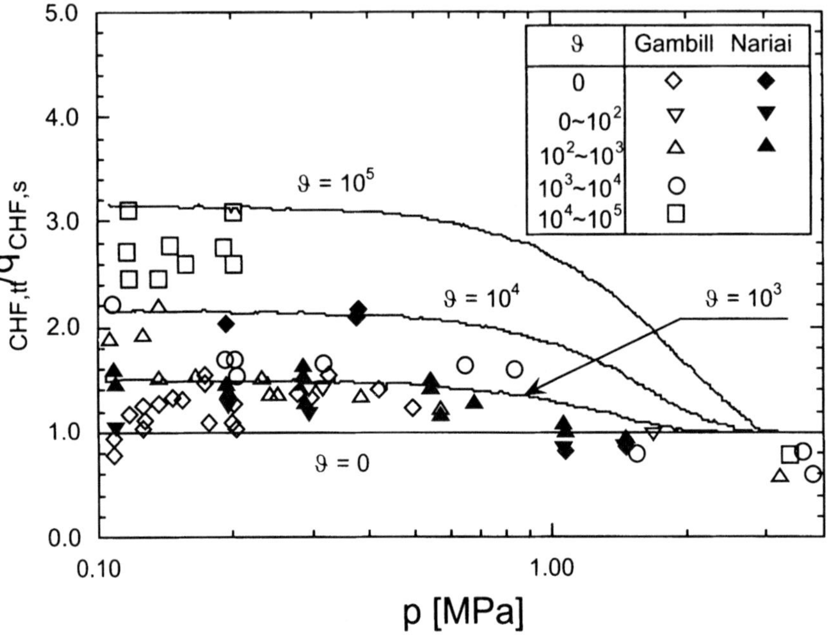

Figure 4.1. Swirl flow CHF data using twisted tapes, Gambill et al. (1961), Nariai et al. (1992).

Extended surfaces. Kovalevev (1976) investigated three different fin design and obtained up to a factor of 10 increase in the CHF for low velocity (0.021 to 0.14 m/s) subcooled flow in an annulus. A thorough review on the use of fins has been given by Boyd (1985a). A very intriguing technique using fins, but placed perpendicular to the fluid flow (in subcooled flow boiling), is the so-called *hypervapotron* technique. From a physical viewpoint, the hypervapotron effect consists of the following succession of events. The liquid inside two adjacent fins of high conductivity material and in contact with the heated wall starts boiling while the fluid bulk outside the fins is under subcooled conditions. Once the slot is full of steam, this latter undergoes a quick condensation in the subcooled liquid bulk, emptying the slots and making their replenishment with cold liquid easier. The heated wall is rewetted until the wall temperature during the uncovered phase is below the Leidenfrost temperature. The base of the fin is allowed to operate at a temperature greater than the CHF temperature while the remaining portion operates near the temperature for the onset of stable nucleate boiling. This continuous boiling and condensation sequence (frequency between 10 and 40 Hz) allows to get a high CHF, essentially on the basis of the transport of the latent heat extracted from the heated wall during boiling and transferred to the coolant outside the fins during condensation. Cattadori et al. (1993) obtained a maximum CHF of 29.4 MW/m^2. A typical picture from visualized tests is reported in Figure 4.3.

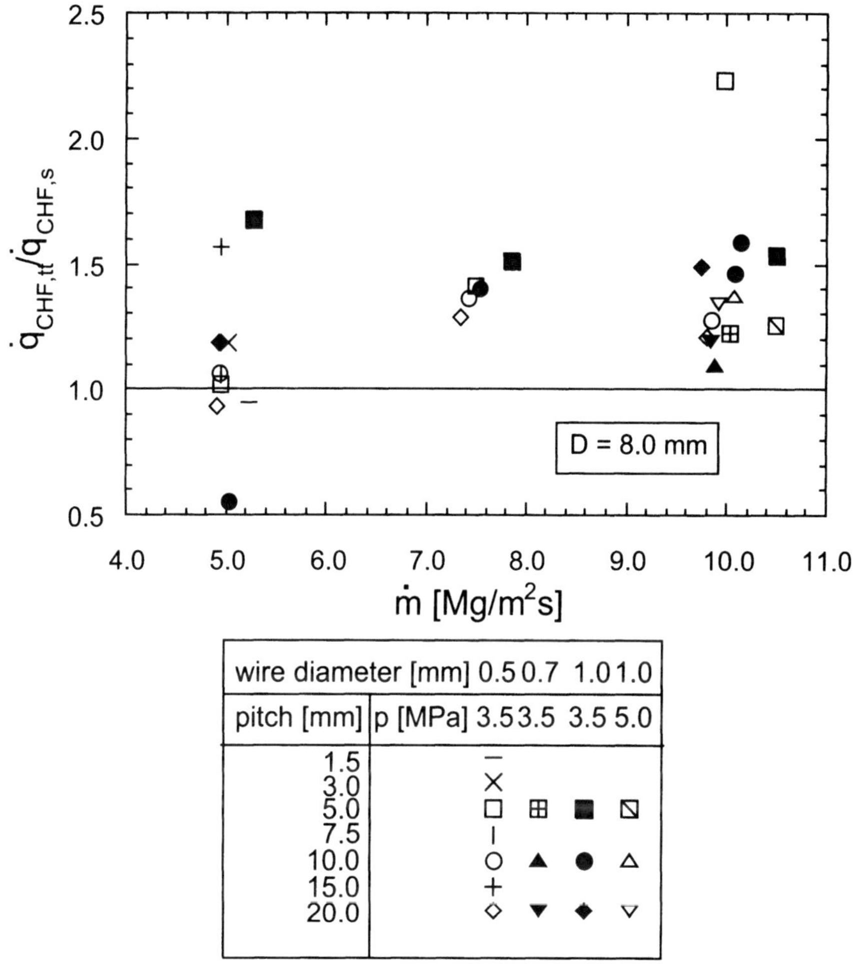

Figure 4.2. Swirl flow CHF data using helically coiled wires, Celata et al. (1994b).

Helically coiled tubes. Use of helically coiled tubes to get higher CHF values has been experienced by various researchers, such as Jensen & Bergles (1981), Berthoud & Jayanti (1990), Kaji et al. (1995) among the others. In the coiled tube, the liquid film thickness distribution is nonuniform around the tube circumference. But, due to the secondary flow caused by centrifugal forces, the entrainment rate of liquid droplets from the inside to the outside of the coil is large and the liquid film flows around the circumference. This may cause the dryout quality and therefore the critical heat flux to increase in the coiled tube.

4.2 Post CHF Heat Transfer Enhancement Techniques

Swirl flow promoters such as twisted tapes are also an effective enhancement technique for the augmentation of heat transfer in the post-CHF region of two-phase flow. Here the mechanism for the augmentation includes the effect of the radial velocity concentrating liquid, from the center of the flow stream, at the heat transfer surface. The first experiment of swirl flow post-CHF heat transfer was conducted by Bergles et al. (1971). They proposed a correlation, which is very complex, for the heat transfer coefficient:

$$\alpha = C\,Re_g^{0.8}\,Pr_g^{0.4}\left(\frac{T_b}{T_w}\right)^{0.32}\left(1+0.25\,\frac{Gr_g^{0.5}}{Re_g}\right)\frac{\lambda_g}{D_h} +$$

$$1.1\left(\frac{\lambda_g^3 h_{lg}^* g \theta \rho_l \rho_g}{\Delta T \eta_g (\pi/6)^{1/3}}\right)^{1/4}\left[\frac{(1-x)(6/\pi)}{x(\rho_l/\rho_g-1)+1}\right]^{2/3}\left(\frac{\pi}{4}\right)Z \qquad (4\text{-}5)$$

where

$$Re_g = \frac{\dot{m}D}{\eta_g}\left[x+(1-x)\left(\frac{\rho_g}{\rho_l}\right)\right]$$

$$C = 0.021\left[1+\frac{0.035\,\pi^2}{D(TTR)^2\,(1+\pi^2/4(TTR)^2)}\right]$$

$$h_{lg}^* = h_{lg}\left[1+\frac{0.35\,c_{p,g}\,(T_w - T_{sat})}{h_{lg}}\right]^{-3}$$

Gr_g is the Grashof number of the gas phase, and Z is an empirically determined constant related to droplet size, while D is in feet. For film boiling of nitrogen at low mass velocity and at a reduced pressure of about 0.045, a value of $Z = 7$ gave satisfactory results.

The post-CHF heat transfer in helical coiled tubes has been studied by Chen & Zhou (1986). Here the heat transfer coefficient in the helical coiled tube is higher than in the straight tube, also because of the action of the secondary flow and the deposition of liquid droplets in the vapour core which continuously we the place of dryout.

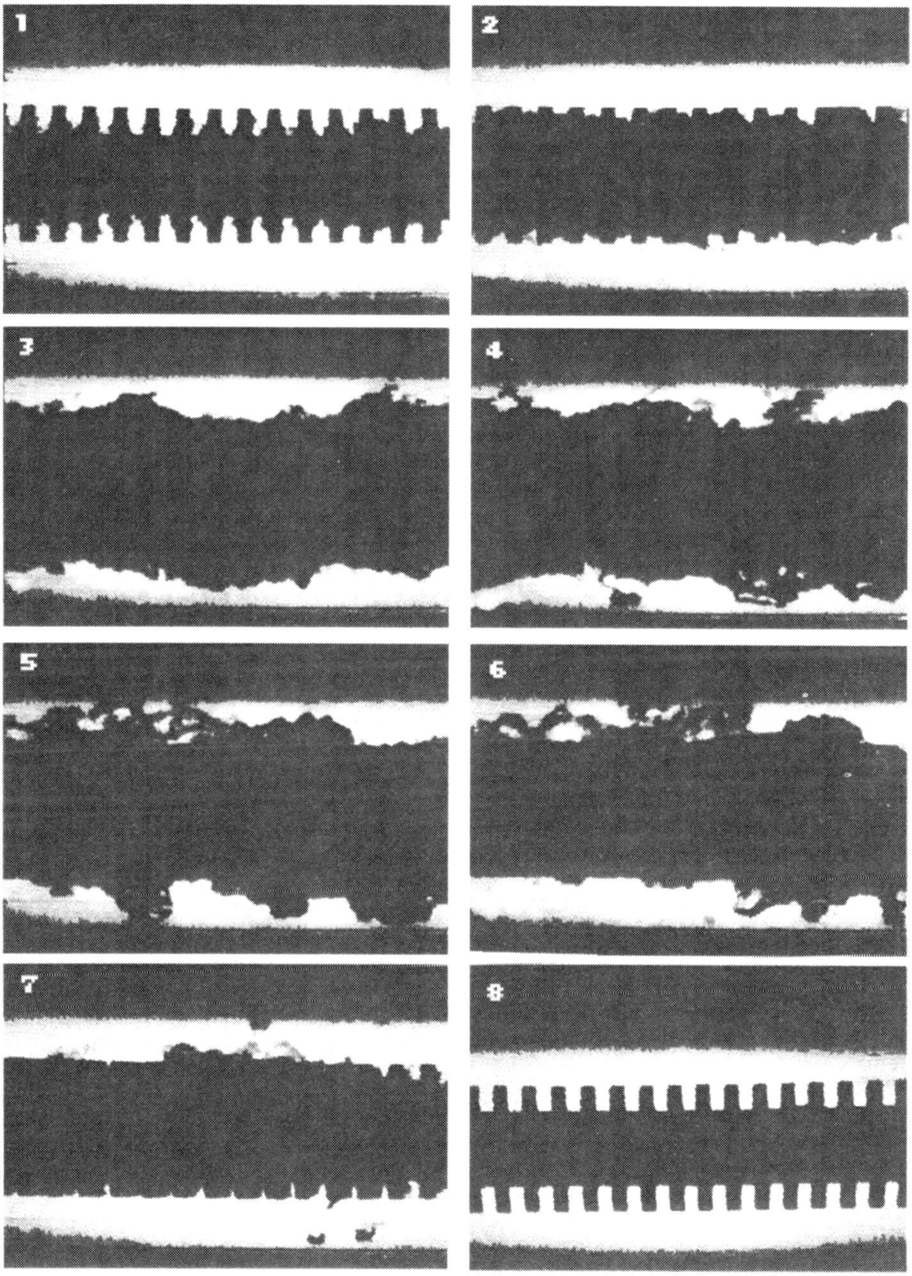

Figure 4.3. Visualization of the hypervapotron effect, Cattadori et al. (1993).

Nomenclature

A	heated surface
a_t	centrifugal acceleration
C_D	drag coefficient
CHF	critical heat flux
c_p	specific heat
D	diameter
D_h	equivalent hydraulic diameter
DR	deposition rate
ER	entrainment rate
f	friction factor
F_c	shape factor
g	gravitational acceleration
h, Δh	enthalpy, entalpy difference
h_{lg}	latent heat of vaporization
$I.D.$, $O.D.$	inlet and outlet diameter in annulus
L	length, heated length
l_b	boiling length
$\dot{M}$	mass flow rate
$\dot{m}$	mass flux
$\dot{m}_{lf}$	liquid film mass flux
p	pressure
p_c	critical pressure
$\dot{Q}$	critical power
$\dot{q}$	heat flux
R	radius
s^*	superheated layer
T, ΔT	temperature, temperature difference
TTR	twisted tape ratio
u	velocity
u_r	resultant water velocity, with twisted tape
u_τ	friction velocity
x	steam quality
X	mixture composition in the liquid phase
Y	mixture composition in the vapour phase
z	axial distance

Non-dimensional numbers

Bo	Boiling number
Fr	Froude number
Gr	Grashof number
Ja	Jakob number

Nu	Nusselt number
Pr	Prandtl number
Re	Reynolds number

Greek symbols

α	heat transfer coefficient
β	contact angle
δ	liquid layer thickness
ε	void fraction
ϕ	angle
η	dynamic viscosity
ϑ	centrifugal acceleration (non dimensional)
λ	thermal conductivity
θ	radial acceleration
ρ	density
σ	surface tension

Subscripts

B	pertaint to the blanket
b	bulk
cal	calculated
CHF	critical heat flux
ex	exit condition
exp	experimental
g	vapor
in	inlet
l	liquid, fluid
m	mean value
nu	non uniform
sat	saturated value
st	smooth tube
sub	subcooled condition
tt	twisted tape
u	uniform
w	wall

References

Achilli A., Cattadori G. and Gaspari G.P. (1993), Subcooled Burnout in Uniformly and Non-Uniformly Heated Tubes, Paper C2 presented at the European Two-Phase Flow Group Meeting, Stockholm, June

Aladyev I.T., Miropolsky Z.L., Doroshchuk V.E. and Styrikovich M.A. (1961), Boiling Crisis in Tubes, Int. Developments in Heat Transfer, Vol. II, Paper 28, University of Colorado, Boulder

Alekseev G.V., Zenkevitch B.A., Peskov O.L., Sergeev N.D. and Subbotin V.I. (1965), Burnout Heat Fluxes under Forced Water Flow, Teploenergetika, Vol. 12, n. 3, pp. 47-51

Andrews D.G., Hooper F.C., and Butt P. (1968), Velocity, Subcooling and Surface Effects in the Departure from Nucleate Boiling of Organic Binaries, Can. J. Chem. Engng., Vol. 46, pp. 194-199

Auracher H. and Marroquin A. (1995), Critical Heat Flux and Minimum Heat Flux of Film Boiling of Binary Mixtures Flowing Upwards in a Vertical Tube, Engineering Foundation Conference on Convective Flow Boiling, Paper V-3, Banff, May

Bahr A., Herkenrath H. and Mork-Morkenstein P. (1969), Anomale Druck-abhängigkeit der Wärmeübertragung im Zweiphasengebeit bei Annäherung an der Kritischen Druck, Brennstoff-Wärme-Kraft, Vol. 21, n. 12, pp. 631-633

Bailey N.A. (1971), Film Boling on Sumberged Vertical Cylinders, AEEW-M1051

Barnett P.G. (1963), An Investigation into the Validity of Certain Hypotheses Implied by Various Burnout Correlations, AEEW-R 214

Becker K. (1971), Measurements of Burnout Conditions for Flow of Boiling Water in Horizontal Round Tubes, AERL-1262

Bennet A.W., Hewitt G.F. and Keeys R.K.F. (1967), Heat Transfer to Steam-Water Mixtures Flowing in Uniformly Heated Tubes in Which the Critical Heat Flux Has Been Exceeded, Paper 27 presented a tthe Thermodynamics and Fluid Mechanics Convention, IMechE, Bristol, March, 1968 (Also AERE-R 5573)

Bennet A.W., Hewitt G.F., Kearsey H.A., Keeys R.K.F. and Pulling D.J. (1966), Studies of Burnout in Boiling Heat Transfer to Water in Round Tubes with Non-Uniform Heating, AERE-R 5076

Bergel'son B.R. (1980), Burnout Under Conditions of Subcooled Boiling and Forced Convection, Thermal Engineering, Vol. 27, n. 1, pp. 48-50

Bergles A.E. (1963), Subcooled Burnout in Tubes of Small Diameter, ASME Paper 63-WA-182

Bergles A.E. (1977), Burnout in Boiling Heat Transfer, Part II: Subcooled and Low-Quality Forced Convection Systems, Nuclear Safety, Vol. 18, n. 2, p. 154

Bergles A.E. (1992), Heat Transfer Enhancement-Second Generation Heat Tranfer Technology, Proc. 10th UIT National Heat Transfer Conference, pp. 3-21, Genoa, June

Bergles A.E. and Scarola L.S. (1966), Effect of a Volatile Additive on the Critical Heat Flux for Surface Boiling of Water in Tubes, Chem. Engng. Science, Vol. 21, pp. 721-723

Bergles A.E., Collier J.G., Delhaye J.M., Hewitt G.F. and Mayinger F. (1981), Two-Phase Flow and Heat Transfer in the Power and Process Industries, Hemisphere Publishing Corporation, New York

Bergles A.E., Fuller W.D. and Hynek S.J. (1971), Dispersed Flow Film Boiling of Nitrogen with Swirl Flow, Int. J. Heat Mass Transfer, Vol. 14, pp. 1343-1354

Berthoud G. and Jayanti S. (1990), Characterization of Dryout in Helical Coils, Int. J. Heat Mass Transfer, Vol. 33, n. 7, pp. 1451-1463

Bertoletti S., Gaspari G.P., Lombardi C., Peterlongo G., Silvestri M. and Tacconi F.A. (1965), Heat Transfer Crisis with Steam-Water Mixtures, Energia Nucleare, Vol. 12, n. 3, pp. 121-172

Bertoni R., Cipriani R., Cumo M. and Palazzi G. (1976), Upflow and Downflow Burnout, CNEN Report RT/ING(76)24

Bowring R.W. (1972), A Simple but Accurate Round Tube Uniform Heat Flux, Dryout Correlation over the Pressure Range 0.7-17 MN/m^2 (100-2500 psia), AAEW-R 789

Boyd R.D. (1985a), Subcooled Flow Boiling Critical Heat Flux (CHF) and its Application to Fusion Energy Components. Part I: A Review of Fundamentals of CHF and Related Data Base, Fusion Technology, Vol. 7, pp. 7-30

Boyd R.D. (1985b), Subcooled Flow Boiling Critical Heat Flux (CHF) and its Application to Fusion Energy Components. Part II: A Review of Microconvective, Experimental, and Correlational Aspects, Fusion Technology, Vol. 7, pp. 31-52

Boyd R.D. (1988), Subcooled Water Flow Boiling Experiments Under Uniform High Flux Conditions, Fusion Technology, Vol. 13, pp. 131-142

Boyd R.D. (1989), Subcooled Water Flow Boiling at 1.66 MPa Under Uniform High Heat Flux Conditions, ASME Winter Annual Meeting, S. Francisco, December 10-15 (HTD - Vol. 119, pp. 9-15)

Boyd R.D. (1990), Subcooled Water Flow Boiling Transition and the L/D Effect on CHF for a Horizontal Uniformly Heated Tube, Fusion Technology, Vol. 18, pp. 317-324

Bromley L.A., LeRoy N.R. and Robbers J.A. (1953), Heat Transfer in Forced Convection Film Boiling, Ind. and Engng. Chem., Vol. 45, n. 12, pp. 2639-2646

Burdunin M.N., Zvonarev Yu.A., Komendatov A.S. and Kuzma-Kichta Yu.A. (1987), Investigation of Post-Dryout Heat Transfer in Channel of Complex Shape, Heat Transfer-Soviet Research, Vol. 19, n. 1, pp. 115-121

Cardella A., Celata G.P., Dell'Orco G., Gaspari G.P., Cattadori G. and Mariani A. (1992), Thermal Hydraulic Experiments for the NET Divertor, Proc. 17th Symposium on Fusion Technology, Vol. 1, pp. 206-210, Rome, September

Carne M. (1963), Studies of the Critical Heat Flux for some Binary Mixtures and their Components, Can. J. Chem. Engng., pp. 235-240

Cattadori G., Gaspari G.P., Celata G.P., Cumo M., Mariani A. and Zummo G. (1993), Hypervapotron Technique in Subcooled Flow Boiling CHF, Experimental Thermal and Fluid Science, Vol. 7, pp. 230-240

Celata G.P. (1996), Critical Heat Flux in Water Subcooled Flow Boiling: Experimentation and Modelling, keynote lecture, Proc. 2nd European Thermal-Sciences Conference, Vol. I, pp. 27-40, Edizioni ETS, Pisa, May

Celata G.P. (1997), Modelling of Critical Heat Flux in Subcooled Flow Boiling, keynote lecture, Convective Flow and Pool Boiling Conference, Irsee, 18-23 May 1997

Celata G.P. and Cumo M. (1996), Forced Convective Boiling of Refrigerant Binary Mixtures, keynote lecture, Proc. 4th International Symposium on Heat Transfer, pp. 70-80, Beijing, September

Celata G.P. and Mariani A. (1993), A Data Set of Critical Heat Flux in Water Subcooled Flow Boiling, presented at the 3rd Specialists' Workshop on the Thermal-Hydraulics of High Heat Flux Components in Fusion Reactors, J. Schlosser Ed., Cadarache, September

Celata G.P., Cumo M. and Mariani A. (1993a), Burnout in Highly Subcooled Water Flow Boiling in Small Diameter Tubes, Int. J. Heat Mass Transfer, Vol. 36, n. 5, pp. 1269-1285

Celata G.P., Cumo M., Inasaka F., Mariani A. and Nariai H. (1993b), Influence of Channel Diameter on Subcooled Flow Boiling Burnout at High Heat Fluxes, Int. J. Heat Mass Transfer, Vol. 36, n. 13, pp. 3407-3410

Celata G.P., Cumo M. and Mariani A. (1994a), Assessment of Correlations and Models for the Prediction of CHF in Subcooled Flow Boiling, Int. J. Heat Mass Transfer, Vol. 37, n. 2, pp. 237-255

Celata G.P., Cumo M. and Mariani A. (1994b), Enhancement of CHF Water Subcooled Flow Boiling in Tubes using Helically Coiled Wires, Int. J. Heat Mass Transfer, Vol. 37, n. 1, pp. 53-67

Celata G.P., Cumo M., Mariani A., Simoncini M. and Zummo G. (1994c), Rationalization of Existing Mechanistic Models for the Prediction of Water Subcooled Flow Boiling Critical Heat Flux, Int. J. Heat Mass Transfer, Vol. 37, n. 7, Suppl. 1, pp. 347-360

Celata G.P., Cumo M. and Setaro T. (1994d), Critical Heat Flux in Upflow Convective Boiling of Refrigerant Binary Mixtures, Int. J. Heat Mass Transfer, Vol. 37, n. 7, pp. 1143-1153

Celata G.P., Cumo M., Mariani A. and Zummo G. (1995a), Preliminary Remarks on Visualization of High Heat Flux Burnout in Subcooled Water Flow Boiling, Proc. International Symposium on Two-Phase Flow Modelling and Experimentation, Vol. 2, pp. 859-866, Rome, October

Celata G.P., Cumo M., Mariani A. and Zummo G. (1995b), The Prediction of Critical Heat Flux in Water Subcooled Flow Boiling, Int. J. Heat Mass Transfer, Vol. 38, n. 6, pp. 1111-1119

Celata G.P., Cumo M. and Mariani A. (1997), Geometrical Effects on the Subcooled Flow Boiling Critical Heat Flux, Proc. 4th World Conference on Experimental Heat Transfer, Fluid Mechanics, and Thermodynamics, Vol. II, pp. 867-872, Bruxelles, June

Chang S.H., Baek W.P. and Bae T.M. (1991), A Study of Critical Heat Flux for Low Flow of Water in Vertical Round Tubes under Low Pressure, Nuclear Engineering and Design, Vol. 132, pp. 225-237

Chen J.C., Sundaram R.K. and Ozkaynak F.T. (1977), A Phenomenological Correlation for Post-CHF Heat Transfer, Lehigh University, NUREG-0237

Chen X.J. and Zhou F.D. (1986), Forced Convection Boiling and Post Dryout Heat Transfer in Helical Coiled Tube, Proc. 8th International Heat Transfer Conference, Vol. 6, pp. 2221-2226, S. Francisco, August

Collier J.G. and Thome J.R. (1994), Convective Boiling and Condensation, Third Edition, Clarendon Press, Oxford

Costigan G., Holmes A.W and Ralph J.C. (1984), Steady-State Post-Dryout Heat Transfer in a Vertical Tube with Low Inlet Quality, Proc. 1st UK National Heat Transfer Conference, Vol. 1, pp. 1-11 (IChemE Symp. Ser. n. 86)

Cumo M., Fabrizi F. and Palazzi G. (1978), The Influence of Inclination on CHF in Steam Generators Channels, CNEN Report, RT/ING (78)11

Cumo M., Palazzi G., Urbani G. and Frazzoli F.V. (1980), Full Scale Tests on Axial Profile Heat Flux Influence on the Critical Quality in PWR Steam Generators, CNEN Report RT/ING (80)5

Denham M.K. (1984), Inverted Annular Flow Film Boiling and the Bromley Model, Proc. 1st UK National Heat Transfer Conference, Vol. 1, pp. 13-23 (IChemE Symp. Ser. n. 86)

Doroshchuk V.E., Levitan L.L., Lantzman E.P., Nigmatulin R.I. and Borevsky L.Ya. (1978), Investigation into Burnout Mechanism in Steam-Generating Tubes, Proc. 6th International Heat Transfer Conference, Vol. 1, pp. 393-398

Doroshchuk V.E., Levitan L.L. and Lantzman F.P. (1975), Investigation into Burnout in Uniformly Heated Tubes, ASME Publication 75-WA/HT-22

Dougall R.S. and Rohsenow W.M. (1963), Film Boiling on the Inside of Vertical Tubes with Upward Flow of the Fluid at Low Qualities, Mech. Engng. Dept. Engineering Project Laboratory, MIT Report 9079-26

Fiori M.P. and Bergles A.E. (1968), Model of Critical Heat Flux in Subcooled Flow Boiling, MIT Report-DSR 70281-56

Fiori M.P. and Bergles A.E. (1970), Model of Critical Heat Flux in Subcooled Flow Boiling, Proc. 4th International Heat Transfer Conference, Vol. VI, p. B6.3, Hemisphere, New York

Fung K.K., Gardiner S.R.M. and Groeneveld D.C. (1979), Subcooled and Low Quality Flow Boiling of Water at Atmospheric Pressure, Nuclear Engineering and Design, Vol. 55, pp. 51-57

Gambill W.R. (1968), Burnout in Boiling Heat Transfer, Part II: Subcooled Forced-Convection Systems, Nuclear Safety, Vol. 9, n. 6, p. 467

Gambill W.R. and Greene N.D. (1958), Boiling Burnout with Water in Vortex Flow, Chem. Eng. Prog., Vol. 54, n. 10, pp. 68-76

Gambill W.R., Bundy R.D. and Wansbrough R.W. (1961), Heat Transfer, Burnout and Pressure Drop for Water in Swirl Flow through Tubes with Internal Twisted Tapes, Chem. Eng. Symp. Ser., Vol. 57, n. 32, pp. 127-137

Ganic E.N. and Rohsenow W.M. (1976), Dispersed Flow Heat Transfer, Int. J. Heat Mass Transfer, Vol. 20, pp. 855-866

Gaspari G.P. (1993), Comparison Among Data of Electrically and E-Beam Heated Tubes, Proc. 3rd International Workshop on High Heat Flux Components Thermal Hydraulics in Fusion Reactors, J. Schlosser, Ed., Cadarache, September

Glickstein M.R. and Whitesides R.H. (1967), Forced Convection Nucleate and Film Boiling of Several Aliphatic Hydrocarbons, ASME Paper 67-HT-7, presented at the ASME-AIChE Heat Transfer, Seattle

Govan A.H. (1984), Comparison of the Harwell Annular Flow Model with Critical Heat Flux Data, AERE-R 11298

Govan A.H., Hewitt G.F., Owen D.G. and Bott T.R. (1988), An Improved CHF Modelling Code, Proc. 2nd UK National Conference on Heat Transfer, Vol. 1, pp. 33-48, IMechE, 14-16 September

Groeneveld D.C. (1972), The Thermal Behaviour of a Heated Surface at and Beyond Dryout, Atomic Energy of Canada Report, AECL-4309

Groeneveld D.C. (1973), Post-Dryout Heat Transfer at Reactor Operating Conditions, AECL-4513

Groeneveld D.C. (1981), Heat Transfer Phenomena Related to the Boiling Crisis, AECL-7239, Chalk River National Laboratory

Groeneveld D.C. and Delorme G.G.J. (1976), Preciction of the Thermal Non-Equilibrium in the Post-Dryout Regime, Nuclear Engineering and Design, Vol. 36, pp. 17-26

Groeneveld D.C., Cheng S.C. and Doan T. (1986), AECL-UO Critical Heat Flux Look-Up Table, Heat Transfer Engineering, Vol. 7, pp. 46-62

Groeneveld D.C., Leung L.K.H., Kirillov P.L., Bobkov V.P., Smogalev I.P., Vinogradov V.N., Huang X.C. and Royer E. (1996), The 1995 Look-up Table for Critical Heat Flux in Tubes, Nuclear Engineering and Design, Vol. 163, pp. 1-23

Gunther F.C. (1951), Photographic Study of Surface-Boiling Heat Transfer to Water with Forced Convection, Trans. ASME, Vol. 73, n. 2, pp. 115-123

Hancox W.T. and Nicoll W.B. (1973), On the Dependence of the Flow-Boiling Heat Transfer Crisis on Local Near-Wall Conditions, 73-HT-38, ASME

Hebel W., Detavernier A. and Decreton M. (1981), A Contribution to the Hydrodynamics of Boiling Crisis in a Forced Flow of Water, Nuclear Engineering and Design, Vol. 64, pp. 433-445

Hein D. and Köhler W. (1984), A Simple-To-Use Post-Dryout Heat Transfer Model Accounting for Thermal Non-Equilibrium, Report USNRC-NUREG/CP-0060, pp. 369-372

Herkenrath H., Mork-Morkenstein P., Jung U. and Weckermann F.J. (1967), Heat Transfer in Water with Forced Circulation in 140-150 bar Pressure Range, EUR 3658d

Hewitt G.F. (1978), Critical Heat Flux in Flow Boiling, Proc. 6th International Heat Transfer Conference, Toronto

Hewitt G.F. (1980), Burnout, in Handbook of Multiphase Systems, Hetsroni G., Ed. McGrawHill, pp. 6.66-6.141

Hewitt G.F. and Govan A.G. (1989), Phenomenological Modelling of Non-Equilibrium Flows with Phase Change, Proc. EUROTHERM Seminar 7, Thermal Non-Equilibrium in Two-Phase Flow, pp. 7-40, ENEA, 23-24 March

Hewitt G.F. and Hall-Taylor N.S. (1970), Annular Two-Phase Flow, Pergamon Press, New York

Hewitt G.F., Kearsey H.A., Lacey P.M.C. and Pulling D.J. (1963), Burnout and Nucleation in Climbing Film Flow, Int. J. Heat Mass Transfer, Vol. 8, p. 793

Hewitt G.F., Kearsey H.A., Lacey P.M.C. and Pulling D.J. (1965), Burnout and Film Flow in the Evaporation of Water in Tubes, Proc. Inst. Mech. Eng., Vol. 80, Part 3C, p. 206

Hino R. and Ueda T. (1985), Studies on Heat Transfer and Flow Characteristics in Subcooled Flow Boiling - Part 2: Flow Characteristics, Int. J. Multiphase Flow, Vol. 11, pp. 283-298

Hsu Y.Y and Graham R.W. (1986), Transport Processes in Boiling and Two-Phase Systems, American Nuclear Society, La Grange Park, Il, USA

Iloeje O.C., Plummer D.N., Rohsenow W.M. and Griffith P. (1974), A Study of Wall Rewet and Heat Transfer in Dispersed Vertical Flow, Mech. Engng. Dept. MIT, Report 72718-92, September

Inasaka F. and Nariai H. (1996), Evaluation of Subcooled Critical Heat Flux Correlations for Tubes with and without Internal Twisted Tapes, Nuclear Engineering and Design, Vol. 163, pp. 225-239

Jensen M.K. (1984), A Correlation for Predicting the Critical Heat Flux Condition with Twisted-Tape Swirl Generators, Int. J. Heat Mass Transfer, Vol. 27, pp. 2171-2173

Jensen M.K. and Bergles A.E. (1981), Critical Heat Flux in Helically Coiled Tubes, Trans. ASME, Vol. 103, n. 4, pp. 660-666

Jones O.C. and Zuber N. (1977), Post-CHF Heat Transfer - A Non-Equilibrium Relaxation Model, ASME Paper 77-HT-79 presented at the 17th National Heat Transfer Conference, Salt Kake City, August

Kaji M., Mori K., Nakanishi S., Hirabayashi K. and Ohishi M. (1996), Dryout and Wall-Temperature Fluctuations in Helically Coiled Evaporating Tubes, Heat Transfer-Japanese Research, Vol. 24, n. 3, pp. 239-254

Katto Y. (1986), Forced-Convection Boiling in Uniformly Heated Channels, in Handbook of Heat and Mass Transfer, Vol. 1: Heat Tranfer Operations, Cheremisinoff N.P., Ed., Gulf Publishing Company, Houston, Chapter 9, pp. 303-325

Katto Y. (1990a), A Physical Approach to Critical Heat Flux of Subcooled Flow Boiling in Round Tubes, Int. J. Heat Mass Transfer, Vol. 33, n. 4, pp. 611-620

Katto Y. (1990b), Prediction of Critical Heat Flux of Subcooled Flow Boiling in Round Tubes, Int. J. Heat Mass Transfer, Vol. 33, n. 9, pp. 1921-1928

Katto Y. (1992), A Prediction Model of Subcooled Water Flow Boiling CHF for Pressure in the Range 0.1-20 MPa, Int. J. Heat Mass Transfer, Vol. 35, n. 5, pp. 1115-1123

Katto Y. (1994), Critical Heat Flux, Int. J. Multiphase Flow, Vol. 20, Suppl., pp. 563-90

Katto Y. (1995), Critical Heat Flux Mechanisms, Proc. Eng. Foundation Convective Flow Boiling Conference, Keynote Lecture V, Banff, May

Katto Y. and Ohno H. (1984), An Improved Version of the Generalized Correlation of Critical Heat Flux for the Forced Convective Boiling in Uniformly Heated Vertical Tubes, Int. J. Heat Mass Transfer, Vol. 26, n. 8 pp. 1641-1648

Keeys R.K.F., Ralph J.C. and Roberts D.N. (1972), Post Burnout Heat Transfer in High Pressure Steam-Water Mixtures in a Tube with Cosine Heat Flux Distribution, Progress in Heat and Mass Transfer, Vol. 6, pp. 99-118

Kefer V., Köhler W. and Kastner W. (1989), Critical Heat Flux (CHF) and Post-CHF Heat Transfer in Horizontal and Inclined Evaporator Tubes, Int. J. Multiphase Flow, Vol. 15, n. 3, pp. 385-392

Kirby G.J., Staniforth R. and Kinneir J.H. (1967), A Visual Study of Forced Convective Boiling. Part II: Flow Patterns and Burnout for a Round Test Section, AEEW - R506

Kirillov P.L., Kahcheyev V.M., Muranov Yu.V. and Yuriev Yu.S. (1987), A Two-Dimensional Mathematical Model of Annular-Dispersed and Dispersed Flows - Parts I and II, Int. J. Heat Mass Transfer, Vol. 30, n. 4, pp. 791-806

Kovalev S.A. (1976), Heat Transfer Crisis of Boiling of Subcooled Water on a Finned Surface Under Forced Convection Conditions, Heat Transfer-Soviet Research, Vol. 8, n. 4, p. 73

Kramer T.J. (1976), Fluid Flow and Convective Heat Transfer in Square Capillary Ducts Subjected to Nonuniform High Heat Flux, ASME Paper 76-WA-HT-29

Kutateladze S.S. and Leontiev A.I. (1966), Some Applications of the Asymptotic Theory of the Turbulent Boundary Layer, Proc. 3rd International Heat Transfer Conference, Vol. 3, pp. 1-6, Chicago, Il, August

Lautenschlager G. and Mayinger F. (1986), Post-Dryout Heat Transfer to R 12 in a Circular 90-Deg-Tube-Bend, Proc. 8th International Heat Transfer Conference, Vol. 6, pp. 2373-2378

Lee C.H. and Mudawwar I. (1988), A Mechanistic Critical Heat Flux Model for Subcooled Flow Boiling Based on Local Bulk Flow Conditions, Int. J. Multiphase Flow, Vol. 14, pp. 711-728

Lee D.H. (1965), An Experimental Investigation of Forced Convection Burnout in High Pressure Water. Part 4. Large Diameter Tubes at About 1600 psia, AEEW-R 479

Lee D.H. (1977), Prediction of Burnout, in Two-Phase Flow and Heat Transfer, Butterworth D. and Hewitt G.F., Eds., Oxford University Press, Oxford, pp. 295-322

Lee D.H. and Obertelli J.D. (1963), An Experimental Investigation of Forced Convection Boiling in High Pressure Water. Part I, AEEW-R 213

Lee S., Inone A. and Takahashi M. (1995), Critical Heat Flux Characteristics of R 113 Boiling Two-Phase Flow in Twisted Tape Inserted Tubes, Heat Transfer-Japanese Research, Vol. 24, n. 3, pp. 272-287

Lin W.S., Lee C.H. and Pei B.S. (1989), An Improved Theoretical Critical Heat Flux Model for Low-Quality Flow, Nuclear Technology, Vol. 88, pp. 294-306, December

Liu Q.S., Shiotsu M. and Sakurai A. (1992), A Correlation for Forced Convection Film Boiling Heat Transfer from a Hot Cylinder under Subcooled Conditions, Fundamentals of Subcooled Flow Boiling, HTD-Vol. 217, pp. 21-32

Lombardi C. and Mazzola A. (1998), A Criterion Based on Independent Parameters for Distinguishing Departure from Nucleate Boiling and Dryout in Water Cooled Systems, Revue Generale de Thermique, Vol. 37, n. 1, pp.31-38

Mattson R.J., Hammit F.G. and Tong L.S. (1973), A Photographic Study of the Subcooled Flow Boiling Crisis in Freon-113, ASME Paper 73-HT-39

Matzner B. (1963), Basic Experimental Studies of Boiling Fluid Flow and Heat Transfer at Elevated Pressures, TID 18978

Mazzola A. (1997), Integrating Artificial Neural Networks and Empirical Correlations for the Prediction of Water Subcooled Critical Heat Flux, Revue Generale de Thermique, Vol. 36, n. 11, pp. 799-806

Merilo M. (1977), Critical Heat Flux Experiments in a Vertical and Horizontal Tube with Both Freon-12 and Water as Coolant, Nuclear Engineering & Design, Vol. 44, n. 1, pp. 1-16

Mishima K. (1984), Boiling Burnout at Low Flow Rate and Low Pressure Conditions, Ph.D. Thesis, Kyoto University, Japan

Moon S.K. and Chang S.H. (1994), Classification and Prediction of the Critical Heat Flux using Fuzzy Clustering and Artificial Neural Networks, Nuclear Engineering and Design, Vol. 150, pp. 151-161

Moon S.K., Baek W.P. and Chang S.H. (1996), Parametric Trends Analysis of the Critical Heat Flux Based on Artifical Neural Networks, Nuclear Engineering and Design, Vol. 163, pp. 29-49

Moose R.A. and Ganic E.N. (1982), On the Calculation of Wall Temperatures in the Post-Dryout Heat Transfer Region, Int. J. Multiphase Flow, Vol. 8, n. 5, pp. 525-542

Mori H., Yoshida S., Ohno M., Kusumoto K. and Itoh T. (1990), Critical Heat Flux for Non-Azeotropic Binary Mixtures at High Pressures, Proceedings of JSME, No. 908-2, Saga, pp. 210-214

Motte E.I. and Bromley L.A. (1957), Film Boiling of Flowing Subcooled Liquids, Ind. and Engng. Chem., Vol. 49, n. 11, pp. 1921-1928

Naboichenko K.V., Kiryutin A.A. and Gribov B.S. (1965), A Study of Critical Heat Flux with Forced Flow of Monoisopropyldeiphenyl-Benzene Mixture, Teploenergetika, Vol. 12, n. 11, pp. 81-86

Nariai H. and Inasaka F. (1992), Critical Heat Flux and Flow Characteristics of Subcooled Flow Boiling with Water in Narrow Tubes, in Dynamics of Two-Phase Flows, Jones O.C. and Michiyoshi I. Eds., CRC Press, pp. 689-708

Nariai H., Inasaka F. and Shimura T. (1987), Critical Heat Flux of Subcooled Flow Boiling in Narrow Tube, ASME-JSME Thermal Engineering Joint Conference, Honolulu, March

Nariai H., Inasaka F., Fujisaki W. and Ishiguro H. (1992), Critical Heat Flux of Subcooled Flow Boiling in Tubes with Internal Twisted Tapes, Proc. ANS Winter Meeting (THD), pp. 38-46, San Francisco, November

Nariai H., Inasaka F., Ishikawa A. and Fujisaki W. (1992), Critical Heat Flux of Subcooled Flow Boiling in Tube with Internal Twisted Tape Under Non-Uniform Heating Conditions, Proc. 2nd JSME-KSME Thermal Engineering Conference, Vol. 3, pp. 285-288

Newbold F.J., Ralph J.C. and Ralph J.A. (1976), Post-Dryout Heat Transfer under Low Flow and Low Quality Conditions, AERE-R 8390

Nishikawa K., Yoshida S., Mori H. and Takamatsu H. (1986), Post-Dryout Heat Transfer to Freon in a Vertical Tube at High Subcritical Pressures, Int. J. Heat Mass Transfer, Vol. 29, n. 8, pp. 1245-1251

Obot N.T. and Ishii M. (1988), Two-Phase Flow Regime Transition Criteria in Post-Dryout Region Based on Flow Visualization Experiments, Int. J. Heat Mass Transfer, Vol. 31, n. 12, pp. 2559-2570

Papell S.S. (1970), Buoyancy Effects on Liquid Nitreogen Film Boiling in Vertical Flow, Advances in Cryogenic Engng., Vol. 16, pp. 435-444

Papell S.S. (1971), Film Boiling of Cryogenic Hydrogen during Upward and Downward Flow, Paper NASA-TMX-67855 presented at the 13th Int. Congress on Refrigeration, Washington

Papell S.S., Simoneau R.J. and Brown D.D. (1966), Buoyancy Effects on Critical Heat Flux of Forced Convective Boiling in Vertical Flow, NASA-TND-3672

Plummer D.N., Griffith P. and Rohsenow W.M. (1977), Post-Critical Heat Transfer to Flowing Liquid in a Vertical Tube, J. Heat Transfer, Vol. 4, pp. 151-158

Purcupile J.C. and Gouse S.W. Jr. (1972), Reynolds Flux Model of Critical Heat Flux in Subcooled Forced Convection Boiling, ASME Paper 72-HT-4

Rohsenow W.M. (1988), Post-Dryout Heat Transfer Prediction Method, Int. Comm. Heat Mass Transfer, Vol. 15, pp. 559-569

Schlosser J., Cardella A., Massmann P., Chappuis P., Falter H.D., Deschamps P. and Deschamps D.H. (1991), Thermal Hydraulic Tests on NET Divertor Targets Using Swirl Tubes, Proc. ANS Winter Meeting. (THD), pp. 26-31, San Francisco, CA, November

Schmidt KR. (1959), Wärmetechnische Untersuchungen and Hoch Belasteten Kesselheizflächen, Mitteinlungen der Vereinigung der Grosskessel-bezitzer, December, pp. 391-401

Slaughterback D.C., Veseley E.W., Ybarrondo L.J., Condie K.G. and Mattson R.J. (1973a) Statistical Regression Analyses of Experimental Data for Flow Film Boiling Heat Transfer, Paper presented at the ASME-AIChE Heat Transfer Conference, Atlanta, August

Slaughterback D.C., Ybarrondo L.J. and Obenchain C.F. (1973b), Flow Film Boiling Heat Transfer Correlations - Parametric Study with Data Comparison, Paper presented at the ASME-AIChE Heat Transfer Conference, Atlanta August

Smith R.A. (1986), Boiling Inside Tubes: Critical Heat Flux for Upward Flow in Uniformly Heated Tubes, ESDU Data Item No. 86032, Engineering Science Data Unit International Ltd., London

Smogalev I.P. (1981), Calculation of Critical Heat Fluxes with Flow of Subcooled Water at Low Velocity, Thermal Engineering, Vol. 28, n. 4, pp. 208-211

Staub F.W. (1968), The Void Fraction in Subcooled Boiling - Prediction of Vapour Volumetric Fraction, J. Heat Transfer, Vol. 90, pp. 151-157

Sterman L., Abramov A. and Checheta G. (1968), Investigation of Boiling Crisis at Forced Motion of High Temperature Organic Heat Carriers and Mixtures, Int. Symposium on Research into Co-current Gas-Liquid Flow, Univ. of Waterloo, Ontario, Canada, Paper E2

Styrikovich M.A., Newstrueva E.I. and Dvorina G.M. (1970), The Effect of Two-Phase Flow Pattern on the Nature of Heat Transfer Crisis in Boiling, Proc. 4th International Heat Transfer Conference, Vol. 9, pp. 360-362, Hemisphere, New York

Swenson H.S., Carver J.R. and Szoeke G. (1961), The Effects of Nucleate Boiling versus Film Boiling on Heat Transfer in Power Boiler Tubes, ASME Paper 61-W-201, presented at ASME Winter Annual Meeting, New York, 26 November-1 December

Theofanous T.G. (1996), Introduction to a Round Table Discussion on Reactor Power Margius, Nuclear Engineering and Design, Vol. 163, pp. 213-282

Thom J.R.S., Walker W.W., Fallon T.A. and Reising G.F.S. (1965), Boiling in Subcooled Water During Flow Up Heated Tubes or Annuli, Symposium on Boiling Heat Transfer in Steam Generating Units and Heat Exchangers, Paper 6, Manchester, IMechE, September

Thome J.R. (1990), Enhanced Boiling Heat Transfer, Hemisphere Publ. Corp., New York
Thorgerson E.J., Knoebel D.H. and Gibbons J.G. (1974), A Model to Predict Convective Subcooled Critical Heat Flux, J. Heat Transfer, Vol. 96, pp. 79-82
Tippets F.E. (1962), Critical Heat Fluxes and Flow Patterns in High Pressure Boiling Water Flows, Paper 62-WA-162 presented at the ASME Winter Annual Meeting, New York, 25-30 November
Tolubinsky V.I. and Matorin P.S. (1973), Forced Convective Boiling Heat Transfer Crisis with Binary Mixtures, Heat Transfer-Soviet Research, Vol. 5, n. 2, pp. 98-101
Tong L.S. (1966), Boundary Layer Analysis of the Flow Boiling Crisis, Proc. 3rd International Heat Transfer Conference, Vol. III, pp. 1-6, Hemisphere, New York
Tong L.S. (1968), Boundary-Layer Analysis of the Flow Boiling Crisis, Int. J. Heat Mass Transfer, Vol. 11, pp. 1208-1211
Tong L.S. (1969), Critical Heat Fluxes in Rod Bundles, Proc. Symp. on Two-Phase Flow and Heat Transfer in Rod Bundles, ASME Winter Annual Meeting, Los Angeles, CA, pp. 31-46
Tong L.S. (1975), A Phenomenological Study of Critical Heat Flux, ASME Paper 75-HT-68
Tong L.S. and Hewitt G.F. (1972), Overall View Point of Film Boiling CHF Mechanisms, ASME Paper No. 72-HT-54
Tong L.S., Currin H.B. and Larsen T.S. (1966), Influence of Axially Non-Uniform Heat Flux on DNB, WCAP-2767, Published in CEP Symp. Ser., Vol. 62
Tong L.S., Currin H.B. and Thorp A.G. (1968), An Evaluation of the Departure from Nucleate Boiling in Bundles of Reactor Fuel Rods, Nuclear Science Engineering, Vol. 33, pp. 7-15
Tong L.S., Currin H.B., Larsen P.S. and Smith D.G. (1966), Influence of Axially Non-Uniform Heat Flux on DNB, Chem. Eng. Prog. Symp. Ser., Vol. 62, n. 64, p. 35
Tong L.S., Currin H.B., Larsen P.S. and Smith O.G. (1965), Influence of Axially Non-Uniform Heat Flux on DNB, AIChE Symposium Series, Vol. 64, pp. 35-40
Unal C., Tuzla K., Badr O., Neti S. and Chen J.C. (1988), Parametric Trends for Post-CHF Heat Transfer in Rod Bundles, J. Heat Transfer, Vol. 110, pp. 721-727
van der Molen S.B. and Galjee F.W.B.M. (1978), The Boiling Mechanism during Burnout Phenomena in Subcooled Two-Phase Water Flow, Proc. 6th International Heat Transfer Conference, Vol. 1, pp. 381-385, Hemisphere, New York
Vandervort C.L., Bergles A.E. and Jensen M.K. (1992), The Ultimate Limits of Forced Convective Subcooled Boiling Heat Transfer, RPI Interim Report HTL-9 DE-FG02-89ER14019
Wang M.J. and Mayinger F. (1995), Post-Dryout Dispersed Flow in Circular Bends, Int. J. Multiphase Flow, Vol. 21, n. 3, pp. 437-454
Wang S.W. and Weisman J. (1983), Post-Critical Heat Flux Heat Transfer: A Survey of Current Correlations and their Applicability, Progress in Nuclear Energy, Vol. 12, n. 2, pp. 149-168
Weatherhead R.J. (1963), Nucleate Boiling Characteristics and the Critical Heat Flux Occurrence in Subcooled Axial Flow Water Systems, ANL 6675
Weisman J. (1992), The Current Status of Theoretically Based Approaches to the Prediction of the Critical Heat Flux in Flow Boiling, Nuclear Technology, Vol. 99, pp. 1-21, July
Weisman J. and Ileslamlou S. (1988), A Phenomenological Model for Prediction of Critical Heat Flux under Highly Subcooled Conditions, Fusion Technology, Vol. 13, pp. 654-659 (Corrigendum in Fusion Technology, Vol. 15, p. 1463-1989)
Weisman J. and King S.H. (1983), Theoretically Based CHF Prediction at Low Qualities and Intermediate Flows, Transaction American Nuclear Society, Vol. 45, pp. 832-833
Weisman J. and Pei B.S. (1983), Prediction of Critical Heat Flux in Flow Boiling at Low Qualities, Int. J. Heat Mass Transfer, Vol. 26, pp. 1463-1477
Wesman J. and Ying S.H. (1983), Theoretically Based CHF Prediction at Low Qualities and Intermediate Flows, Transactions American Nuclear Society, Vol. 45, pp. 832-833
Whalley P.B. (1987), Boiling, Condensation, and Gas-Liquid Flow, Clarendon Press, Oxford, pp. 163-166

Whalley P.B., Azzopardi B.J., Hewitt G.F. and Owen R.G. (1982), A Physical Model of Two-Phase Flow with Thermodynamic and Hydrodynamic Non-Equilibrium, Proc. 7th International Heat Transfer Conference, Vol. 5, pp. 181-188, Munich, August

Whalley P.B., Hutchinson P. and Hewitt G.F. (1974), The Calculation of Critical Heat Flux in Forced Convection Boiling, Proc. 5th Int. Heat Transfer Conference, Tokyo, Paper B6.11

Yagov V.V., Puzin V.A. and Kudryavtsev A.A. (1987), Investigation of the Boiling Crisis and Heat Transfer in Dispersed-Film Boiling of Liquids in Channels, Heat Transfer-Soviet Research, Vol. 19, n. 1, pp. 1-8

Yapo T., Embrechts M.J., Cathey S.T. and Lahey R.T. (1992), Prediction of Critical Heat Fluxes using a Hybrid Kohonen-backpropagation Neural Networks, in Topics in Intelligent Engineering Systems Through Artificial Neural Networks, Dali C.H. et al., Eds., Vol. 2, ASME Press, New York

Yin S.T., Liu T.J., Huang Y.D. and Tain R.M. (1988), An Investigation of the Limiting Quality Phenomenon of Critical Heat Flux, in Particulate Phenomena and Multiphase Transport, Veziroglu T.S. Ed., Hemisphere Publishing Corporation, Washington, Vol. 2, pp. 157-173

Yoo S.J. and France D.M. (1996), Post-CHF Heat Transfer with Water and Refrigerants, Nuclear Engineering and Design, Vol. 163, pp. 163-175

Interaction between Turbulence Structures and Inertial Particles in Boundary Layer: Mechanisms for Particle Transfer and Preferential Distribution

Cristian Marchioli, Maurizio Picciotto and Alfredo Soldati*

Centro di Fluidodinamica e Idraulica and Dipartimento di Energetica e Macchine, University of Udine, 33100, Udine, Italy

Abstract. Particle transfer in the wall region of turbulent boundary layers is dominated by the coherent structures which control the turbulence regeneration cycle. Coherent structures bring particles toward the wall and away from the wall and favour particle segregation in the viscous region. In this work we examine turbulent transfer of heavy particles to the wall and away from the wall in connection with the coherent structures of the boundary layer.

First, a detailed analysis of wall turbulence phenomena in a boundary layer will be provided. We will focus on the evolutionary dynamics of the structures populating the boundary layer: according to Schoppa & Hussain (1996, 1997), we will identify the following turbulence regeneration cycle: (i) low-speed streaks generate quasi-streamwise vortices, (ii) quasi-streamwise vortices generate sweeps and ejections, (iii) sweeps and ejections contribute to maintain the low-speed streaks.

We will then examine the behaviour of a dilute dispersion of heavy particles – flyashes in air – in a vertical channel flow, using pseudo-spectral direct numerical simulation to calculate the turbulent flow field at a shear Reynolds number $Re_\tau = 150$, and Lagrangian tracking to describe the dynamics of particles. Drag force, gravity and Saffman lift are used in the equation of motion for the particles, which are assumed to have no influence on the flow field. Particles interaction with wall is fully elastic. As reported in several previous investigations, we found that particles are transferred by sweeps in the wall region, where they preferentially accumulate in the low-speed streak environments, whereas ejections transfer particles from the wall region to the outer flow. We quantify the efficiency of the instantaneous realizations of the Reynolds stresses – sweeps and ejections – in transferring different size particles to the wall and away from the wall, respectively. Our findings confirm that sweeps and ejections are efficient transfer mechanisms for particles. However, the efficiency of the transfer mechanisms is conditioned by the presence of particles to be transferred. In the case of ejections, particles are more rarely available since, when in the viscous wall layer, they are concentrated under the low-speed streaks. Even though the low-speed streaks are ejection-like environments, particles remain trapped for a long time. Following the parent-offspring regeneration cycle for near-wall quasi-streamwise vortices, suggested by Brooke & Hanratty (1993), we find some evidence that the coupling of mature vortices with associated newly-born vortices is responsible for particle trapping in a sediment layer confined under the low-speed streak, between the offspring vortex and the wall. This mechanism may help to explain the existence of net particle fluxes toward the wall (turbophoretic drift).

Further analysis on particle distribution in the viscous sublayer confirmed that particles build-up under the low-speed streaks is due to the trapping action of the near-wall coherent

* Corresponding author. Email: soldati@uniud.it; Ph.: +390432558020; Facs: +390432558027; www: 158.110.32.35

structures. It is apparent that particles are not entrained in the coherent structures but rather accumulate in the proximity of a source point at the wall, located well below the low-speed streak.

1 Introduction

The physics of particle transfer in a turbulent boundary layer is of great importance in a number of applications, from environmental systems to industrial processes. Turbulent particle transfer mechanisms in proximity of a wall are characterized by complex interactions between turbulence structures and dispersed phase. Despite several decades of extensive experimental and numerical studies (Friedlander & Johnstone 1957), exhaustive explanations of particle transfer mechanisms have still to be produced. For instance, it is widely accepted that heavy particles have a tendency to migrate toward the wall (Caporaloni *et al.* 1975; Reeks 1983; McLaughlin 1989; Brooke *et al.* 1992) and that, when in the wall layer, they segregate preferentially in regions characterized by streamwise velocity lower than the mean (Pedinotti, Mariotti & Banerjee 1992; Eaton & Fessler 1994; Niño & Garcia 1996; Pan & Banerjee 1996), and yet explanations which offer physical mechanisms to justify and, possibly, link these observations appear incomplete.

The answer to particle behaviour in turbulent boundary layer is to be found in the relationships between turbulence structures and particle dynamics. This will explain the relationship between turbulence structure and particle number density (Rouson & Eaton 2001), which is the relevant information sought, and it could suggest ways to size and control transfer rates, mixing processes and reaction rates. A brief literature review will help to clarify some of the currently open issues on particle behaviour in turbulent boundary layers.

Cleaver & Yates (1975) proposed a sub-layer model for the deposition of small solid particles from a turbulent gas stream, which emphasizes the role of sweeps and ejections, which are instantaneous realizations of the Reynolds stresses, in determining the deposition rates. Once a particle is entrained in a sweep, *i.e.* fluid downwash toward the wall, it is expected to continue within the sweep and to approach the wall. The possibility of contacting the boundary depends on particle inertia and on the position at which the particle is entrained by the sweep. Only those particles entrained in the sweeps are likely to reach the wall: the others will be under the influence of outward fluid motions which will drive them away from the wall into the outer region. This and several other findings by a number of researchers who reported and demonstrated accumulation of particles close to the wall (Caporaloni *et al.* 1975; Reeks 1983; Sun & Lin 1986; Kallio & Reeks 1989) led to the conclusion that the mechanisms which transfer particles to the wall are more efficient than those which entrain particles into the outer flow.

After the first detailed Direct Numerical Simulation (DNS) of turbulent channel flow (Kim, Moin & Moser 1987), it has been possible to examine accurately the role of time-dependent turbulence structures on particle behaviour in the boundary layer. Among the subsequent DNS driven works, McLaughlin (1989) was the first to exploit a direct simulation of turbulence to simulate a three-dimensional time-dependent vertical channel flow, in which rigid spherical particles were released. The author observed that particles tend to accumulate in the viscous sublayer by virtue of inward turbulent motions in the buffer region, i.e. sweeps. He observed also that particles have a large residence time in the viscous sublayer.

From similar simulations, Brooke *et al.* (1992) found that particles appear to be driven toward the wall by coherent eddies which impart strong spanwise, wall parallel motions to the particle trajectories before they deposit.

These coherent eddies are the very same quasi-streamwise structures which play a fundamental role in the wall turbulence regeneration cycle (Guezennec, Piomelli & Kim 1989; Lyons, Hanratty & McLaughlin 1991; Brooke & Hanratty 1993; Schoppa & Hussain 1996, 1997; Jeong *et al.* 1997; Jimenez & Pinelli 1999; Zhou *et al.* 1999; Soldati & Marchioli 2001) and which control turbulence scalar transfer mechanisms at the wall (De Angelis *et al.* 1997). The mechanism for particle trapping developed by Brooke *et al.* (1992) suggests that particles undergo a long lasting sideways wandering in the wall normal direction until they are trapped in the coherent quasi-streamwise vortex which brings them directly to the wall. In the viscous sublayer, the spanwise motion of such vortex overwhelms the motion toward the wall so that particles experience a wall parallel transverse drift which acts to concentrate them in low-speed streaks.

This scenario is similar to that proposed by Banerjee and co-workers: their simulations (see Pedinotti *et al.* 1992; Pan & Banerjee 1996) and experiments (Kaftori, Hetsroni & Banerjee 1995a, 1995b) indicate that particle motion, as well as entrainment and deposition processes, are controlled by the action of coherent wall structures. The behaviour of particles is consistent with the motion and the effects of such structures, which appear to have a triple effect: i) they cause the formation of particle streaks in the low-speed regions near the wall, ii) they create suitable conditions for particle entrainment and iii) they assist in deposition, by conveying particles from the outer flow to the wall region through downsweeps, and eventually in particle resuspension, by conveying them from the wall region to the outer flow through ejections. Further analyses on particle fluxes (Kaftori *et al.* 1995b) showed that particles which tend to concentrate in low-velocity regions are ascending ones, indicating that the low-velocity environment is preferred by particles with an off-the-wall (rather than downward) trajectory. Other recent direct numerical simulations of particle dispersion in turbulent channel flow (see Ounis, Ahmadi & McLaughlin 1993; Ujittewaal & Oliemans 1996; van Haarlem, Boersma & Nieuwstadt 1998; Rouson & Eaton 2001) and in a three-dimensional mixing layer (Wei Ling *et al.* 1998) confirm that the dynamics of turbulent dispersion phenomena are strongly influenced by coherent wall structures and related sweep-ejection events cycle.

Despite the general consensus on several features of particle behaviour in the boundary layer, there are still many open issues concerning particle transfer mechanisms and particle segregation. In particular, even though we may hypothesize that sweeps and ejections control particle transfer to the wall and away from the wall, it is not entirely clear why particles tend to accumulate at the wall (Young & Leeming 1997; Cerbelli, Giusti & Soldati 2001), or why, once at the wall, particles remain trapped in the low streamwise velocity regions. In this work, we will try to address these issues.

We present results from a direct numerical simulation of the passive transport of solid particles – flyashes – in a fully developed upward turbulent channel flow. Our aim is to identify the role of turbulence structures which are responsible for particle motion and distribution in the buffer region, and which are also responsible for particle entrainment and deposition. We started from a well established background on wall turbulence (see Brooke *et al.* 1992; Brooke & Hanratty 1993; Schoppa & Hussain 1996, 1997; Jeong *et al.* 1997) focusing first on a simplified, though still complex and fully relevant, theoretical case in which the feedback of particles onto the flow field is ignored. Since our calculations involved large swarms of particles – $O(10^5)$

– the one-way coupling allowed us to reduce computing time required. We believe, nevertheless, that our results are of general validity for dilute dispersion. Previous experiments (Kulick, Fessler & Eaton, 1994; Kaftori *et al.* 1995a, 1995b) demonstrate that turbulence modifications due to the presence of small particles at low enough concentration are negligible. Even though particle concentration in the wall region may be locally large, turbulence structures appear modified only from a quantitative – small modifications of the intensities – viewpoint.

In the boundary layer, the most statistically common quasi-streamwise coherent structures are single streamwise-oriented vortices, generally centred within the buffer layer, slightly tilted upward – about 9° average – with streamwise dimension of 200 – 400 wall units; clockwise and counter-clockwise rotating vortices are also slightly tilted 4° left and right respectively (Schoppa & Hussain 1996, 1997). Quasi-streamwise vortices generate strongly coherent sweeps on the downwash side and strongly coherent ejections on the upwash side.

In the outer region, several recent investigations suggest that the most common vortex structures appear like hairpins whose legs are the counter-rotating quasi-streamwise vortices in the near-wall region (Robinson 1991). Neither these hairpins usually possess perfect spanwise symmetry nor the counter-rotating vortices have equal strength. Spanwise axisymmetric one-sided hairpins are also observed (Guezennec & Choi 1999). These new models revise and improve the classical concept of Ω-shaped horseshoe vortices and are widely, though not totally, accepted. In particular, Zhou *et al.* (1999) proposed a mechanism for generating coherent packets of hairpin vortices in the mean turbulent field of a low Reynolds number channel flow. This mechanism is similar to that proposed by Brooke & Hanratty (1993) and by Bernard, Thomas & Handler (1993).

The purpose of this work is to establish a physical link between the large-scale streamwise structures which populate the wall region and control momentum transfer to the wall and particle transfer fluxes. Further, we will try to put in evidence the mechanisms which trap particles at the wall. First, we will establish from a quantitative viewpoint the role of sweeps and ejections in determining particle fluxes to the wall and away from the wall. Second, we will focus on the mechanisms which prevent particles from being entrained by ejections and trap them under the low-speed streaks. In this context, we will examine the role of statistically probable secondary quasi-streamwise vortices, the characteristics of which were described by Brooke & Hanratty (1993) and by Bernard *et al.* (1993).

This paper is organized as follows. First, we will resume some of the results presented in Marchioli & Soldati (2002) in order to characterize turbulence events in the wall layer from a statistical viewpoint and to describe the dynamics of the wall structures which control the Reynolds stresses. Second, we will provide details about the computational methodology used in the numerical simulations. Finally, we will discuss new results, which provide quantitative information on particle preferential distribution in the wall region of a turbulent boundary layer.

2 Wall Turbulence Phenomena in a Boundary Layer

We consider in this section the boundary layer which occurs in the 2D fully developed turbulent channel flow. In a turbulent boundary layer, momentum, heat and mass transfer are controlled by the Reynolds stresses. Of the convective, correlation events which characterize the Reynolds stresses, strong local motions of fluid which are called *ejections* and *sweeps* are the most efficient transfer agents and control turbulent mixing close to the wall. Specifically, ejections bring

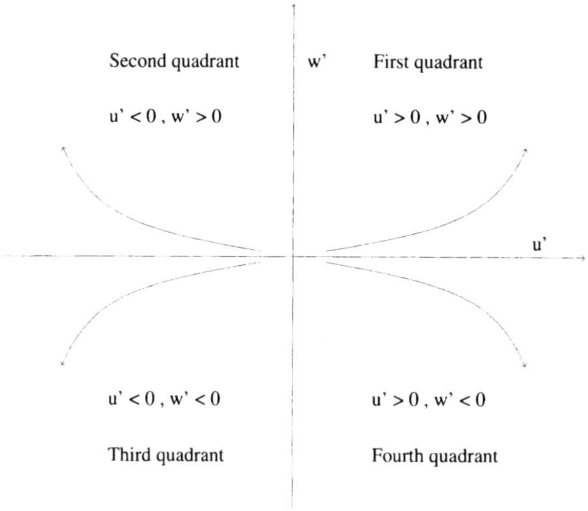

Figure 1. Identifications of $u'w'$ Reynolds stresses: $u' - w'$ plane.

the low-momentum fluid close to the wall into the outer region whereas sweeps bring the high-momentum fluid from the outer flow into the wall region. Ejections and sweeps control momentum transfer at the wall and are also well correlated to heat transfer (Pavassiliou & Hanratty 1995, Kasagi & Iida 1999) and mass transfer (De Angelis *et al.* 1997) at the wall. A complete characterization of sweeps and ejections and of their generation mechanisms is thus fundamental in order to gain insight into the physics of turbulence structure at the wall.

2.1 Statistics of Turbulence Phenomena at the Wall

Reynolds stresses may be conveniently examined by employing the quadrant analysis for the wall-normal (w') and the streamwise (u') component of the fluctuating velocity field (Willmarth & Lu 1972). Considering the events in the $u' - w'$ plane shown in Figure 1, the Reynolds stress is produced by four types of events: first quadrant events (I), characterized by outward motion of high-speed fluid, with $u' > 0$ and $w' > 0$, second quadrant events (II), characterized by outward motion of low-speed fluid, with $u' < 0$ and $w' > 0$, which are usually called *ejections*, third quadrant events (I), characterized by inward motion of low-speed fluid, with $u' < 0$ and $w' < 0$; and finally, fourth quadrant events (IV), which represent motions of high-speed fluid toward the wall, with $u' > 0$ and $w' < 0$, and are usually called *sweeps*. The drag generating events fall in the second and in the fourth quadrant. Both sweeps and ejections contribute to the negative Reynolds stress, i.e. to increase turbulence production, thus their increase corresponds in general to an increase in drag. First and third quadrant events contribute to positive Reynolds stress, i.e. to decrease turbulence production and their increase corresponds in general to a decrease in drag.

As in previous DNS works (Kim *et al.* 1987), we performed the quadrant analysis (Wallace, Eckelmann & Brodkey 1972) in the case of a plane boundary layer by time-averaging and space-

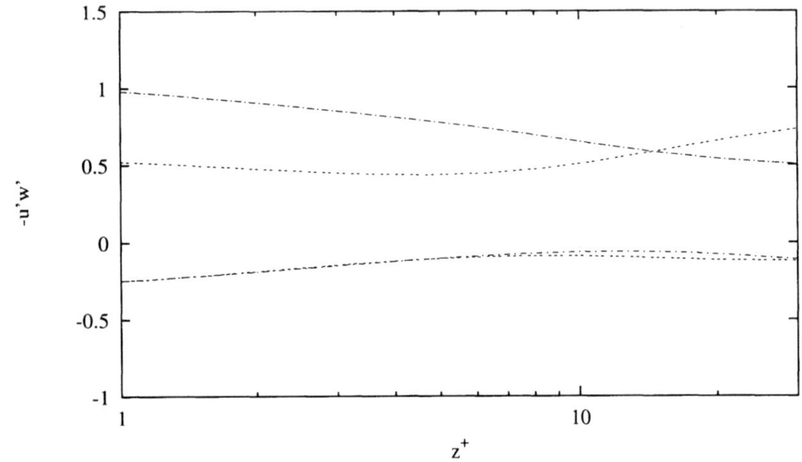

Figure 2. Quadrant analysis of $u'w'$.

averaging over the $x - y$ plane the mean fractional contribution to the Reynolds stress from each quadrant. Results are shown in Figure 2, plotted as a function of the non-dimensional wall distance [1], and normalized by the mean value of $u'w'$ at the wall distance considered.

Very close to the wall, the sweep events (IV) predominate over the ejection events (II). In other words, the fraction of the Reynolds stress associated with the sweep event is larger than that associated with the ejection event. Farther from the wall, ejection events dominate, with a crossover point at $z^+ \simeq 12$.

In order to examine the spatial relationship of the different events and the shear stress at the wall, we calculated the probability distributions of first, second, third and fourth quadrant events as functions of the shear stress at the wall. We followed the procedure used by Lombardi *et al.* (1996) and we considered only those events with significant spatial coherence. The probability density functions are shown in Figure 3. Sweeps and ejections are separated by the level of wall shear stress they generate: sweeps clearly correspond to high shear stress region, whereas ejections correspond to regions of low shear stress. Considering the average value of the shear stress at the wall, about 86% of the sweeps correspond to value of the shear stress above the average, whereas 98% of the ejections induce shear stress values lower than the average.

2.2 Dynamics of Turbulence Structures at the Wall

In the previous section, we examined the statistical relationship between the Reynolds stress and the shear stress at the wall. We will try to visualize the single events and the dynamics of the structures which control their occurrence. The correlations "sweep-high shear stress" and

[1] Turbulence analyses and data are usually reported in terms of a special set of dependent and independent variables. In particular, distance from the wall is made dimensionless as $z^+ = zu_\tau/\nu$, where $u_\tau = (\tau_w/\rho)^{0.5}$ is the *shear velocity* (ratio of the wall shear stress to fluid density) and ν is the kinematic viscosity of the fluid.

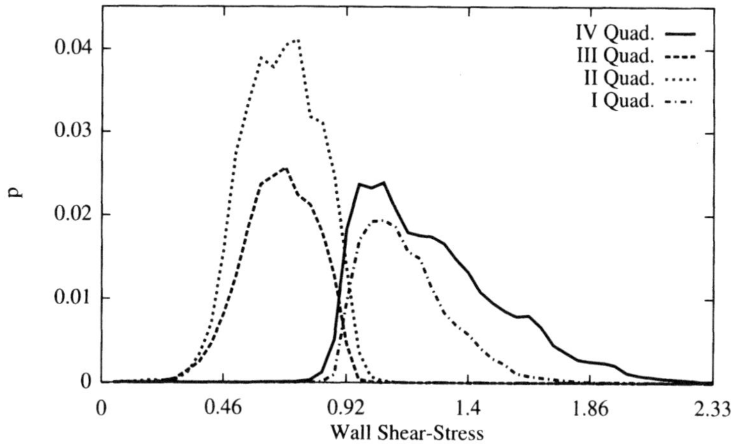

Figure 3. Probability function of coherent $u'w'$ events versus wall shear stress.

"ejection-low shear stress", demonstrated in Figure 3, may be appreciated in the snapshot shown in Figure 4, where the footprint of the shear-stress at the wall is shown together with sweep and ejection events. In this figure, flow is from left to right along the x direction. Sweeps and ejections are identified with the same isocontour of the instantaneous surface at $u'w' = -3$ in wall units. As discussed before, low shear-stress regions, shown in blue, correspond to the ejections (blue lumps), whereas high shear-stress regions, in red, correspond to the sweeps (gold lumps). Sweeps and ejections are just the final outcome of the dynamics of turbulence structures in the wall layer, and there is still some uncertainty about the mechanisms which generate and maintain the sweep/ejection events. They appear to be generated by the quasi-streamwise vortices which populate the near wall region. Quasi-streamwise vortices have been the object of a number of works starting with Kline *et al.* (1967). There is a general agreement about their characteristics: quasi-streamwise vortices appear to have a characteristic length of about 200 wall units and a spacing of about 400 wall units (Kline & Robinson 1990, Jeong *et al.* 1997, Schoppa & Hussain 1997, Jimenez & Pinelli 1999). These vortices are slightly tilted away from the wall and are responsible for pumping fluid towards and away from the wall. Quasi-streamwise vortices are slightly tilted upward - about 9° average (Schoppa & Hussain 1997); clockwise and counterclockwise rotating vortices are also slightly tilted about 4° left and right respectively. The streamwise vortices may be identified by using pressure, vorticity or other indicators (Bonnet & Delville 1996, Hunt, Wray & Moin 1998, Dubief & Delcayre 2000). A convenient method for identification is based on calculating the eigenvalues of the strain rate tensor $\mathbf{\Omega}$ (Perry & Chong 1987, Chong, Perry & Cantwell 1990, Jeong & Hussain 1995). The vector $\mathbf{\Omega}$ represents the strength and direction of the rotation of the streamlines and is based on the identification of flow regions where the rate-of-deformation tensor $\partial u_i/\partial x_j$ exhibits complex eigenvalues (Perry & Chong 1987; Chong, Perry & Cantwell 1990; Jeong & Hussain 1995). For further details, see Appendix A.

In Figure 5, two counterrotating vortices, identified by one isosurface of $\mathbf{\Omega}$, are shown together with the ejections and sweeps they generate. The elongated red and pale blue structures are two isosurfaces with the same absolute value of $\mathbf{\Omega}$ (and opposite sign) and indicate clockwise

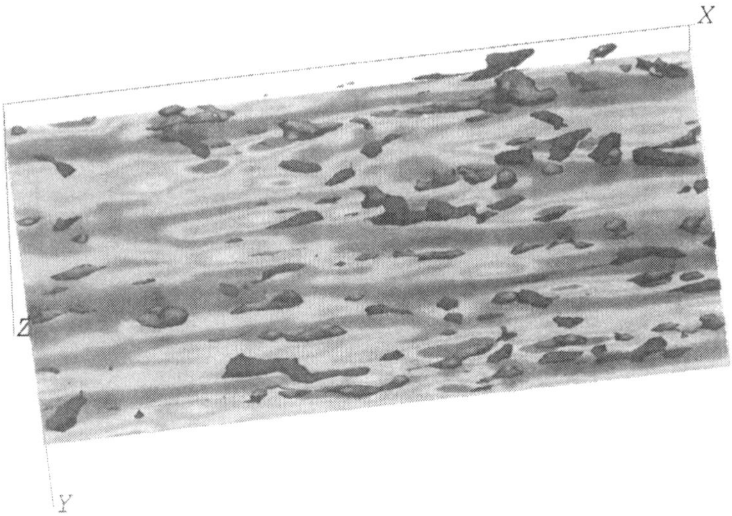

Figure 4. Snapshot footprint of the wall shear-stress with corresponding sweep and ejection events in the whole computational domain. At the wall, red indicates high shear-stress; blue indicates low shear-stress. Gold 3D regions are isosurfaces characterizing sweeps whereas blue 3D regions characterize ejections. 3D isosurfaces are traced at $u'w' = -3$ in dimensionless units.

rotating (red) and counterclockwise rotating (pale blue) vortices. Flow is going from bottom left to top right in figure and vortices appear tilted away from the wall by the mean strain rate. The blue lumps of fluid in between the two vortices are ejections and the green lumps of fluid outside the two vortices are sweeps. Ejections and sweeps also appear to be affected by the mean strain rate. Owing to the continuous action of the quasi-streamwise vortices in generating sweeps and ejections, regions between two vortices such as those shown in Figure 5 are characterized by a streamwise velocity lower than the mean, whereas the regions outside the two vortices are characterized by a streamwise velocity higher than the mean. Specifically, the regions with velocity lower than the mean are called *low-speed streaks*, whereas the regions with velocity higher than the mean are called *high-speed regions*. Many quasi-streamwise vortices are usually associated with one single low-speed streak. Low-speed streaks are sinuous regions about 1000 wall units long and are more coherent than the high-speed regions.

In Figure 6, a 450 wall units long piece of one low-speed streak is shown, flanked by two counter-rotating quasi-streamwise vortices. In this figure, the action of the quasi-streamwise vortices in lifting up the low-speed streak is clear. Streamwise vortical structures overlap in x as a staggered array, and this is clearly shown in Figure 7, where quasi-streamwise vortices are shown associated with one low-speed streak - a 650 wall unit-long piece of one low-speed streak. One single low-speed streak has a longer life than quasi-streamwise vortices and survives a number of vortex generations. In Figure 8, one single low-speed streak is shown for a length of about 1200 wall units. The isosurface indicates a streamwise velocity value of 0.56 U_c where U_c is centerline velocity. The streak is sinuous and, in time, oscillates with a meandering motion in the spanwise direction. The isosurface of the streak has different slope and shape at different locations.

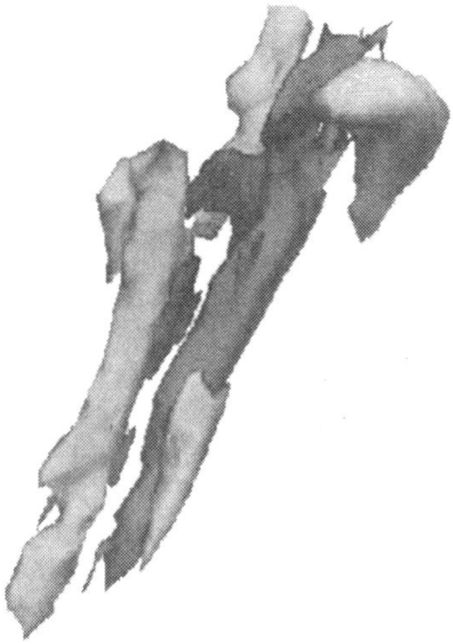

Figure 5. Quasi-streamwise counterrotating vortices together with ejections and sweeps. Quasi-streamwise vortices extend for about 200 – 300 wall units. Two isosurfaces of the same absolute value of Ω indicate clockwise rotating (red) and counterclockwise rotating (pale blue) vortices. Sweeps and ejections are indicated by green and blue respectively.

Figure 6. Two counterrotating quasi-streamwise vortices onto a single low-speed streak (red). Green isosurface of Ω indicates clock-wise rotating vortex, blue isosurface of Ω indicates counter clock-wise rotating vortex.

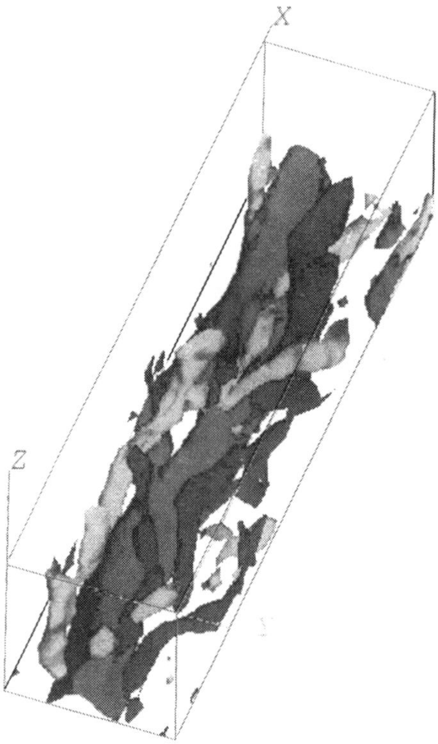

Figure 7. One low-speed streak (red) with counterrotating quasi-streamwise vortices in a box about 150 wall units wide and 650 wall units long. Clockwise rotating vortices (blue) are on the right side of the low-speed streak and are slightly tilted right; counterclockwise rotating vortices (green) are on the left side and are slightly tilted left.

It has been shown (Kim & Hussain 1993) that the generation of the quasi-streamwise vortices is associated with changes in the shape of the low-speed streak surface (Soldati 2000). The Lagrangian evolution of the shape of a low-speed streak is shown in Figure 9. The section shown in Figure 9(a) is followed downstream with the velocity $0.56\ U_c$ which is the advection velocity of the low-speed streak (Kim & Hussain 1993). This enables a Lagrangian tracking of the evolution of the chosen section of the streak. During the sequence shown in Figure 9, the wall layer undergoes a quiescent phase, Figures 9(a) and (b), during which there is no evidence of quasi-streamwise vortices, and an active phase, Figure 9(c), (d) and (e), during which there is an alternate appearance of quasi-streamwise vortices of opposite sign, and again a quiescent phase, Figure 9(f). It is interesting to observe that clockwise and counterclockwise rotating vortices do not appear at the same time in the same location, as discussed by Schoppa & Hussain (1997). Over the whole cycle, the section of the streak was advected downstream for 640 wall units.

It is clear that changes in the shape of the cross section of the low-speed streaks are associated to the occurrence of quasi-streamwise vortices. Since quasi-streamwise vortices control turbulence transfer mechanisms at the wall, the key to control transfer rates is to control quasi-streamwise vortices.

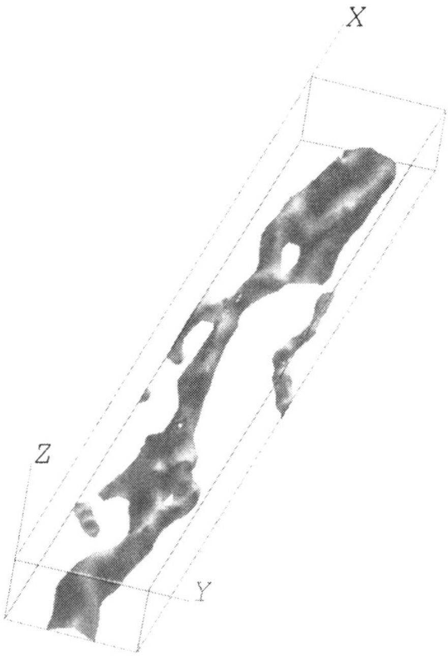

Figure 8. Velocity isosurface describing one isolated low-speed streak. Isosurface velocity value is $0.56\ U_c$.

In recent papers, Schoppa & Hussain (1997, 2000) suggested that wall turbulence is dominated by a cycle in which low-speed streaks generate quasi-streamwise vortices, which in turn generate ejections and sweeps. These finally contribute to maintain the low-speed streaks. A crucial question is how the low-speed streaks can generate the quasi-streamwise vortices. Schoppa & Hussain (1997) answer this by examining the behavior of a single low-speed streak and finding that it is subject to lateral, sinuous instability. Stability analysis of an idealized low-speed streak showed it to be unstable to lateral perturbations. Basing their examination also on the DNS of a minimal channel-flow unit (Jimenez & Moin 1991), the authors were able to demonstrate that even a vortex-less streak generates new streamwise vortices.

With respect to previous pictures of the dynamics of turbulence structures in the wall layer (see Kline *et al.* 1967, Brooke & Hanratty 1993, Choi 2001), somehow the view of evolutionary dynamics of the structures in the boundary layer changes perspective in that streaks are considered responsible for the initial generation of quasi-streamwise vortices. Based on this view, Schoppa & Hussain (1997) conclude by suggesting different strategies for turbulence control. The most interesting strategy seems to be stabilization of the low-speed streaks by means of large-scale forcing motions, given the obvious technological advantages. In practice, a low-speed streak which is more stable to spanwise perturbations would reduce its meandering (see Figure 8) and reduce the tripping frequency of quasi-streamwise vortices (see Figure 9), eventually reducing the frequency and the intensity of turbulence production events - i.e. sweeps and ejections. This suggestion would also explain why turbulent boundary layers loaded with small particles may exhibit reduced drag (Pan & Banerjee 1996). Small particles characterized by a suitable

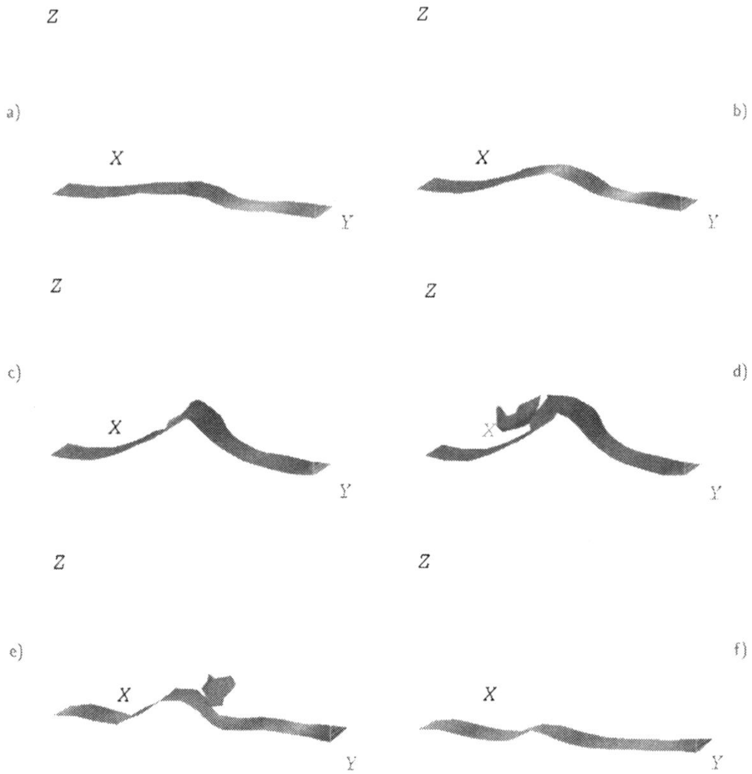

Figure 9. Lagrangian evolution of low-speed streak surface - identified by the $0.56\ U_c$ isosurface - associated with the presence of streamwise vortices - identified by the Ω isosurface: red is clockwise rotating and blue is counterclockwise rotating. The streamwise location is advected downstream with velocity $0.56\ U_c$. The wall units time of the different snapshots is (a) $t^+ = 0$, (b) $t^+ = 24$, (c) $t^+ = 52$, (d) $t^+ = 94$, (e) $t^+ = 131$, and (f) $t^+ = 181$. The wall region undergoes modifications from quiescent phase (a) and (b), to active phase, (c) and (d) in which the presence of a counterclockwise rotating streamwise vortex is noted, and (e) in which the presence of a clockwise rotating streamwise vortex is observed, back to quiescent phase, (f). During the time sequence, the initial cross-section advanced about 640 wall units downstream.

inertia parameter tend to segregate into the low-speed streaks. Higher concentrations of particles in the low-speed streaks increase the inertia of the low-speed streaks making them more stable to lateral perturbations.

3 Methodology

3.1 Channel flow simulation

The flow into which particles are introduced is a turbulent channel flow of air, assumed to be incompressible and Newtonian. The flow is driven upward by a pressure gradient. The reference geometry, shown in Figure 10, consists of two infinite vertical flat parallel walls: the origin of the coordinate system is located at the centre of the channel and the x, y and z axes point in

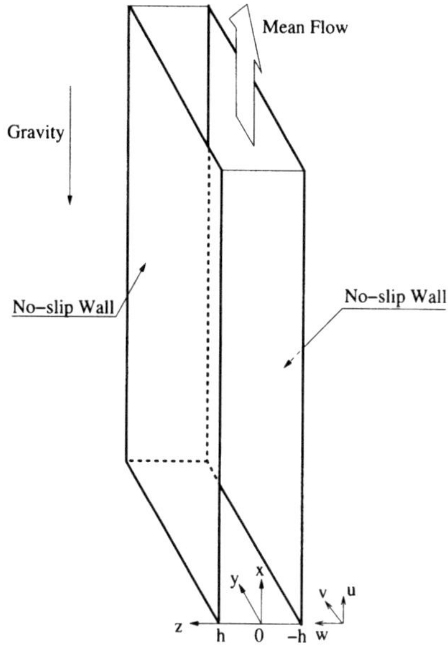

Figure 10. Schematic of vertical channel flow configuration

the streamwise, spanwise and wall-normal directions respectively. Gravity is directed along the negative x direction. Periodic boundary conditions are imposed on the fluid velocity field in both streamwise and spanwise directions and no-slip boundary conditions are enforced at the walls. We assume that particle number density and particle size are both small, and that there is no feedback of the particles onto the gas flow. The flow field was calculated by integrating mass and momentum balance equations in dimensionless form, obtained using the duct half-width, h, and the shear velocity, u_τ, defined as:

$$u_\tau = \left(\frac{\tau_w}{\rho}\right)^{1/2} \tag{1}$$

where τ_w is the shear at the wall and ρ is fluid density. Therefore, mass and momentum balance equations in dimensionless form are:

$$\frac{\partial u_i}{\partial x_i} = 0 \quad , \tag{2}$$

and

$$\frac{\partial u_i}{\partial t} = -u_j \frac{\partial u_i}{\partial x_j} + \frac{1}{Re} \frac{\partial^2 u_i}{\partial x_j \partial x_j} - \frac{\partial p}{\partial x_i} + \delta_{1,i} \quad , \tag{3}$$

where u_i is the ith component of the dimensionless velocity vector, p is the fluctuating kinematic pressure (pressure divided by density), $\delta_{1,i}$ is the mean dimensionless pressure gradient that drives the flow, and $Re_\tau = hu_\tau/\nu$ is the shear Reynolds number. Equations (2) and (3) were solved directly using a pseudo-spectral method previously used in different types of flow (Pan & Banerjee, 1996; Soldati & Banerjee, 1998) and similar to that used by Kim *et al.* (1987) to solve the turbulent, closed-channel flow problem. Equation (3) may be recast as:

$$\frac{\partial u_i}{\partial t} = S_i + \frac{1}{Re_\tau} \frac{\partial^2 u_i}{\partial x_j \partial x_j} - \frac{\partial p}{\partial x_i} \tag{4}$$

where S_i includes the convective term and the mean pressure gradient (Kim *et al.* 1987; Lam & Banerjee 1992). The pseudo-spectral method is based on transforming the field variables into wavenumber space, using Fourier representations for the streamwise and spanwise directions and a Chebyshev representation for the wall-normal (non-homogeneous) direction. A two-level explicit Adams-Bashforth scheme for the nonlinear terms S_i and an implicit Crank-Nicolson method for the viscous terms, were employed for time advancement. Details of the method have been published previously (Lam & Banerjee 1992).

In the present study, we consider air with density of $1.3\ kg\ m^{-3}$ and kinematic viscosity of $15.7 \cdot 10^{-6}\ m^2\ s^{-1}$. Since the pressure gradient is equal for all simulations, the shear velocity is $11.775 \cdot 10^{-2}\ m\ s^{-1}$, and the shear Reynolds number, Re_τ, is equal to 150. The mean velocity is $1.65\ m\ s^{-1}$ and the Reynolds number based on mean velocity and half duct width is ≈ 2110. Our calculations have been performed in dimensionless units, the Reynolds number being the one parameter to scale the flow. The computational domain was $1885 \times 942 \times 300$ wall units in x, y and z with $64 \times 64 \times 65$ nodes. The spacing of collocation points in the streamwise and spanwise directions was $\Delta x^+ \approx 30$, $\Delta y^+ \approx 15$ in wall units. The first collocation point away from the wall is at $z^+ = 0.18$: this grid resolution is sufficient to describe the significant length scales in the channel flow. The time step used was $\Delta t^+ = 0.35325$ in wall time units. We will not show here the statistics of the flow field which match closely with those obtained by Lyons *et al.* (1991) for the same Reynolds number.

Even though the grid is slightly less refined compared to other DNS databases (Kim *et al.* 1987), the large-scale wall structures are well resolved. We compared the 64^3 results against results obtained with a 128^3 grid (twice the resolution in each direction). From a statistical viewpoint, the results obtained with the two different grids match closely, both collapsing on the results obtained by Lyons *et al.* (1991) for the same Reynolds number. We examined in detail the evolution of the wall structures below $z^+ = 80$. We found hardly any difference in the shape, extent and duration of the structures which dominate wall transfer mechanisms – large-scale quasi-streamwise vortices, low-speed streaks, sweeps and ejections.

An accurate calculation of the forces acting on the particle requires accurate evaluation of the instantaneous fluid velocity at the particle location. Balachandar & Maxey (1989) used a fourth-order Hermitian scheme in two directions followed by a Fourier interpolation in the third direction. Yeung & Pope (1988) tested both a third-order Taylor-series interpolation scheme and a cubic-spline scheme, concluding that the first scheme gives higher interpolation accuracy with adequate spatial resolution. Pan & Banerjee (1996) used cubic splines as well as a hybrid interpolation scheme that employs cubic-splines in the homogeneous directions followed by a Chebyshev summation in the non-homogeneous direction. This approach is similar to that proposed by Kontomaris, Hanratty & McLaughlin (1992), who employed Lagrange polynomials in the homogeneous directions and Chebyshev polynomials in the non-homogeneous direction. This procedure showed to be highly accurate and the computational work requirement was smaller by roughly the factor $(6/N_x)(6/N_y)$ than the computational work requirement for a fully spectral evaluation of the fluid velocity field at the centre of the particle, which involves summing the Fourier-Chebyshev series (Soldati *et al.* 1997).

In recent papers, several authors who investigated the behaviour of large swarms of particles used time-efficient lower-order interpolation schemes, proving they were accurate enough to maintain statistical accuracy (van Haarlem *et al.* 1998) and to preserve local resolution for the small scales of the boundary layer (Rouson & Eaton 2001). Among the others, Rouson & Eaton (2001) used a three-dimensional linear interpolation, whereas van Haarlem *et al.* (1998) and Uijttewaal & Oliemans (1996) used a quadratic interpolation scheme. In the present work, we used a Lagrange interpolation of order three.

3.2 Particle equation of motion

Particles are injected into the flow at concentration low enough for particle-particle interaction due to their inertial force to be negligible – dilute system conditions. Furthermore, particles are assumed to be pointwise, rigid, spherical and to obey the following vectorial Lagrangian equation of motion:

$$\frac{d\mathbf{v}}{dt} = \frac{C_d}{\tau_P}(\mathbf{u}-\mathbf{v}) + \left(1 - \frac{\rho}{\rho_P}\right)\mathbf{g} - \xi(\epsilon)\frac{6.46}{12\pi}\frac{d_P}{\tau_P}\left|\frac{\partial u_x}{\partial z}\right|^{\frac{1}{2}}(v_x - u_x)\mathbf{e_z}, \tag{5}$$

in which $\mathbf{v}$ is particle velocity vector, $\mathbf{u}$ is fluid velocity vector at particle location, $C_d = 24(1+0.15Re_P^{0.687})/Re_P$ is the Stokes drag coefficient, $\tau_P = d_P^2\rho_P/18\rho$ is the particle relaxation time (d_P and ρ_P being particle diameter and density respectively), $\mathbf{g}$ is gravitational acceleration and $\mathbf{e_z}$ is the unit vector in wall-normal direction. All physical quantities have been made dimensionless in terms of wall units based on shear velocity, kinematic viscosity and fluid density. The lefthand side of (5) represents particle inertia, and the terms on the righthand side of (5) represent the effects of Stokes drag, gravity and Saffman lift force. The Saffman lift force reproduces the effects of the local gradient of the fluid velocity field $\partial u_x/\partial z$ acting on a rigid spherical particle in a streamwise-oriented, time-dependent shear flow that is a function of wall-normal direction (Saffman 1965). In (5), the lift force term is written according to Saffman (1965) with an additional correction factor $\xi(\epsilon)$ to account for situations where the velocity difference between the particle and the continuous phase becomes larger (McLaughlin 1991). The correction factor is computed as follows (McLaughlin 1991):

$$\begin{cases} \xi(\epsilon) = \frac{1}{K}[-32\pi^2\epsilon^5\ln(\epsilon^{-2})],\ \epsilon \le 0.025 \\ \xi(\epsilon) = \frac{1}{K}[1.418\ \arctan(2.8\epsilon^{2.44})],\ 0.025 > \epsilon \le 20 \\ \xi(\epsilon) = \frac{1}{K}(K - 0.6463\,\epsilon^{-2}),\ \epsilon > 20. \end{cases} \tag{6}$$

where $K = 2.225$ and $\epsilon = |\partial u_x^+/\partial z^+|^{0.5}\, d_P^+(Re_P)^{-1}$. We have that $\xi(0) = 0$ whereas $\xi(\epsilon)$ becomes equal to unity for high values of ϵ.

Other forces acting on the particle, such as hydrostatic force, Magnus effect, Basset history force and added mass force are not taken into account since they are assumed to be negligible (orders of magnitude smaller than the three effects considered) because of the specific set of physical parameters of our simulations (Rizk & Elghobashi 1985; Armenio & Fiorotto 2001).

In the present simulations, 48^3 flyash particles, characterized by a particle to fluid density ratio equal to 769.23, have been released at randomly chosen locations within the computational

Run	$\tau_p(ms)$	τ_p^+	$d_p(\mu m)$	d_p^+
R1	4.35	3.8	40	0.3
R2	32.9	29.1	110	0.825
R3	131.7	116.3	220	1.65

Table 1. Parameters relative to the simulation of particle dispersion. The superscript + identifies dimensionless variables.

box. During the simulation, particles go toward either wall. Comparing the statistics of particle dynamics obtained for the two walls, we found no significant difference, thus indicating that our particle swarm is large enough to ensure meaningful quantitative analysis. Using samples of 48^3 particles for each diameter, particles Lagrangian velocity statistics (not shown here) become stationary in time after a few particle time constants. The trajectories of the particles were tracked individually through integration of (5) by an explicit method, using the channel flow DNS code to supply the fluid velocity field at each time step. The initial velocities of the particles were set equal to the interpolated fluid velocities at each particle location. In Table 1, dimensionless and dimensional parameters characterizing the tracked particles are reported. The three particle sets considered here are large compared to those examined by other authors, who considered particles in the range $\tau_P^+ = 0 - 3$ (McLaughlin 1989; Brooke *et al.* 1992; Pan & Banerjee 1996 and Ujittewaal & Oliemans 1996). Particle behaviour changes dramatically over such range of particle relaxation times. In particular, particles with τ_P^+ close enough to zero will behave like fluid parcels showing no preferential concentration. This is required to agree with mass conservation for the continuous phase (Brooke *et al.* 1992; Ujittewaal & Oliemans 1996).

Particles with finite inertia will respond to the coherent vortical structures of the boundary layer in different ways. Following Pan & Banerjee (1996), the particle samples which seem to respond "at best" to the specific size of the structures are characterized by a dimensionless relaxation time around $\tau_P^+ = 1.0$. Notice that (5) does not include wall effects: when the distance of the particle from the closest wall becomes small compared to particle size, the actual mechanism of deposition is complicated by the possible rise of different surface related phenomena (McLaughlin 1991). In our calculations, these phenomena were not taken into account and we simply considered that a particle is elastically reflected away from the wall when its centre is less than a distance $d_P/2$ from the boundary. Perfect elastic reflection, i.e. where no dissipation occurs during the collision, is at the other extreme with respect to the perfectly absorbing wall model, in which particle kinetic energy is completely lost during the collision. Real cases usually fall between these limiting situations.

4 Results and Discussion

4.1 Particle distribution in the boundary layer

We investigated turbulent transfer mechanisms and segregation of inertial particles in a boundary layer over a flat vertical wall. The response of rigid, pointwise particles to turbulence is related to the Stokes number, defined as the ratio between the particle relaxation time τ_P and some representative time scale of the turbulent fluid motion τ_f (Crowe, Gore & Troutt 1988): thus, the

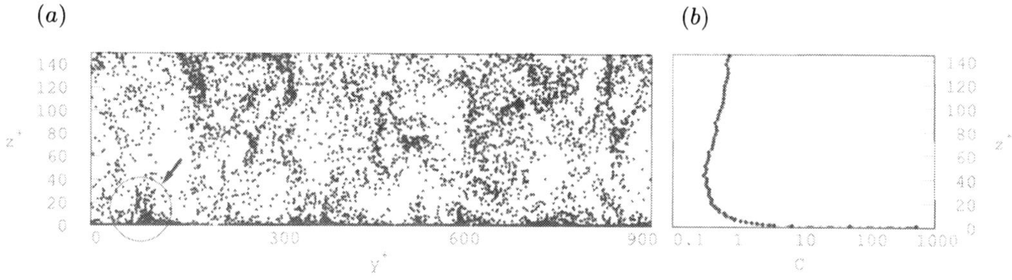

Figure 11. Instantaneous distribution of particles characterized by $\tau_p^+ = 116.3$ at time $t^+ = 2700$. View of particle position in the yz-plane for $700 < x^+ < 1000$ (a) and corresponding xy-plane average number density distribution as a function of the wall normal direction (b).

controlling variable was τ_P (or diameter d_P, since all the other parameters were kept constant during the simulations).

A representative view of particle dynamics and distribution in a turbulent boundary layer is provided in Figure 11a, where a cross-sectional view of particle instantaneous distribution for upward turbulent channel flow is shown. The snapshot is taken at the dimensionless simulation time $t^+ = 2700$. For clarity of presentation, we show only one-sixth of the channel length (from $x^+ = 700$ to $x^+ = 1000$) and one-half of the channel wall-to-wall distance (from $z^+ = 0$ to $z^+ = 150$). Recall that particles are uniformly distributed in the entire computational domain at time $t^+ = 0$. Figure 11a shows a number of features which can classify the process of particle dispersion and transfer in turbulent boundary layer. First, we can observe that particles are not homogeneously distributed along the channel, but they tend to cluster. In particular, particles tend to cluster around the large vortical structures. From these clusters, particles are transported toward the wall, where they accumulate in specific "reservoirs" (one of these is indicated by the black circle) where concentration build-up occurs. These accumulation regions are classified by flow streamwise velocity lower than the mean. Particles tend to stay long times in these low-speed regions so that eventually particle concentration in the wall region increases. Since, in the present upflow configuration, gravity does not affect particle deposition and segregation significantly, this behaviour (which is qualitatively similar for $\tau_P^+ = 3.8$ and $\tau_P^+ = 29.1$ particles, not shown here). is probably due to turbulence structures and has also been observed in pipe flow boundary layer (Cerbelli *et al.* 2001, Marchioli *et al.* 2003).

To quantify near-wall accumulation, the particle number density distribution is plotted as a function of the non dimensional distance z^+ from the wall in Figure 11b. A logarithmic scale is used to capture the detail of particle behaviour as they approach the near-wall region. The concentration profile was computed at fixed time intervals by subdividing the channel into 129 slabs (through Chebyshev polynomials) and counting the fraction of particles that fell within each slab, i.e. by averaging over the streamwise and spanwise coordinates, x and y, respectively. As already observed in pipe flow geometry (Cerbelli *et al.* 2001, Marchioli *et al.* 2003), the concentration profile is developing with time, and at the instant captured in Figure 11b, we observe that particle number density profile has developed a maximum well into the near-wall region ($0 < z^+ < 20$).

This behaviour can be viewed as the consequence of non-uniform turbulence advection mechanisms, whose intensity decreases to very low values in the near wall region. In other words, from a macroscopic, engineering viewpoint, particle transport toward the wall can be roughly envisioned as a two-stage process characterized by different time-scales. Particles are driven away from the outer flow region by energetic turbulent convective mechanisms, accumulate close to the wall, and are then slowly transported toward the wall.

Particle migration to the wall in turbulent boundary layer has been observed previously (Caporaloni *et al.* 1975; Reeks 1983) and the name turbophoretic drift was given to this phenomenon, which is extremely relevant for a number of industrial and environmental applications. The mechanisms leading to these strong particle net fluxes to the wall are precisely the aim of this work and we will try to investigate this issue by examining particle dynamics in connection with the dynamics of wall coherent structures.

4.2 Particle distribution in the wall region

Even when segregated in the wall region, particles do not attain uniform distribution in the spanwise wall parallel direction. As reported previously (Pedinotti *et al.* 1992; Eaton & Fessler 1994; Pan & Banerjee 1996; Zhang & Ahmadi 2000) particle positions tend to correlate with the instantaneous location of the lower values of the streamwise velocity – low-speed streaks. In Figure 12 the instantaneous distribution of particles in the region between the wall and $z^+ = 1$ is shown. As reported in previous works (Pedinotti *et al.* 1992; Kaftori *et al.* 1995a, 1995b; Pan & Banerjee 1996; van Haarlem *et al.* 1998), we can observe that particles tend to line-up along streamwise oriented streaks.

To quantify the correlation between the instantaneous location of the low-speed regions and particle streaks, in Figure 13, the normalized joint probability density function (PDF) between the particle position and the fluctuation of the streamwise velocity, u', in the near-wall region ($5 \leq z^+ \leq 15$) is shown. The joint PDF profiles were computed as follows: i) we calculated the average streamwise velocity $\overline{u}_{slab}(z)$ of the fluid in the region $5 \leq z^+ \leq 15$, ii) we subdivided this region in 10 equally spaced slabs, iii) we determined the slab containing the particle, iv) we computed the local streamwise velocity fluctuation of the fluid $u'(x, y, z) = u(x, y, z) - \overline{u}_{slab}(z)$ in the position of the particle, v) we counted the number of particles associated with each value of $u'(x, y, z)$ and normalized it by the total number of particles located into each slab.

From Figure 13, it appears that particles tend to attain a preferential distribution in the regions of lower-than-mean fluid velocity. There is also evidence of an effect of particle relaxation time: the correlation between particle position and regions of lower negative u' is better for $\tau_P^+ = 29.1$ particles (intermediate size) rather than for $\tau_P^+ = 3.8$ and $\tau_P^+ = 116.3$ particles.

The tendency of inertial particles to accumulate in the low-speed regions may support a possible use of particles as smart roughness (Pan & Banerjee 1996). In real situations, characterized by a two-way coupling between particles and fluid, the presence of particles would increase the inertia of the low-speed streaks. Since low-speed streak stability to lateral perturbation has an impact on the wall turbulence regeneration cycle (Schoppa & Hussain 1996, 1997), the presence of specific inertia or size particles in turbulent boundary layer might be exploited to tune wall transfer mechanisms.

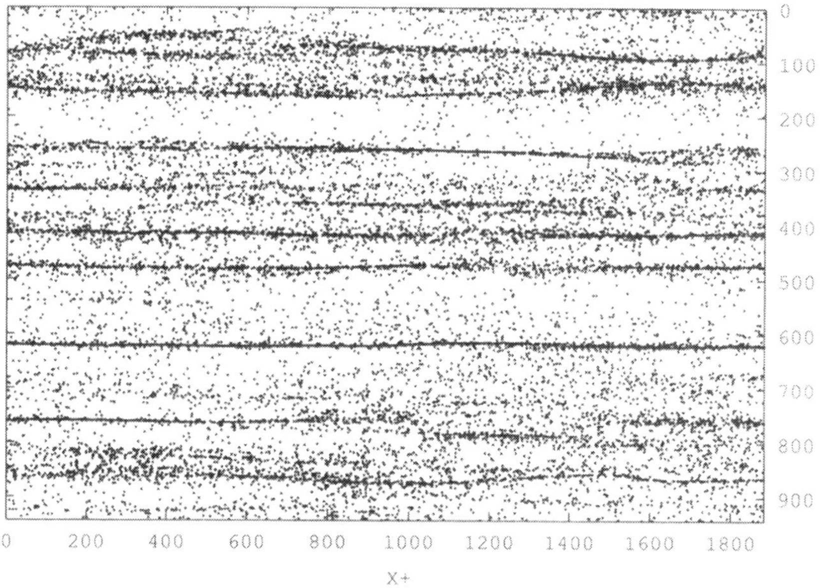

Figure 12. Distribution of $\tau_P^+ = 116.3$ particles in the wall region. Particles comprised between $z^+ = 1$ and the wall are shown.

4.3 Particle transfer fluxes

As observed before (Marchioli and Soldati 2002), particle transfer to the wall is a remarkably efficient phenomenon. In Figure 11, we observed pathways of particles striking the wall. These pathways correspond to coherent advective motions which scale with the buffer layer and are the instantaneous realizations of the Reynolds stresses. The relationships between these advective motions and particle transfer to the wall may be elucidated through quadrant analysis (Wallace *et al.* 1972). As mentioned in Section 2.1, the presence of a sweep corresponds to a local increase of the shear stress at the wall whereas the presence of an ejection corresponds to a local decrease of the shear stress at the wall.

Here, we have two aims: the first is to verify whether sweeps and ejections are the actual mechanisms by which particles are transferred toward the wall and are entrained into the outer flow. The second is to quantify the role of sweeps and ejections in determining particle fluxes to and off the wall layer.

In Figure 14, we highlight the spatial correlation of the location of sweeps and ejections and the preferential locations where particles penetrate and exit the wall layer. In Figure 14(a), we show the probability density functions of $u'w'$ events plotted as a function of the local wall shear stress, which is normalized to its average value. As can be observed from Figure 14(a), sweeps and ejections are separated by a crossover level of the wall shear stress, the sweeps corresponding to high shear stress regions and the ejections corresponding to low shear stress regions. A slight overlapping between the two distributions exists and the value 1.0 of the normalized shear stress separates high and low shear stress regions. A similar trend is obtained for the distributions of first and third quadrant events, which, however, are much smaller in area. In Figures 14(b-d), we

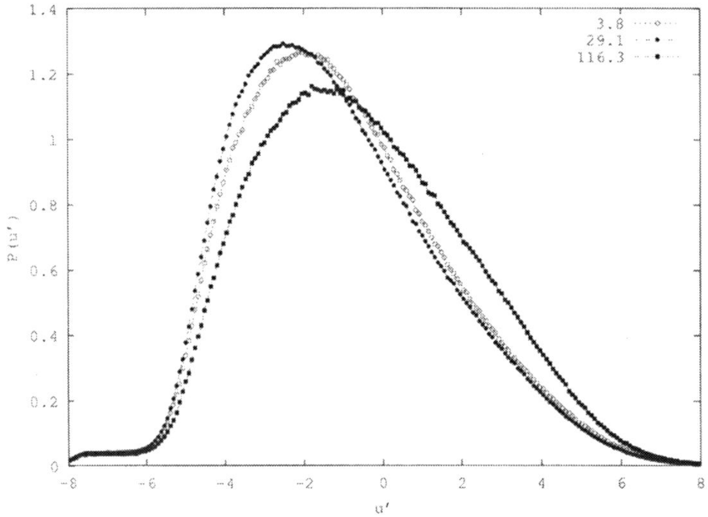

Figure 13. Correlation between particle number density in the wall region with the streamwise fluctuating velocity u' for $\tau_P^+ = 3.8$, $\tau_P^+ = 29.1$ and $\tau_P^+ = 116.3$ particles.

show the probability density function of particles having positive wall-normal velocity – toward the wall – and negative wall-normal velocity – toward the outer flow – plotted as a function of the local wall shear stress.

Particle distribution along the wall-normal coordinate changes with time, and particle fluxes toward the wall and away from the wall change therefore with time. We tried to quantify the fluxes toward and away from the wall counting the instantaneous number of particles that cross a specific monitor slab calculated at two different simulation times, $t_1^+ = 1342$ and $t_2^+ = 2048$. We counted the particles having positive or negative wall-normal velocity instantaneously present in the monitor slab of 10 wall units (from $z^+ = 5$ to $z^+ = 15$ from the wall). The normalized profiles reported in Figures 14(b-d) were computed by:

$$N_{in/out} = \frac{\sum_{i=1}^{20} n_i}{TA} \tag{7}$$

where n_i is the number of particles with negative/positive wall-normal velocity w_P counted at the measuring points per unit time, T is the length of the time averaging period and A is the measuring area (Kaftori *et al.* 1995b). To have a larger particle set for calculating particle fluxes, we averaged fluxes over a short time of length $\Delta t^+ = 142$ (20 instantaneous realizations of the flow field), centred around t_1^+ and t_2^+. For sake of brevity, we will refer to the instantaneous number of particles that cross a specific monitor slab as particle flux (Kaftori *et al.* 1995b).

For $\tau_P^+ = 3.8$ and $\tau_P^+ = 29.1$ particles, fluxes to the wall and away from the wall are similar. Particle fluxes toward the wall decrease a little with time for the smaller particles, but increase for the larger particles. Particle flux away from the wall decrease with time for smaller particles and remains roughly constant for the larger particles.

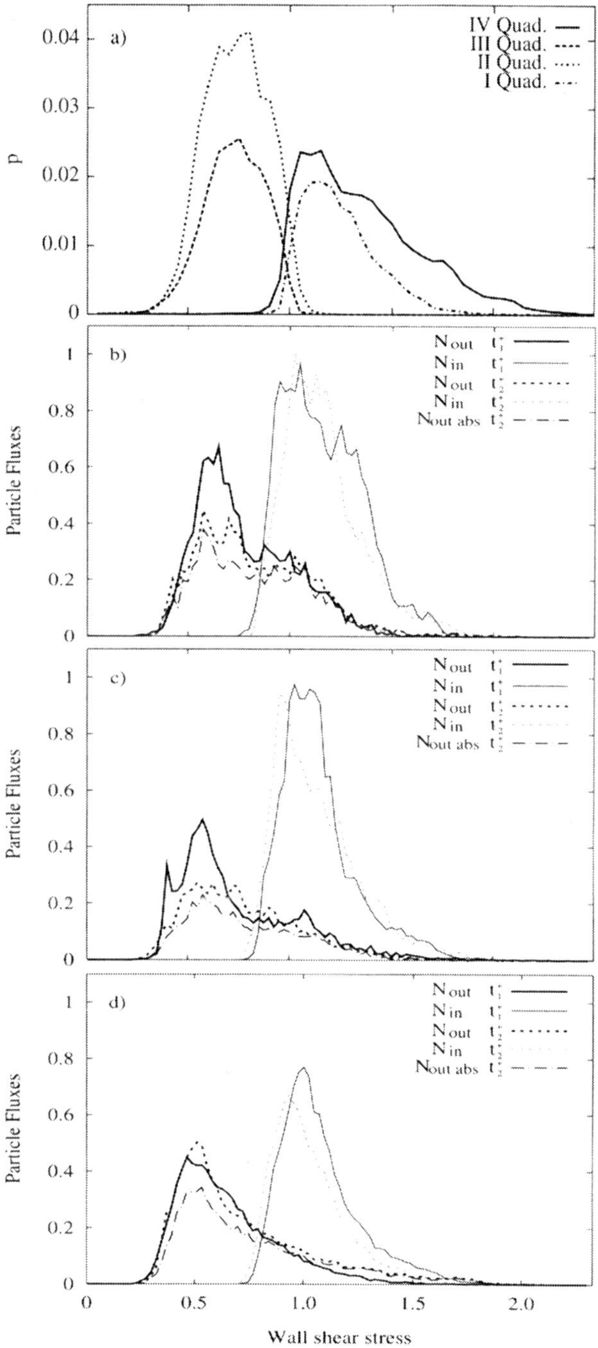

Figure 14. Correlation between particle fluxes in and out the wall layer and wall shear stress distribution. (a) Probability distribution of $u'w'$ events in I, II, III and IV quadrant events versus wall shear stress. (b), (c) and (d) Correlation of the normalized particle fluxes in and out the wall layer with wall shear stress for $\tau_P^+ = 3.8$, $\tau_P^+ = 29.1$ and $\tau_P^+ = 116.3$ particles, respectively.

Two main conclusions can be drawn from Figures 14(b-d). First, it is confirmed that, regardless of particle size, a strong correlation exists between particle fluxes to the wall (N_{in}) and high wall shear stress regions, which correspond to sweep events; low wall shear stress regions correspond to ejection events and are well correlated with off-the-wall particle fluxes (N_{out}). Second, fluxes to the wall always have a greater intensity compared with fluxes toward the outer flow, this trend being enhanced when particle size is smaller. This suggests that, particularly in the case of smaller particles, ejections are somehow unable to lift up all the particles that sweeps drive toward the wall, i.e. particles tend to settle in a sediment layer at the wall, which roughly corresponds to the viscous sublayer. This by no means implies that larger inertia particles are more efficiently transferred far from the wall by ejections. It merely indicates that such particles may find other ways to exit the wall layer. This is apparent on considering a perfectly elastic particle-wall interaction: larger particles gain greater momentum in the sweeps and do not always need organized structures to be driven out from the wall region. They may simply bounce off the wall. To support this observation, we computed the normalized particle flux profiles ($N_{out\ abs}$) in the case of perfectly absorbing wall (i.e. particles do not bounce off the wall) at time t_2^+. The overall intensity of upward particle flux – i.e. the area under the curve – for $\tau_P^+ = 116.3$ particles undergoes a 17.4% decrease compared with the case of perfectly reflecting wall. This decrease is smaller for smaller particles (9.8% for $\tau_P^+ = 29.1$ particles and 7.2% for $\tau_P^+ = 3.8$ particles).

4.4 Instantaneous particle transfer mechanisms

As discussed, particle fluxes are dominated by the same coherent flow structures which control momentum transfer to the wall. The shear stress at the wall is governed by the strongly coherent flow structures generated by the quasi-streamwise vortices, i.e. sweeps and ejections. From the previous results, it appears that particles are driven to the wall by the sweeps and are entrained away from the wall by the ejections. However, since particle distribution in the wall normal direction changes over time, increasing in the wall layer and decreasing in the outer layer, fewer particles will be available to be transferred to the wall, the reverse being true for particle transfer away from the wall. The efficiency of a coherent convective motion is thus conditioned by the presence of particles. Here, we will try to establish the efficiency of particle transfer mechanisms to the wall and away from the wall.

By analogy with momentum transfer (Orlandi & Jimenez 1994), mass transfer (De Angelis *et al.* 1997) and heat transfer (Lu & Hetsroni 1995), we focused on the effects of strongly coherent sweep and ejection events on particle transfer and considered the wall layer up to $z^+ = 12$ (Lombardi, De Angelis & Banerjee 1996). Within this layer, we chose five monitor $x-y$ planes (the heights away from the wall of these planes are $z^+ = 4$, $z^+ = 6$, $z^+ = 8$, $z^+ = 10$, $z^+ = 12$). At each time step considered, we recorded the velocity components on the five monitor planes at each collocation point. To count only those events with substantial spatial coherence, an event is recorded, at some point $(\overline{x}, \overline{y})$, when on at least four of the five monitor planes, $u'w'$ belongs to the same quadrant (Lombardi *et al.* 1996). Considering only sweeps and ejections, which contribute to the negative turbulence-producing part of the Reynolds stress, sweeps contribute to a larger fraction of the negative Reynolds stress for $z^+ < 11$ whereas ejections contribute to a larger fraction of the negative Reynolds stress for $z^+ > 11$ (Willmarth & Lu 1972; Wallace *et al.* 1972; Kim *et al.* 1987).

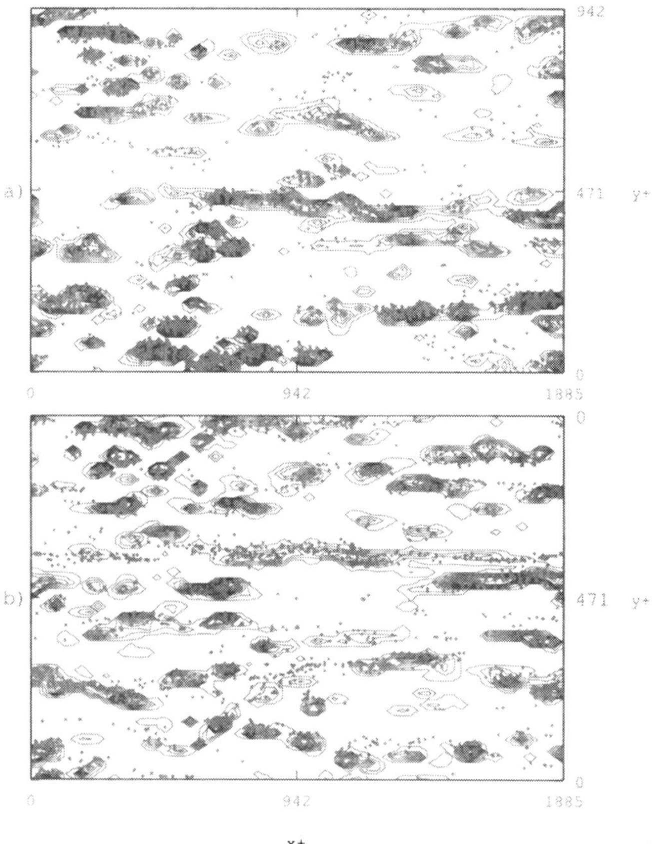

Figure 15. Instantaneous correlation between strongly coherent sweeps and ejections and $\tau_P^+ = 3.8$ particle fluxes toward the wall and away from the wall: (a) instantaneous correlation between strongly coherent sweep events and particle fluxes toward the wall at $z^+ = 11$, (b) instantaneous correlation between strongly coherent ejection events and particle fluxes away from the wall at $z^+ = 11$.

We therefore focused on the crossover plane between sweep/ejection dominance, at $z^+ = 11$, and we characterized all particles in the region between 10 and 12 z^+ from the wall with their velocity w_P, which is negative toward the wall and positive toward the outer flow. In Figure 15(a), we show the instantaneous position of strongly coherent sweeps, characterized by the isocontours corresponding to $u'w' < 0$, and the instantaneous position of all the $\tau_P^+ = 3.8$ particles directed toward the wall. There is evidence of a strong correlation between particles with negative w_P and sweeps, since only a small fraction of the particles falls out of sweep regions.

In Figure 15(b), the position of the particles with positive w_P is shown together with the instantaneous position of strongly coherent ejections, characterized again by the isocontours of negative $u'w'$. There is evidence of a strong correlation between the position of an ejection and the location from which particles exit the wall layer. The same type of visualizations for the larger particle sets (not shown here) confirm the qualitative results of Figure 15.

Our aim now is to quantify the visual observations of Figure 15 and, in particular, to determine the fraction of particle fluxes toward the wall and away from the wall actually driven by the sweeps and the ejections, respectively. Thus, we computed, at fixed time intervals, the correlation between the wall-normal velocity of particles located in the wall layer up to $z^+ = 12$ and the quadrant events occurring at each particle position. We evaluated the average area on the monitor plane and, sampling 50 instantaneous realizations, we found that about 60% of the monitor plane is occupied by the strongly coherent sweeps and ejections.

However, if we count the number of particles that cross the monitor plane and observe where they cross it, we find that, for $\tau_P^+ = 3.8$ particles, on the overall number of 42280 particles crossing the plane in both directions, 98.05% of the particles are located in regions where we find either a strongly coherent sweep or a strongly coherent ejection. For the $\tau_P^+ = 29.1$ particles, this percentage is 98.01% on the overall number of 41090 particles transferred, whereas for the $\tau_P^+ = 116.3$ particles, the percentage is 91.51% on the overall number of 28961 particles transferred. Even for the largest particles, it is apparent that the strongly coherent events control particle fluxes to the wall and away from the wall, and represent almost exclusively inlet and outlet channels to the wall region and away from the wall region.

We quantified the efficiency of the transfer mechanisms as shown in Figure 16, where negative values of $|u'|w'$ are associated with strongly coherent sweeps and positive values of $|u'|w'$ are associated with strongly coherent ejections. To include only events with statistically significant occurrence, calculations were made over 250 instantaneous flow realizations (covering an overall time from $t^+ = 353$ to $t^+ = 2118$). For clarity of presentation, we chose to show only 1 over 5 points to represent the quadrant point distribution in Figure 16.

For $\tau_P^+ = 3.8$ particles (Figure 16a), we find that 30456 out of 31482 particles with negative wall-normal velocity w_P are located within $|u'|w' < 0$ regions whereas only 1026 are located in $|u'|w' > 0$ regions. This indicates that almost all the particles with a downward trajectory are engulfed in a sweep (the probability for a particle to be entrained in a sweep conditioned by having wall-directed velocity is $P_S^d = P(|u'|w' < 0 \mid w_P < 0) = 96.74\%$) and only a small proportion is involved in an ejection ($P_E^d = P(|u'|w' > 0 \mid w_P < 0) = 3.26\%$). Considering the semi-plane $w_P > 0$, 8820 out of 9974 particles are located in $|u'|w' > 0$ regions whereas 954 are located in $|u'|w' < 0$ regions. This indicates that the percentage of ejection-entrained particles with a trajectory away from the wall is equal to $P_E^u = P(|u'|w' > 0 \mid w_P > 0) = 90.24\%$ whereas the percentage of such particles which are entrained in a sweep is equal to $P_4 = P(|u'|w' < 0 \mid w_P > 0) = 9.76\%$.

For the $\tau_P^+ = 29.1$ particle case (see Figure 16b), the plotted points are subdivided as follows: 30769 particles characterized by $w_P < 0$ (28388 located in $|u'|w' < 0$ regions, corresponding to $P_S^d = 92.26\%$, and 2381 located in $|u'|w' > 0$ regions, corresponding to $P_E^d = 7.74\%$) and 9503 particles characterized by $w_P > 0$ (7996 located in $|u'|w' > 0$ regions, corresponding to $P_E^u = 84.14\%$ and 1507 with $w_P < 0$, corresponding to $P_4 = 15.86\%$).

In Figure 16(c), the correlation between particle wall-normal velocity and sweep/ejection events is shown for $\tau_P^+ = 116.3$ particles: 15966 particles have $w_P < 0$ (12972 located in $|u'|w' < 0$ regions, corresponding to $P_S^d = 81.22\%$, and 2994 located in $|u'|w' > 0$ regions, corresponding to $P_E^d = 18.78\%$) and 10536 have $w_P > 0$ (6936 located in $|u'|w' > 0$ regions, corresponding to $P_E^u = 65.82\%$ and 3600 located in $|u'|w' < 0$ regions, corresponding to $P_4 = 34.18\%$).

These results are summarized in Table 2. It is interesting to observe that the overall particle number driven toward the wall is similar for the smaller particle size (31482 and 30769 for the $\tau_P^+ = 3.8$ and $\tau_P^+ = 29.1$ particles respectively), whereas it decreases abruptly for the largest particles (15966 for the $\tau_P^+ = 116.3$ particles). This may be due to the influence of the stronger

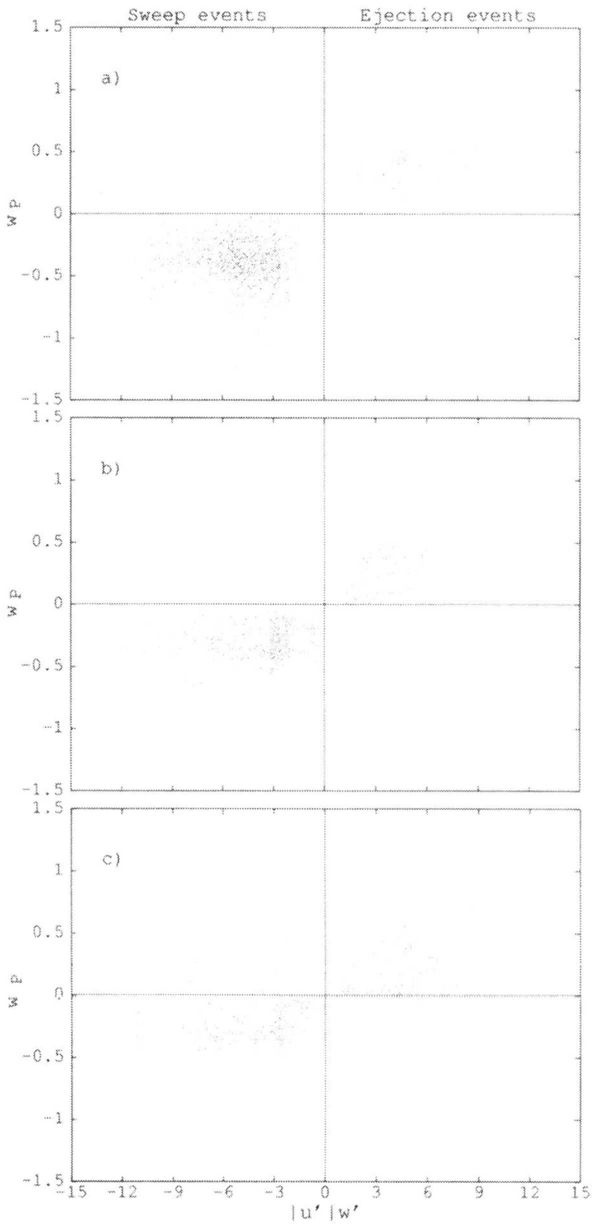

Figure 16. Time-averaged correlation between strongly coherent events and particle fluxes toward the wall and away from the wall for (a) $\tau_P^+ = 3.8$ particles, (b) $\tau_P^+ = 29.1$ particles, (c) $\tau_P^+ = 116.3$ particles.

	w_P	$\lvert u' \rvert w'$	Event Type	$\tau_P^+ = 3.8$	$\tau_P^+ = 29.1$	$\tau_P^+ = 116.3$
P_S^d	< 0	< 0	Sweep	96.74 %	92.26 %	81.22 %
P_E^d	< 0	> 0	Ejection	3.26 %	7.74 %	18.78 %
P_E^u	> 0	> 0	Ejection	90.24 %	84.14 %	65.82 %
P_S^u	> 0	< 0	Sweep	9.76 %	15.86 %	34.18 %
P_N^d	< 0	-	None	1.54 %	1.56 %	7.8 %
P_N^u	> 0	-	None	3.23 %	3.36 %	9.52 %

Table 2. Probabilities representing the correlation between particle wall-normal velocity w_P and sweep events ($\lvert u' \rvert w' < 0$) or ejection events ($\lvert u' \rvert w' > 0$) at plane $z^+ = 11$. Probabilities are defined as follows: $P_S^d = P(\lvert u' \rvert w' < 0 \mid w_P < 0)$, $P_E^d = P(\lvert u' \rvert w' > 0 \mid w_P < 0)$, $P_E^u = P(\lvert u' \rvert w' > 0 \mid w_P > 0)$, $P_E^u = P(\lvert u' \rvert w' < 0 \mid w_P > 0)$. The fraction of particles with negative or positive wall-normal velocity which are located in non-sweep/non-ejection environments is represented by P_N^d and P_N^u respectively.

structures in the region beyond $z^+ = 11$. These structures impart considerable momentum to particles and may contribute to decorrelate larger particles from the sweeps. Particles behave as a low-pass filter and respond to perturbations of the appropriate time-scale. The larger the particle, the larger the structure has to be to modify particle trajectory. Rather surprising, we observe that he overall particle number driven away from the wall layer is almost independent of particle size (9974, 9503 and 10536 for the $\tau_P^+ = 3.8$, $\tau_P^+ = 29.1$ and $\tau_P^+ = 116.3$ particles respectively). From the data in Table 2, we also notice that smaller particles follow closely sweeps and ejections and their transfer rates are well correlated with the instantaneous realizations of turbulent Reynolds stresses. The larger the particle size, the weaker the correlation and a larger fraction of particles is found to travel to the wall in an ejection or to travel away from the wall in a sweep. It is particularly relevant to observe that a large fraction – 34.18% – of the larger particle set travels toward the outer flow in sweep environments. As observed before, larger particles may return toward the outer flow simply bouncing elastically on the wall by exploiting the large momentum gained.

Results similar to those presented here on particle fluxes were previously obtained by Kaftori *et al.* (1995b), who ascribed particle behaviour to the gravitational pull experienced by the particles in a horizontal turbulent boundary layer. In our simulations, the force due to gravity can not directly cause particle deposition at the wall and the buildup of particles in a sediment layer must be explained, from a physical viewpoint, by different mechanisms, in which near-wall turbulent coherent structures play a crucial role.

4.5 Turbulent structures and particle trapping mechanism

In the previous sections, we presented both qualitative and quantitative results to put in evidence the mechanisms by which particles are transferred to the wall by the sweeps and are eventually re-entrained into the outer flow by the ejections. We also determined how strongly coherent $u'w'$ events influence particle fluxes. The picture is that particles enter the wall layer advected by the strongly coherent sweeps and exit the wall layer advected by the strongly coherent ejections. However, exit fluxes are much weaker than inlet fluxes. We will try to address that issue in this section. Specifically, our aim is to explain, from a physical viewpoint, the mechanisms for

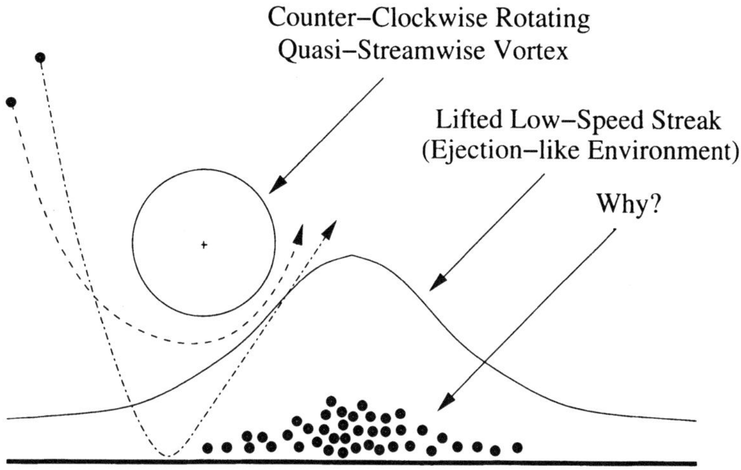

Figure 17. Schematic mechanisms for particle transfer to the wall and away from the wall.

particle segregation within the boundary layer. For this aim, we have to link all the phenomena observed to the dynamics of the near wall turbulent vortical structures.

The vortical structures which dominate the wall layer dynamics are the aforementioned quasi-streamwise vortices, which generate sweeps on the downwash side, and ejections on the upwash side. In turn, ejections contribute to the maintenance of the lifted low-speed streaks on the upwash side of the quasi-streamwise vortices. Recent results (Schoppa & Hussain 1996, 1997; Soldati & Marchioli 2001) show that clockwise and counterclockwise, quasi-streamwise vortices appear flanking the low-speed streak as a staggered array in most of the cases. Only rarely do a clockwise and a counterclockwise quasi-streamwise vortices appear together. Considering the particle transfer mechanisms described in the previous sections, we can describe particle transfer as illustrated schematically in Figure 17. We can envision the following cycle for particles initially in the outer flow: if a particle is captured by a sweep, it moves along a curved trajectory around the quasi-streamwise vortex generating the sweep, approaches the wall and moves between the vortex and the wall. During this phase, the particle may touch the wall or not. Then, the particle is on the upwash side of the vortex and is subject to the influence of the ejection.

The next step involves trespassing the lifted low-speed streak and exiting from the wall layer. Considering in particular the conservative conditions of our simulations, with perfectly elastic rebound, particles should migrate toward the surface of the lifted low-speed streak, which is an ejection-like environment, and find an ejection strong enough to drive them into the outer flow. Yet most of the particles remain trapped under the lifted low-speed streak.

To investigate on the mechanisms leading to particle accumulation under the lifted low-speed streaks, we examined a large number of snapshots showing the action of quasi-streamwise vortices on particle transfer in the wall region. We show one of these in Figure 18(a), which focuses on a $y - z$ window of the computational domain extruded for the length of one streamwise cell (dimensions are $30 \times 58 \times 108$ in dimensionless wall units) at time $t^+ = 1412$. The main character in this picture is the green counterclockwise-rotating quasi-streamwise vortex characterized by positive ω_x vorticity, centred at $z^+ = 36$. We also show, for the $\tau_P^+ = 3.8$ case, the particles with negative w_P – directed to the wall – with black circles, the particles with positive w_P – directed away from the wall – with blue circles and the particles with w_P almost zero – $|w_P| < 10^{-3}$ in wall units – with empty circles. The action of the large vortex in transferring the black particles to the wall and the blue particles away from the wall is apparent. Particles with negligible wall-normal velocity accumulate under the lifted low-speed streak, which we described with a blue isosurface indicating a streamwise velocity value of $0.56\ U_C$ (Kim *et al.* 1987), where $U_C = 16.76$ is the centreline velocity. The low-speed streak appears lifted by the counterclockwise quasi-streamwise vortex.

The counterclockwise quasi-streamwise vortex is visualized by the streamline rotation vector $\mathbf{\Omega}$, previously defined in Section 2.2. We drew an isosurface plot of $\mathbf{\Omega}$ selecting a value equal to the 25% of the instantaneous maximum of $\mathbf{\Omega}$, which is high enough to capture only the strong vortices (Lombardi *et al.* 1996; Soldati & Marchioli 2001).

If we characterize all the coherent structures present in the area with the same $|\mathbf{\Omega}|$ isosurface but with both signs, a secondary, but relevant, character in this picture appears as a red isosurface identifying a smaller counter-rotating quasi-streamwise vortex of negative ω_x vorticity, centred at $z^+ = 9$ and extending well into the viscous wall layer. The presence of such small vortices was also put in evidence by Brooke & Hanratty (1993), who found that each turbulence-producing quasi-streamwise vortex in the viscous wall region is created in time in the downwash of another flow-oriented vortex. Brooke & Hanratty (1993) proposed a turbulence regeneration cycle in which each mature quasi-streamwise vortex, *parent vortex* (the green vortex in Figure 18a), produces a small quasi-streamwise vortex of opposite sign, *offspring vortex*. Following other interpretations, the offspring vortices may be interpreted as the rear, wall-touching end of a counter-rotating quasi-streamwise vortex farther downstream (Schoppa & Hussain 1996, 1997).

It is not our object here to focus on the turbulence regeneration cycle. However, we aim at verifying and at quantifying the action of the offspring vortices in trapping the particles in the wall layer. From Figure 18(a), it is apparent that particles which enter the field of the offspring vortex may not easily escape and go under the influence of the ejection maintained by the mature vortex. We verified if the situation just described has statistical relevance and used a visual criterion to determine whether the structure dynamics shown in Figure 18(a) is statistically more probable than others. We examined 50 flow fields spaced over time in order to have a large set of uncorrelated realizations. We observed an average sample of about 50 vortices taken from the same instantaneous flow field realization and we found single flow-oriented mature vortices coupled with secondary counter-rotating newly-born vortices in more than 70% of the observations. Occasionally, this coupling is not present. We detected pairs of quasi-streamwise vortices with equal strength – i.e. characterizing the legs of the so-called *horseshoe vortices* (Zhou *et al.* 1999; Choi 2001) – in just 25% of the observations. Our observations agree well with previous results (Schoppa & Hussain 1996, 1997), in which it was pointed out that quasi-streamwise vortices preferably line-up flanking the low-speed streaks as a staggered array.

We followed the evolution in time of the structures in Figure 18(a): Figure 18(b) shows the same box of Figure 18(a) approximately 30 wall time units later and 150 wall units downstream. We followed the evolution of the newly-born vortex with a convection velocity equal to $0.5\ U_C$, which corresponds to the $z^+ = 10$ location of the newly-born vortex. The parent vortex previ-

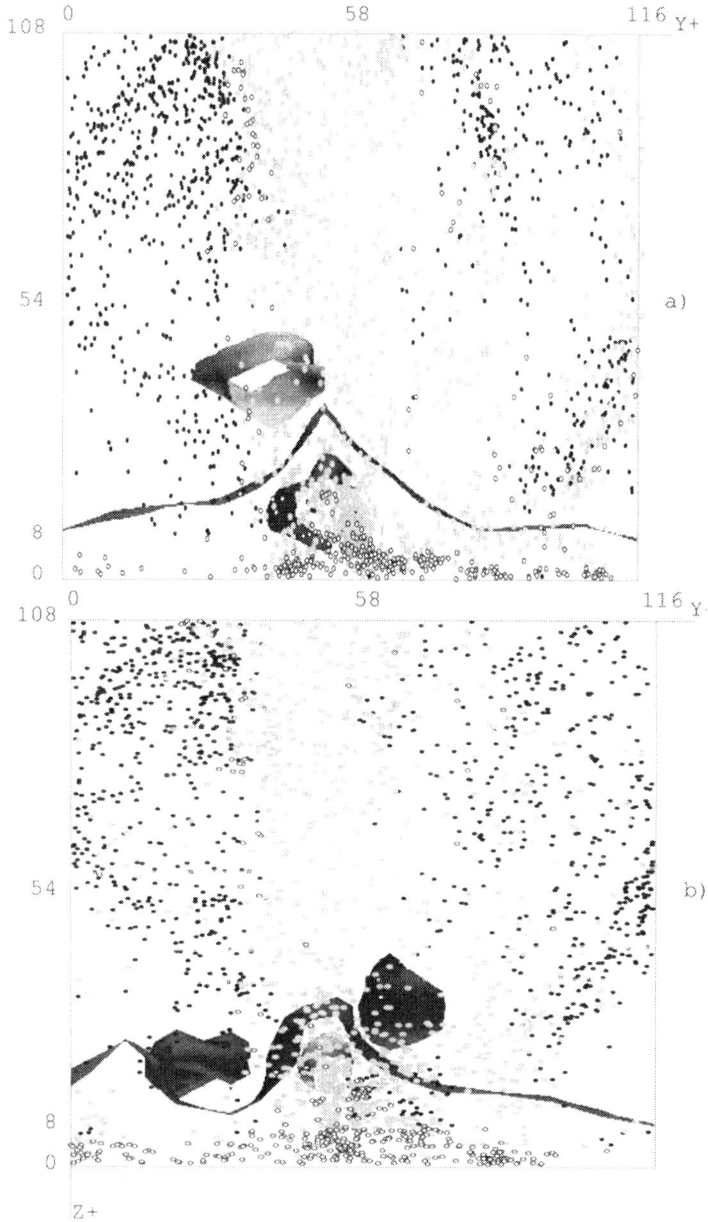

Figure 18. Snapshots of particle distribution and turbulent coherent structures in near-wall region at (a) $t^+ = 1412$ and (b) $t^+ = 1450$ respectively. Green isosurfaces indicate counterclockwise-rotating vortices, red isosurfaces indicate clockwise-rotating vortices, blue isosurfaces indicate low-speed streaks. Most of the ascending particles (black dots) are located in the region between a mature vortex and a newly-born vortex. Most of the descending particles (blue dots) are concentrated on one side of the mature vortex. Particle trapped in the wall layer (particles with $|w_P| < 10^{-3}$) are shown with empty circles.

ously shown is no longer visible whereas the pocket of negative ω_x has grown both in length and size and has lifted from the wall. Under the cusp of the lifted low-speed streak, the green Ω isosurface indicates the presence of a patch of positive ω_x vorticity, that will later become a third generation vortex (Brooke & Hanratty 1993). Colours and symbols in Figure 18(b) have the same meaning as in Figure 18(a). The clockwise-rotating vortex on the left, pertaining to the influence area of the low-speed streak on the left of the figure, is not discussed in this context.

As is apparent from Figure 18, the role of the secondary vortex is crucial in preventing particles from being entrained in the outer flow. As discussed in the previous section, energetic, strongly coherent ejections correlate well with particle fluxes away from the wall. In turn, strongly coherent ejections are generated by strong, mature vortices, which at the same time, are associated with offspring vortices. The combined action of the newly-born vortex and the mature vortex is such as to reduce the width of the 'ejection avenue' through which particles in the wall layer have to pass to reach the outer flow.

Particles with $|w_P| < 10^{-3}$ (empty circles) are mostly settled under the low-speed streak, in a wall layer confined between the offspring vortex and the wall. These *trapped* particles were pushed toward the wall by previous downsweeps but no ejection has yet occurred sufficiently energetic to re-entrain them. This behaviour is probably due to the above mechanism for near-wall vortex regeneration. The birth of new vortices is associated with strong spanwise motions which counteract the wall-normal pulls due to local turbulence gradients or fluid ejections and prevent some particles from being re-entrained. The overall effect is to concentrate and keep particles within the sediment layer in elongated streaks that may be viewed as low-stress regions associated with a stagnation flow.

It is important now to underline the timing at which the different events contributing to bring and segregate particles in the wall layer occur. The low-speed streaks, under which particles are accumulated, are long-lived wall structures, The time duration of quasi-streamwise vortices is much shorter. Strongly coherent sweeps, capable of driving particles toward the wall, and ejections strong enough to drive particles away from the wall are generated simultaneously by the forward end of the quasi-streamwise vortex. As observed in previous works (De Angelis *et al*. 1997), the frequency of ejections is about 80 t^+ and it is the only way for a particle to be driven away from the wall layer (see previous section). Thus, when a mature quasi-streamwise vortex generates a strongly coherent sweep, it simultaneously generates a strongly coherent ejection. However, the contemporary presence of the newly-born vortex – or rear end of a counter-rotating streamwise vortex farther downstream – prevents a large fraction of the particles from accessing to the ejection area. Thus, the newly-born vortex acts to enhance the energy level required to carry particles in the outer flow region and plays a primary role in reducing overall particle mixing.

A Lagrangian description of the local trajectory of particles when under the influence of the structures just described may help to elucidate further the particle trapping mechanisms. In Figure 19(a), we show the mature vortex and the offspring vortex at the same time-step of Figure 18(a), together with the trajectory of a number of particles.

We chose several particles in interesting positions, i.e. trapped or ejected away from the wall layer, and we tracked their trajectory backward and forward. The dot-to-dot distance on particle trajectory is $\Delta t^+ = 0.7065$ in wall time units. During this time, we can hypothesize that turbulence structures change slightly, their average life being more than 100 wall time units (Brooke & Hanratty 1993; Schoppa & Hussain 1997, 2000). In this figure, we also show the instantaneous flow streamlines calculated at the same time at which we visualized the quasi-

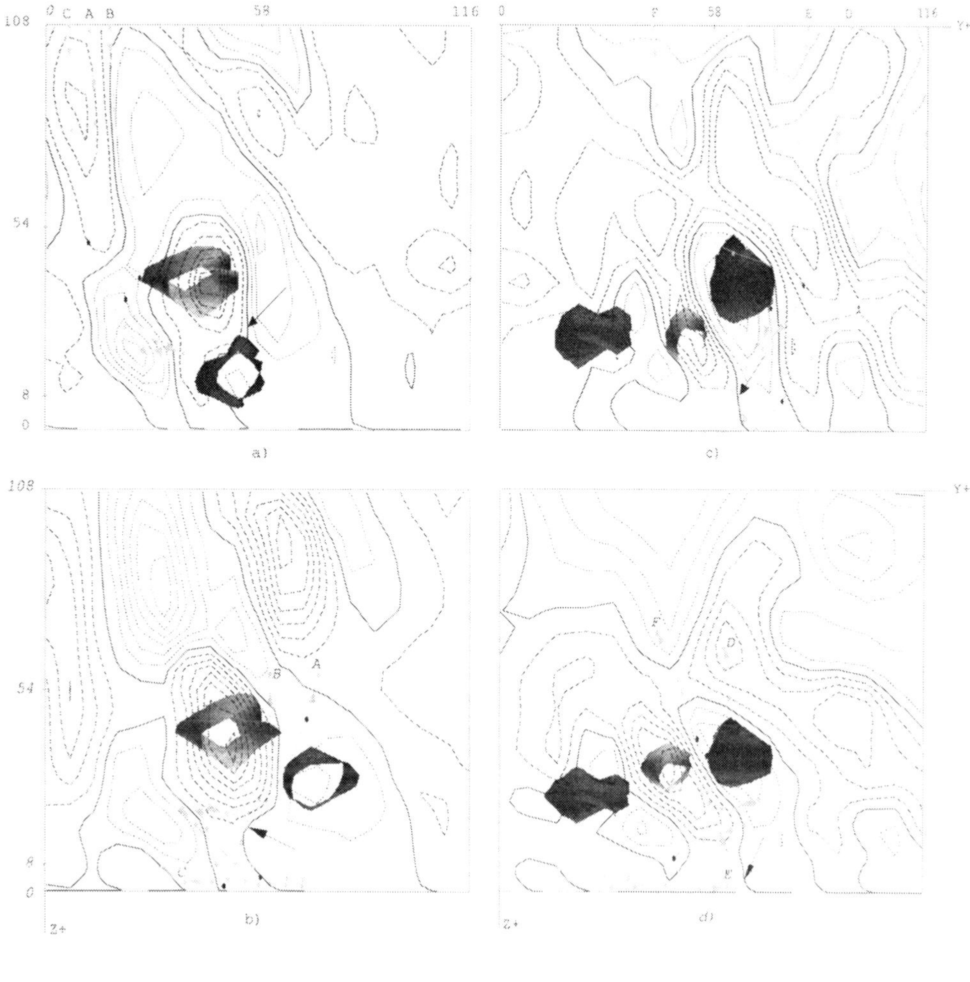

Figure 19. Near-wall turbulent structures with superimposed streamlines of channel flow at different times of the simulation: (a) $t^+ = 1412$,(b) $t^+ = 1414.8$, (c) $t^+ = 1450$, (d) $t^+ = 1452.8$. Dashed lines represent streamlines with positive values of the stream function ($\Psi > 0$), dotted lines represent streamlines with negative values of Ψ. Streamlines with $\Psi = 0$ are plotted as solid lines. Sample trajectories of few particles approaching the wall (labelled with letters A to F) are also plotted to evidence the influence of counter-rotating vortices on particle dynamics at the wall. Black dots on sample trajectories indicate the particle position at the same time step at which the flow structures are visualized. Particle positions are tracked backward and forward around this time step – dot-to-dot time is 0.7065.

streamwise vortices. Positive values for the stream function Ψ (dashed lines) are associated with counterclockwise-rotating vortices whereas negative values of Ψ (dotted lines) are associated with clockwise-rotating vortices. Streamlines with $\Psi = 0$ are plotted as solid lines. Consider the three particles labelled A, B and C in Figure 19(a): their position at the time of the figure is identified by the black dot. Tracking their trajectory backward, we observe that the three particles left from the same fluid environment and, tracking their trajectory forward, we see that they end up in the same neighbourhood. However, and this is important for their future destiny, they have different trajectory curvatures.

Going now to Figure 19(b), we can see the wall structure 2.8 wall time units later, approximately 11 wall units downstream. After this short time, the large vortex changed slightly, and the smaller vortex moved farther from the wall. The position of the three particles is again identified by the black dot. We shall now consider the overall trajectory of the three particles: the particle labelled A follows a neat path around the mature quasi-streamwise vortex in green, and after being swept toward the wall, enters the outer flow driven by the ejection. The particle labelled B follows a path similar to that of particle A but, before being entrained by the ejection, it bounces elastically on the wall. Particle C goes under the offspring vortex, very close to the wall, where it finds an adverse flow which pushes it backward parallel to the wall. This particle will be confined longer in the viscous wall layer.

In Figure 19(c), we examined a similar situation generated by the clockwise-rotating quasi-streamwise vortex in red, with all symbols and positioning of letters maintaining the same meaning. Again, we chose three particles with different destiny. Particle D, after being entrained by the sweep, is able to pass between the mature vortex and the offspring vortex, along the black streamline – $\Psi = 0$ – which identifies the only escape way from the wall region. Particle E is driven too far under the offspring vortex and is not able to escape from the wall region. Particle F bounces off the wall and is able to follow the ejection to the outer flow. If we now consider the streamlines patterns in Figure 19, the action of the offspring vortex in trapping the particles in the wall layer is evident. The flow regions bordering the $\Psi = 0$ streamlines, indicated with the black arrows in Figure 19, are source flows from the wall region. The presence of the offspring vortex associated with the effects of the mature vortex contributes to squeeze these regions (increasing contour density) thus reducing the probability for a particle to be entrained to the outer flow.

The aim of this Section was to elucidate the mechanisms by which inertial particles are trapped in the near wall region by the sincronicity among the turbulent transfer mechanisms, namely strongly coherent sweeps and ejections, and the regeneration cycle of the quasi-streamwise vortices. To this purpose, we focused on the smaller particles in our set ($\tau_P^+ = 3.8$). A sample analysis conducted for the larger particle cases gave similar results from a qualitative viewpoint, indicating that, for the particle time-scales investigated interacting with the wall structures characteristic of a channel flow at $Re_\tau = 150$, particles segregation mechanisms are similar. Quantitative figures relative to the different particles are different due to the different particle inertia, as previously pointed out in Sections 3.2, 3.3 and 3.4.

4.6 Turbulent structures and particle distribution

In previous sections, we have put in evidence the non-random fashion of particle distribution in the boundary layer, due to the interactions between dispersed phase and near-wall coherent flow structures. In Figure 13, we gave a quantitative measure of particle preferential distribution in the low-speed streaks. Particles correlate well with regions of lower-than-mean fluid velocity. Since low-speed streaks are ejection-like environments, particles should leave soon the wall region driven by the flow but this does not seem the most probable case. Thus, we want to explore deeper this correlation.

To this aim we relate particle distribution to coherent flow structures, according to the general classification scheme proposed by Chong *et al.* (1990) and Blackburn *et al.* (1996). This classification scheme groups all elementary three-dimensional flow patterns and is based on the rate-of-deformation invariants (P, Q and R). Details on the classification scheme and on P, Q and R are provided in Appendix A. For incompressible flow (as in the present case), the first invariant P vanishses so that all possible topologies can be represented using the (Q, R)-plane (Blackburn *et al.* 1996). Four regions can be identified: two vortical flow regions (the so-called stable focus/stretching and unstable focus/compressing) and two convergence regions (the so-called stable node/saddle/saddle and unstable node/saddle/saddle critical nodes).

Following Rouson & Eaton (2001), we applied this classification system to 400 instantaneous realizations of the flow field. Thus, we computed Q and R at each fluid grid point in the viscous sublayer ($z^+ < 5$). We also conditionally sampled Q and R at the positions of solid particles in order to determine if the particles show any preference for or against any of the aforementioned four topologies. Figure 20 shows the resulting joint probability density function (PDF) of Q and R. Note that Q and R are both zero for the mean flow and that zero is the most probable value for all instantaneous realizations considered.

The PDF sampled at the fluid grid points (Figure 20a) shows that the preferred quadrants correspond to the stable focus/stretching and the unstable node/saddle/saddle topologies. The points associated with the latter topology are found along the positive-R side of the $D = 0$ line: this indicates the predominance of vortex stretching in turbulence.

The PDF sampled at $\tau_p^+ = 116.3$ particle positions (Figure 20b) occupies a relatively small area, centered around the origin of the (Q,R)-plane, indicating that larger particles tend to avoid the strongest vortical regions (I and II quadrants in Figure 20b). The particles also avoid the strongest vortex-stretching regions along the positive-R, zero-D line (IV quadrant of Figure 20b). Convergence regions (III quadrant in Figure 20b) seem to be little affected by the preferential sampling of these particles. Topologies in this region have two axis along which the flow converges toward a sink point, providing a mechanism for particle accumulation. The efficiency of such mechanism is limited, however, by the incompressibility constraint, which requires outflow along the third axis. The unstable node/saddle/saddle region in the fourth quadrant of the (Q, R)-plane has no such limiting factor. Topologies in this region have two outflow directions (source point) and thus can be expected to induce low particle number densities. In the following, we will show that this is not true.

The PDF sampled at $\tau_p^+ = 29.1$ particle positions (Figure 20c) occupies a broader area, suggesting a behaviour very similar to that of fluid particles in the proximity of the wall. The PDF sampled at $\tau_p^+ = 3.8$ particle positions (Figure 20d) indicates an intermediate behaviour.

Results obtained by Rouson & Eaton (2001) are qualitatively similar but there are significant differencies from a quantitative viewpoint. Discrepancies may be due to several causes: Rouson & Eaton (2001) simulated a vertical turbulent downflow (not upflow, as in the present case) channel with $Re_\tau = 180$; also, the authors did not include the effect of Saffman lift force on particle tracking and considered particles with relaxation times different from those investigated in the present work.

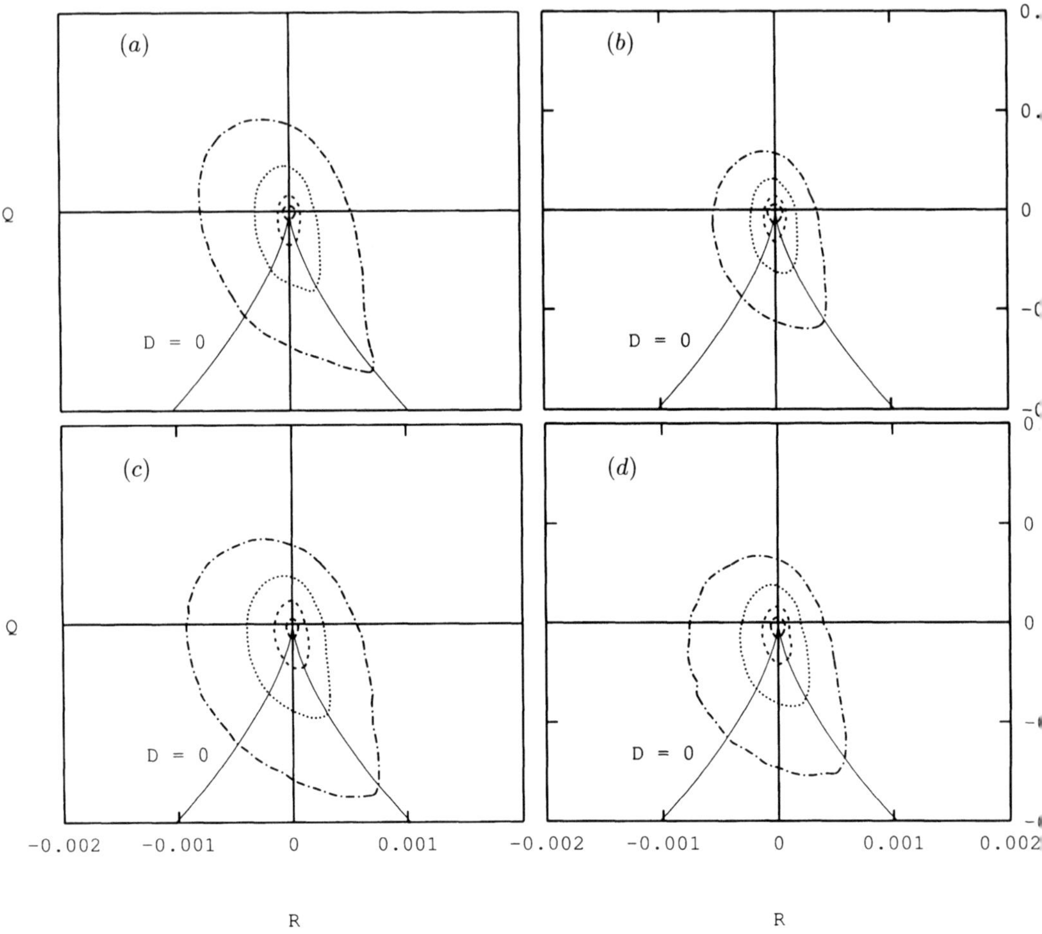

Figure 20. Viscous sublayer ($z^+ < 5$) joint PDF of Q, R conditionally sampled at fluid grid points (a), $\tau_p^+ = 116.3$ particle positions (b), $\tau_p^+ = 29.1$ particle positions (c), and $\tau_p^+ = 3.8$ particle positions (d). Plotted values are: — — — PDF = 10, - - - PDF = 1, · · · PDF = 0.1, — · — PDF = 0.01.

The topological analysis seems to suggest that, very near the wall, "strongly coherent vortical structures are depleted of particles as would be if the local flow is producing particle nonhomogeneous concentration" (Rouson & Eaton 2001).

This result is confirmed by Figure 21, in which the correlation between particle number density distribution in the viscous sublayer and the magnitude of the streamline rotation vector, Ω (Chong *et al.* 1990), is shown. We represent Ω in the range from 0 to 0.12, which is the threshold value we use to visualize all coherent vortices (Marchioli & Soldati 2002). Regardless of particle size, particle number density profile peaks at $\Omega = 0$, which is the most probable value for all the 400 fields studied (the time-step between two fields is $\Delta t^+ = 1.76625$).

According to the definition of streamline rotation vector (Chong *et al.* 1990), Figure 21 demonstrates that coherent vortical structures tend to avoid the $\Omega = 0$ regions of the viscous

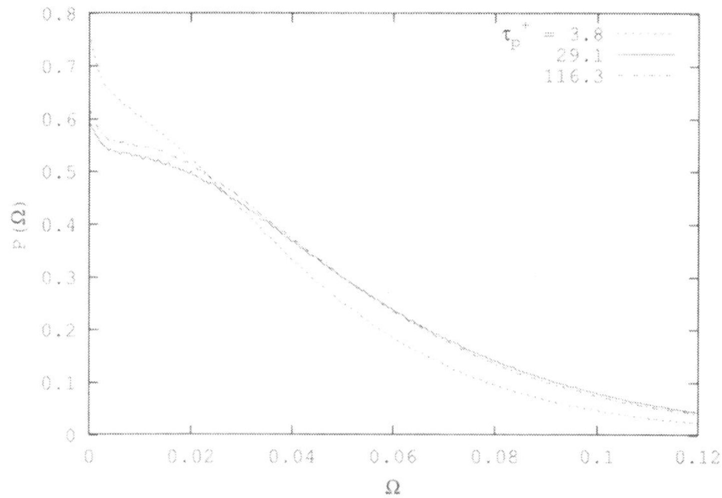

Figure 21. Viscous sublayer ($z^+ < 5$) particle number density distribution as a function of streamline rotation vector magnitude for $\tau_p^+ = 3.8$, $\tau_p^+ = 29.1$ and $\tau_p^+ = 116.3$ particles.

sublayer where particles tend to attain a preferential distribution. In other words, the coherent structures responsible for the non-random particle distribution under the low-speed streaks are not collocated with particle clusters they induce at the wall. This observation is consistent with results shown in Figure 20.

These results are relevant and yet provide no specific information about the wall regions where particle build-up occurs. To identify these regions we need a different identification criterion. To this aim, in Figure 22 we show particle density distribution as a function of the fluctuating components of vorticity. Particle density distribution profiles were computed averaging over 400 instantaneous realizations of the flow field to include only those events with significant statistical occurrence. The distribution profile appears almost perfectly symmetric and centered around zero with respect to ω_x' and ω_z' (see Figure 22a and 22b respectively). As expected, ω_z' is one order of magnitude smaller than the other two components. Symmetry is lost with respect to ω_y' and the distribution profile is centered around $\omega_y' = -0.3$ and $\omega_y' = -0.4$, depending on particle relaxation time. Thus, in the viscous sublayer of a non-homogeneous flow, particles collect in regions with zero vorticity fluctuations along the streamwise and the wall-normal directions and negative vorticity fluctuations along the spanwise direction, differently from homogeneous turbulent flows, where particles are known to collect in regions of low vorticity and high strain (Squires & Eaton 1991). This tendency can be explained considering the biasing effect of the wall shear stress, which is related to the near-wall fluid streaky structure.

For incompressible flow, the fluctuating part, $u_{i,j}'$, of the velocity gradient tensor at the wall has the following form:

$$u_{i,j}'|_w = \left.\begin{pmatrix} 0 & 0 & \partial u'/\partial z \\ 0 & 0 & \partial v'/\partial z \\ 0 & 0 & 0 \end{pmatrix}\right|_w .$$

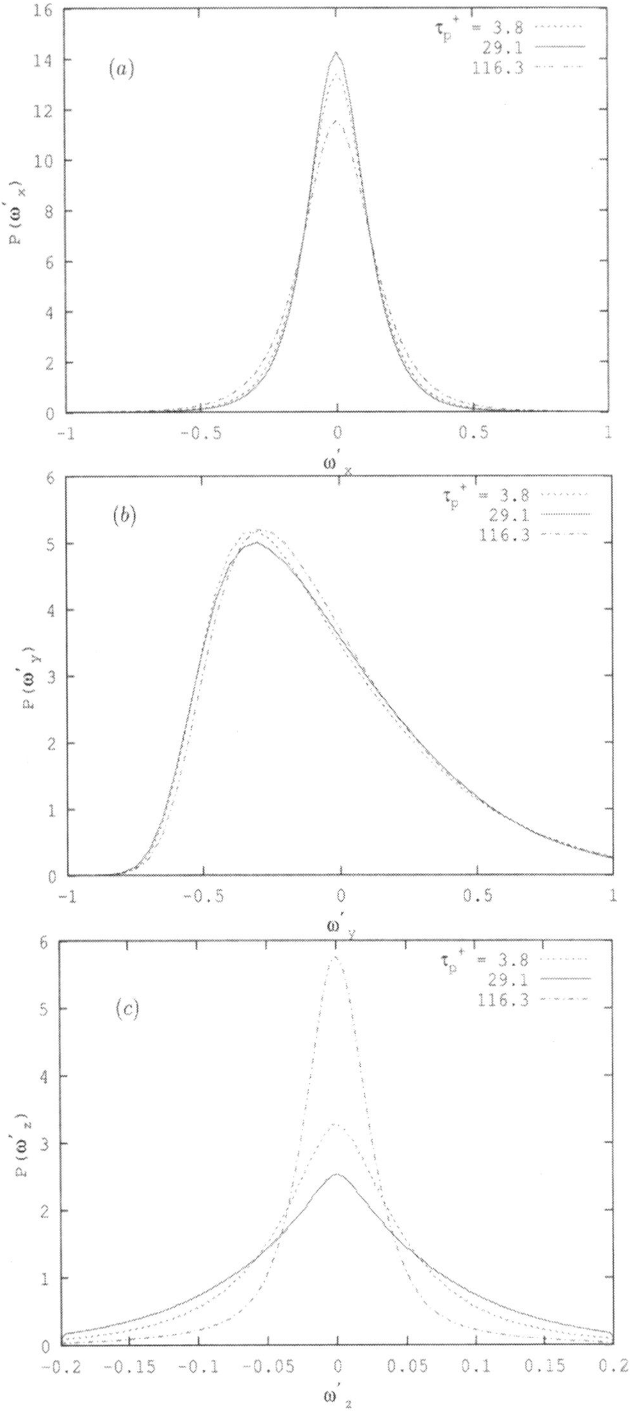

Figure 22. Particle number density distribution as a function of fluctuating vorticity components: (a) $\omega_x^{'}$, (b) $\omega_y^{'}$ and (c) $\omega_z^{'}$.

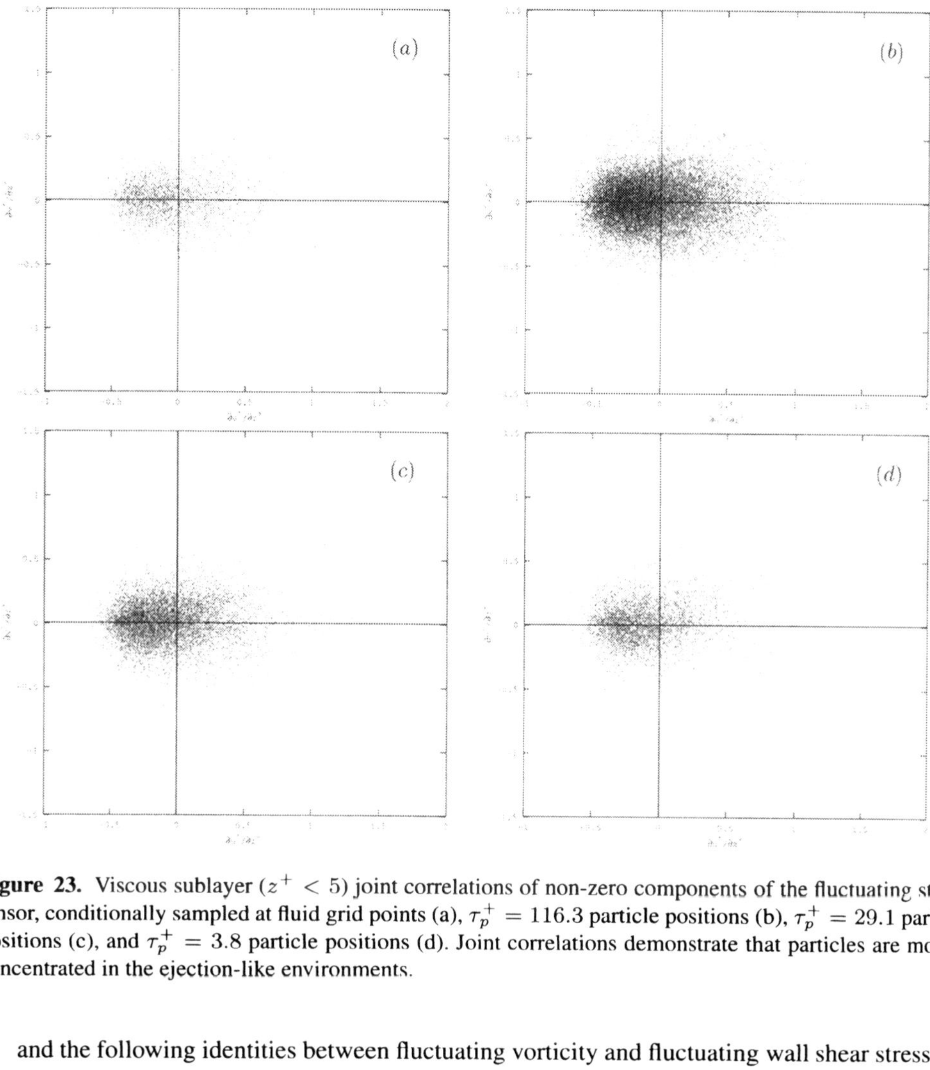

Figure 23. Viscous sublayer ($z^+ < 5$) joint correlations of non-zero components of the fluctuating strain tensor, conditionally sampled at fluid grid points (a), $\tau_p^+ = 116.3$ particle positions (b), $\tau_p^+ = 29.1$ particle positions (c), and $\tau_p^+ = 3.8$ particle positions (d). Joint correlations demonstrate that particles are mostly concentrated in the ejection-like environments.

and the following identities between fluctuating vorticity and fluctuating wall shear stress are true:

$$\mu \cdot \omega_x'|_w = -\mu \cdot \partial v'/\partial z = -\tau_{yz}'|_w \,, \tag{8}$$

$$\mu \cdot \omega_y'|_w = \mu \cdot \partial u'/\partial z = \tau_{xz}'|_w \,, \tag{9}$$

where μ is the (constant) dynamic fluid viscosity.

We already pointed out that the low speed streaks are ejection-like environments that correlate with lower-than-mean wall shear stress regions, where $\tau_{xz}'|_w < 0$ i.e. $\omega_y'|_w < 0$, and appear much wider than high speed streaks, associated with higher-than-mean wall shear stress regions, where $\tau_{xz}'|_w > 0$ i.e. $\omega_y'|_w > 0$. Thus, particle positions necessarily sample $\omega_y'|_w < 0$ regions more often than $\omega_y'|_w > 0$ regions.

We exploited the relationship between the wall shear stress and the elements of the velocity gradient computed at the wall to describe the near-wall flow regions where particle build-up occurs. In Figure 23 we show the instantaneous joint correlations of non-zero components of

	Upflow		*No Gravity*	
τ_p^+	$\partial u'/\partial z\|_w < 0$	$\partial u'/\partial z\|_w > 0$	$\partial u'/\partial z\|_w < 0$	$\partial u'/\partial z\|_w > 0$
3.8	61%	39%	60%	40%
29.1	57%	43%	59%	41%
116.3	59%	41%	59%	41%

Table 3. Probabilities representing the correlations between particle position and $\partial u'/\partial z|_w$. Probabilities are averaged over 400 instantaneous instants of the flow field.

$u'_{i,j}|_w$. We considered the correlations for the fluid (sampled at grid points) and for $\tau_p^+ = 3.8$, 29.1 and 116.3 particles (sampled at particle positions). Correlations were computed as follows: (i) we considered only the position of particles located in the viscous sublayer, (ii) we projected each particle position onto the wall, (iii) we computed the components of $u'_{i,j}$ at the projected wall location. We can observe two distinct near-wall flow regions: with respect to the wall-normal direction, we can identify a sweep-like inflow region, characterized by $\tau'_{yz}|_w = \mu \cdot \partial v'/\partial z|_w = 0$ and by $\tau'_{xz}|_w = \mu \cdot \partial u'/\partial z|_w > 0$, and an ejection-like outflow region, characterized by $\tau'_{yz}|_w = 0$ and by $\tau'_{xz}|_w < 0$.

Correlations shown in Figures 23b, 23c and 23d demonstrate that, regardless of particle size, particles in the viscous sublayer tend to accumulate in proximity of wall regions characterized by negative values of $\partial u'/\partial z|_w$: 61% of the $\tau_p^+ = 3.8$ particles, 57% of the $\tau_p^+ = 29.1$ particles and 59% of the $\tau_p^+ = 116.3$ particles fall in the $\partial u'/\partial z|_w < 0$ quadrants. Even fluid particles (see Figure 23a) show preferential distribution in such quadrants. This behavior can be explained by considering the mechanisms by which the low-speed streaks are generated: a jet of fluid which is directed to the wall generates the sweep and also the high speed region; then the jet of fluid, by continuity, is deflected by the wall and generates the low-speed ejection. Due to the entrainment of surrounding fluid, the sweep is more intense and concentrated and the following ejection spreads over a wider cross-section and has lower momentum. Low-speed low-shear regions, where $\tau'_w < 0$ i.e. $\frac{\partial u'}{\partial z^+} < 0$, appear much wider than high-speed high-shear regions, where $\tau'_w > 0$ i.e. $\frac{\partial u'}{\partial z^+} > 0$. Thus, fluid points necessarily sample $\frac{\partial u'}{\partial z^+} < 0$ regions more often than $\frac{\partial u'}{\partial z^+} > 0$ regions. However, in the latter regions, fluid points distribution appears more scattered covering a wider range of values of $\frac{\partial u'}{\partial z^+}$.

The correlations shown in Figure 23 are typical of what was observed in all the 400 instants of the flow field. This result, together with the fact that the preferred wall regions are also characterized by values of $\partial v'/\partial z|_w$ nearly equal to zero, indicates that particle concentration build-up in the viscous sublayer occurs preferentially in the proximity of a near-wall outflow region. This result is rather surprising since the outflow region would be expected to induce low particle number densities. It can be explained, however, considering the crucial role played by quasi-streamwise vortices in trapping particles very near the wall (Marchioli & Soldati 2002).

Similar conclusions were drawn from further analysis on particle behaviour in absence of gravity. In Table 3, we show the probabilities representing the correlation between particle position and $\partial u'/\partial z|_w$ for two different flow configurations: upflow (gravity directed against the streamwise direction) and channel flow without gravity acting on particles. Again, $\partial u'/\partial z|_w < 0$ represents near-wall ejection-like regions whereas $\partial u'/\partial z|_w > 0$ represents near-wall sweep-like regions. For both flow configurations, particles tend to attain a preferential distribution in the ejection-like outflow regions: this is particularly true for the smaller inertia particles. The comparison of the upflow case data with those for the no-gravity case seems to indicate that preferential accumulation of different size particles in the near-wall region is not much affected by body forces applied on particles, at least for the cases considered.

To provide a unifying pictorial view of the mechanisms discussed in this paper by statistical means, in Figure 24 the instantaneous snapshot of particle distribution and turbulent coherent structures in the near-wall region of the channel is shown, superposed to the 2D footprint of the wall shear-stress. The perspective view and the inset top view focus on a window of the computational domain (dimensions are $180 \times 135 \times 55$ in dimensionless wall units) and give a clear rendering of the physical phenomenon. Using the $\Omega = 0.03$ isosurface with both signs, we have identified two quasi-streamwise vortices: for clarity of presentation, we only show the rear-end of the mature counterclockwise-rotating vortex (blue isosurface) together with the forward-end of the secondary clockwise-rotating vortex (red isosurface). As expected, the two vortices are separated by the $\Psi = 0$ streamsurface (green isosurface), cut at $z^+ = 20$. The environment bordering the streamsurface is an ejection-like outflow region, characterized by low values of the wall shear-stress (in blue). As reported by Marchioli & Soldati (2002), this region is squeezed by the two vortices, which act to reduce the probability for particles to be resuspended into the outer flow. Most of the particles (plotted as light grey spheres) accumulate in the ejection-like region, avoiding the high shear-stress sweep-like environment at the wall (in red). It is also apparent that coherent vortices are not able to entrain particles. Events shown in Figure 24 have been observed in a large number of instants at different locations of the channel.

5 Conclusions

The identification of the mechanisms leading to particle transfer in the wall region and to particle segregation in the viscous sublayer in regions where the streamwise fluid velocity is lower than the mean is of fundamental significance for a number of technological and environmental applications. A sound understanding of these mechanisms requires analysis of the interactions between the coherent structures controlling the turbulent wall transfer and particle dynamics.

In this work, we examined the dynamics of large swarms of heavy particles – flyashes in air – dispersed in a vertical upward channel flow. The particle to fluid density ratio was 769.23 and particle size was $d_P = 40\ \mu m$, $d_P = 110\ \mu m$ and $d_P = 220\ \mu m$ in the three cases investigated. The fluid was driven by a pressure gradient and the shear Reynolds number was $Re_\tau = 150$ which, for a channel 4 cm wide, gave an average velocity of $1.65\ m\ s^{-1}$. We had two objects in this research: the first was to identify and quantify the turbulent convective mechanisms which transfer particles toward the wall and toward the outer flow. The second was to determine why particle transfer toward the wall is more efficient than particle transfer away from the wall.

Since our aim was to examine the influence of turbulence structures on particle behaviour, we considered no feedback of particles on the flow field. As demonstrated in the experiments

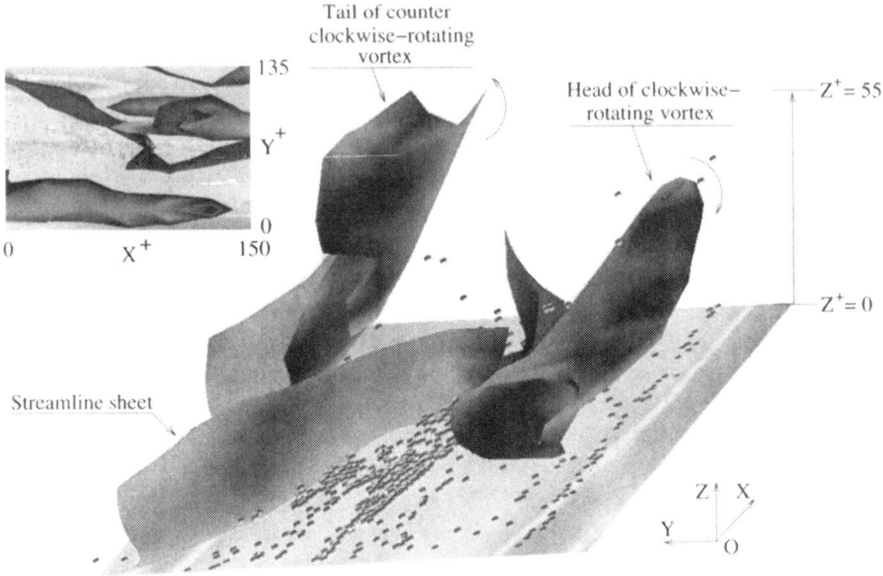

Figure 24. Instantaneous snapshot of particle distribution and turbulent coherent structures in the near-wall region. Blue isosurfaces indicate counterclockwise-rotating vortices, red isosurfaces indicate clockwise-rotating vortices. Vortices are separated by the $\Psi = 0$ streamsurface, in green. Particles, represented as blue spheres, and structures are superposed to the footprint of the wall shear stress: red indicates high shear-stress, blue indicates low shear-stress.

by Kaftori *et al.* (1995a, 1995b), turbulence characteristics change slightly for dilute dispersion, thus permitting us to obtain results with general relevance.

In the first place, we examined the relationship between particle fluxes in and out the wall layer and momentum fluxes at the wall. We found that a strong correlation exists between sweep events – i.e. coherent downwash of outer fluid to the wall – and particle flux toward the wall, and between ejection events – i.e. coherent upwash of wall fluid toward the outer flow – and particle flux toward the outer flow. This correlation is almost perfect for smaller particles and a little weaker for larger particles. We found that particles are transferred almost exclusively by strongly coherent sweeps and ejections.

We also tried to quantify the efficiency of these coherent local convective motions, calculating the probability that a particle will go toward the wall or away from the wall conditioned by the presence of a sweep or an ejection. We found that sweeps and ejections are extremely efficient for transferring small particles of $\tau_P^+ = 3.8$. In other words, if a small particle travels toward the wall, it is engulfed in a sweep whereas if the particle travels away from the wall, it is driven by an ejection. For larger particles, $\tau_P^+ = 116.3$, we found that most of the particles are still transferred by sweeps and ejections but a higher proportion of particles with positive wall normal velocity appears in fluid environments characterized by negative wall-normal velocity, and viceversa. In particular, for the larger set of particles, the fraction of particles travelling toward the wall in a non-sweep environment is smaller than the fraction of particles travelling away from the wall in a non-ejection environment.

A possible explanation is to be found in the ratio of particle time-scale to fluid structure time-scale. The characteristic time-scale of turbulent structures decreases progressively as the structures lie closer to the wall. Larger particles have a larger time-scale and filter out the effects of the smaller fluid scales. Thus, the larger momentum gained by the large particles through the interactions with the large scales in the buffer layer which are able to drive particles to the wall may be sufficient to let the particle bounce elastically off the wall toward the outer flow, crossing the smaller scale structures in the vicinity of the wall, unable to further modify particle trajectory.

Having established quantitatively the action of the instantaneous realizations of the Reynolds stress on particle transfer, our object was then to determine why a large fraction of the particles entering the wall region are unable to escape and remain trapped in the low streamwise velocity environment. From previous works (Schoppa & Hussain 1996, 1997, 2000; Jeong *et al.* 1997; Soldati & Marchioli 2001), it was possible to establish that: i) low-speed streaks are long-lived structures, ii) low-speed streaks are flanked by clockwise and counterclockwise-rotating quasi-streamwise vortices distributed mostly (in about 75% of the cases) as a staggered array, iii) quasi-streamwise vortices generate strongly coherent sweeps and ejections. In this work, we were able to verify that iv) particles are driven toward the wall and toward the outer flow only by the strongly coherent sweeps and ejections. Thus, examining in detail the dynamics of the wall structures in connection with the dynamics of the particles entering and exiting the wall layer, we were able to appreciate fully the relevance of a secondary wall structure which was described by Brooke & Hanratty (1993) and by Bernard *et al.* (1993). In particular, the downstream part of the quasi-streamwise vortex, usually centred in the buffer layer, is associated to a smaller streamwise oriented vortex, centred much closer to the wall. The local flow structure produced by this couple prevents a number of the particles entered in the wall layer from being entrained toward the outer flow. In particular, even though the strongly coherent sweep events required to drive particles to the wall are associated with strongly coherent ejections capable of driving the particles toward the outer flow, the simultaneous presence of the offspring vortex acts as to reduce the width of the 'ejection channel'. In practice, only particles which enter the wall-layer with specific trajectory curvature may be able to be entrained into the outer flow.

Further statistical analysis were performed to clarify the link between the non-random particle distribution in the viscous sublayer and the near -wall coherent flow structures. Results confirmed that particles build-up under the low-speed streaks is due to the trapping action of the near-wall coherent structures. It is apparent that particle are not entrained in the coherent structures but rather accumulate in the proximity of a source point at the wall, located well below the low-speed streak.

The identification of these trapping mechanisms is relevant for the future direction of mathematical modelling of wall-bounded particle-laden flows, including advection-diffusion type field approaches (Slater & Young 2001; Young & Leeming 1997; Cerbelli *et al.* 2001) and Lagrangian approaches based on Large-Eddy Simulations (Armenio, Piomelli & Fiorotto 1999).

Acknowledgements

The authors would like to thank Giorgio Michelazzo for performing some calculations and visualizations. Financial support from the Regional Authority of Friuli Venezia Giulia under the

Fluidodinamica e Analisi delle Dispersioni nella Bassissima Atmosfera del Friuli Venezia Giulia Grant are gratefully acknowledged.

References

Armenio, V. & Fiorotto, V. (2001). The importance of the forces acting on particles in turbulent flows. *Phys. Fluids*, **13**, 2437-2440.

Armenio, V., Piomelli, U. & Fiorotto, V. (1999). Effect of the subgrid scales on particle motion. *Phys. Fluids*, **11**, 3030-3042.

Balachandar, S. & Maxey, M. R. (1989). Methods for evaluating fluid velocities in spectral simulations of turbulence. *J. Comput. Phys.*, **83**, 96-125.

Bernard, P. S., Thomas, J. M. & Handler, R. A. (1993). Vortex dynamics and the production of Reynolds stress. *J. Fluid Mech.*, **253**, 385-419.

Blackburn, H. M., Mansour, N.N. & Cantwell, B. J. (1996). Topology of fine-scale motions in turbulent channel flow. *J. Fluid Mech.*, **310**, 269-292.

Bonnet, J. P. & Delville, J. (1996). General concepts on structure identification. In Bonnet, J. P., ed., *Eddy Structure Identification*, Springer-Verlag, Wien, 1-59.

Brooke, J. W., Kontomaris, K., Hanratty, T. J. & McLaughlin, J. B. (1992). Turbulent deposition and trapping of aerosols at a wall. *Phys. Fluids A*, **4**, 825-834.

Brooke, J. W. & Hanratty, T. J. (1993). Origin of turbulence-producing eddies in channel flow. *Phys. Fluids A*, **5**, 1011-1022.

Caporaloni, M., Tampieri, F., Trombetti, F. & Vittori, O. (1975). Transfer of particles in nonisotropic air turbulence. *J. Atmos. Sci.* (Boston), **32**, 565-568.

Cerbelli, S., Giusti, A. & Soldati, S. (2001). ADE approach to predicting dispersion of heavy particles in wall bounded turbulence. *Int. J. Multiphase Flow*, **27**, 1861-1879.

Choi, K. S. (2001). Turbulent drag reduction mechanisms: strategies for turbulence management. In *Turbulence Modulation and Control* (ed. A. Soldati & R. Monti), CISM Courses and Lectures, vol. 415, pp. 161-211, Springer.

Chong, M. S., Perry, A. & Cantwell, B. J. (1990). A general classification of three-dimensional flow fields. *Phys. Fluids A*, **2**, 765-777.

Cleaver, J. W. & Yates, B. (1975). A sub layer model for the deposition of particles from a turbulent flow. *Chemical Engineering Science*, **30**, 983-992.

Crowe, C. T., Chung, J. N. & Troutt, T. R. (1988). Particle mixing in free shear flows. *Prog. Energy Combust. Sci.*, **14**, 171-194.

De Angelis, V., Lombardi, P., Andreussi, P. & Banerjee, S. (1997). Microphysics of scalar transfer at air-water interfaces. Invited Paper, IMA Conference on *Wind over Wave Couplings*, Salford, UK, 8-10 April, 1997, Oxford University Press.

Dubief, Y. & Delcayre, F. (2000). On coherent-vortex identification in turbulence. *Journal of Turbulence*, **1**, 11–32. Retrieved at *http://jot.iop.org*.

Eaton, J. K. & Fessler, J. R. (1994). Preferential concentration of particles by turbulence. *Int. J. Multiphase Flow*, **20**, 169-209.

Friedlander, S. K. & Johnstone, H. F. (1957). Deposition of suspended particles from turbulent gas streams. *Ind. Eng. Chem.*, **49**, 1151-1156.

Guezennec, Y. G. & Choi, W. C. (1989). Stochastic estimation of coherent structures in turbulent boundary layers. In *Proc. Zoran P. Zaric Memorial International Seminar on Near Wall Turbulence*, May 1988 (ed. S. J. Kline & N. H. Afghan), pp. 420-436, Hemisphere.

Guezennec, Y. G., Piomelli, U. & Kim, J. (1989). On the shape and dynamics of wall structures in turbulent channel flow. *Phys. Fluids A*, **1**, 764-766.

Hunt, J. C. R., Wray, A. A. & Moin, P. (1998). Eddies, stream and covergence zones in turbulent flows. *Center of Turbulence Research Rep.*, CTR-S88, 193.

Jimenez, J. & Moin, P. (1991). The minimal flow unit in near-wall turbulence. *Journal of Fluid Mechanics*, **225**, 213–233.

Jimenez, J. & Pinelli, A. (1999). The autonomous cycle of near-wall turbulence. *J. Fluid Mech.*, **389**, 335-359.

Jeong, J. & Hussain, F. (1995). On the identification of a vortex. *J. Fluid Mech.*, **285**, 69-83.

Jeong, J., Hussain, F., Schoppa, W. & Kim, J. (1997). Coherent structures near the wall in a turbulent channel flow. *J. Fluid Mech.*, **332**, 185-214.

Kaftori, D., Hetsroni, G. & Banerjee, S. (1995a). Particle behavior in the turbulent boundary layer. I. Motion, deposition, and entrainment. *Phys. Fluids*, **7**, 1095-1106.

Kaftori, D., Hetsroni, G. & Banerjee, S. (1995b). Particle behavior in the turbulent boundary layer. II. Velocity and distribution profiles. *Phys. Fluids*, **7**, 1107-1121.

Kallio, G. A. & Reeks, M. W. (1989). A numerical simulation of particle deposition in turbulent boundary layers. *Int. J. Multiphase Flow*, **15**, 433-446.

Kasagi, N. & Iida, O. (1999). Progress in direct numerical simulation of turbulent heat transfer. Keynote Paper, *5th ASME/JSME Joint Thermal Engineering Conference*, San Diego, CD-ROM Publication, ASME, March, 1999.

Kim, J. & Hussain, F. (1993). Propagation velocity of perturbations in turbulent channel flow. *Physics of Fluids A*, **5**, 695–706.

Kim, J., Moin, P. & Moser, R. (1987). Turbulence statistics in fully developed channel flow at low Reynolds number. *J. Fluid Mech.*, **177**, 133-166.

Kline, S. J., Reynolds, W. C., Schraub, F. A. & Runstadler, P. W. (1967). The structure of turbulent boundary layer. *Journal of Fluid Mechanics*, **70**, 741–773.

Kline, S. J. & Robinson, S. K. (1990). Quasi-coherent structures in the turbulent boundary layer:Part 1. Status report on community-wide survey of the data. In Kline, S. J., Afgan, N. H., eds., *Near-Wall Turbulence*. Hemisphere, New York.

Kontomaris, K., Hanratty, T. J. & McLaughlin, J. B. (1992). An algorithm for tracking fluid particles in a spectral simulation of turbulent channel flow. *J. Comput. Phys.*, **103**, 231-242.

Kulick, J. D., Fessler, J. R. & Eaton, J. K. (1994). Particle response and turbulence modification in fully developed channel flow. *J. Fluid Mech.*, **277**, 109-134.

Lam, K. & Banerjee, S. (1992). On the condition of streak formation in bounded flows. *Phys. Fluids A*, **4**, 306-320.

Lombardi, P., De Angelis, V. & Banerjee, S. (1996). Direct numerical simulation of near-interface turbulence in coupled gas-liquid flow. *Phys. Fluids*, **8**, 1643-1665.

Lu, D. M. & Hetsroni, G. (1995). Direct numerical simulation of a turbulent channel flow with heat transfer. *Int. J. Heat Mass Transfer*, **38**, 3241 - 3251.

Lyons, S. L., Hanratty, T. J. & McLaughlin, J. B. (1991). Large-scale computer simulation of fully developed turbulent channel flow with heat transfer. *Int. J. Numer. Methods Fluids*, **13**, 999-1028.

Marchioli, C. & Soldati, A. (2002). Mechanisms for particle transfer and segregation in turbulent boundary layer. *J. Fluid Mech.*, **468**, 283-315.

Marchioli, C., Giusti, A., Salvetti, M. V. & Soldati, A. (2003). Direct numerical simulation of particle wall transfer and deposition in upward turbulent pipe flow. *Int. J. Multiphase Flow*, **29**, 1017-1038.

McLaughlin, J. B. (1989). Aerosol particle deposition in numerically simulated channel flow. *Phys. Fluids*, **1**, 1211-1224.

McLaughlin, J. B. (1991). Inertial migration of a small sphere in linear shear flows. *J. Fluid Mech.*, **224**, 261-274.

Narayanan, C., Lakehal, D., Botto, L., Soldati, A. (2003). Mechanisms of particle deposition in a fully-developed turbulent open channel flow. *Phys. Fluids*, **15**, 763-775.

Niño, Y. & Garcia, M. H. (1996). Experiments on particle-turbulence interactions in the near-wall region of an open channel flow: implications for sediment transport. *J. Fluid Mech.*, **326**, 285-319.

Orlandi, P. & Jimenez, J. (1994). On the generation of turbulent wall friction. *Phys. Fluids*, **6**, 634-641.

Ounis, H., Ahmadi, G. & McLaughlin, J. B. (1993). Brownian particle deposition in a directly simulated channel flow. *Phys. Fluids*, **5**, 1427-1432.

Pan, Y. & Banerjee, S. (1996). Numerical simulation of particle interactions with wall turbulence. *Phys. Fluids*, **8**, 2733-2755.

Papavassiliou, D. V. & Hanratty, T. J. (1995). The use of Lagrangian methods to describe turbulent transport of heat from a wall. *Industrial and Engineering Chemistry Research*, **34**, 3359–3367.

Pedinotti, S., Mariotti, G. & Banerjee, S. (1992). Direct numerical simulation of particle behavior in the wall region of turbulent flow in horizontal channels. *Int. J. Multiphase Flow*, **18**, 927-941.

Perry, A. & Chong, M. S. (1987). A description of eddying motions and flow patterns using critical point concepts. *Annu. Rev. Fluid Mech.*, **9**, 125-148.

Reeks, M. W. (1983). The transport of discrete particles in inhomogeneous turbulence. *J. Aerosol Sci.*, **14**, 729-739.

Rizk, M. A. & Elghobashi, S. E. (1985). The motion of a spherical particle suspended in a turbulent flow near a plane wall. *Phys. Fluids*, **28**, 806-817.

Robinson, S. K. (1991). Coherent motions in the turbulent boundary layer. *Ann. Rev Fluid Mech.*, **23**, 601-639.

Rouson, D. W. I. & Eaton, J. K. (2001). On the preferential concentration of solid particles in turbulent channel flow. *J. Fluid Mech.*, **428**, 149-169.

Saffman, P. G. (1965). The lift on a small sphere in a slow shear flow. *J. Fluid Mech.*, **22**, 385-400. [Corrigendum **31**, 624 (1968)].

Schoppa, W. & Hussain, F. (1996). New aspects of vortex dynamics relevant to coherent structures in turbulent flows. In *Eddy Structure Identification* (ed. J. P. Bonnet), CISM Courses and Lectures, vol. 353, pp. 61-143, Springer.

Schoppa, W. & Hussain, F. (1997). Genesis and dynamics of coherent structures in near-wall turbulence. In *Self-sustaining Mechanisms of Wall Turbulence* (ed. R. Panton), Advances in Fluid Mechanicsi, vol. 15, pp. 385-422, Computational Mechanics Publications.

Schoppa, W. & Hussain, F. (2000). Coherent structure dynamics in near-wall turbulence. *J. Fluid Dynamics Research*, **26**, 119-139.

Slater, S. A. & Young, J. B. (2001). The calculation of inertial particle transport in dilute gas-particle flows. *Int. J. Multiphase Flow*, **27**, 61-87.

Squires, K. D. & Eaton, J. K. (1991). Preferential concentration of particles by turbulence. *Phys. Fluids A*, **3**, 1169-1178.

Soldati, A., Casal, M., Andreussi, P. & Banerjee, S. (1997). Lagrangian simulation of turbulent particle dispersion in Electrostatic Precipitators. *AIChE J.*, **43**, 1403-1413.

Soldati, A. & Banerjee, S. (1998). Turbulence modification by large scale organized electrohydrodynamic flows. *Phys. Fluids*, **10**, 1742-1756.

Soldati, A. (2000). Modulation of turbulent boundary layer by EHD flows. *ERCOFTAC Bulletin*, **44**, 50–56.

Soldati, A. & Marchioli, C. (2001). Prospects for modulation of turbulent boundary layer by EHD flows. In *Turbulence Structure and Modulation* (ed. A. Soldati & R. Monti), CISM Courses and Lectures, vol. 415, pp. 119-160, Springer.

Sun, Y. F. & Lin, S. P. (1986). Aerosol concentration in a turbulent flow. *J. Colloid Interface Sci.*, **113**, 315-320.

Tchen, C. M. (1947). Mean value and correlation problems connected with the motion of small particles suspended in a turbulent fluid. Ph.D. Thesis, Delft.

Uijttewaal, W. S. J. & Oliemans, R. V. A. (1996). Particle dispersion and deposition in direct numerical and large eddy simulations of vertical pipe flows. *Phys. Fluids*, **8**, 2590-2604.

van Haarlem, B., Boersma, B. J. & Nieuwstadt, F. T. M. (1998). Direct numerical simulation of particle deposition onto a free-slip and no-slip surface. *Phys. Fluids*, **10**, 2608-2620.

Wallace, J. M., Eckelmann, H., Brodkey, R. S. (1972). The wall region in turbulent shear flow. *J. Fluid Mech.*, **54**, 39-48.

Wang, Q. & Squires, K. D. (1996). Large eddy simulation of particle deposition in a vertical turbulent channel flow. *Int. J. Multiphase Flow*, **22**, 667-683.

Wei Ling, Chung, J. N., Troutt, T. R. & Crowe, C. T. (1998). Direct numerical simulation of a three-dimensional temporal mixing layer with particle dispersion. *J. Fluid Mech.*, **358**, 61-85.

Wells, M. R. & Stock, D. E. (1983). The effect of crossing trajectories on the dispersion of particles in turbulent flow. *J. Fluid Mech.*, **136**, 31-62.

Willmarth, W. W. & Lu, S. S. (1972). Structure of the Reynolds stress near the wall. *J. Fluid Mech.*, **55**, 65-92.

Yeung, P. K. & Pope, S. B. (1988). An algorithm for tracking fluid particles in numerical simulations of homogeneous turbulence. *J. Comput. Phys.*, **79**, 373-416.

Young, J. & Leeming, A. (1997). A theory of particle deposition in a turbulent pipe flow. *J. Fluid Mech.*, **340**, 129-159.

Zhang, H. & Ahmadi, G. (2000). Aerosol particle transport and deposition in vertical and horizontal turbulent duct flows. *J. Fluid Mech.*, **406**, 55-80.

Zhou, J., Adrian, R. J., Balachandar, S. & Kendall, T. M. (1999). Mechanisms for generating coherent packets of hairpin vortices in channel flow. *J. Fluid Mech.*, **387**, 353-396.

A Streamline Rotation Vector

The recognition of coherent structures in turbulence is a puzzling issue since no widely accepted definiton of coherent structure exists. Most definitions involve subjective choices of cut-off values or reference frames. For example, local pressure minima below some threshold value are often used to identify vortex cores. One scheme for rational selection of the cut-off values was developed by Chong, Perry & Cantwell (1990). Their scheme involves the examination of every point in the flow domain from a reference frame translating with the local fluid velocity. Each point is then a critical point of the dynamical system obtained by Taylor expansion of the instantaneous fluid velocity $\mathbf{u}$ about a point $\mathbf{x}$. For all case of practical interest, the phase-space trajectory of such dynamical system, and hence the physical-space trajectory of fluid elements, is governed by a three-dimensional set of first-order differential equations, written as:

$$\dot{x} = Ax, \tag{10}$$

where:

$$\dot{x} = \begin{pmatrix} \dot{x}_1 \\ \dot{x}_2 \\ \dot{x}_3 \end{pmatrix}, \quad A = \begin{pmatrix} a_{11} & a_{12} & a_{13} \\ a_{21} & a_{22} & a_{23} \\ a_{31} & a_{32} & a_{33} \end{pmatrix}, \quad x = \begin{pmatrix} x_1 \\ x_2 \\ x_3 \end{pmatrix}.$$

Elements a_{ij} of matrix A are real constants. In the case of a fluid flow, the a_{ij} are the elements of the rate-of-deformation tensor $\partial \dot{x}_i / \partial x_j = \partial u_i / \partial x_j = u_{i,j}$, where u_i and x_i are the ith Cartesian components of the vectors $\mathbf{u}$ and $\mathbf{x}$, the subscript comma symbolizes partial differentation with respect to the subsequent index, and Einstein summation convention applies.

Let us now split the rate-of-deformation tensor into a symmetric S_{ij} and antisymmetric part Ω_{ij} defined as follows:

$$S_{ij} = \frac{1}{2}(u_{i,j} + u_{j,i}), \quad \Omega_{ij} = \frac{1}{2}(u_{i,j} - u_{j,i}). \tag{11}$$

S_{ij} is the rate-of-strain tensor and describes the rate at which an infinitesimal control volume of fluid changes shape. Ω_{ij} is the rotation (or spin) tensor: S_{ij} and Ω_{ij} represent the local balance between shear strain rate and vorticity magnitude. Shear strain and vorticity have the same magnitude at a stationary wall and thus the rate-of-deformation tensor vanishes at the wall. If λ_1,λ_2,λ_3 are the eigenvalues of matrix A, then:

$$[A - \lambda I]\mathbf{e} = 0, \tag{12}$$

where $\mathbf{e}$ is the eigenvector. The eigenvalues can be determined by solving the characteristic equation $det[A - \lambda I] = 0$ which, for a 3×3 matrix, can be written as $\lambda^3 - P\lambda^2 + Q\lambda - R = 0$, where:

$$P = a_{11} + a_{22} + a_{33} = -tr[A] = -S_{ii}, \tag{13}$$

$$Q = \begin{vmatrix} a_{11} & a_{12} \\ a_{21} & a_{22} \end{vmatrix} + \begin{vmatrix} a_{11} & a_{13} \\ a_{31} & a_{33} \end{vmatrix} + \begin{vmatrix} a_{22} & a_{23} \\ a_{32} & a_{33} \end{vmatrix}$$

$$= \frac{1}{2}(P^2 - tr[A^2]) = \frac{1}{2}(P^2 + \Omega_{ij}\Omega_{ij} - S_{ij}S_{ij}), \tag{14}$$

and

$$R = \begin{vmatrix} a_{11} & a_{12} & a_{13} \\ a_{21} & a_{22} & a_{23} \\ a_{31} & a_{32} & a_{33} \end{vmatrix} = det[A] = \frac{1}{3}(P^3 - 3PQ + tr[A^3])$$

$$= \frac{1}{3}(P^3 - 3PQ + S_{ij}S_{jk}S_{ki} + 3\Omega_{ij}\Omega_{jk}\Omega_{ki}), \tag{15}$$

are the first, second and third invariants of A. Recall that the characteristic equation can have (i) all real roots which are distinct, (ii) all real roots where at least two roots are equal, or (iii) one real root and a conjugate pair of complex roots.

Chong *et al.* (1990) classified all possible local streamline patterns around any point in a flow (in a reference frame moving with the velocity of that point) according to the values of the three invariants of the rate-of-deformation tensor (P, Q and R), *i.e.* according to the values of eigenvalues λ_i. They proposed that a vortex core is a region with complex eigenvalues of $\partial u_i/\partial x_j$: from a physical viewpoint, complex eigenvalues imply a local swirling motion so that the local streamline pattern is closed or spiral in a reference frame moving with the point. In other words, a vortex core is a region where the vorticity is sufficiently strong to cause the rate-of-deformation tensor to be dominated by the rotation tensor.

For incompressible flow, $P = 0$ due to mass conservation, $Q = -\frac{1}{2}u_{i,j}u_{j,i}$ and $R = det(u_{i,j})$. Thus complex eigenvalues will occur when the discriminant Δ of the characteristic equation is positive:

$$\Delta = (\frac{1}{3}Q)^3 + (\frac{1}{2}R)^2 > 0. \tag{16}$$

For visualization purposes, the vector $\boldsymbol{\Omega}$ is computed only at those point where $\Delta > 0$ by means of the following equation:

$$\boldsymbol{\Omega} = Im(\lambda_c)\frac{\mathbf{e}_{\lambda_r}}{|\mathbf{e}_{\lambda_r}|}\frac{\mathbf{e}_{\lambda_r}\cdot[Re(\mathbf{e}_{\lambda_c})\times Im(\mathbf{e}_{\lambda_c})]}{|\mathbf{e}_{\lambda_r}\cdot[Re(\mathbf{e}_{\lambda_c})\times Im(\mathbf{e}_{\lambda_c})]|}, \tag{17}$$

where $Im(\lambda_c)$ is the imaginary part of the pair of complex eigenvalues, $Re(\mathbf{e}_{\lambda_c})$ and $Im(\mathbf{e}_{\lambda_c})$ are the real part and the imaginary part of the conjugate complex eigenvectors corresponding to the complex eigenvalues, $\mathbf{e}_{\lambda_r}$ is the eigenvector corresponding to the real eigenvalue λ_r.

The rate-of-deformation invariants can be used to provide a classification of elementary three-dimensional flow patterns defined by instantaneous streamlines for flow at and away from no-slip boundaries (Chong *et al.* 1990, Blackburn *et al.* 1996). This scheme provides an objective classification that eliminates the arbitrary choice of threshold values used in other schemes.

For incompressible flow, all possible topologies can be classified in the (Q, R)-plane, as shown in figure 25. Considering the $D = 0$ curve, the (Q, R)-plane is divided into four regions: the so-called stable focus/stretching, unstable focus/compressing, stable node/saddle/saddle and unstable node/saddle/saddle critical points. The first two are vortical flow regions, the last two are essentially convergence zones.

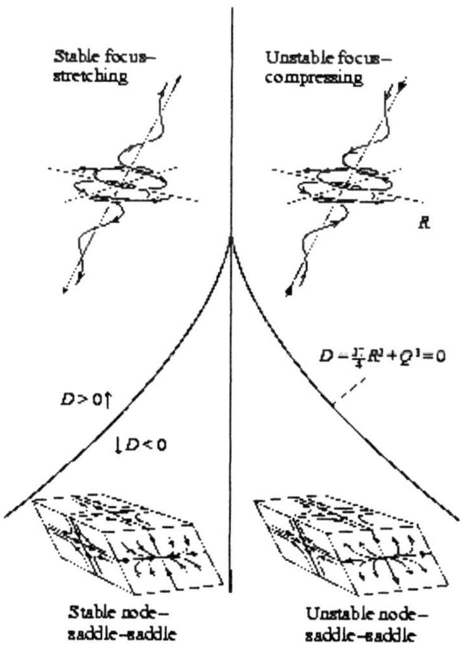

Figure 25. Incompressible flow critical point topologies, counterclockwise from upper right: stable focus/stretching, unstable focus/compressing, stable node/saddle/saddle and unstable node/saddle/saddle (figure courtesy of J. Chacin).

Breinigsville, PA USA
10 February 2011
255262BV00001BA/15/P

9 783211 207574